Cyanotoxins – Occurrence, Causes, Consequences

Springer

Berlin
Heidelberg
New York
Barcelona
Hong Kong
London
Milan
Paris
Singapore
Tokyo

Ingrid Chorus (Ed.)

Cyanotoxins

Occurrence, Causes, Consequences

With 128 Figures

Springer

EDITOR:

Dr. Ingrid Chorus
Umweltbundesamt (Federal Environmental Agency)

Corrensplatz 1
14195 Berlin
Germany

Many of the research results presented here were funded by the German Federal Ministry for Education, Science, Research and Technology (grant no. 0339547)

ISBN-13: 978-3-642-64004-9 Springer-Verlag Berlin Heidelberg New York

Library of Congress Cataloging-in-Publication Data
Cyanotoxins: occurrence, causes, consequences / Ingrid Chorus (ed.).- Berlin; Heidelberg; New York; Barcelona; Hong Kong; London; Milan; Paris; Singapore; Tokyo: Springer 2001

ISBN-13: 978-3-642-64004-9 e-ISBN-13: 978-3-642-59514-1
DOI: 10.1007/978-3-642-59514-1

Springer-Verlag Berlin Heidelberg New York
a member of BertelsmannSpringer Science+Business Media GmbH
http://www.springer.de

Softcover reprint of the hardcover 1st edition 2001

Cover Design: *design & production*, Heidelberg
Cover photograph by Dr. Hartmut Rönicke
Typesetting: Camera-ready by Heike Bickel

SPIN: 10683397 31/3130xz – 5 4 3 2 1 0

To Professor Johannes-Günter Kohl
(04. 04. 1936 – 03. 03. 2000),
who initiated the first investigations of
cyanobacterial toxicity in Germany in 1982
and had a key role in establishing
the research results presented here.

Beyond this, his competent guidance
as mentor and research partner
is appreciated with deep gratitude.

In particular, his absolutely contagious enthusiasm
for understanding how biological mechanisms work
in aquatic ecosystems will remain unforgotten.

Contents

Contributions marked with an asteric were supported by the German Federal Ministry for Education and Research, grant no. 0339547.

Contributors

Postal addresses refer to the institution in which the contributions were prepared; e-mail addresses are current ones.

Annadotter, Helene: Inst. Of Ecology, Dept. of Limnology, Lund University, 223 62 Lund, Sweden. Helene.Annadotter@limnol.lu.se

Baganz, Daniela: Institute of Freshwater Ecology and Inland Fisheries, Mueggelseedamm 301, 12561 Berlin, Germany.

Bickel, Heike: National Institute of Public Health, P.O. Box 4404 Torshov, 0403 Oslo, Norway, and Albert-Ludwigs-Universität, Inst. für Biologie II, Schänzlestr. 1, 79104 Freiburg, Germany. mail@heikebickel.de

Bláha, Ludek: RECETOX-Research Center for Environmental Toxicology, Masaryk University, Veslarská 230B, 637 00 Brno, Czech Republic. blaha@vri.cz

Börner, Thomas: Inst. of Biology/Genetics, Humboldt University Berlin, Chausseestrasse 117, Berlin, Germany. thomas=boerner@biologie.hu-berlin.de

Böttcher, Gabriele: Umweltbundesamt, P.O. Box 330022, 14191 Berlin, Germany. gabriele.boettcher@uba.de

Bumke-Vogt, Christiane: Umweltbundesamt, P.O. Box 330022, 14191 Berlin, Germany.

Chorus, Ingrid: Umweltbundesamt, P.O. Box 330022, 14191 Berlin, Germany. ingrid.chorus@uba.de

Cronberg, Gertrud: Inst. of Ecology, Dept. of Limnology, Lund University, 223 62 Lund, Sweden. gertrud.cronberg@limnol.lu.se

Dewender, Marion: Humdoldt University Berlin, FG Ökologie, Luisenstr. 53, 10117 Berlin, Germany. marion=dewender@biologie.hu-berlin.de

Dittmann, Elke: Inst. of Biology/Genetics, Humboldt University Berlin, Chausseestrasse 117, Berlin, Germany. Elke=dittmann@biologie.hu-berlin.de

Erhard, Marcel: AnagnosTec, Im Biotechnologiepark TGZ II 14943 Luckenwalde, Germany. www.AnagnosTec.de

Ewald, Silvia: Umweltbundesamt, P.O. Box 330022, 14191 Berlin, Germany.

Fastner, Jutta: Umweltbundesamt, P.O. Box 330022, 14191 Berlin, Germany. jutta.fastner@uba.de

Fitzgerald, D. James: Environmental Health Branch, South Australian Health Commission, P.O. Box 6, Rundle Mall, Adelaide, SA 5000, Australia. jim.fitzgerald@dhs.sa.gov.au

Flury, Thomas: Novartis Crop Protection Inc., Lead Discovery, WRO-1060 4.06, 4002 Basel, Switzerland. thomas.flury@cp.Novartis.com

Forchert, Ayako: Umweltbundesamt, P.O. Box 330022, 14191 Berlin and Max Vollmer Institute, Technical University, Franklinstr. 29, 10587, Berlin, Germany. forchert@sispa.chem.tu-berlin.de

Giddings, Michele: Head, Drinking Water Section, Safe Environments Programme, Healthy Environments & Consumer Safety Branch, 123 Slater St. (A.L. 3505A) Ottawa, Canada. michele_giddings@hc-sc.gc.ca

Gjolme, Nina: National Institute of Public Health, P.O Box 4404 Torshov, 0403 Oslo, Norway. nina.gjolme@folkehelsa.no

Göthe, Ulf: Office for Public Health, 241 80 Eslöv, Sweden. ulf.gothe@eslov.se

Gröbe, Kerstin: Eurawasser Aufbereitungs- und Entsorgungs GmbH, Blüscherstr. 54, 18055 Rostock, Germany. k.groebe@eurawasser.de

Gupta, Shalini. Management of Toxics Substances Division, Safe Environments Programme, Healthy Environments & Consumer Safety Branch, Environmental Health Centre, Tunney's Pasture, Ottawa, Canada. shalini_gupta@hc-sc.gc.ca

Hansson, Hans-Bertil: Regional Center for Infectious Diseases, Malmö Univ. Hospital, SE-205 02 Malmö, Sweden. hb.hansson@pop.skane.se

Heinze, Rita: Umweltbundesamt, P.O. Box 330022, 14191 Berlin, Germany. rita.heinze@uba.de

Henning, Manfred: Humdoldt University Berlin, FG Ökologie, Luisenstr. 53, 10117 Berlin, Germany. manfred=henning@biologie.hu-berlin.de

Henriksen, Peter: National Environmental Research Institute, Dept. of Marine Ecology and Microbiology, P.O. Box 358, 4000 Roskilde, Denmark. PET@dmu.dk

Hesse, Karina: Humdoldt University Berlin, FG Ökologie, Luisenstr. 53, 10117 Berlin, Germany. karina=hesse@biologie.hu-berlin.de

Hinze, Thomas: Institute of Freshwater Ecology and Inland Fisheries, Mueggelseedamm 301, 12561 Berlin, Germany. hintze@igb-berlin.de

Hübner, Conrad: Nordthüringer Wasserversorgung und Abwasserbehandlung GmbH, Haarbergstraße 37, 99097 Erfurth, Germany.

Jaime, Elke: Friedrich Schiller Universität Jena, Inst. of Nutrition, Dornburger Str. 25, 07743 Jena, Germany. jaime@mampf.ieu.uni-jena.de

Jones, Gary: CSIRO Land and Water, Resources Sciences Centre, 80 Meiers Rd., Building B, Indooroopilly, Old, 4068, Australia. gjones@enterprise.canberra.edu.au

Jungblut, Peter: Max-Planck-Institut für Infektionsbiologie, Monbijoustr. 2, 10117 Berlin, Germany. jungblut@mpiib-berlin.mpg.de

Kohl, Johannes-Günter: Humdoldt University Berlin, FG Ökologie, Luisenstr. 53, 10117 Berlin, Germany.

Kruschwitz, Christine: Landeshygieneinst. Rostock, Dez. Wasser Bodenhygiene, Gerturdenstr. 9, 18057 Rostock, Germany.

Kruspe, Ute: Landestalsperrenverwaltung des Freistaates Sachsen, Labor Radeburg, Am Stausee, 01471 Radeburg, Germany.

Lawton, Linda: Faculty of Science and Technology, School of Applied Sciences, St. Andrew Street, Aberdeen, AB1 1HG, United Kingdom.apsll@app-science.rgu.ac.uk

Lindenschmidt, Karl-Erich: Umweltbundesamt, P.O. Box 330022, 14191 Berlin, Germany. kel@gm.ufz.de

Lyck, Susanne: Freshwater Biological Laboratory, University of Copenhagen, Helsingørgade, 51 3400 Hillerød, Denmark. slyck@zi.ku.dk

Maršálek, Blahoslav: Institute of Botany, Czech Academy of Sciences, Kvetna 8, 603 65 Brno, Czech Republic. Marsalek@brno.cas.cz

Neca, Jirí: Veterinary Research Institute, Hudcova 70, 621 32 Brno, Czech Republic. neca@vri.cz

Neilan, Brett A.: The University of New South Wales, School of Microbiology and Immunology, 2052, Sydney, NSW, Australia. b.neilan@unsw.edu.au

Neumann, Uwe: Albert-Ludwigs-Universität, Inst. für Biologie II, Schänzlestr. 1, 79104 Freiburg, Germany.

Niesel, Verena: Umweltbundesamt, P.O. Box 330022, 14191 Berlin, Germany. verena.niesel@uba.de

Nixdorf, Brigitte: Brandenburgische Technische Universität, Lehrstuhl für Gewässerschutz, Seestr. 45, 15526 Bad Saarow, Germany. b.nixdorf@t-online.de

Oberemm, Axel: Institute of Freshwater Ecology and Inland Fisheries, Mueggelseedamm 301, 12561 Berlin, Germany. a.oberemm@bgvv.de

Papendorf, Olaf: Inst. for Pharmaceutical Biology, Technical Univ. of Braunschweig, Mendelsohnstr. 1, 38106 Braunschweig, Germany.

Park, Ho-Dong: Shinshu University, Dept. of Environm. Sciences, Matsumoto 390, Japan. pparkhd@gipac.shinshu-u.ac.jp

Pflugmacher, Stephan: Institute of Freshwater Ecology and Inland Fisheries, Mueggelseedamm 301, 12561 Berlin, Germany. pflug@igb-berlin.de

Pietsch, Jörg: DVGW-Technologiezentrum Wasser, Scharfenberger Str. 152, 01139 Dresden, Germany. DVGW-TZW.DRESDEN@t-online.de

Pütz, Klaus: Landestalsperrenverwaltung des Freistaates Sachsen, Wassergütebewirtschaftung, P.O. Box 94, 01782 Pirna, Germany.

Rabe, Carol: Fernwasserzweckverband Nord- und Ostthüringen, Am Wasserwerk, 07407 Zeigerheim, Germany.

Rohrlack, Thomas: Institute of Biology/Ecology, Humboldt University Berlin, Luisenstrasse 53, 10117 Berlin, Germany. TRohrlack@fbl.zi.ku.dk

Schlag, Gertrud: Umweltbundesamt, P.O. Box 330022, 14191 Berlin, Germany. gertrud.schlag@uba.de

Schmidt, Wido: DVGW-Technologiezentrum Wasser, Scharfenberger Str. 152, 01139 Dresden, Germany. schmidt@tzw-dresden.de

Sheffer, Marla: Scientific Editor, Ottawa, Canada. yaeger@sympatico.ca

Skulberg, Olav M.: Nowegian Institute for Water Research, P.O. Box 173 Kjelsås, Oslo, Norway. olav.skulberg@niva.no

Skulberg, Randi: Nowegian Institute for Water Research, P.O. Box 173 Kjelsås, Oslo, Norway.

Staaks, Georg: Institute of Freshwater Ecology and Inland Fisheries, Mueggelseedamm 301, 12561 Berlin, Germany. oki@igb-berlin.de

Steinberg, Christian: Institute of Freshwater Ecology and Inland Fisheries, Mueggelseedamm 301, 12561 Berlin, Germany. stein@igb-berlin.de

Tillett, Daniel: The University of New South Wales, School of Microbiology and Immunology, 2052, Sydney, NSW, Australia. d.jacobs@unsw.edu.au

Turánek, Jaroslav: Veterinary Research Institute, Hudcova 70, 621 32 Brno, Czech Republic. turanek@vri.cz

Underdal, Bjarne: The Norwegian Collge of Veterinary Medicine, Oslo, Norway.

Utkilen, Hans: Nat. Inst. Public Health, P.O. Box 4404 Torshov, 0403 Oslo, Norway. hans.utkilen@folkehelsa.no

Vasconcelos, Vitor: Porto University, Dept. of Zoology and Antropology, Praca Gomes Teixeira, 4050 Porto, Portugal. vmvascon@fc.up.pt

Von Döhren, Hans: Technical University Berlin, Fränklinstr. 29, 10587 Berlin, Germany. doehren@chem.tu-berlin.de

Walz, Norbert: Institute of Freshwater Ecology and Inland Fisheries, Mueggelseedamm 301, 12561 Berlin, Germany. walz@igb-berlin.de

Weckesser, Jürgen: Albert-Ludwigs-Universität, Inst. für Biologie II, Schänzlestr. 1, 79104 Freiburg, Germany. jweckess@uni-freiburg.de

Welker, Martin: Institute of Freshwater Ecology and Inland Fisheries, Mueggelseedamm 301, 12561 Berlin, Germany. Martin.Welker@chem.TU-Berlin.de

Wiedner, Claudia: Brandenburgische Technische Universität, Lehrstuhl für Gewässerschutz, Seestr. 45, 15526 Bad Saarow, Germany. Claudia.wiedner@chem.uva.nl

Wiegand, Claudia: Institute of Freshwater Ecology and Inland Fisheries, Mueggelseedamm 301, 12561 Berlin, Germany. cwiegand@igb-berlin.de

Willmitzer, Hartmut: Thüringer Talsperrenverwaltung, Talsperrenstr. 25-27, 99897 Tambach-Dietharz, Germany. hartmut.willmitzer.ttv@t-online.de

Wirsing, Birgit: Albert-Ludwigs-Universität, Inst. für Biologie II, Schänzlestr. 1, 79104 Freiburg, Germany. Wirsing@uni-bonn.de

Woitke, Peter: Humdoldt University Berlin, FG Ökologie, Luisenstr. 53, 10117 Berlin, Germany. peter.woitke@uba.de

Frequently used Abbreviations

ADP	Adenosine diphosphate
AMP	Adenosine monophosphate
A. spp.	*Anabaena* species
AT_{ex}	Anatoxin-a, external, i.e. dissolved
AT_{int}	Anatoxin-a, internal, i.e. cell-bound
ATP	Adenosine triphosphate
Chl.-a	Chlorophyll-a
DOC	Dissolved organic carbon
DOM	Dissolved organic matter
DW	Dry weight
FFA	Fuhrberger fulvic acid
ELISA	Enzyme Linked Immuno Sorbent Assay
Fe	Iron
GSH	Glutathione, reduced
GST	Glutathione-S-transferases
HPLC	High performance liquid chromatography
i.p.	Intra-peritoneal
LD_{50}	Lethal dose for 50 % of a population
M. a.	*Microcystis aeruginosa*
MALDI-TOF	Matrix assisted laser desorption ionisation-time-of flight
MC	Microcystin(s)
MC_{ex}	Microcystins, external, i.e. dissolved
MC_{intra}	Microcystins, internal, i.e. cell-bound
MCYST	Microcystin(s)
M. i.	*Microcystis ichthyoblabe*
M. n.	*Microcystis novaceckii*
M. spp.	*Microcystis* species
M. w.	*Microcystis wesenbergii*

μ	Growth rate
μ_m	Maximal growth rate
N	Nitrogen
P	Phosphorus
P. a.	*Planktothrix agardhii*
PAR	Photosynthetically active radiation
P. r.	*Planktothrix rubescens*
PSP-toxins	Paralytic shellfish poisoning toxins
Q_{max}	Maximal cell quota
q_o	Cell quota
TN	Total nitrogen
TP	Total phosphorus
WHO	World Health Organisation
Z_{eu}	Euphotic depth
Z_{max}	Maximal depth
Z_{mean}	Mean depth

1 Introduction: Cyanotoxins – Research for Environmental Safety and Human Health

How dangerous are toxic cyanobacteria? Numerous lethal animal poisonings worldwide and a number of cases of human illness have drawn the attention of the World Health Organization, the public and an increasing scientific community to cyanobacteria and their toxins. Gradually, a clearer picture is emerging on the occurrence, causes and consequences of the proliferation of toxic cyanobacteria in surface waters.

Through the centuries, isolated incidents of cyanobacterial mass developments have been reported, often conspicuous as so-called water blooms on the surface of lakes, reservoirs or slowly flowing rivers, and the first incident of animal poisoning attributed to toxic cyanobacteria was, in fact, published in 1878 in Nature (Francis 1878). Reports start to become more frequent in the second half of the 20th century (Carmichael et al. 1985). Two factors triggered intensification of research on cyanobacterial toxins, or cyanotoxins in the 1980s: one was the increasing availability of analytical techniques that enabled structural elucidation and identification of several groups of cyanotoxins. The other was the increased occurrence of mass developments of cyanobacteria in consequence of accelerated eutrophication of many waterbodies worldwide, enhancing concern of health risks due to the occurrence of these natural toxins in unnatural frequency and concentrations.

Cyanobacteria are microscopically small, unicellular or filamentous bacteria with some characteristics of algae such as cell wall structure, pigments and the ability to perform oxygenic photosynthesis. For this reason they are commonly termed blue-green algae. Many cyanobacteria occur suspended in water, i.e. they are planktonic (as part of the phytoplankton), others attach themselves on surfaces or sediments, i.e. are benthic. While some occur as single cells, others form filaments or colonies.

For health-risk assessments, monitoring and surveillance, an important feature to take into account is buoyancy regulation of some planktonic cyanobacteria. Through gas vesicles which adjust the specific weight of the cells to that of water, they are prevented from sinking into deep, dark water layers where photosynthesis is not possible. However, a few genera may perform vertical migrations in water bodies, thus spending some time in water layers with ample light for photosynthesis and some time in deep layers which are too dark for photosynthesis, but may contain high nutrient concentrations. The mechanism currently assumed to be responsible for vertical migration of cyanobacteria is a passive one: if photosynthesis assimilates more energy than the cells can use immediately, excess energy is stored as carbohydrates. These are heavier than water and act as ballast, causing the cells to sink until these carbohydrates have been consumed (Visser et al. 1995). Sometimes this mechanism is not effective, and cells are trapped at the water surface, where they form surface scums, commonly termed blooms. Scums are then easily swept together by wind. Accumulations of cyanobacterial blooms along downwind shores may reach extremely high densities, more than 1000-fold above those found in the open water.

Cyanotoxins cover a wide range of chemical structures and mechanisms of toxicity, a comprehensive review of which is given by Sivonen and Jones (1999). Cyanotoxins comprise cyclic peptides which are chiefly hepatotoxic (microcystins and nodularins), alkaloids and an organophosphate with strong neurotoxic effects (anatoxin-a, anatoxin-a(S) and saxitoxins), a cyclic guanidine alkaloid which inhibits protein synthesis (cylindrospermopsin), lipopolysaccharides with pyrogenic properties (often referred to by their abbreviation LPS), and in marine cyanobacteria also dermatoxic alkaloids (aplysiatoxins and lyngbyatoxins).

The majority of these substances are potent toxins, with LD_{50} (concentration acutely lethal to half of a population of mice in the bioassay) in the range between 10 and 200 $\mu g\ kg^{-1}$ (determined from intraperitoneal injection into mice). However, their occurrence in water is usually very dilute, unless cells accumulate. Although numerous reports of lethal animal poisoning through cyanobacteria are published, to date human fatalities have only been reported after exposure through renal dialysis (see Chap. 5.1, this Vol. and Kuiper-Goodman et al. 1999). This contradiction may be explained by the fact that dose-response curves for most cyanotoxins are steep: a dose only five to ten-fold lower than one that was barely lethal may show no apparent effect. Animals will drink enough cyanobacteria-infested water for a lethal dose if they have no access to clear water, but humans are less likely to swallow a sufficiently large amount of cyanoabacterial suspension to be exposed to a lethal dose.

Dermal irritations and/or allergic reactions to cyanobacteria are probably due to components other than the above-mentioned cyanotoxins, and have scarcely been studied. There is ample evidence for further bioactive, potentially toxic compounds in cyanobacteria which are currently poorly understood, and cyanobacteria are being intensively studied in pharmacological research in quest for new pharmaceuticals (Moore et al. 1996). Thus, health risk assessments of cyanobacterial toxicity must take into account that these organisms may contain harmful substances not yet classified as cyanotoxins and thus not subject to analytical screening methods.

As reports on cyanobacterial toxicity were increasingly published from Australia, but also from North America, Britain, Scandinavia and the Netherlands, it became evident that cyanotoxins occur worldwide, regardless of climate and latitude. However, most investigations were either triggered by human health concerns or by scientific curiosity in the fields of microbiology and biochemistry, and most publications centred around the occurrence of toxicity or structural elucidation of toxins. Studies on ecological aspects of cyanobacterial toxicity, with respect both to factors causing or enhancing cyanotoxin occurrence and to cyanotoxin impacts on aquatic biota, were in the minority.

For Germany, very little information was available on cyanotoxins. Consequently, a comprehensive program on occurrence, causes and consequences of toxic cyanobacteria in German water bodies was conducted from 1995 to 1997 with extensive support from the Federal Ministry for Education, Science, Research and Technology (BMBF), the results of which are presented in this book. A screening program to assess the extent of cyanotoxin occurrence was a major component of the program. This survey included 16 waterbodies intensively studied by three of the five partners of the cooperative cyanotoxin project. For broader coverage to include a larger variety of water bodies from different parts of the country, collaboration was established with ongoing sampling programs from 15 other institutions (9 public authorities responsible for resource monitoring, 4 research institutes, 2 university institutes). In return for information on the toxin content of their samples, these partners provided samples and limnological background data. Thus, this survey could include more than 670 samples from a total of 133 water bodies, 38 of which were studied more intensively. Almost all of the samples were analyzed for microcystins, and the majority of the samples was also tested for hepatotoxicity with a cellular assay using primary rat hepatocytes. More than half of the samples were further analyzed for anatoxin-a, and occurrence of PSP toxins was qualitatively assessed in a selected subset of these samples. Chapter 2.1 is a comprehensive report of these survey results.

Further, we are grateful for contributions of recent cyanotoxin survey results from five other countries (Chap. 2.2). Taken together, these data allow a first assessment of the relative importance of microcystins and cyanobacterial neurotoxins, and they indicate geographic gradients of cyanotoxin concentrations to be further elucidated in future studies.

The issue whether or not microcystin content of cyanobacterial strains is influenced by environmental factors is addressed in Chapter 3 with three different experimental approaches using continuous cultures. Chapter 4 adresses the mechanisms determining toxin content in natural populations of cyanobacteria.

Chapter 5 adresses the relevance of cyanotoxins for human exposure, beginning with an overview of current toxicological knowledge (Chap. 5.1) contributed from Australia. Chapter 5.2 relates data on microcystin concentrations at recreational sites sampled within the German survey to a general worst-case exposure estimate. Chapter 5.3 presents data on cyanotoxins in drinking water with case reports from Sweden, Canada and the Czech Republic and four pilot studies from Germany.

Chapter 6 sheds some light on the impact of cyanotoxins on aquatic biota. Chapter 6.1 demonstrates effects of cyanotoxins on early life stages of fish and amphibians, Chapter 6.2 shows uptake and detoxification mechanisms, and Chapter 6.3 reports behavioral changes induced in fish by microcystins. Chapter 6.4 shows how different the impact on *Daphnia galeata* may be if they are exposed to different strains of *Microcystis aeruginosa*, some of which contain microcystins and others which do not.

Chapter 7 collates information on further bioactivity of cyanobacterial components, beginning with an overview of peptides and depsipeptides in cyanobacteria (Chap. 7.1). It further describes the substantial toxic impact of many cyanobacterial samples beyond the effects that can be explained by the toxins identified in these samples, which became evident in the various bioassays conducted in the context of the German cyanotoxin survey (Chap. 7.2), and it reports a variety of cyanobacterial metabolites identified in the cyanobacterial samples (Chap. 7.3).

Chapter 8 assesses applicability of different methods for detection of cyanobacterial toxicity, i.e. cellular asssays (Chap. 8.1 and 8.2) as well as methods for sensitive analysis of low concentrations of microcystins (Chap. 8.3). Chapter 8.4 describes the application of the early life-stage fish test for cyanotoxin assessment. Finally, Chapter 8.5 introduces MALDI-TOF mass spectrometry as a newly available, highly sensitve technique for identifying cyanobacterial compounds at a sensitivity sufficient for analysis of 1 µg of dry material and even of larger individual *Microcystis* colonies. Routine analytical procedures applied in the German cyanotoxin survey are given in Chapter 9.

Acknowledgements

Particular thanks are due to the German Federal Ministry for Education, and Research for 3 years of funding for five research groups (grant no. 0339547), and to Ingo Fitting for continuous patient support of the project. We are also thankful to Günter Klein for initiating the proposal, to Gabi Böttcher, Gertrud Schlag, Inge Flieger and Bettina Schiers for editorial, accounting, secretarial and laboratory support, to Elke Pawlitzky for polishing up the figures, to Joachim Borchert and Heike Bickel for lay-out, to Jean von dem Bussche for language editing, to numerous referees and their suggestions for some chapters, and to the Association for Water, Soil and Air Hygiene for supporting a taxonomic workshop on cyanobacteria held in 1997. It was a great pleasure and benefit to have enjoyed support and taxonomical training from Jiri Komárek, Jaroslava Komárkova and Judit Padisák at this workshop. Last, but by no means least, the data set could not have been compiled without the active and engaged support of all those external partners who provided samples and background data.

References

Carmichael W, Hones C, Mahmood N, Theiss W (1985) Algal toxins and water-based diseases. Critical Rev Environ Contr 15:275–313

Francis G (1878) Poisonous Australian lakes. Nature 18:11–12

Kuiper-Goodman T, Falconer I, Fitzgerald J (1999) Human health aspects. In: Chorus I, Bartram J (eds) Toxic cyanobacteria in water: a guide to their public health consequences, monitoring and management. Published on the behalf of WHO by E & FN Spon, London, pp. 113–153

Moore R, Corbett T, Patterson GML, Valeriote FA (1996) The search for new antitumor drugs from blue-green algae. Curr Pharm Design 2:317–330

Sivonen K, Jones G (1999) Cyanobacterial toxins. In: Chorus I, Bartram J (eds) Toxic cyanobacteria in water: a guide to their public health consequences, monitoring and management. Published on the behalf of WHO by E & FN Spon, London, pp. 41–111

Visser PM, Ibelings BW, Mur LR (1995) Autumnal sedimentation of *Microcystis* spp. as a result of an increase in carbohydrate ballast at reduced temperature. J Plankton Res 17:919–933

2 Cyanotoxin Occurrence in Freshwaters

How likely is a given cyanobacterial bloom to be toxic? This question was addressed in a number of countries already in the 1980s, even before the structures of many cyanotoxins were known. Thus, early surveys were conducted using the mouse bioassay, and results generally showed toxicity in more than half of the cyanobacterial samples tested (e.g. Leeuwangh et al. (1983) in The Netherlands, Törökné-Kozma and Gábor (1988) in Hungary, and the NRA Report (1990) in England). In the wake of the development of chemical methods of cyanotoxin identification and quantification, these have been employed in a number of countries for more extensive surveys of cyanotoxin occurrence in relation to dominant cyanobacterial taxa. In some cases they were combined with bioassays for toxicity. Further, limnological aspects are increasingly being included in study designs and data interpretation.

The following chapter compiles current knowledge from recent surveys on the occurrence cyanotoxins. Chapter 2.1 presents the results of the survey in Germany. Contributions from five other countries summarize results of national surveys in Chapter 2.2 to enable international comparison along a north-south gradient from southern Norway (Chapter 2.2.1), Denmark (Chapter 2.3.2), Germany (Chapter 2.1.2), the Czech Republic (Chapter 2.2.3), Portugal (Chapter 2.2.4) and Korea (Chapter 2.2.5). These studies differ in approach and in the data base they produced. Nonetheless, in combination with the literature they discuss, a number of observations are emerging about the occurrence of cyanotoxins and the concentration ranges to expect. In Chapter 2.3, these are proposed as seven hypotheses to be further assessed for potential generalization.

Further, an important aspect of cyanotoxin occurrence is release of toxins from cyanobacterial cell material and persistence of dissolved toxins in water. Chapter 2.3 therefore concludes this chapter with a contribution on release, persistence and degradation of microcystins.

2.1 Cyanotoxin Occurrence in Germany

As in many countries, eutrophication is a problem for a large fraction of Germany´s lakes and slowly flowing rivers, particularly in lowland regions where nutrients from agriculture and treated sewage have accumulated. As a result, massive proliferation of cyanobacteria is widespread. Measures to reduce nutrient loading from sewage have been increasingly implemented since the mid-1980s, and better practices of fertilizer use in agriculture are also becoming more widespread. However, due to phosphorus stored in soils, inputs from farmland into waterbodies have declined only slightly. Further, in a number of waterbodies, internal loading with phosphorus accumulated in the sediments is maintaining a high trophic level even after external loads have substantially been reduced. In consequence, the measures taken are currently beginning to show success with respect to the reduction of phosphorus concentrations in many waterbodies, but total phosporus concentrations are often still too high to limit phytoplankton biomass and cyanobacterial blooms.

In spite of the widespread occurrence of cyanobacterial blooms, only two studies investigated toxic cyanobacteria in German waterbodies in the 1970s and 1980s (Kalbe et al. 1976; Henning and Kohl 1981). These described field observations of animal intoxications and demonstrated toxicity of some cyanobacterial blooms using the mouse bioassay. Almost two decades passed until the issue was addressed again in Germany, now with chemical detection of the most prominent cyanotoxins, the microcystins (Fastner 1994; Jakobi et al. 1996; Jungmann et al. 1996; Ueno et al. 1996b, Fastner et al. 1999). All of these investigations indicated a high incidence of microcystin detection, but did not address cyanobacterial neurotoxicity.

The results of the comprehensive screening program presented here cover 133 waterbodies, 38 of which were sampled regularly or at least three times during the cyanobacterial season for microcystins and hepatotoxicity and often also for anatoxin-a. The occurrence of PSP-toxins was qualitatively assessed in a selected subset of these samples.

Chapter 2.1.1 provides an overview of these water-bodies and their key limnological characteristics. Chapter 2.1.2 presents the results of the survey for occurrence of microcystin and hepatotoxicity, and Chapter 2.1.3 shows neurotoxin occurrence.

2.1.1
The Waterbodies Surveyed for Cyanotoxins in Germany

Claudia Wiedner, Ingrid Chorus and Jutta Fastner

with support from Hans Güde (LAU-BW), Lutz Hoehne (LUA-BB), Antje Köhler (ITOX), Swantje Koch (Tm-UP), Rainer Koschel and Lothar Krienitz (IGB-N), Bernd Melzer (TmFZ), Barbara Meyer (MPI-Pl), Brigitte Nixdorf (BTU-C), Ernst Nusch (Ruhr-V), Martina Oehms and Liane Portugal, (LWW-RP), Hartmut Rönicke (UFZ-M), Hans-Heinrich Schuster (TH-H), Sonja Stottmeister (TmZmW), Ekkehard Vareschi (Uni-O), Norbert Walz (IGB-B), Rosemarie Wanka (StAU-H), Heidrun Wege and Ute Kruspe (Tm-GW), Hartmut Willmitzer (Thü-TV), (legend for abbreviations see Table 2.1)

The waterbodies included in the German cyanotoxin survey were selected to cover a wide variety of types with respect to morphology, mixing behavior, trophic state, cyanobacterial dominance and regional distribution. Limnological characteristics are given in detail for all lakes and reservoirs in Table 2 at the end of this chapter together with the concentrations of cyanotoxins found in each waterbody and the institutes that provided samples and background data. The following overview summarizes the characteristic patterns of morphometry, mixing, trophic state and cyanobacterial taxa which were dominant among the diverse waterbodies sampled. It also provides information on their regional distribution.

Morphology of the waterbody ranges from small, shallow lakes with an area from 0.02 km^2 over large shallow lakes such as Steinhuder Meer with an area of 29 km^2 to deep, large Lake Constance with 472 km^2. More than half of the waterbodies were larger than 1 km^2 and deeper than 5 m maximal depth (Fig. 1A, B). The eight largest of these have an area of more than 5 km^2 and a maximal depth of more than 25 m. A large number of small (< 1 km^2) and shallow (< 5 m maximal depth) waterbodies was also included, particularly among those sampled only once or twice. Mixing conditions split roughly half and half between water-bodies with thermal stratification and dimictic turnover, and polymictic shallow systems (Fig. 1C).

The trophic gradient covered by the survey ranged from oligotrophic to polytrophic (i.e. hypertrophic) as classified by the laboratories providing the samples; however, the majority was classified as eutrophic to poly- or hypertrophic (Fig. 1D). This high share of very eutrophic water-bodies is reflected by frequent cyanobacterial dominance in the samples.

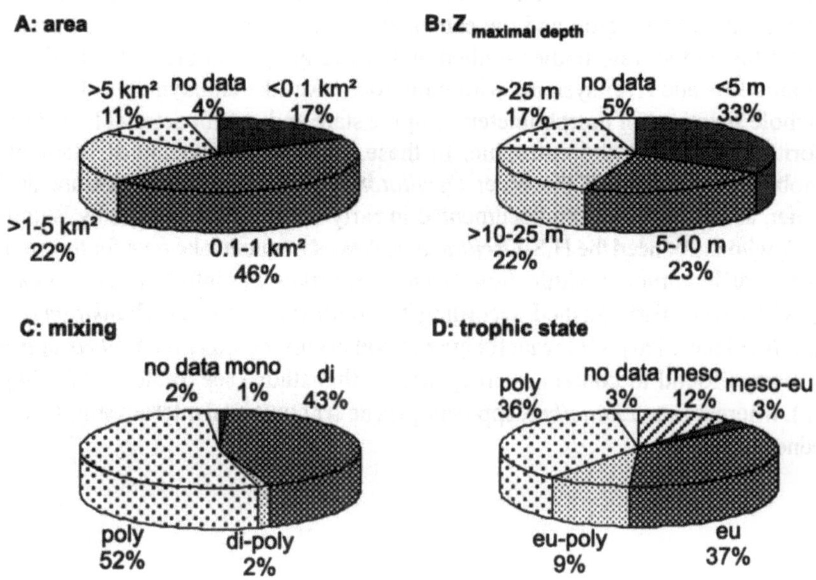

Fig. 1 A-D. Features of the 133 waterbodies surveyed: area (**A**), maximum depth (**B**), mixing characteristics (**C**) and trophic state (**D**)

For 60% of the waterbodies, the share of cyanobacteria in total phytoplankton could be estimated for samples studied. Among these, cyanobacteria contributed 1/3 of total phytoplankton biovolume in 9% of the waterbodies, 2/3 in 14%, and they constituted almost all of the phytoplankton population in 77% of all waterbodies surveyed (Table 2). Detailed information on taxonomic composition of the cyanobacterial populations was available for 73% of all lakes. This shows a wide variety of species and genera. The most widespread genera were *Planktothrix agardhii, Limnothrix* spp., *Microcystis* spp., *Anabaena* spp., *Aphanizomenon* spp. and *Planktothrix rubescens*. Taxa of other cyanobacterial genera were observed in 12 waterbodies (summarized as other cyanobacteria in Table 2).

The occurrence only of some species appeared to be influenced by lake morphometry and trophic state: *Plantothrix argardhii* and *Planktothrix rubescens* show obvious patterns of occurrence in relation to lake morphometry: *Planktothrix agardhii* formed substantial populations in the more shallow and eutrophic waterbodies with not more than 13 m maximal and 6.3 m mean depth, and *Planktothrix rubescens* was recorded only in deeper dimictic lakes and reservoirs (Z_{max} of 12-49 m and Z_{mean} of 6-29 m) with stable thermal stratification which allows the formation of metalimnetic population maxima (Table 2). Further, *P. rubescens* was found only in those classified as meso- or as eutrophic, but never in hyper- or polytrophic waters. In contrast, populations of *Aphanizomenon, Anabaena, Microcystis, Limnothrix* could not be directed to a certain lake type.

The geographic distribution of the waterbodies included in the cyanotoxin survey is shown in Fig. 2. One regional focal point is the Brandenburg and Berlin area with the Scharmützelsee region and the catchment area of the rivers Dahme, Spree and Havel. Most of the waterbodies studied in this area are part of these slowly flowing, eutrophic lowland river systems with numerous lake-like embayments. They cover the whole spectrum of morphometery, trophic state and cyanobacterial taxa, but the majority are shallow and eutrophic. In these, mass developments of filamentous cyanobacteria, mainly of the order *Oscillatoriales*, are a common feature during summer, as has already been documented in early limnological studies by Wundsch (1940), who introduced the H_2S-*Oscillatoria* lakes as frequent lake type for this region. A further differentiation within these *Oscillatoria* lakes was introduced by Rücker et al. (1997), who distinguished according to dominance between *Planktothrix* and *Limnothrix* lakes. This is relevant for cyanotoxin occurrence, as *Planktothrix agardhii* was always found to contain microcystins in this study (see Table 2 and Chapter 2.1.2), whereas pure *Limnothrix* spp.-samples never contained microcystins (see also Wiedner 1999).

Microcystis spp. were dominant in a smaller number of the lakes of this region, often accompanied by minor populations of Aphanizomenon spp., e.g. in the large and polymictic lakes Wannsee and Müggelsee. In Wannsee, these populations are followed by dominance of *Planktothrix agardhii* during autumn, and sometimes also during spells of windy and rainy weather in summer (Fastner 1999). In Mueggelsee, *Aphanizomenon* often became dominant over *Microcystis* (Teubner et al. 1999). The stratified lakes Schlachtensee and Tegeler See contained substantially smaller populations of cyanobacteria as a consequence of restoration (Chorus and Schlag 1993), but when they occurred, *Microcystis* spp. was also the dominant cyanobacterial taxon.

Most of the lakes studied at the northern fringe of Brandenburg and in the neighboring state of Mecklenburg-Vorpommern are small, stably stratifying, dimictic and of considerable depth, classified as meso- to eutrophic. Their cyanobacterial populations typically are quite diverse, with a variety of filamentous species belonging to the genera *Anabaena, Anabaenopsis, Aphanizomenon, Limnothrix, Pseudanabaena, Planktolyngbia,* and sometimes also *Planktothrix agardhii/ rubescens.*

The six reservoirs studied in Thuringia and Saxonia ranged from eutrophic to polytrophic. In Thuringia, the Zeulenroda Reservoir feeds into the Weida Reservoir, and both are still affected by wastewater, through restoration measures are under way. The Weida Reservoir is directly used for drinking-water abstraction, as is the Bautzen Reservoir in Saxonia. In the Pöhl and Weida Reservoirs, *Planktothrix rubescens* was the key cyanobacterial species, whereas the Bautzen Reservoir was dominated by *Microcystis* spp. at the time of the investigation. The Zeulenroda Reservoir harboured only minor populations of cyanobacteria, none of which ever contained detectable concentrations of microcystins or anatoxin-a.

In the state of Schleswig Hollstein, a large number of 47 lakes was sampled once or twice. Many of these calcium-rich hardwater lakes are part of the Schwentine drainage system (see Hofmann 1989 for further information). Due to their location in a lowland region with intensive agriculture, most of these are eutrophic or polytrophic. The lakes are of glacial origin with considerable variation in size and shape, but the the majority are small, shallow and polymictic. Cyanobacteria were present in most of them at the time of the sampling but dominant only in eight. *Microcystis, Anabaena* and *Aphanizomenon* were the genera most frequently observed.

Fig. 2. Location of the water-bodies surveyed and range of microcystin concentrations (in seston dry weight and/or cell-bound concentrations per L^{-1} water, as data are available) *Open triangle* No MC detected; *shaded triangle* 0.1-1 µg L^{-1} and/or 1-100 µg g^{-1} DW; *solid triangle* 1-10 µg L^{-1} and/ or 100-1000 µg g^{-1} DW; *solid circle* > 10 µg L^{-1} and/or > 1000 µg g^{-1} DW

Similarly, a large number of 35 lakes in the state of Baden-Württemberg were also sampled once or twice; however, most of these are less eutrophic and contain a high variety of algal and cyanobacterial populations. With the exception of Buchsee and Bodensee, all of these were small (<1 km^2) and many were polymictic. Not all of them contained cyanobacterial populations at the time of sampling, but where they occurred, *Microcystis* was the most frequent genus.

Arendsee in Sachsen-Anhalt is one of the deepest lakes in the survey (maximal depth of 49 m), and its populatons of *Planktothrix rubescens* contained the highest microcystin concentrations (up to 5600 $\mu g\ g^{-1}$ dry weight) found in the entire survey.

Most of the waterbodies surveyed were intensively used, usually for several purposes: fishing was the usage most frequently given, followed by swimming and other water sports (Table 1). Nature conservation was mentioned as usage of roughly 20% of the waterbodies, whereas only 11 of the total of 133 waterbodies was used as drinking-water resource, either through direct abstraction or through bank filtration. Thus, the key human health issue in most of these waterbodies is risk assessment for exposure through swimming. Although fishing is a use frequently reported, the risk assessment discussed by Vasconcelos in Chapter 2.2.4 for microcystin exposure through consumption of fish indicates that the dose thus acquired is likely to be low, even if the fish are from waterbodies with heavy blooms of cyanobacteria containing high microcystin concentrations.

The following two chapters present results for microcystin concentrations and for neurotoxins in samples collected in the pelagial (i.e. the open water) and not in scums accumulated along shorelines. Thus, they provide a general measure for cyanobacterial population densities and concomitant toxin concentrations, rather than describing worst-case situations. Microcystin concentrations in shoreline scums were sampled in a separate subprogram in 1997, the results of which are reported in Chapter 5.2.

Table 1. Usage of the water-bodies surveyed
Information is available from 103 waterbodies; many of them are used for multiple purposes

Usage	N	%
Fishing	94	91
Swimming	62	60
Other water sports	44	43
Nature conservation	18	17
Drinking-water	11	11

Table 2. (p. 12–20) Waterbodies included in the cyanotoxin survey (1995-1997), prevalent cyanobacteria, cyanotoxins analyzed and their concentration ranges, trophic state, mixing characteristics, morphometric features, utilization, and laboratory providing toxin samples and limnological data. Legend: see p. 21

Lake[a)] and gridlines for northing and easting	Cyanobacteria		Microcystin[d)]				Anatoxin[d)]		PSP				Morphometry				
	Share cyano[b)]	Taxa[c)]		µg g^{-1} DW		µg L^{-1}		µg L^{-1}	Qual. Data only		Trophic state[e)]	Mix -ing[f)]	A	Z_{max}	Z_{mean}	Utilisa- tion[g)]	Studied by[h)]
			N	Mean Range	N	Mean Range	N	Mean Range	N				(km^2)	(m)	(m)		
Alter Weiher 353950 / 531110		L** Pa***	1	0	1	0,0					poly	poly	0.12	3,2	0,8	F S	LAU-BW
Altglobsower See 4575040 / 5889068	***	Pa*** Ab*	1	19			1	0,00	1	-	eu	poly	0.15	3,2	2,8	F S	IGB-N
Arendsee 4455041 / 5851969	**	Ab** Ap** Pr***	17	2298 11-5595			5	0,00	6	+	eu	di	5,14	48,7	28,6	A F N S	UFZ-M
Argensee 356960 / 529950		oC** M**	1	0	1	0,0					meso	poly	0,27	6,4	2,3	A S	
Badsee 357480 / 529060		Ab*** Ap** L*** M*	1	55			1	0,09			poly	di	0,42	4,5		A S	LAU-BW
Behler See 3595820 / 6004410					1	0,0					poly	di	3,30	43,0	13,0	F S	MPI-PI
Belauer See 3581950 / 5997350					1	0,0					eu	di	1,13	25,6	9	F S	MPI-PI
Bibersee 3541 / 5303		oC**	1	0	1	0,0					meso	di	0,03	6,7			LAU-BW
Biggetalsperre Olper Vorbecken 3418114 / 5652154	***	M***	1	282	1	2,0	1	0,00			eu	di	0,88	15,5	6,0	A S	Ruhr-V
Blankensee 4414700 / 5964700	***	M***	5	73 0-147	4	3,6 1,9-5,8	6	0,00			poly	poly	2,50	1,2	0,8	F	LUA-BB
Bodensee Obersee 6528 / 5276		Ap**	1	0	1	0,0					meso	mono	472	252	101	A F S	LAU-BW
Bothkamper See 3574080 / 6009120	***	M*** Ab* oC*			1	6,8					poly	poly	1,23	2,5	1,3	F S	MPI-PI
Brammer Teich 3581750 / 6024250	***	Ap***			1	0,0					poly	poly	0,24	2,5	1,5	F	MPI-PI

Breiter Luzin 4593986 / 5630106	**	L*** Ab* Ap*	18	116 0-570	18	0,5 0,0-2,8	12	0,00	1	+	meso-eu	di	2,68	58,5	25,2	A F S	IGB-N
Buchsee 354280 / 530460		Ab*** Ap** L*	2	312 231-393			3	0,00	2	-	poly	poly	6,06	1,6	0,9	F	LAU-BW
Dagowsee 4570296 / 5891051			1	0			1	0,00	1	+	eu	di	0,30	9,5	5,0	F S	IGB-N
Dannauer See 44044560 / 6012940	***	M*** Ab* Ap*			1	4,7			1	-	poly	poly	0,19	1,3	0,8	F	MPI-Pl
Degersee 354880 / 527500		M*	1	0	1	0,0					meso	di	0,29	17,0	6,0	F S	LAU-BW
Dieksee 4402610 / 6005100					1	0,0					poly	di	3,86	38,0	14,0	A F S	MPI-Pl
Dobersdorfer See 358540 / 6021670	***	M*** Ab*** Ap*			3	0,67 0,0-1,1					eu	poly	3,12	18,8	5,4	F S	MPI-Pl
Dolgowsee 4555811 / 5885467		M***	1	584	1	15,0	1	0,00			eu	di	0,67	10,0			LUA-BB
Ebenweiler See 353770 / 530720			1	0	1	0,0					eu	poly	0,07	3,2	1,8	S	LAU-BW
Edebergsee 3594990 / 6003370					2	0,0					poly	di	0,09	11,0	6,9	F	MPI-Pl
Egelsee		M**	1	0	1	0,0					eu	poly				S	LAU-BW
Elitzer Weiher 3561 / 5281		Ab* Ap* L*** M**	1	0	1	0,0					eu	poly	0,03	1,9	1,1		LAU-BW
Federsee 354700 / 532760		Ab*** M***	1	21			1	0,00			poly	poly				N	LAU-BW
Fuhlensee	**	Ab***			2	0,0					eu	poly	0,14	5,3	2,8	F	MPI-Pl
Glörtalsperre 3395 / 5679		M**	1	1859	1	55,0	1	0,00	1	-	meso	di	0,22	28,0	9,6	A S	Ruhr-V
Gödfeldteich 3591690 / 6014140	***	Ab***			1	0,2					poly	poly	0,51	2,0	1,5	F	MPI-Pl
Görnitzer See 3596550 / 6008810					1	0,0					poly	poly	0,11	2,5	1,8		MPI-Pl
Gr. Binnensee 4411200 / 6022470					1	0,0					poly	poly	5,28	3,0	2,0	F	MPI-Pl
Gr. Bornhorster See 3451600 / 58940000			1	19							meso	di	0,41	23,5	11,4	A S	Uni-O

Lake[a)] and gridlines for northing and easting	Cyanobacteria		Microcystin[d)]				Anatoxin[d)]		PSP				Morphometry				
	Share cyano[b)]	Taxa[c)]		µg g^{-1} DW		µg L^{-1}		µg L^{-1}	Qual. Data only		Trophic state[e)]	Mix -ing[f)]	A	Z_{max}	Z_{mean}	Utilisa-tion[g)]	Studied by[h)]
			N	Mean Range	N	Mean Range	N	Mean Range	N				(km^2)	(m)	(m)		
Gr. Glubigsee 4636937 / 5786324	***	Ap* L***	6	139 0-515	6	7,6 0,0-25,5	5	0,00			eu	di	0,64	12,0	4,0	A F S	TU C
Gr. Großhartmanns-dorfer Teich	***	Ab* Ap*** M**	3	329 79-670	1	1,0	1	0,00			poly	poly	0,61	7,5	2,8	N	Tm FmZ
Gr. Madebrökensee 3595450 / 6003300					1	0,0					eu	di	0,09	10,5	4,0		MPI-PI
Gr. Müggelsee 5408309 / 5812411	**	M** Ab** Ap**	35	224 0-688	98	1,2 0,0-47,6	12	0,00	18	-	eu	poly	7,30	7,5	4,9	A F S	IGB-B
Gr. Plöner See 3592100 / 5997130		M** Ab**			2	0,0					eu	di	29,10	60,5	16	A F S	MPI-PI
Gr. Ursee 357670 / 529090		Ab*** L*** Pa**	1	145			1	0,10			eu-poly	di	0,19	8,0		N	LAU-BW
Gr. Wannsee 4580000 / 5812000	***	M*** Ap** Pa***	41	466 0-1562	39	6,0 0,0-32,4	9	0,00	20	+	poly	poly	2,82	9,8	5,5	A D F S	UBA
Gransee 4567128 / 5874483		M	2	76 53-98	2	4,6 3,6-5,6	2	0,55 0,30-0,80			poly	poly	0,80	2,0		F	LUA-BB
Grebiner See 4401940 / 6009320					1	0,0					eu	di	0,29	25,3	8,3	S	MPI-PI
Grimnitzsee 4621 / 5874	**	Ab*** Ap* M*	8	160 13-511			8	0,01 0,00-0,02	5	+	eu	poly	7,73	9,9	4,6	F S	IGB-B
Gülper See 4522531 / 5840686	**	M**	1	148	1	2,2	1	0,00			poly	poly	6,16	1,5	0,6	F N	LUA-BB
Halener See 5860050 / 3442500		M**	3	0							meso	di	0,12	13,0	6,4	A S F	Uni-O
Haussee 4588881 / 5908227		M* Pa* Pr* Ab* L***	18	144 0-1019	6	0,0 0,0-0,1	16	0,01 0,00-0,05	1	+	eu	di	1,30	12,0	6,3	A F S	IGB-N

Heiligensee 4582 / 5831	***	L***	1	7							poly	di	0,30	9,5	5,9	F S	IGB-B
Höftsee 3595240 / 6003910					1	0,0					poly	di	0,20	19,0	9,0	A F	MPI-Pl
Holzsee 3577380 / 6003830					1	0,0					eu	poly	0,20	9,0	4,7	F	MPI-Pl
Honigsee 3577410 / 6011140					1	0,0					eu-poly	poly	0,05	3,3	1,6	S	MPI-Pl
Hüttensee 355360 / 527920		M*** oC**	1	0	1	0,0					eu	poly	0,02	2,5	2,0	F S	LAU-BW
Illmensee 352850 / 530170		Ab* oC**			1	0,0					meso	di	0,70	16,0	8,4	A S	LAU-BW
Karsee 356050 / 529020		L	3	0	3	0,0					eu	di	0,03	6,0	4,1	F S	LAU-BW
Kirchsee 3583800 / 6012000	***	L*** Ab*** Ap***			1	0,0					poly	poly	0,10	2,5	1,8	A F	MPI-Pl
Kl. Bornhorster See 3451200 / 5895000			1	0							meso	di	0,21	19,2	9,2	A S	Uni-O
Kl. Glubigsee 4637990 / 5787001	***	Ap* L***	6	338 56-813	5	8,5 1,5-19,9	5	0,00			poly	poly	0,06	4,5	3,5	A F	TU C
Kl. Madebrökensee 3595230 / 6002870	***	M*** Ab*** Ap***			1	1,4					poly	poly	0,04	2,0	1,3		MPI-Pl
Kl. Plöner See 3591370 / 6004280	***	oC*** M***			2	0,2 0,0-0,3					eu-poly	poly	2,39	31,4	9,0	F S	MPI-Pl
Kolksee 3586690 / 6009600					1	0,0					eu	di	0,03	9,7	4,1	F	MPI-Pl
Kronsee 3586170 / 6006700	***	Ab***			2	0,0					poly	poly	0,23	7,3	3,4	F	MPI-Pl
Kühreiner Teich 3584020 / 6007450					1	0,0					poly	poly	0,40	2,0	1,5	F	MPI-Pl
Langensee 355270 / 527860		M*** oC***	1	0	1	0,0					eu-poly	poly	0,06	2,0	1,0	F	LAU-BW
Langer See 5426309 / 5721512	***	Ab*Ap** L** Pa***	28	1501 31-2860	22	77,9 0,0-454,9	29	0,21 0,00-1,20	3	-	poly	poly	1,55	3,8		A F S	TU C
Lanker See 3585000 / 6008600	***	M*** Ap* Ab*			3	0,0 0,0-0,1					eu-poly	di	3,80	20,5	3,6	A F S	MPI-Pl

Lake[a)] and gridlines for northing and easting	Cyanobacteria		Microcystin[d)]				Anatoxin[d)]		PSP		Trophic state[e)]	Mix-ing[f)]	Morphometry			Utilisa-tion[g)]	Studied by[h)]
	Share cyano[b)]	Taxa[c)]		µg g^{-1} DW		µg L^{-1}		µg L^{-1}	Qual. Data only				A	Z_{max}	Z_{mean}		
			N	Mean Range	N	Mean Range	N	Mean Range	N				(km^2)	(m)	(m)		
Lebbiner See 5420244 / 5785644	***	L**	3	0	3	0,0	3	0,00								F	TU C
Lebrader Teich 3594040 / 6010900					1	0,0					poly	poly	0,90	1,3	0,5	F N	MPI-Pl
Löptiner See 3579900 / 6005800	***	M*** Ab*** Ap**			1	0,6					poly	poly	0,12	4,5	2,5	S	MPI-Pl
Melangsee 4500000 / 5874015	***	Ab**Ap** L*** M* Pa**	17	236 19-910	17	8,6 0,0-42,7	17	0,01 0,00-0,07	2	+	poly	poly	0,14	3,0	1,2	F	TU C
Metzisweiler			1	0	1	0,0					eu-poly	poly	0,27	2,5	1,6		LAU-BW
Muttelsee 355040 / 5275700		M*** oC***	1	11			2	0,00	1	-	eu	di	0,08	5,5		F S	LAU-BW
Muxaller Teich 3581900 / 6042600					1	0,0					poly	poly	0,08	2,5	1,5	F	MPI-Pl
Ob. Ausgrabensee 3594580 / 6006190		M*** Ap* Ab*			1	11,1					poly	poly	0,02	2,0		F S	MPI-Pl
Obersee 3566 / 5296		Ab L M	2	31 0-61							eu-poly	di	0,25	8,8		F S	LAU-BW
Oestertalsperre 3406618 / 5663475	***	Pa***	1	3603	1	6,7	1	0,00	1	-	meso	di	0,25	30,0	12,4	A S	Ruhr-V
Osterwischteich 3594000 / 6009000					1	0,0					poly	poly					MPI-Pl
Passader See 3585710 / 6024670	***	M*** Ab*** Ap***			2	0,4 0,1-0,6					eu	di	2,73	10,7	4,9	A F S	MPI-Pl
Petersdorfer See 5451308 / 5805812	***	Ap* L***	6	17 0-48	5	2,0 0,0-7,7	4	0,00			poly	poly	0,23	2,6		A F S	TU C
Plußsee 3594310 / 6006720	***	oC*** Ab*			1	0,1					eu-poly	di	0,14	28,0	9,4		MPI-Pl

Postsee 3582100 / 6012320	***	M*** Ap***			2	2,1 0,4-3,8					poly	poly	2,76	9,1	3,3	A F S	MPI-PI
Postweiher 3416 / 5603	***	Ab* M***	3	445 382-569					3	-	poly	poly	0,13	4,5	2,5	A F S	LWW-RP
Rangsdorfer See 4596068 / 5795845	*		2	0	2	0,0	2	0,00			poly	poly	0,03	2,5	1,5		LUA-BB
Rixdorfer Teich 3591710 / 6009370	***	Ap***			1	0,0					poly	poly	0,42	2,5	1,5	F	MPI-PI
Rohrsee 3563 / 5304		oC*	1	0	1	0,0					eu	poly	0,50	2,5	1,0		LAU-BW
Rosensee 3582110 / 6017930					1	0,1					poly	poly	0,38	6,0	3,3	A F	MPI-PI
Rößler Weiher 355110 / 529600			1	0	1	0,0					eu	poly	0,14	4,0	2,5	F S	LAU-BW
Rudower See 1131250 / 5366000		M***	1	24					1	-	eu	poly	2,13	5,0	4,0	F S	UFZ
Ruschweiler See 352760 / 530360		M** L*	1	0	1	0,0					meso	di	0,20	16,0	8,5	F	LAU-BW
Schaarsee 3585670 / 6011200					1	0,0					poly	poly	0,36	2,9	1,6	F	MPI-PI
Scharmützelsee 5431638 / 5785473	***	Ap* L***	10	4 0-32	8	0,0	6	0,00	1	-	eu	di	12,05	28,0	8,8	A F S	TU C
Schierensee Greb 3597470 / 6007920	***	Ap***			1	0,0					eu	di	0,15	20,1	8,0		MPI-PI
Schierensee Wank 3580390 / 5997780					1	0,0					eu	poly	0,27	5,5	3,6	F S	MPI-PI
Schlachtensee 4579397 / 5804895	*	M** Ab*	31	269 0-1047	37	0,8 0,0-4,8	7	0,00	9	+	meso	di	0,42	9,6	4,6	A D S F	UBA
Schloßsee 355580 / 530960		L* M*	1	0	1	0,0					eu	di	0,04	5,8	3,5	F	LAU-BW
Schluensee 35958700 / 6007140		Ab*** Ap*			2	0,0					meso-eu	di	1,27	45	16,3	F S	MPI-PI
Schmaler Luzin / Carwitzer B. 4597868 / 5908922		Ab* Ap** L***	15	130 0-572			10	0,00			eu	di	0,58	33,0	18,1	A F N S	IGB-N
Schmaler Luzin / Mittelbecken 5909000 / 4596000		Ab* Ap** L***	15	192 0-674	21	1,1 0,0-7,9	12	0,00			eu	di	0,76	34,0	13,3	A F N S	IGB-N
Schreckensee 3542 / 5305			1	0	1	0,0					eu	di	0,30	11,3	6,1		LAU-BW

Lake[a] and gridlines for northing and easting	Cyanobacteria		Microcystin[d]				Anatoxin[d]		PSP				Morphometry				
	Share cyano[b]	Taxa[c]		µg g^{-1} DW		µg L^{-1}		µg L^{-1}	Qual. Data only		Trophic state[e]	Mix -ing[f]	A	Z_{max}	Z_{mean}	Utilisa-tion[g]	Studied by[h]
			N	Mean Range	N	Mean Range	N	Mean Range	N				(km^2)	(m)	(m)		
Schwielowsee 4563469 / 5798002	***	Ab* M***	10	699 70-1387	9	8,6 0,2-21,7	9	0,00	2	-	poly	poly	8,50	6,5	3,0		LUA-BB
Selenter See 3593870 / 6020000	***	M*** Ab*** Pa***			2	0,2 0,0-0,4					meso	poly	22,39	35,8	13,2	A F S	MPI-PI
Speicher Radeburg 2 5406618 / 5663475	***	Ap* M***	8	1161 741-2763	26	12,3 26-43,6	10	0,06 0,00-0,25	2	-	poly	poly	2,38	6,0	2,7	D F N	Tm-GW
Springsee 4459193 / 5715417	***	Ap* L***	16	3 0-34	14	0,0	11	0,01 0,00-0,02			eu	di	0,59	18,0	13,2	A F S	TU C
Stadtsee 3592220 / 6004330					1	0,0					eu-poly	poly	0,07	7,6	4,1	F	MPI-PI
Stausee Rötha 4523296 / 5673806	**	Ab*	8	49 0-162	2	0,0	13	0,00	2	-	poly	poly	0,71	2,9	1,5	A F N	Tm-UP
Steinhuder Meer 3522685 / 5807306	***	Ab* Ap** L*** M* Pa*	2	164 101-226			9	0,06 0,00-0,20	2	-	eu-poly	poly	29,10	2,8	1,4	A F N S	TH-H
Stocksee 3589140 / 5996420					1	0,0					eu	di	2,11	30,2	12,3	F	MPI-PI
Stockweiher 353250 / 5299880		Ab*	1	88			1	0,00			eu	di	0,07	4,2	1,9	F	LAU-BW
Stolper See 3580870 / 600460					2	0,0					eu	poly	1,40	14,6	6,9	F S	MPI-PI
Storkower See / Nord 5420244 / 5785645	***	Ab* Ap** L*** Pa*	8	2 0-13	14	0,0	5	0,04 0,00-0,16				di-poly	1,39	11,4	4,7	A F S	TU C
Storkower See / Süd 5420244 / 5785643	***	Ab* Ap** L*** Pa*	8	9 0-70			4	0,1 0,00-0,31				di	2,33	8,0	3,8	A F S	TU C

Suhrer See 4400130 / 6002490					1	0,0					meso	di	1,37	24,7	8,3	S	MPI-PI
Tegeler See 4584709 / 5826800	*	Ab** M**	36	169 0-490	42	0,6 0,0-4,0	5	0,00	7	-	meso-eu	di	4,00	15,9	6,6	A D F S	UBA
Tiefer See 4535170 / 5938706	*	Ab* Pr**	7	11 0-69	6	0,1 0,0-0,5	5	0,00			meso	di	0,63	23,0		F N S	TU C
Tollensesee 4577408 / 5930287	*	Ab**	7	47 0-141	3	1,5 0,6-3,3	2	0,00			eu	di	17,40	33,2	17,2	A F S	IGB-N
Trammer See 3592120 / 6005290	***	L***			1	0,0					eu	di	1,63	33,4	11,1	F S	MPI-PI
Trenthorster See 3588310 / 6008930					1	0,0					poly	poly	0,13	3,0	1,5	F S	MPI-PI
Trentsee 3592350 / 6004620					1	0,0					eu	poly	0,09	6,3	4,8		MPI-PI
TS Bautzen 5453308 / 5662841	**	M***	6	836	5	19,3	5	0,00	2	-	eu	poly	5,33	13,5	7,4	A D F S N	Tm GW
TS Pöhl 451350 / 560120	**	Pr***	3	2091 1722-2677			1	0,00			eu	di	3,65	42,5	15,6	A F S	Tm ZmE
TS Weida 4501297 / 5618570	**	Pr***	31	1638 84-5089	32	3,0	13	0,00	3	-	meso-eu	di	0,90	17,0	10,0	D	Thür-TV
TS Zeulenroda 4488202 / 5607026	*	oC*	29	0	31	0,0	13	0,00			eu	di	2,28	25,0	11,6	D	Thür-TV
Unt. Ausgrabensee 3594030 / 6005920					1	0,0					poly	poly	0,05	6,0	3,3	F	MPI-PI
Vierer See 3594270 / 5999960	***	Ap***			2	1,2 0,0-2,4					eu	poly	1,32	18,8	7,7	F S	MPI-PI
Wielandsee 354840 / 527590		Ab* oC*	1	0	1	0,0					eu	poly	0,02			F S	LAU-BW
Wielener See 3586540 / 6008050	***	Ap***			1	0,3					eu-poly	di	0,25	15,4	6,3	F S	MPI-PI
Wolziger See 5420709 / 5793312	***	Ap** L** Pa***	23	532 0-2205	21	10,2	20	0,05 0,00-0,22			eu-poly	di-poly	5,79	13,0		A F S	TU C
Zeuthener See 5397688 / 5797188	***	Ab* Ap** L* Pa**	16	840 128-2169	15	6,2	17	0,01 0,00-0,05	8	-	poly	poly	2,31	4,0	2,5	A F S	ITOX
Zeuthensee 4555811 / 5885467		M***	1	1617	1	35,7	1	0,00	1	-	poly	poly	0,18	2,0		F	LUA-BB

Lake[a)] and gridlines for northing and easting	Cyanobacteria		Microcystin[d)]				Anatoxin[d)]		PSP				Morphometry				
	Share cyano[b)]	Taxa[c)]		µg g^{-1} DW		µg L^{-1}		µg L^{-1}	Qual. Data only		Trophic state[e)]	Mix -ing[f)]	A	Z_{max}	Z_{mean}	Utilisa-tion[g)]	Studied by[h)]
			N	Mean Range	N	Mean Range	N	Mean Range	N				(km^2)	(m)	(m)		
sampled only for anatoxin-a																	
Gr. Seddiner See 4570440 / 5793815	***	L*** Pa*					1	0,00			poly	poly	2,18	7,2	3,1		UBA
Gr. Vätersee 4605 / 5876	**	oC					1	0,00			meso	di	0,12	11,5	5,2	S	IGB-B
Gröbener See	***						1	0,02			poly	poly	0,41	2,5	1,5		UBA
Schleinsee 354780 / 527250		Ab*					1	0,00			meso	di	0,14	11,6	6,4	F	LAU-BW
Siethener See 4582435 / 5795965		Ab L Pa					1	0,01									UBA
Stechlinsee 4567 / 5890	*	Ab*					1	0,00			oligo	di	4,25	68,0	22,8	A F N S	IGB-N
Süßer See		M					1	0,00			eu	poly				A F N S	StAU-H
Teupitzer See 4610474 / 5780280							2	0,25 0,20-0,29									UBA

a) locations of lakes given as Gauß-Krüger-coordinates (northing / easting)
b) share of cyanobacteria of total phytoplanktonbiovolume (refers to samples analysed for toxins): * = 1 third, ** = 2 thirds, *** = more than 2 thirds
c) most frequent cyanobacteria (refers to samples analysed for toxins): Ab = *Anabaena* spp. and *Anabaenopsis* spp., Ap = *Aphanizomenon* spp., L = *Limnothrix* spp., *Pseudanabaena* spp. and *Planktolynghya* spp., M = *Microcystis* spp., Pa = *Planktothrix agardhii*, Pr = *Planktothrix rubescens*, oC = other cyanobacteria; making with *, **, *** indicates * 1 third, ** 2 thirds, *** more than 2 thirds of total cyanobacteria biovolume.
d) values for the mean and the range (min – max) of concentrations of Microcystins and Anatoxin a; N = number of samples analyzed, DW = dry wight.
e) oligo = oligotrophic, meso = mesotrophic, eu = eutrophic, poly = polytrophic (i.e. hypertrophic if classified after OECD, 1982).
f) poly = polymictic, di = dimictic, mono = monomictic
g) D = drinkingwater, F = fishing, N = nature conservation area, S = swimming, A = other aquatic sports
h) **BTU-C**: Brandenburgische Technische Universität Cottbus, Lehrstuhl Gewässerschutz, Seestr. 45, 15526 Bad Saarow, Prof. Dr. B. Nixdorf
IGB-B: Institut für Gewässerökologie und Binnenfischerei, Müggelseedamm 260, 12562 Berlin, Dr. N. Walz
IGB-N: Institut für Gewässerökologie und Binnenfischerei, Alte Fischerhütte 2, 16775 Neuglobsow, Prof. R. Koschel
ITOX: Umweltlabor ITOX, Chausseestr. 48, 10115 Berlin, A. Köhler
LAU-BW: Landesanstalt für Umweltschutz, Baden-Württemberg, Institut für Seenforschung, Postfach 4953, 88081, Dr. H. Güde
LUA-B: Landesumweltamt Brandenburg, Abteilung Ökologie und Umweltanalytik, Berlinerstr. 21-25, 14467 Potsdam, L. Hoehne
LWW-RP: Landesamt für Wasserwirtschaft, Rheinland-Pfalz, Am Zollhafen 9, 55118 Mainz, M. Oehms
MPI-Pl: Max-Planck-Institut für Limnologie, Postfach 165, 24302 Plön, Dr. B. Meyer
Ruhr-V: Ruhr River Basin Management Association, Chemical and Biological Laboratory, Kronprinzenstr. 37, 45128 Essen, Dr. E. Nusch
StAU-H: Staatliches Amt für Umweltschutz Halle (Saale), Postfach 113, 06036 Halle (Saale), R. Wanka
TH-H: Botanisches Institut, Tierärztliche Hochschule Hannover, Bünteweg 17d, 30559 Hannover, Prof. Dr. A. Schmidt
Thür-TV: Thüringer Talsperrenverwaltung, Talsperrenstr. 25-27, 99897 Tambach-Dietharz, H. Willmitzer
Tm FmZ: Landestalsperrenverwaltung des Freistaates Sachsen, Talsperrenmeisterei Freiberger Mulde/Zschopau, Labor Saidenbach, An der Talsperre 2, 09514 Reifland, B. Meltzer
Tm ZmW: Landestalsperrenverwaltung des Freistaates Sachsen, Talsperrenmeisterei Zwickauer Mulde/Weiße Elster, Labor Plauen, Bärenstr. 46, 08523 Plauen, S. Stottmeister
TmGW: Landestalsperrenverwaltung des Freistaates Sachsen, Talsperrenmeisterei Gottleuba-Weißeritz, Labor Radeburg. Am Stausee, D-01471 Radeburg, H. Wege
TmUP: Landestalsperrenverwaltung des Freistaates Sachsen, Talsperrenmeisterei Untere Pleisse, Gartenstr. 34, 04571 Rötha, S. Koch
UBA: Umweltbundesamt, Institut für Wasser- Boden- und Lufthygiene, Corrensplatz 1, 14191 Berlin, Dr. I. Chorus
UFZ-M: Umweltforschungszentrum, Leipzig-Halle GmbH, Sektion Gewässerforschung Magdeburg, Am Biederitzer Busch, 39114 Magdeburg, Dr. H. Rönicke
Uni-O: Universität Oldenburg, ICBM, AG aquatische Ökologie, Postfach 2503, 26111 Oldenburg, Prof. Dr. E. Vareschi

2.1.2 Microcystins and Hepatocyte Toxicity

Jutta Fastner, Birgit Wirsing, Claudia Wiedner, Rita Heinze, Uwe Neumann and Ingrid Chorus

Introduction

A comprehensive microcystin screening program to assess the extent of the problem was conducted from 1995 to 1997 including more than 100 lakes sampled by a range of institutions (see Chapter 2.1). Targets of this program were to assess

- both microcystin concentrations (determined with HPLC analysis) and hepatotoxicity (determined in a bioassay with primary rat hepatocytes),
- microcystin concentrations in relation to different parameters for cyanobacterial biomass – seston dry weight, chlorophyll-a, and biovolume,
- microcystin concentrations in water, both cell-bound (intracellular) and dissolved (extracellular), and
- differences in toxicity and microcystin concentration of frequent cyanobacterial taxa.

Complementary analysis of microcystin concentrations and hepatotoxic effects aimed at assessing the toxicity of the microcystin mixture in the samples. Additionally, it enables detection of further toxicity not accounted for by microcystins; this is highlighted in Chapter 7.

Estimating human exposure requires toxin concentrations in relation to water volumes potentially ingested as drinking water or when swimming. Whereas swimmers will ingest water including cyanobacterial cells, toxin dissolved in water is relevant for drinking-water, which is either filtered through bank filtration or in the drinking water treatment plant. This chapter addresses both fractions for samples from the pelagic zone (i.e. the open water outside of shoreline scum areas), whereas maximal microcystin accumulations in scums are addressed in Chapter 5.

Sampling and Analysis

As described in Chapter 2.1.1, 124 lakes and reservoirs were sampled for microcystin analysis from 1995 to 1997, most of which were included in the study because they were known to contain cyanobacteria, at least in subdominant quantities. The frequency of sampling differed between water bodies and ranged from weekly over 2 years to single sampling within the 3 years.

Samples were either collected by net hauls for direct determination of microcystin in relation to dry weight, or they were enriched by filtration for direct determination of microcystin in relation to water volume. In order to relate microcystin concentrations to the other parameters addressed by the program, conversions were calculated as follows:

1. Total microcystin concentrations per chlorophyll-a (µg µg^{-1}). For results obtained from net samples, this ratio was calculated from microcystin concentrations per seston biomass (µg g^{-1} DW) and chlorophyll-a concentrations determined in aliquots of the concentrated seston samples. For results obtained from samples on filters, this ratio was calculated from the results for total microcystin L^{-1} and from chlorophyll-a concentrations L^{-1}, which were available for a large number of the waterbodies.
2. Total microcystin concentration per unit cyanobacterial biovolume (µg mm^{-3}). For results obtained from net samples, this ratio was calculated in analogy to (1) from microcystin concentrations per seston biomass and from biovolume determined within the net samples. For results obtained from samples on filters, this ratio could be calculated from lake water biovolume determinations, where these were available.
3. Total intracellular (cell-bound) microcystin concentrations per litre lake water (µg L^{-1}) were either determined directly from filters or were calculated from the microcystin concentrations per chlorophyll-a (µg µg^{-1}) and chlorophyll-a concentrations in the lake water (µg L^{-1}).

The results of these conversions must be considered in the light of bias caused by net sampling, as larger species and colonies (e.g. *Microcystis*) may be selectively enriched in relation to small phytoplankton, which passes through the net. This effect shifts net sample species composition in relation to original seston in favour of larger organisms, which may be the microcystin-containing cyanobacteria. Further, it must be considered that the relation of the microcystin concentrations to biomass parameters such as dry weight and chlorophyll-a is most reliable during pronounced dominance of toxin-producing cyanobacteria. If cyanobacteria are subdominant dry weight is influenced by varying shares of seston other than cyanobacteria. While chlorophyll-a is then more specific, excluding zooplankton and detritus, it still includes other phytoplankton. Thus, the relation to cyanobacterial biovolume is most specific because it directly reflects biomass of the toxin-containing taxa.

Microcystin extraction, microcystin analysis with HPLC for cell-bound and with ELISA for dissolved microcystins, determination of dry weight, chlorophyll-a and biovolume were performed as described in Chapter 9.1. Significance of differences between the mircocystin concentrations in different cyanobacterial taxa was assessed with the t-test based on log-transformed data.

Toxicity was tested using the assay with primary rat hepatocytes developed in the context of this program as presented in Chapter 8.1. Different solvents were tested for extraction (see Chapter 9.2): samples from 1995 were extracted with 100% methanol, and samples from 1996 and 1997 were extracted with 75% aqueous methanol.

Dominant Cyanobacterial Taxa

Information on species composition was available for most of the samples analyzed. Based on results of quantitative and qualitative phytoplankton analysis and the knowledge about potentially microcystin-producing species, samples were grouped by dominance of (1) *Microcystis* spp., (2) *Planktothrix agardhii*, (3) *Planktothrix rubescens*, (4) *Anabaena* spp. and (5) a mixture of these species, (6) samples in which other cyanobacteria than these potential microcystin producers dominated and (7) samples with dominance of other phytoplankton such as diatoms, dinoflagellates etc. Dominance was determined as biovolume of the respective taxon amounting to more than 50% of total biovolume in samples which were quantified. For some samples dominance was determined semiquantitatively in the net samples.

Within the *Microcystis* group the species *Microcystis aeruginosa* and *Microcystis flos-aquae* dominated in most samples, often accompanied by minor quantities of *M. viridis*, *M. wesenbergii*, *M. novacekii* or *M. ichtyoblabe*. A few samples were dominated by *M. viridis*. Among *Anabaena* species *Anabaena circinalis*, *A. flos-aquae*, *A. spiroides*, *A. macrospora* and *A. crassa* were identified. Other cyanobacteria – to date not known to synthesize microcystins – such as *Aphanizomenon flos-aquae*, *Aphanizomenon gracile*, *Limnothrix* and *Pseudanabaena* species dominated in group (6). In these samples as well as in group (7) with samples dominated by other phytoplankton, *Microcystis*, *Planktothrix* or *Anabaena* species were often present in small numbers or even as subdominant populations.

The total of 642 samples was heterogeneous with respect to the species composition. 59 (9% of all samples) were dominated by *Microcystis* spp., 39 (6%) by *Planktothrix agardhii*, 36 (6%) by *Planktothrix rubescens* and only 10 (2%) by *Anabaena* species (Table 2). In 44 samples (7%) a mixture of different potentially microcystin-containing genera (*Microcystis, Planktothrix, Anabaena*) dominated the phytoplankton.

Table 1. Frequency of hepatocyte toxicity in all samples and in samples dominated by different species/genera of cyanobacteria and other phytoplankton

		samples dominated by cyanobacteria potentially producing microcystins					Other cyano-bacteria	Other phyto-plankton or phytoplankton not determined
		One species/genus				Several species		
		M. spp.	P.a.	P.r.	A.spp.	M.spp., P.a., P.r., A.spp		
Number of samples (n)	443	41	36	24	6	28	146	162
Samples toxic to primary rat hepatocytes	185	36	36	24	2	14	25	48
Frequency (%)	42	88	100	100	33	50	17	30

M.spp. = different species of *Microcystis*. P.a. = *Planktothrix agardhii*. P.r. = *Planktothrix rubescens*. A.spp. = different species of *Anabaena*

The majority of all samples was either dominated by other cyanobacteria (240 samples, 37%) or by other phytoplankton (214 samples, 33%).

Frequency of Hepatocyte Toxicity

Of the total of 443 samples tested, 42% were toxic to primary hepatocytes (Table 1). Notably, toxicity was detected in all of the samples dominated by *Planktothrix agardhii* or *P. rubescens* and in 88% of the samples dominated by *Microcystis* spp., but only 33% of the samples dominated by *Anabaena* spp. were toxic, and samples dominated by a mixture of any of these four groups showed toxicity in 50% of the cases studied. Less than 20% of the samples dominated by other cyanobacteria or other phytoplankton were toxic to the hepatocytes (Table 1), and their toxicity may be due to small amounts of *Microcystis, Planktothrix or Anabaena* in these samples.

Frequency of Microcystin Detection

Microcystins were detected in 66% of all samples (Table 2). This higher frequency of microcystin detection in relation to the frequency of toxicity detection (Table 1) can be due to the fact that only about half of all samples were subjected to the hepatocyte assay, but also due to the lower sensitivity of the bioassay, which has a limit of detection of 200 $\mu g\ g^{-1}$ DW (see Chapter 8.1) as compared to 1 $\mu g\ g^{-1}$ DW of HPLC analysis. Thus, samples containing microcystins in the range of 200 $\mu g\ g^{-1}$ DW or less showed no toxicity in the primary rat hepatocyte assay.

Table 2. Frequency of microcystins in all samples and in samples dominated by different species/ genera of cyanobacteria and other phytoplankton

		samples dominated by cyanobacteria potentially producing microcystins					Other cyano-bacteria	Other phyto-plankton
		One species/genus				Several species		
		M. spp.	P.a.	P.r.	A.spp.	M.spp., P.a., P.r., A.spp		
Number of samples (n)	642	59	39	36	10	44	240	214
Samples toxic to primary rat hepatocytes	422	57	39	36	5	36	149	100
Frequency (%)	66	97	100	100	50	93	62	47

M.spp. = different species of *Microcystis*. P.a. = *Planktothrix agardhii*. P.r. = *Planktothrix rubescens*. A.spp. = different species of *Anabaena*.

Table 3. Examples of microcystin concentrations (μg g^{-1} DW) reported in field samples

Country	Samples Toxic/Total	Microcystin	Microcystin concentration [μg g^{-1} DW]	References
Australia	4/?	LR	8000	Jones & Orr 1994
Brazil	6/8	Total	270-1121	Yunes et al. 1998
Canada	168/226	LR	1-1552	Kotak et al. 1995
Canada	37/50	LR	4-610	Kotak et al. 1993
Denmark	198/296	Total	10-1695	Henriksen & Mostrup 1997
Finland	17/20	LR	< 10-900	Lahti et al. 1997
France	16/22	Total	70-3970	Vezie et al. 1997
Germany	8/15	LR	36-364	Fastner 1994
Germany	1 lake	LR	0.4-14700	Jungmann et al. 1996
Ireland	4/4	Total	1.6-168	Sherlock et al. 1997
Japan	11/17	RR, YR, LR	2.5-2120	Park et al. 1993
Japan	60/?	RR, YR, LR	0.02-2260	Park et al. 1998a
Korea	34/47	RR, YR, LR	19-1489	Park et al. 1998b
Portugal	12/12	Total	1600-7100	Vasconcelos et al. 1996
South Afrika	?	Total	5-415	Wicks & Thiel 1990
United Kingdom	11/25	Total	20-3430	Ward et al. 1997
USA	9/?	Total	1865-12800	McDermott et al. 1995

Table 4. Number of samples per waterbody and frequency of microcystin detection

		Number of samples per lake		
	Total	1	2 - 10	> 10
No. of lakes (n)	124	73	34	17
Microcystins detected in lakes (n)	64	24	24	16
Frequency (%)	52	33	71	94

Microcystin detection of 66% on a per-sample basis is within the range of results from other surveys (Table 3). Ueno and co-workers (1996b) detected microcystins by ELISA in 74% of all samples ($n = 38$) from some German lakes, and 70 and 74% of all samples from France and Canada contained microcystins as determined by HPLC (Kotak et al. 1995; Vezie et al. 1997). A monitoring program in Denmark detected microcystins in 66% of almost 300 samples (Henriksen and Moestrup 1997).

As observed for hepatocyte toxicity, microcystins were found in all of the samples dominated *Planktothrix* spp. (Table 2); 97% of the samples dominated by *Microcystis* spp. and 93% of those dominated by a mixture of *Microcystis*, *Planktothrix* and/or *Anabaena* species contained microcystins. By contrast, microcystins were detected in only 50% of all *Anabaena*-dominated samples microcystins (Table 2). About 60% of the samples which were dominated by other cyanobacteria, and about 50% of those consisting mainly of other phytoplankton, were found to contain microcystins (Table 2). The detection of microcystins in these samples could in all cases be attributed to small numbers or subdominant populations of *Microcystis*, *Planktothrix* and/or *Anabaena*.

In 52% of the 124 waterbodies investigated microcystins were detected at least once within the investigation period (Table 4). However, the likelihood of detecting microcystins in a given lake depends to some extent on the frequency of sampling: microcystins were found in only 33% of the waterbodies that were sampled once, but in 71% of those sampled two to ten times and in 94% of those sampled more than ten times within the 3-year period (Table 4).

Levels of Hepatocyte Toxicity and Microcystin Concentrations

Toxicity to Heptatocytes

The median LC_{50} of the 86 toxic samples from 1995 tested after extraction with pure methanol amounted to 0.24 mg DW ml^{-1}, whereas the 94 toxic samples tested from 1996 after extraction with 75% aqueous methanol were found to be almost twice as toxic with a median LC_{50} of only 0.13 mg DW ml^{-1} (Fig. 1A). Since it has been shown that pure methanol does not extract microcystins quantitatively from lyophilized cells and 75% aqueous methanol may extract substantially more microcystins (Fastner et al. 1998), the lower toxicity detected for the 1995 samples may be attributed to a lower microcystin content of this set of samples (extracted with pure methanol) as compared to the 1996 samples (extracted with 75% aqueous methanol). Further, relationships between hepatocyte toxicity and microcystin content of the samples discussed in Chapter 7 provide evidence that the choice of extraction solvents affects toxicity, probably due to solvent-specific extraction of additional bioactive substances.

Microcystin Concentrations per Seston Dry Weight

Four hundred and six net samples were available to study microcystin concentrations in relation to seston dry weight (Fig. 1B). If microcystins were detected, the minimum concentration found was 11 µg g^{-1} DW (at a detection limit of 1 µg g^{-1} DW); 76% of the samples contained less than 1000 µg microcystins g^{-1} DW (Fig.1B), and more than 1000 µg g^{-1} DW were detected in 24% of the samples. Only few samples showed high concentrations of up to 6000 µg g^{-1}. The overall microcystin concentrations per seston biomass found during this study were comparable with findings from other countries (Table 3). They were within the range of results reported from Canada, France, Korea, the United Kingdom and Japan, although occasionally other studies from Germany, Japan and South Africa reported lower concentrations, and another German study, as well as Portuguese and Australian studies, showed higher maximal concentrations (Table 3).

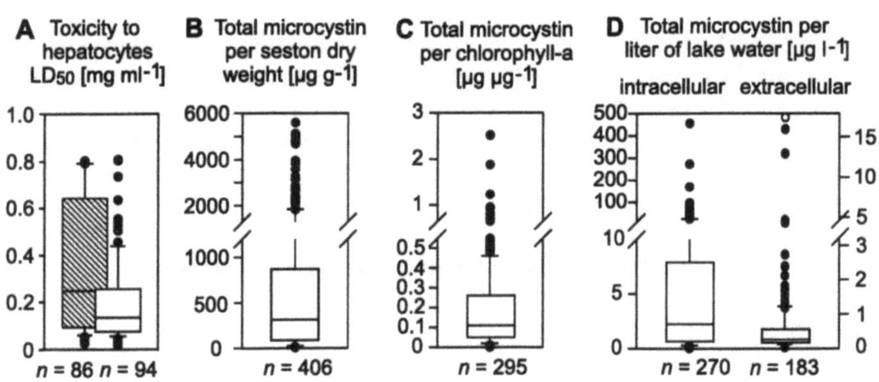

Fig. 1 A-D. Box-Whiskers plots of **A** toxicity of seston dry mass to primary rat hepatocytes after extraction with 100% methanol (*shaded box*) and after extraction with 75% aqueous methanol (*white box*) [LC $_{50}$ in mg DW ml^{-1} test solution]. **B** Microcystin concentrations per seston biomass (µg g^{-1} DW). **C** Microcystin to chlorophyll-a ratios (µg $µg^{-1}$); **D** Concentrations of cell-bound and dissolved microcystin L^{-1} [µg]. *Solid dots* represent data outside the 10th and 90th percentile. *Open circle* in **D** represents five samples with concentrations > 16 µg L^{-1}. Except for A all other results are based on extraction with 75% aqueous methanol

Microcystin to Chlorophyll-a Ratios

No data on microcystin to chlorophyll-a ratios were found in the literature. For field situations dominated by potentially microcystin-producing genera such ratios can be useful for preliminary estimation of microcystin concentrations. Chlorophyll-a concentrations and microscopic determination of taxonomic composition of phytoplankton, routine parameters in many sampling programs, can be used for this purpose. As shown in Fig. 1C, in 94% of 295 samples the ratio of microcystin to chlorophyll-a was lower than 0.5 $\mu g\ \mu g^{-1}$. A factor of 0.5 may thus be used for such preliminary estimates of potential microcystin hazard. Only exceptionally (9% of the samples) were higher ratios of 1-3 $\mu g\ \mu g^{-1}$ found during pronounced dominance of either *Microcystis* spp., *Planktothrix agardhii* or *Planktothrix rubescens* (Figs. 1C, 2C). Furthermore, during monospecific mass developments of *Microcystis* spp., bloom-specific ratios may persist both spatially and temporally within the same waterbody with rather little variation (see Chapter 5.1). Thus, microcystin to chlorophyll-a ratios established for a given population may be applied to interpolate microcystin concentrations from chlorophyll-a concentrations for sampling sites on occasions between microcystin analyses.

Microcystin Concentrations in Water

The total intracellular (cell-bound) microcystin concentration per liter was assessed from 270 pelagic samples by calculation or direct measurement as described above. In 79% of all cases the total cell-bound microcystin concentration in the pelagic zone amounted to less than 10 $\mu g\ L^{-1}$, and the median value was 2.3 $\mu g\ L^{-1}$ (Fig. 1D). Concentrations above 100 were detected only occasionally. The maximum amounted to almost 500 $\mu g\ L^{-1}$.

Extracellular microcystin per liter ($\mu g\ L^{-1}$) was found in 183 of 289 samples tested with the ELISA at a detection limit of 0.1 $\mu g\ L^{-1}$. Concentrations were substantially

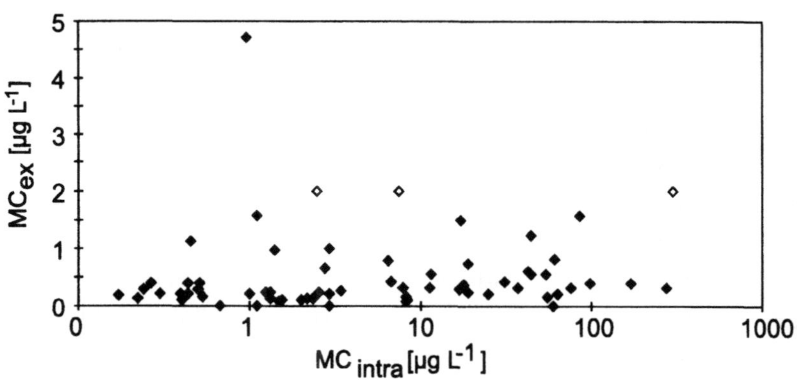

Fig. 2. Extracellular versus intracellular concentrations of total microcystins ($\mu g\ L^{-1}$). Open diamonts: > 1.6 $\mu g\ L^{-1}$, actual concentration not determined

lower than for intracellular microcystins. 75% of the positive samples contained less than 0.5 µg L^{-1} extracellular microcystin, and only 10% contained concentrations between 1 and 16 µg L^{-1} (Fig. 1D). Five samples showed extracellular concentrations above 16 µg L^{-1}, but the exact amount could not be determined, thus these data were not included in Fig. 1D. Three of these contained more extracellular microcystin than cell-bound microcystin L^{-1}, while in all of the other samples, extracellular microcystin concentrations were lower than intracellular concentrations.

Extracellular microcystins showed little relation to the concentration of intracellular microcystins per liter: extracellular concentrations above 0.5 µg L^{-1} were observed over the entire range of intracellular concentrations observed (Fig. 2). These results confirm other observations that microcystin is largely found within the cells and only to a small extent dissolved in water (for an overview see Sivonen and Jones 1999).

In a screening study conducted near Frankfurt, Germany, Ueno et al. (1996b) found total microcystin concentrations from 0.07-32 µg L^{-1} (sum of intra- and extracellular microcystin). Comparison of our data with findings from other countries revealed similar intra- and extracellular microcystin concentrations per liter in the pelagic zone (Table 5).

Table 5. Examples of cell-bound and dissolved microcystin concentrations (µg L^{-1}) in field samples from different countries

		Microcystin concentration (µg L^{-1})		
Country	Toxin	cell-bound	extracellular	References
Canada	MC (LR)	<1-6		Kotak et al. (1995)
Finland	MC	37		Lindholm et al. (1989)
Finland	MC (LR)	< 0.01-3.2	0.001-0.21	Lahti et al. (1997)
Germany	MC (total)	< 0.01-454	<0.1-16	this study
Germany	MC (total)	1,5-36*		Ueno et al. (1996b)
Japan	MC (RR/YR/RR)	0.02-378	0.02-2.64	Tsuji et al. (1996)
Japan	MC (total)	0.052-1300*	0.07-5.6	Ueno et al. (1996b)
Japan	MC (total)	0.06-94*		Nagata et al. (1997)
Japan	MC (total)	0.05-184	0.05-3.61	Park et al. (1998a)
Korea	MC (total)	0.7-856		Park et al. (1998b)
Portugal	MC (total)	0.1-37*		Ueno et al. (1996b)
Thailand	MC (total)	0.082-0.354*		Ueno et al. (1996b)

Hepatocyte Toxicity and Microcystin Concentrations in Relation to the Biomass of Dominant Cyanobacterial Taxa

Differentiated evaluation of the data according to the dominance of different cyanobacterial taxa shows that some taxa are substantially more toxic and contain more microcystin than others.

Toxicity to Heptatocytes

Both *Planktothrix* species were significantly ($p < 0.001$) more toxic compared to the other taxa, showing very low median LC_{50} values of 0.07 and 0.08 mg ml^{-1} test solution. In contrast, the samples dominated by *Microcystis* spp. and those dominated by a mixture of *Microcystis*, *Planktothrix* and *Anabaena* were substantially less toxic, as is reflected by the higher LC_{50} values (Fig. 3A). Differences between solvents (i.e. 100 and 75% aqueous methanol) used for extraction were far less pronounced than differences between taxa.

Microcystin Concentrations per Seston Dry Weight

Total microcystin concentrations per biomass varied significantly between taxa (Fig. 3B; Table 6) and inversely reflect hepatocyte toxicity (Fig. 3A), i.e. taxa with high toxicity (indicated by low LC_{50}) contained high concentrations of microcystins. The highest microcystin concentrations within the 3 years studied were found in samples dominated by *Planktothrix rubescens* (Fig. 3B, Table 6). Concentrations varied from about 3000 to 5595 µg g^{-1} DW when *Planktothrix rubescens* amounted from 80 to 90% of the total phytoplankton. In samples dominated by *Planktothrix agardhii* microcystin concentrations were also high (Fig. 3B, Table 6). The highest amount found in an almost monospecific field sample of *Planktothrix agardhii* was 3603 µg g^{-1} DW, and the median microcystin concentration of samples dominated by this species was 1788 µg g^{-1} DW (Table 6), similar to that of *Planktothrix rubescens*.

Table 6. Total Microcystin concentrations detected within different dominant groups of cyanobacteria

Dominant genus/species	Number of samples	Total microcystin concentrations [µg g^{-1} dry weight] Median	Minimum	Maximum
Microcystis	51	770	21	2763
Planktothrix agardhii	39	1788	19	3603
Planktothrix rubescens	32	1885	415	5595
Anabaena	4	107	11	500
Mixed genera	37	294	11	944

Although microcystin concentrations have been published for isolated strains of *Planktothrix agardhii and Planktothrix rubescens* (Luukainen et al. 1993), reports on microcystin concentrations in field samples dominated by this taxon are scarce. Similar to our results, *Planktothrix agardhii*-dominated samples from Danish

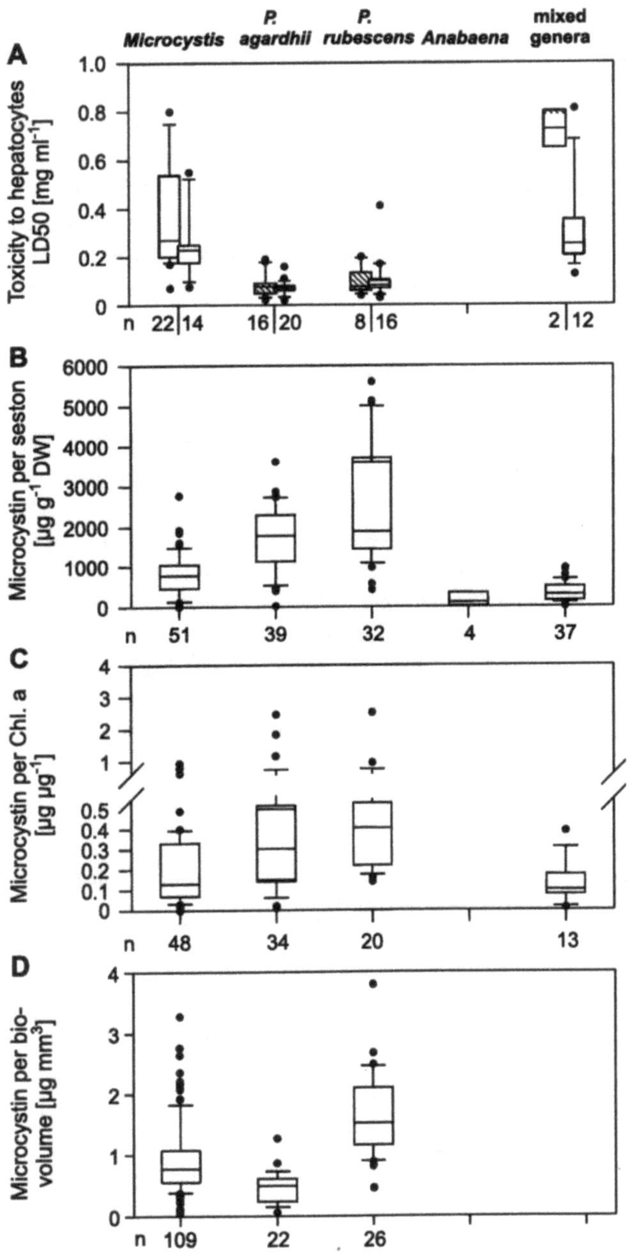

Fig. 3A-D. Box-Whiskers plots of

A toxicity of seston dry mass to primary rat hepatocytes after extraction with 100% methanol (*shaded boxes*) and after extraction with 75% aqueous methanol (*white boxes*) (LC_{50} in mg DW ml^{-1} test solution)

B Total microcystin concentration per biomass (µg g^{-1} DW)

C Total microcystin to chlorophyll-a ratio (µg $µg^{-1}$)

D total microcystin concentrations per bio-volume [µg mm^{-3}] *Solid dots* represent data outside the 10th and 90th percentile. *Boxes* show values during dominance of either *Microcystis* spp.,*Planktothrix agardhii, Planktothrix rubescens, Anabaena* spp. or a mixture of these three genera. Except for **A** all other results are based on extraction with 75% aqueous methanol

waterbodies also contained relatively high microcystin concentrations (Henriksen 1996), with the red-pigmented variants of this species showing the highest concentrations within this group (Henriksen; Chapter 2.4). In contrast, Ueno et al. (1996b) reported substantially lower microcystin concentrations (368 $\mu g\ g^{-1}$) from one sample of *Planktothrix rubescens* from Germany.

During dominance of *Microcystis* spp., concentrations of microcystin were significantly lower, ranging from about 1000 to 2763 $\mu g\ g^{-1}$ DW (Fig. 3B; Table 6) as compared to samples dominated by *Planktothrix agardhii* or *Planktothrix rubescens* ($p < 0.001$). Essentially the same result is reported by Henriksen (1996; Chapter 2.3) for the comparison of Danish populations of *Planktothrix agardhii* and *Microcystis* spp.

The lowest median microcystin concentrations were observed during dominance of different genera/species or during dominance of *Anabaena* species (Table 6); this difference to the other groups was found to be statistically significant ($p < 0.001$). Substantially lower total microcystin concentrations compared to *Microcystis* and *Planktothrix* were also observed during dominance of *Anabaena* spp. or a mixture of several species in Danish waterbodies (Henriksen 1996).

Microcystin to Chlorophyll-a Ratios

Microcystin concentrations in relation to chlorophyll-a showed species-specific patterns for dominance of the different taxa (Fig. 3C) similar to those observed for microcystins in relation to seston dry weight (Fig. 3B): The median ratios for *Planktothrix agardhii* and *Planktothrix rubescens* of 0.3 and 0.4 were significantly higher ($p < 0.001$) than that for *Microcystis* spp. of 0.13 and for the mixed genera group, which amounted to only 0.1. Higher ratios from 1 up to 2.5 could be observed only in some samples during dominance of either *Microcystis* and *Planktothrix* (Fig. 3C).

Microcystin Concentrations per Biovolume

Microcystin concentrations were related to cyanobacterial biovolume only for samples with presence or dominance of only one microcystin-producing taxon, thus allowing allocation of microcystins to that taxon and determination of species-specific microcystin/biovolume ratios. Microcystin to biovolume ratios were found to range from 0.05 to 3.8, with significant differences ($p < 0.001$) between the taxa studied (Fig. 3D): *Planktothrix rubescens* populations showed the highest absolute (3.8 $\mu g\ mm^{-3}$) and median (1.5 $\mu g\ mm^{-3}$) microcystin concentration per biovolume. This coincides with high microcystin concentrations in seston samples of this species. Maximal values found for *Microcystis* spp. were similar to those for *Planktothrix rubescens;* however, the median of *Microcystis* spp. (0.8 $\mu g\ mm^{-3}$) was considerably lower, as was to be expected from the substantially lower microcystin concentrations in seston samples for this taxon (Fig. 3B). Surprisingly, microcystin concentrations per biovolume of *Planktothrix agardhii* showed the lowest median ratio of the species studied, substantially lower than expected from their relatively high seston concentrations. This was due to the different set of samples used: microcystin/biovolume ratios for *Planktothrix agardhii* could be obtained only for a low number of samples which – by coincidence – were those with a relatively low seston content of microcystin; evaluation of a larger number of *P. agardhii* samples for microcystin/biovolume ratios is likely to yield a higher median value.

Microcystin Concentrations in Water

In relation to the lake water volume, concentrations of intracellular microcystin showed differences for the species studied (Fig. 4A): the highest absolute (454 µg L^{-1}) and median (21 µg L^{-1}) concentration observed during dominance of *Planktothrix agardhii* can be explained by the combination of high microcystin concentrations per biomass together with high biovolumes of this species. Although *Microcystis* spp. also formed mass developments, microcystin concentrations of maximally 54 µg L^{-1} in the pelagic samples were significantly lower ($p < 0.001$, median 3.1 µg L^{-1}) due to lower microcystin content of seston biomass. Results for the mixed genera group were similar to those for *Microcystis* spp. (Fig. 4A).

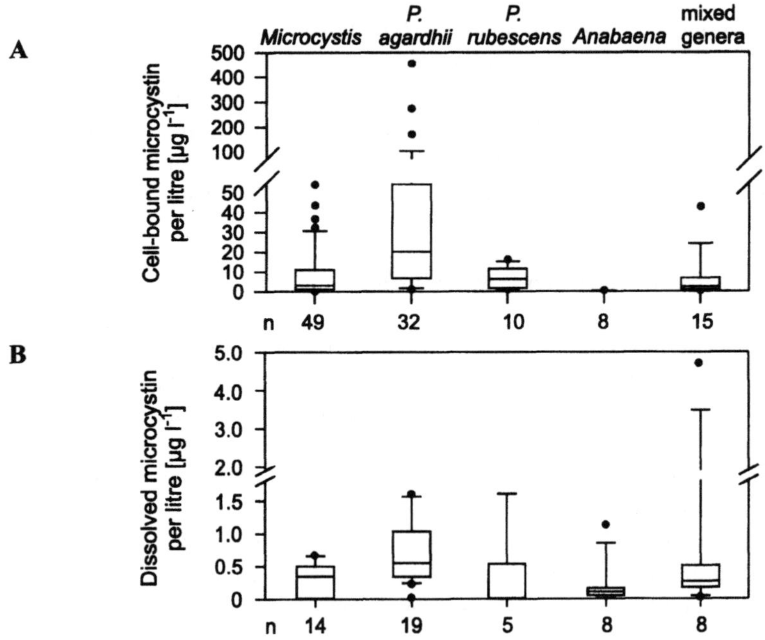

Fig. 4A-B. Box-Whisker plots of concentrations of **A** cell-bound and **B** extracellular microcystins L^{-1} (µg L^{-1}). *Solid dots* represent data outside the 10th and 90th percentile. *Boxes* show values during dominance of either *Microcystis* spp., *Planktothrix agardhii, Planktothrix rubescens, Anabaena* spp. or a mixture of these three genera

Concentrations during dominance of *Planktothrix rubescens* were low (< 16 µg L^{-1}, Fig. 4A) in spite of the high content of biomass of this species (in relation seston dry weight and to biovolume) because this typically mesotrophic species was present in low biovolumes L^{-1} of water.

Although medians of extracellular microcystin concentrations were lower by a factor of 10 and maxima by a factor of 100 as compared to intracellular concentrations, patterns in relation to dominant taxa were similar. The highest median was found during dominance of *Planktothrix agardhii*, and concentrations were lower during dominance of *Microcystis* spp. or *Planktothrix rubescens* (Fig. 4B).

Discussion

This study demonstrates the high prevalence of microcystin-producing cyanobacteria in German freshwaters. A striking result is that almost all blooms of either *Microcystis* spp., *Planktothrix agardhii* or *Planktothrix rubescens* contained microcystins and consequently were toxic to primary rat hepatocytes. Further, toxicity and microcystin content of *Planktothrix* spp. were substantially higher than of *Microcystis* spp. Microcystins were also frequently detected when these taxa were subdominant. In contrast, microcystins and hepatocyte toxicity were less frequent for *Anabaena* spp.; however, only a small number of samples were dominated by this genera.

The fact that about half of all waterbodies investigated contained microcystins at least once during the sampling program reflects the distribution and abundance of these species and their microcystin-producing strains. This frequency of microcystin detection is partially influenced by the specific waterbodies sampled within this study, but also by the sampling frequency. The low percentage of microcystin detection in lakes which were sampled only once may be due to intermediate occurrence of microcystin-producing species or strains. A single sampling of lakes is therefore not likely to give relevant information on the prevalence of microcystin-producing cyanobacteria, and a more frequent sampling of those waterbodies which were sampled only once would have probably increased the percentage of waterbodies in which microcystins could be detected.

Our results on intracellular microcystin concentrations related to the seston dry weight and liter of lake water were similar to those reported from other countries (see above, Tables 3 and 5; Chapter 2.3). The results for extracellular microcystin concentrations confirm reported observations that microcystin is usually largely found within the cells. Extracellular concentrations are generally very low due to the fact that situations of sudden bloom lysis leading to high extracellular concentrations are rare and dilution occurs in the open water.

It must be pointed out that the above-mentioned cell-bound microcystin concentrations L^{-1} were determined in the pelagic zone of the waterbodies, and that microcystin concentrations within lakes may vary substantially due to buoyancy and accumulation of cells by wind. This applies especially to taxa forming surface blooms, such as *Microcystis* and *Anabaena,* which are easily accumulated by several orders of magnitude along shorelines (examples of microcystin concentrations L^{-1} during their accumulation are given in Chapter 5.1). Another species showing concentration gradients is *Planktothrix rubescens* which tends to form metalimnic layers in deep, stratified lakes. By contrast, similar heterogeneity of microcystin concentration within *Planktothrix agardhii* populations is unlikely since this species is usually evenly distributed within the water column.

While early studies published give results in terms of dry weight, recent studies increasingly express microcystin concentrations per liter of water (Table 5), which is required for information on human health concerns. Intracellular microcystin concentration per liter can either be obtained by conversion factors from microcystin concentrations in relation to biomass parameters (such as dry weight, chlorophyll-a or biovolume) and biomass concentrations L^{-1}. However, especially when microcystin-containing species are subdominant, results are more accurate if obtained by measuring microcystin concentrations directly from defined water volumes.

The advantages and disadvantages of measuring microcystins in relation to biomass or to water volume have increasingly been discussed with respect to methods. One obvious advantage of determining toxin in relation to water volume is that filtration and drying of filters is quicker and requires less equipment than freeze-drying net samples. Discussions of the usefulness of results for monitoring and management show that, in practice, two sources of information are needed: toxin concentrations in relation to water volume indicate the actual current exposure hazard at a given site. However, estimation of the potential toxin concentrations developing in the course of bloom formation and bloom accumulation requires a measure of toxin concentration in the cyanobacterial biomass, describing the toxicity of the current cyanobacterial population.

The results reported here give rise to several concerns for the health of users of the waterbodies studied and for aquatic biota: microcystin concentrations may become extremely high if *Microcystis* spp. accumulate to surface blooms, particularly for swimmers and playing children (see also Chapter 5.1). Furthermore, microcystin-producing *Planktothrix rubescens* may occur perennially (Fastner et al. 1999) which may have specific implications: *Planktothrix rubescens* often occurs in meso- to eutrophic deep lakes and reservoirs used for drinking-water abstraction. As this species is difficult to eliminate by conventional flocculation in water-treatment processes (i.e. due to its motility, Lepistö et al. 1994) trichomes may enter drinking water and potentially lead to extended exposition of humans to low levels of microcystin in drinking water. Furthermore, an ecological influence of this potentially perennially persisting species is likely because of possible release of microcystins and other unknown toxic substances in spring when fish eggs are deposited (Feuillade et al. 1996).

A further concern for human health and ecosystem impact of toxic cyanobacteria is indicated by the results of the primary rat hepatocyte toxicity assay which indicate a substantial additional toxic effects in a few samples dominated by *Planktothrix agardhii* (see Chapter 8). Although so far this result is valid only for isolated primary rat hepatocytes and its relevance for entire organisms and ecosystems must be investigated, it does demonstrate the incompleteness of our knowledge of bioactive substances contained in cyanobacteria. While microcystins and hepatotoxicity are likely to be the chief health concern caused by cyanobacteria, risk assessment requires further elucidation of other metabolites (see also Chapter 7).

Sustainable long-term measures to abate these risks will address reduction of eutrophication and, in some cases, hydrophysical management measures to impair growth conditions for the respective taxa. Important immediate measures for reducing public health risks are regular monitoring and raising public awareness regarding the toxic potential of cyanobacteria.

2.1.3
Cyanobacterial neurotoxins

Ingrid Chorus, Christiane Bumke-Vogt, Karl-Erich Lindenschmidt and Elke Jaime

Introduction

Cyanobacterial neurotoxins have not been included in many cyanotoxin surveys. One reason may be that methods applied for cyanotoxin surveys have increasingly shifted from bioassays, which show the integral effect of all toxins, to chemical analyses, which target selected substances. This move has been accompanied by a bias towards assessment of microcystin concentrations. Further, among the cyanobacterial taxa readily accessible for research because they form surface blooms, those containing microcystins, in particular *Microcystis* spp., may occur more widely and frequently than those containing neurotoxins (i.e. *Anabaena* and *Aphanizomenon* spp.). Among the three neurotoxin groups of known chemical structure, anatoxin-a and PSP-toxins have been reported more frequently; a few investigations included anatoxin-a(S). The survey conducted in Germany addressed the occurrence of anatoxin-a and PSP-toxins in various waterbodies with the aim of describing the frequency of occurrence for both neurotoxins. For anatoxin-a, it also aimed at determining ambient concentrations.

Sampling and Analysis

Samples for neurotoxin analysis were taken together with those for microcystin analysis as reported in Chapter 2.1.1.

Anatoxin-a was analyzed from 100-ml samples filtered through cellulose-acetate membranes (pore size 0.45 µm; Schleicher and Schuell, Germany). Intracellular (or cell-bound) anatoxin-a was analyzed from the filters and extracellular anatoxin-a was extracted from the filtrate. Analysis was conducted with the highly sensitive method described by Bumke-Vogt et al. (1996), which is based on derivatization to N-pentafluorobenzyl-anatoxin-a and electron capture detection. By performing gas chromatography with three capillary columns of different polarity, unambiguous identification is possible, and further confirmation with CG-MS in SIM modus was carried out for selected samples. The detection limit of 2.5 pg on column per injection allowed detection of 2.5 ng L^{-1} cell-bound anatoxin-a if 100 ml were filtered and of 50 ng L^{-1} if 5 ml were extracted. The method also detects homo-anatoxin-a which, however, was found only once in a culture of *Anabaena flos-aquae* and never in the field samples.

PSP-toxins were determined only as cell-bound toxin from lyophilyzed samples, in most samples only qualitatively. Analysis was performed as described by Hummert et al. (1995), for some samples modified as described by Yu et al. (1998) and Jaime et al. (1999). Acidic extraction is followed by separation with HPLC. Effective separation of the various PSPs with different toxicity makes use of their different states of ionization in dependence on pH by applying ion pair chromatography for selective effects on hydrophobicity and thus on binding to the stationary phase. For detection, PSPs are transformed to fluorescent derivatives by alkaline oxidation followed by acidification.

As neurotoxins were expected to occur less frequently than microcystins, sample selection for PSP-toxins targeted particularly at those dominated by the cyanobacterial taxa reported to contain neurotoxins (species of *Anabaena, Aphanizomenon* and *Planktothrix*). For anatoxin-a, analysis was more random because for most samples information on dominant taxa was not available at the time of sample selection.

Frequency of Neurotoxin Detection and Concentration Ranges

Both anatoxin-a and PSP-toxins were found in less than 1/4 of all samples analyzed and in 1/4 to 1/3 of the waterbodies investigated (Fig. 1). Thus, it may be concluded

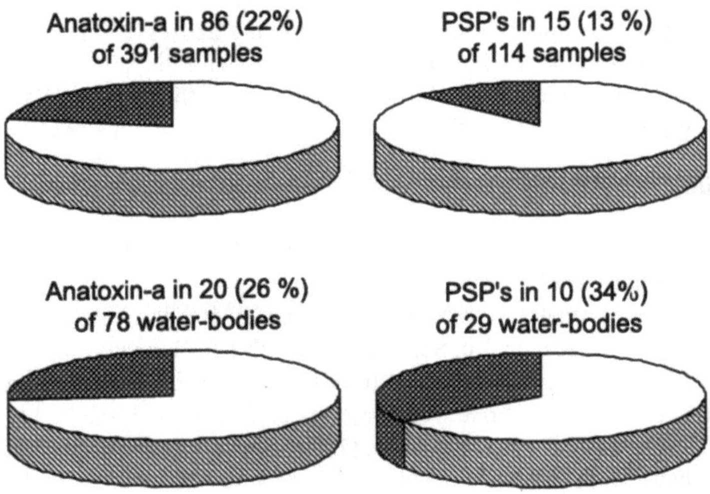

Fig. 1. Frequency of neurotoxin detection in relation to total number of samples and waterbodies

that their frequency of occurrence in Germany is far lower than that of microcystins, which occurred in 66% of all samples analyzed (see Chapter 2.1.2). Extracellular anatoxin-a was not frequently detected. It was found in only 15 of 209 samples analyzed (7%), whereas intracellular anatoxin-a was found in 25% of 328 seston samples. This result contrasts to that for extracellular microcystins, which were found in the majority of the samples analyzed, particularly if intracellular microcystins were detected; see Chapter 2.2).

Neurotoxin concentrations generally were low: the highest level detected was 13 µg L^{-1} for anatoxin-a at the bathing site of the Langer See in July 1997, and the highest concentration found in the pelagic zone was 1.2 µg L^{-1}. Frequency distribution is skewed towards the lower concentration range for both intracellular and extracellular anatoxin-a. Median concentrations were 0.037 µg L^{-1} for intracellular and 0.46 µg L^{-1} for extracellular anatoxin-a (Fig. 2).

Three samples in which PSP-toxins were quantified contained concentrations between 0.0002 and 1 µg g^{-1} DW.

However, in contrast to microcystins, which usually showed substantially lower extracellular than intracellular concentrations, extracellular anatoxin-a concentrations were higher than cell-bound concentrations in all of the nine samples for which both values were analyzed (Fig. 3). This is in agreement with results of a batch culture study with *Anabaena flos-aquae* NRC-44-1, in which extracellular concentrations of anatoxin-a exceeded the intracellular concentrations up to the 7th week of culture growth (Bumke-Vogt et al. 1996). Possibly, exsudation of anatoxin-a from the cells into the medium is more pronounced than for the larger microcystine-molecules, and extracellular anatoxin-a is found in situations where production by cyanobacteria is more efficient than degradation in the medium.

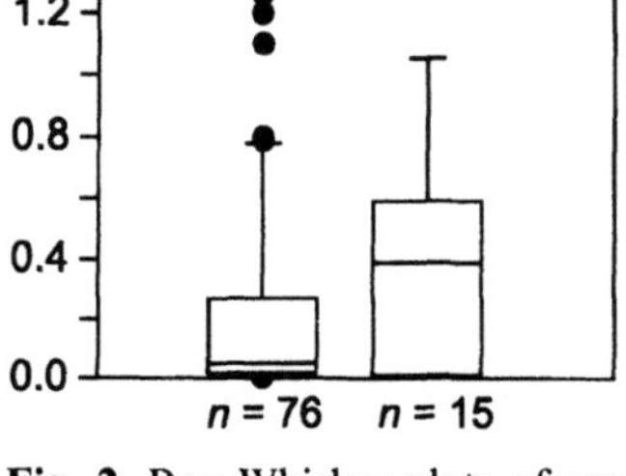

Fig. 2. Box-Whisker plots of concentrations of extracellular (AT_{ex}) and intracellular (AT_{intra}) anatoxin-a (µg L^{-1}). *Solid dots* represent data outside the 10th and 90th percentile

Maximal concentrations in cyanobacterial biomass may be estimated on the basis of several assumptions, the primary one being that the anatoxin-a concentration can be attributed to the biomass of one taxon. However, such samples scarcely occurred in the data set, as most of the samples with anatoixn-a contained several species of cyanobacteria, often including more than one potentially containing anatoxin-a. For one sample, in which *Anabaena* spp. was the only cyanobacterial taxon of significant biomass, the ratio of anatoxin-a to the biovolume of *Anabaena* spp. would amount to 0.045 μg mm^{-3}. Taking 1 mm^3 biovolume to amount to 1 mg fresh weight and dry weight to be 20% of fresh weight, results in 226 μg anatoxin-a g^{-1} DW. Similarly, in another sample 40% of the cyanobacterial biovolume is *Aphanizomenon* spp., and if all of the anatoxin-a in the sample is attributed to this taxon, it would contain 440 μg g^{-1} DW.

Concentrations of anatoxin-a reported in the literature are usually related to total seston dry weight. In cyanobacterial bloom samples, total seston dry weight may be almost equivalent to total cyanobacterial dry weight, but if cyanobacteria are subdominant, other seston particles will 'dilute' the toxin/dry weight ratio. Thus, our figures derived from biovolumes of the toxicogenic species are likely to be skewed to the high end of the range of anatoxin-a/dry weight values reported. In fact, the anatoxin-a concentrations of 226 to 440 μg g^{-1} DW estimated above are in the lower to medium range of concentrations reported in published studies: 0.3-16 μg μg g^{-1} DW found in Japan (Park et al. 1993), 10-100 μg g^{-1} DW reported from Ireland (James et al. 1997), 417-1190 μg g^{-1} DW found in Korea (see Chapter 2.3.8), and up to 4400 μg g^{-1} DW detected in Sweden (Sivonen et al. 1989).

Neurotoxin Occurrence, Species Dominance and Seasonality

Most of the 114 samples analyzed for PSP-toxins proved negative for this toxin group, regardless of the cyanobacterial taxa they contained and regardless of the biovolumes of potentially PSP-producing taxa. Many samples with high biovolumes of more

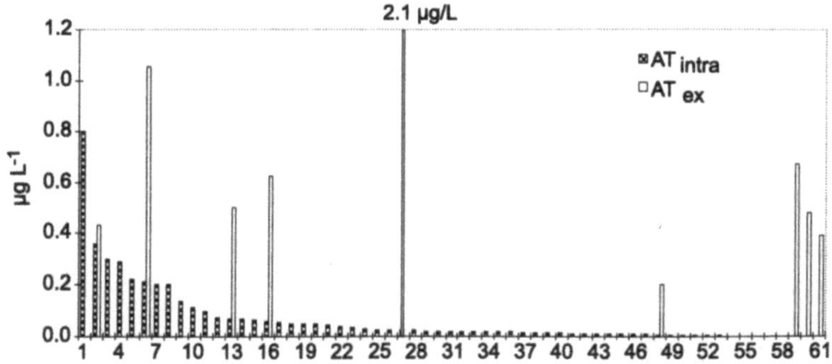

Fig. 3. Intra- and extracellular concentrations of anatoxin-a in 61 samples containing both fractions. (Samples arranged by cell-bound concentrations)

than 10 mm^3 L^{-1} of various cyanobacterial taxa did not contain PSP-toxins, even though their maximal biovolumes were higher than those in the samples with PSP-toxins (Table 1). Among the 15 samples containing PSPs, data on biovolumes of cyanobacterial taxa are available for seven, six of which contained *Aphanizomenon* or *Anabaena*, although in a lower biovolume range than in the samples without PSP-toxins.

Although a higher share of the samples contained anatoxin-a, these also showed no relationship to the biovolumes of any cyanobacterial taxa (Table 2). This may be due to the mixture of taxa in most samples. Particularly the genera often reported to contain anatoxin-a, *Anabaena* spp.and *Aphanizomenon* spp., were dominant only in one and two samples, respectively. *Planktothrix agardhii,* which may also contain anatoxin-a, was dominant in 16 samples, but these also showed no relationship between biovolume and anatoxin-a concentrations.

Occurrence of anatoxin-a showed a clear seasonal pattern in all 3 years studied. Samples taken from April to June and after September contained little or no anatoxin-a. Almost all of the anatoxin-a detected was found from July to September (Fig. 4). Seasonal patterns observed in Langer See and in Zeuthener See coincided with the population development of *Anabaena* spp. and *Aphanizomenon flos-aquae* (Bumke-Vogt et al. 1999).

Table 1. Cyanobacterial taxa, their biovolumes and occurrence of PSP-toxins

Cyanobacterial taxa in the samples analyzed for PSP's	No. of samples with this taxon	Mean biovolumes	% Samples with PSPs; (biovolume range with PSPs)	% Samples without PSPs; (biovolume range without PSPs)
Anabaena spp.	39	1.6 mm^3 L^{-1}	5% (0.05 - 2.1 mm^3 L^{-1})	95% (0.1 - 37 mm^3 L^{-1})
Aphanizomenon spp. (often *A. gracile*)	54	3.3 mm^3 L^{-1}	9% (0.02-7.8 mm^3 L^{-1})	91% (0.1 - 24 mm^3 L^{-1})

Table 2. Cyanobacterial taxa, their biovolumes and occurrence of anatoxin-a

Cyanobacterial taxa in the samples analysed for anatoxin-a	No. of samples	Mean biovolumes	% Samples with anatoxin-a (biovolume range with anatoxin-a)	% Samples without anatoxin-a (biovolume range without anatoxin-a)
Anabaena spp.	51	0.91 mm^3 L^{-1}	48 % (0.01-3.2 mm^3 L^{-1})	52 % (0.003-6.4 mm^3 L^{-1})
Aphanizomenon spp.	70	2.9 mm^3 L^{-1}	44% (0.02-24.3 mm^3 L^{-1})	56% (0.1-14.7 mm^3 L^{-1})
Planktothrix agardhii	44	12.1 mm^3 L^{-1}	57% (0.01-25.8 mm^3 L^{-1})	43% (0.01-12.6 mm^3 L^{-1})

Neurotoxin Risk to Humans

Estimation of human health risks due to the maximum of 13 μg L^{-1} anatoxin-a found in this study can be based on acute toxicity data. No clear evidence for chronic toxicity of cyanobacterial neurotoxins has been reported, and in contrast to microcystins, the neurotoxins' mechanisms of blocking crucial biochemical pathways are reversible, i.e. depuration of toxins occurs if acute toxic effects are not lethal. Thus, concern with regard to chronic toxicity is considered less likely (Kuiper-Goodman et al. 1999).

No guideline values for anatoxin-a in water for human use are available. Fawell and James (1994) found the oral LD_{50} measured over 5 days in a mouse study to be between 5000 and 15000 μg kg^{-1} body weight, where 15000 μg kg^{-1} were immediately lethal within 3 min of application. Fitzgeorge et al. (1994) showed an LD_{50} of 2000 μg kg^{-1} through exposure via nasal mucosa, and exposure of respiratory mucosa is a likely uptake route for swimmers. The LD_{50} was reduced to 500 μg kg^{-1} in the presence of sublethal doses of microcystin. Such a dose could be attained if a child of 10 kg body weight is exposed to 100 ml through aspiration, and this water volume contains 5000 μg of anatoxin-a, or a concentration of 50 mg L^{-1}. Thus, the maximum of 13 μg L^{-1} found in the survey reported here are within a 4000-fold safety margin from acute toxicity.

However, the study includes only a few samples from shorelines and none from scums of *Anabaena* spp. Thus, the survey is not sufficiently comprehensive to exclude risks from such situations. A worst-case scenario may illustrate the potential for lethal intoxications with anatoxin-a. Assuming a concentration of 5000 μg g^{-1} DW (slightly above the 4400 μg g^{-1} DW reported by Sivonen et al. 1989) and a ratio of dry weight to fresh weight for *Anabaena* spp. of 1 to 5, then 5000 μg of anatoxin-a would be contained in 5 g of fresh weight. In theory, if an *Anabaena* concentration were so dense that it displaced all of the water, intranasal exposure to 5 ml of pure *Anabaena* could be in the range of the LD_{50} reported in the presence of microcystins by Fitzgeorge et al. (1994), provided that all of the anatoxin-a in the cells is liberated and comes into contact with the respiratory mucosa and that sublethal concentrations of microcystins (or some other substance enhancing the effect of anatoxin-a) are present in concentrations sufficiently high to cause such a synergism. In practice, such a situation is highly unlikely.

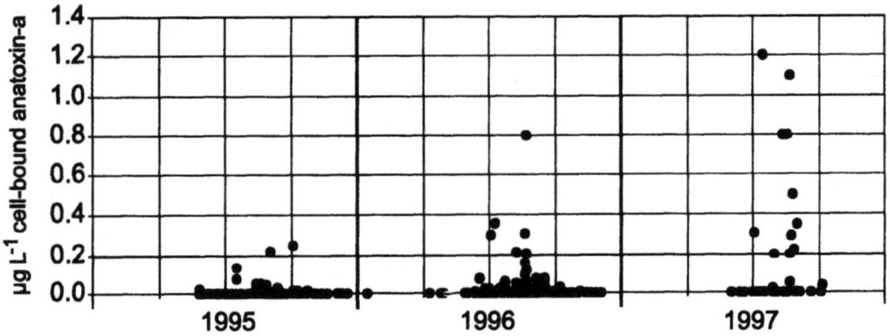

Fig. 4. Concentrations of anatoxin-a (cell-bound) analyzed in summer and autumn in 1995-1997

Liberation of all of the cell-bound toxin is more probable in the digestive tract. If the oral LD_{50} is in the range of 5000 to 15000 µg kg^{-1} body weight, this would be attained by an uptake of 50 - 150 ml of pure *Anabaena* (containing 5000 µg anatoxin-a g^{-1} DW and 1000 µg g^{-1} FW) by a child of 10 kg.

It may be concluded that, although occurrence of anatoxin-a and PSP-toxins is associated with specific cyanobacterial taxa, these only sometimes contain neurotoxins. This contrasts with the result for microcystins, which were usually detected in the presence of *Microcystis* spp. and always in the presence of *Planktothrix* spp. (see Chapter 2.1.2).

References

Bumke-Vogt C, Mailahn W, Chorus I (1999) Anatoxin-a and neurotoxic cyanobacteria in German lakes and reservoirs. Env. Tox. Water Qual. 14:117–126

Bumke-Vogt C, Mailahn W, Rotard W, Chorus I (1996) A highly sensitive analytical method for the neurotoxin anatoxin-a, using GC-ECD, and first application to laboratory cultures. *Phycologia* 35 (6 Supplement):51–56

Chorus I, Schlag G (1993) Importance of intermediate disturbances for the species composition and diversity of phytoplankton in two very different Berlin lakes. Hydrobiologia 249:67–92

Fastner J (1994) First results on the occurence of microcystin-LR in Berlin and Brandenburg lakes. In: Detection methods for cyanobacterial toxins. Codd GA, Jefferies TM, Keevil CW, Potter E (eds) The Royal Society of Chemistry, UK

Fastner J (1999) Microcystins (cyanobacterial hepatotoxins) in German fresh waters: extraction, occurrence, and infuence of environmental factors on microcystin production. PhD Thesis at the Free University Berlin

Fastner J, Flieger I, Neumann U (1998) Optimised extraction of microcystins from field samples – a comparison of different solvents and procedures. Wat Res 32:3177–3181

Fastner J, Neumann U, Wirsing B, Weckesser J, Wiedner C, Nixdorf B, Chorus I (1999) Microcystins (hepatotoxic heptapeptides) in German fresh water bodies. Environ Toxicology 14:13–22

Fawell JK, James HA (1994) (Report No. FR 0434/DoE3728). United Kingdom: Allen House, The Listons, Liston Road, Marlow, Bucks, SL7 IFD

Fitzgeorge RB, Clarc SA, Keevil CW (1994) Routes of intoxication. In: Detection methods for cyanobacterial toxins. (Codd et al., eds) The Royal Society of Chemistry. Cambridge, pp 69–74

Henning M, Kohl JG (1981) Toxic blue-green algae water blooms in some lakes in the German Democratic Republic Int Revue Ges Hydrobiol 66:553–561

Henriksen P (1996) Microcystin profiles and contents in Danish populations of cyanobacteria/ blue-green algae as determined by HPLC. Phycologia 35:102–110

Henriksen P, Moestrup O (1997) Seasonal variation in microcystin contents of danish cyanobacteria. Natural Toxins 5:99–106

Hofmann W (1989) Holstein lakes. In: Lampert W, Rothaupt KO (eds) Limnology in the Federal Republic of Germany, pp 41–45

Hummert C (1995) Der Einsatz chromatographischer Methoden mit selektiver Detektion zur Erfassung von Phycotoxinen aus Algen und Muscheln, Dissertation, Friedrich-Schiller-Universität Jena, Jena, Deutschland

ISO (1992) Water Quality Measurement of biochemical parameters. Spectrometric determination of the chlorophyll-a concentrations. ISO 10260, International Organization for Standardization, Geneva

Jaime E, Hummert C, Luckas B (1999) Eine optimierte HPLC-Methode zur Analyse von PSP-Toxinen in Lebensmitteln mariner Herkunft. 9. Arbeitstagung des Regionalverbandes Südost, Lebensmittelchemie 53:127

Jakobi C, Rinehart KL, Codd GA, Carmienke I, Weckesser J (1996) Occurrence of toxic water blooms containing microcystins in a German lake over a 3-year period. Syst Appl Microbiol 19:249–254

Jones GJ, Orr PT (1994) Release and degradation of microcystin following algicide treatment of a *Microcystis aeruginosa* bloom in a recreational lake, as determined by HPLC and protein phosphatase inhibition assay. Wat Res 28:871–876

Jungmann D, Ludwichowski K-U, Faltin V, Benndorf J (1996) A field study to investigate environmental factors that could affect microcystin synthesis of a *Microcystis* population in the Bautzen reservoir. Int Revue Ges Hydrobiol 81:493–501

Jungmann, D, Benndorf J (1994) Toxicity to Daphnia of a compound extracted from laboratory and natural *Microcystis* spp. and the role of microcystins. Freshwater biol 32:1777–1783

Kalbe L, Duty J, Heise H, Putzke H-P, Schulz E, Tiess D (1976) Zu einer Blaualgen-Wasserblüte (*Nodularia*) der Ostsee im August 1974. Zeitschrift ges Hygiene und Grenzgebiete 22:250–255

Kaya K, Watanabe MM (1990) Microcystin composition of an axenic clonal strain of *Microcystis viridis* and *Microcystis viridis*-containing waterblooms in Japanese freshwaters. J Appl Phycol 2:173–178

Kotak BG, Kenefick SL, Fritz DL, Rousseaux CG, Prepas EE, Hrudey SE (1993) Occurrence and toxicological evaluation of cyanobacterial toxins in Alberta lakes and farm dugouts. Wat Res 27:495–506.

Kotak BG, Lam AK-Y, Prepas EE, Kenefick SL, Hrudey SE (1995) Variability of the hepatotoxin microcystin-LR in hypertrophic drinking water lakes. J Phycol 31: 248–263

Kuiper-Goodman T, Falconer I, Fitzgerald J (1999) Human Health Aspects. In: Chorus I, Bartram J (eds) Toxic cyanobacteria in water: a guide to their public health consequences, monitoring and management. Published on the behalf of WHO by E & FN Spon, London, pp. 133–154

Lahti K, Rapala J, Färdig M, Niemelä M, Sivonen K (1997) Persistence of cyanobacterial hepatotoxin, microcystin-LR in particulate matter and dissolved in lake water. Wat Res 31:1005–1012

Lindholm T, Eriksson JE, JAO Meriluoto (1989) Toxic cyanobacteria and water quality problems – examples from a eutrophicated lake on Aland, South West Finland. Wat Res 23: 481–486

Luukainen R, Sivonen K, Namikoshi M, Färdig M, Rinehart KL, Niemelä SI (1993) Isolation and identification of eight microcystins from thirteen *Oscillatoria agardhii* strains and structure of a new microcystin. Appl Environm Microbiol 59:2204–2209

McDermott CM, Feola R, Plude J (1995) Detection of cyanobacterial toxins (microcystins) in waters of northeastern Wisconsin by a new immunoassay technique. Toxicon 33:1433–1442

Nagata S, Tsutsumi T, Hasegawa A, Yoshida F, Ueno Y (1997) Enzyme immunoassay for the direct determination of microcystins in environmental water. J AOAC Int 80: 408–417

Park H-D, Iwami C, Watanabe MF, Harada K-I, Okino T, Hayashi H (1998a) Temporal variabilities of the concentrations of intra- and extracellular microcystin and toxic *Microcystis* species in a hypertrophic lake, Lake Suwa, Japan (1991–1994). Env Tox Wat Qual 13:61–72

Park H-D, Kim B, Kim E, Okino T (1998b) Hepatotoxic microcystins and neurotoxic anatoxin-a in cyanobacterial blooms from Korean lakes. Env Tox Wat Qual 13:225–234

Park H-D, Watanabe MF, Harada K-I, Nagai H, Suzuki M, Watanabe M, Hayashi H (1993) Hepatotoxin (microcystin) and neurotoxin (anatoxin-a) contained in natural blooms and strains of cyanobacteria from japanese freshwaters. Natural Toxins 1:353–360

Park H-D, Watanabe MF, Harada K-I, Nagai H, Suzuki M, Watanabe M, Hayashi H (1993) Hepatotoxin (microcystin) and neurotoxin (anatoxin-a) contained in natural blooms and strains of cyanobacteria from Japanese freshwaters. Natural Toxins 1:353–360

Rücker J, Wiedner C, Zippel P (1997) Factors controlling the dominance of Planktothrix agardhii and Limnothrix redekei in eutrophic shallow lakes. Hydrobiologia 342/343:107–115

Sherlock IR, James KJ, Caudwell FB, MacKintosh C (1997) First identification of microcystins in Irish lakes aided by a new derivatisation procedure for electrospray mass spectrometric analysis. Natural Toxins 5:247–254

Sivonen K, Himberg K, Luukkainen R, Niemelä SI, Poon GK, Codd GA (1989) Preliminary characterization of neuroroxic cyanobacteria blooms and strains from Finland. Toxicity Assessment 4:339–352

Sivonen K, Jones J (1999) Cyanobacterial toxins. In: Chorus I, Bartram J (eds) Toxic cyanobacteria in water: a guide to their public health consequences, monitoring and management. Published on the behalf of WHO by E & FN Spon, London, pp. 41–112

Teubner, K, Feyerabend R, Henning M, Nicklisch A, Woitke P, Kohl, J-G (1999) Alternative blooming of *Aphanizomenon flos-aquae* or *Planktothrix agardhii* induced by the timing of the critical nitrogen:phosphorus ratio in hypertrophic riverine lakes. Arch Hydrobiol Spec Issues Advanc Limnol 54:325–344

Tsuji K, Setsuda S, Watanuki T, Kondo F, Nakazawa H, Suzuki M, Harada K-I (1996) Microcystin levels during 1992–95 for Lakes Sagami and Tsukui, Japan. Natural Toxins 4:189–194

Ueno Y, Nagata S, Tsutsumi T, Hasegawa A, Watanabe MF, Park H-D, Chen G-C, Chen G, Yu S-Z (1996a) Detection of microcystins, a blue-green algal hepatotoxin, in drinking water sampled in Haimen and Fusui, endemic areas of primary liver cancer in China, by highly sensitive immunoassay. Carcinogenesis 17:1317–1321

Ueno Y, Nagata S, Tsutsumi T, Hasegawa A, Yoshida F, Suttajit M, Krebs D, Pütsch M, Vasconcelos V (1996b) Survey of microcystins in environmental water by a highly sensitive immunoassay based on monoclonal antibody. Natural Toxins 4:271–276

Vasconcelos VM, Sivonen K, Evans WR, Carmichael WW, Namikoshi M (1996) Hepatotoxic microcystin diversity in cyanobacterial blooms collected in Portuguese freshwaters. Wat Res 30:2377–2384

Vezie C, Brient L, Sivonen K, Bertru G, Lefeuvre J-C, Salkinoja-Salonen M (1997) Occurrence of microcystin-containing cyanobacterial blooms of Brittany (France). Arch Hydrobiol 139:401–413

Ward CJ, Beattie KA, Lee EYC, Codd GA (1997) Colorimetric protein phosphatase inhibition assay of laboratory strains and natural blooms of cyanobacteria: comparisons with high-performance liquid chromatographic analysis for microcystins. FEMS Microbiol Lett 153:465–473

Wicks RJ, Thiel PG (1990) Environmental factors affecting the production of peptide toxins in floating scums of the cyanobacterium *Microcystis aeruginosa* in a hypertrophic african reservoir. Environ Sci Technol 24:1413–1418

Wiedner C (1999) Toxische und nicht-toxische Cyanobakterien in Gewässern der Scharmützelseeregion: Ihr Vorkommen in Gewässern unterschiedlicher Trophie und Morphometrie und Steuermechanismen ihrer Dynamik in polymiktischen Flachseen. PhD Thesis at the Brandenburgischen Technischen Universität Cottbus.

Wundsch HH (1940) Beiträge zur Fischereibiologie märkischer Seen, VI. die Entwicklung eines besonderen Seentypus (H2S-Oscillatorien-Seen) im Flußgebiet der Spree und Havel, und seine Bedeutung für die fischereibiologischen Bedingungen in dieser Region. Z. Fischerei XXXVIII: 443–648

Yu RC, Hummert C, Luckas B, Qian PY, Li JI, Zhou V (1998) A modified HPLC method for analysis of PSP toxins in algae and shellfish from China. Chromatographia 48:671–676

Yunes JS, Matthiensen A, Salomon PS, Beattie KA, Ragett SL, Codd GA (1998) Toxic blooms of cyanobacteria in the Patos lagoon Estuary, southern Brazil. J Aquat Ecosys Health 5:223–229

2.2 Comparison of Cyanotoxin Occurrence in Different Countries

2.2.1 Toxic Cyanobacterial Blooms of Inland Waters in Southern Norway 1978-1998

Hans Utkilen, Olav M. Skulberg, Randi Skulberg, Nina Gjølme, Bjarne Underdal

Introduction

In the more densely populated part of southern Norway, cyanobacteria blooms have become a rather common phenomenon in several inland waters. Research on their toxicity was triggered by public complaints about nuisance cyanobacterial blooms in lakes used for multiple purposes, including recreation and a few for municipal water supply.

Phytoplankton was collected at irregular intervals using tow nets samples (25 µm mesh size), water bottles (quantitative analysis, Utermöhl 1958) or membrane filters (0.45 µm; Skulberg 1978). In most cases the sampling was done during summer and autumn when the populations of cyanobacteria were well developed. Taxonomic determination of organisms was performed with an optical microscope on living material and on samples preserved with formaldehyde (Brettum 1989; Skulberg and Skulberg 1985). The regionally most important species were clonally isolated and cultivated. Clone cultures of the toxic strains are kept in the NIVA Culture Collection (Skulberg and Skulberg 1990).

For toxicity testing, material was lyophylilized and then extracted in sterile 0.9% NaCl solution, equivalent to 50 mg DW ml^{-1}. Acute toxicity of the saline suspensions was determined by intraperitoneal (i.p.) injection to mice (standard mouse bioassay, Berg et al. 1987) with two mice to each dosage. Control animals received 1 ml 0.9% NaCl. In addition to observation of the symptoms developed after intraperitoneal injection, gross pathology was performed on mice that died during the experiments, or were killed by spinal dislocation afterwards.

The main types of cyanotoxins involved were classified according to their poisoning effect observed with the mouse bioassay. The responses noticed included three categories: (1) neurotoxins causing rapid death within 30 min after i.p. injection; (2) hepatotoxins causing death within 1-4 h after i.p. injection with massive haemorrage and necrosis of the liver; (3) non-specific toxins causing death within 4-24 h after i.p. injection with no direct observable signs of organ damage.

Table 1. Recognized toxic cyanobacteria in Norwegian inland waters. (Skulberg 1996)

Taxa	Toxins	Type of Poisoning
Chroococcales	Microcystins	Hepatotoxic
Microcystis aeruginosa Kütz.	Microcystins	Hepatotoxic
Microcystis botrys Teil.	Microcystins	Hepatotoxic
Microcystis kützing sp.	Microcystins	Hepatotoxic
Snowella lacustris (Chod.) Kom.et. Hind. [a]	Microcystins	Hepatotoxic
Woronichinia naegeliana (Unger) Elenk. [b]	Microcystins	Hepatotoxic
Nostocales		
Anabaena circinalis Rabenh.	Unknown	Non-specific
Anabaena flos-aquae (Lyngb.) Bréb.	Microcystins	Hepatotoxic
Anabaena lemmermannii P. Richt.	Microcystins	Hepatotoxic
	Anatoxin	Neurotoxic
Anabaena mendotae Trel.	Microcystins	Hepatotoxic
Anabaena solitaria Kleb.	Unknown	Non-specific
Anabaena spiroides Kleb.	Unknown	Non-specific
Aphanizomenon flos-aquae (L.) Ralfs	Unknown	Non-specific
Nostoc Vaucher spp.	Unknown	Non-specific
Oscillatoriales		
Planktothrix agardhii (Gom.) Anagn.et Kom. [c]	Microcystins	Hepatotoxic
	Anatoxin	Neurotoxic
Planktothrix mougeotii (Bory ex Gom.) Anagn.et Kom.	Microcystins	Hepatotoxic
Planktothrix rubescens (DC. ex Gom.) Anagn.et Kom.	Microcystins	Hepatotoxic
Planktothrix prolifica (Gom.) Anagn.et Kom.	Microcystins	Hepatotoxic
Phormidium formosum (Bory ex Gom.) Anagn.et Kom [d]	Microcystins	Hepatotoxic
	Homoanatoxin	Neurotoxic
Phormidium Kützing spp.	Unknown	Non-specific
Trichodesmium lacustre Kleb.	Unknown	Non-specific

Synonyms: [a] *Gomphosphaeria lacustris Chod.*. [b] *Gomphosphaeria naegeliana* (Unger) Lemm.. [c] *Oscillatoria agardhii/rubescens* group. [d] *Oscillatoria formosa Bory ex Gom.*

Results and Discussion

Table 1 shows that cyanobacteria investigated from Norwegian lakes can contain neurotoxins, hepatotoxins and unknown non-specific toxins.

Hepatotoxic cyanobacterial blooms were regularly recorded in 40 lakes between 1978 and 1998. Sixteen of these were dominated by *Anabaena* spp., 11 by *Microcystis* spp., and 9 were dominated by species of *Oscillatoria* (syn. *Planktothrix)*. For a few lakes the abundant toxic cyanobacteria varied during the period investigated, as shown for Lake Frøylandsvatn in Table 2.

Neurotoxic cyanobacterial blooms were recorded in nine lakes between 1978 and 1998. Eight of these blooms were dominated by *Anabaena*, five of them by *A. lemmermannii,* two by *A. flos-aquae* and one by a not identified species of *Anabaena.* One of the neurotoxic blooms was brought about by *Planktothrix agardhii.*

Thus, hepatotoxicity is the most frequent type of cyanobacterial acute toxicity observed in southern Norway, and *Anabaena* is the genus most frequently causing it. In fact, about 62% of the hepatotoxic blooms were due to filamentous cyanobacteria.

A. lemmermannii, a species widely distributed in cyanobacterial blooms in southern Norway, may also contain neurotoxins. The results indicate that species of *Anabaena* were the main producers of both neuro- and hepatotoxins in the lakes examined.

The presence of cyanotoxins of the non-specific category is a regionally widespread event, as indicated by the results of a separate investigation of nuisance blooms of *Aphanizomenon.* The genus *Aphanizomenon* is represented with a few species in Scandinavian inland waters (Cronberg and Komàrek 1994). The species most frequently found are *A. flos-aquae* (L.) Ralfs, *A. gracile* Lemm., *A. klebahnii* (Elenk.) Pechar et Kalina and *A. yezoense* M. Watanabe.

The species referred to here is identified as *A. flos-aquae*, which regularly develops water blooms in eutrophic Norwegian lakes (Brettum 1989). Acute toxicity testing of populations from 8 lakes demonstrated 13 cases with production of non-specific toxin, and non-toxic or only insignificant toxic effects in 10 cases. In no case were neurotoxic responses noticed (Underdal et al. 1999).

These non-specific toxic effects observed in the mouse bioassays with the saline extracts of bloom material from *A. flos-aquae* indicate the presence of secondary

Table 2. Dominating cyanobacteria in hepatotoxic blooms from lake Frøylandsvatn 1978-1998

Dominant species	Date
Microcystis aeruginosa	1978 - 1990
Planktothrix mougeotii *	05.07.1983
Planktothrix mougeotii *	30.07.1984
Planktothrix mougeotii *	27.08.1984
Anabaena lemmermannii	17.06.1991
Anabaena lemmermannii	01.07.1991
Anabaena circinalis, Anabaena lemmermannii	21.07.1993

* *Oscillatoria agardhii* var. *isothrix*

metabolites whose chemical structure so far is unknown. Previous literature states that most strains of the ubiquitous species *A. flos-aquae* are apparently non-toxic, or do not produce cyanotoxins under ambient environmental conditions (Gleason and Wood 1987). In contrast, the results obtained in the Norwegian investigation indicate the frequent occurrence of compounds with a non-specific toxic effect in populations of *A. flos-aquae*.

However, pronounced poisoning of fish eggs due to unknown substances in species of *Aphanizomenon* and other cyanobacterial taxa is reported in Chapter 7.2, this Volum. Together, these results show that chemical structures of further cyanotoxins remain to be elucidated.

2.2.2 Toxic Freshwater Cyanobacteria in Denmark

Peter Henriksen

Introduction

The first Danish record associating fish kills and impoverished health of cattle after consumption of lake water with the green-colouring of the water dates back to 1833 (Moestrup 1996). This has recently been followed by further well-studied animal intoxications in connection with cyanobacterial blooms in lakes (e.g. Henriksen et al. 1997) and in coastal water (Nehring 1993). From a human health perspective, the presence of tumour promoting hepatotoxic microcystins in drinking water reservoirs is a primary concern. Fortunately, Denmark has good groundwater supplies for drinking water, and only two waterworks, in the vicinity of the capital Copenhagen, supplement the groundwater with a maximum of 50% surface water during the summer period (and on an annual basis max. 10-15%). However, the annual recurrence of cyanobacterial blooms in a large number of lakes used for recreational purposes poses a potential acute human health hazard.

The occurrence and toxicity of cyanobacteria in Danish freshwaters was studied from 1993 to 1995. The results presented are based on 309 phytoplankton net samples collected from the surface water of 122 localities. Toxicity of all samples was examined by mouse bioassays and high-performance liquid chromatography (HPLC) for microcystins. In addition, one sample from each locality was analyzed by HPLC for the neurotoxins anatoxin-a and paralytic shellfish poisons (PSP). Acetylcholinesterase (AChE) inhibition assays for anatoxin-a(S) were performed on all samples from localities where one or more samples resulted in a neurotoxic response and salivation in the test mice.

Toxicity of Natural Populations of Cyanobacteria

In 109 localities (89%), one or more samples showed some type of toxicity in the mouse bioassay or contained small amounts of toxins as detected by HPLC or by AChE inhibition. In only 13 localities all samples examined were non-toxic.

Based on survival times of mice, mouse bioassays revealed three types of toxic responses: neurotoxic (< 45 min), hepatotoxic (1-4 h), and protracted toxic (4-24 h) (Table 1). The majority of the samples (58%) were hepatotoxic, 7% neurotoxic, and 17% showed a protracted toxic response. Less than 20% of the samples were non-toxic by this assay. Application of the more sensitive detection methods HPLC and AChE inhibition assay increased the frequency of hepatotoxic and neurotoxic samples to 71 and 9%, respectively, leaving only 13% without detected toxicity/toxins (Table 1).

The most common genera with previously reported toxin production potential were *Anabaena*, *Aphanizomenon*, *Microcystis* and *Planktothrix* (mainly *P. agardhii*, syn. *Oscillatoria agardhii*). In addition, *Woronichinia naegeliana* (syn. *Gomphosphaeria naegeliana*) was very common.

Only 16 samples contained only one single species. Most samples were composed of one or several dominant species and additional subdominant species present in low numbers. Toxicity of samples dominated by the ten most common species in the toxic samples is illustrated in Fig. 1. Samples dominated by *Microcystis* generally comprised several dominant species and showed very high frequencies (89-100%) of hepatotoxicity. They were rarely neurotoxic or protracted toxic. The lack of protracted toxicity in these samples may be due to a masking of this type of toxicity by the hepatotoxins.

Eighty percent of samples dominated by *Planktothrix agardhii* were hepatotoxic and protracted toxicity was found in 26%.

Half of the samples (53%) dominated by *Anabaena flos-aquae* contained hepatotoxins, whereas 33% were protracted toxic and 26% non-toxic in the mouse bioassay. Neurotoxins were detected in only 7% of the samples. In contrast, 71% of

Table 1. Toxicity of cyanobacterial samples determined by mouse bioassay, HPLC and acetylcholinesterase inhibition

Toxicity	Survival time of mice	Frequency by mouse bioassay (%)	Frequency by all detection methods[a] (%)
Hepatotoxic	1-4 h	58	71[b]
Neurotoxic	< 30 (-45) min	7	9[b]
Protracted toxic	4-24 h	17	17[c]
Non-toxic	> 24 h	18	13

[a] Samples with low levels of microcystins or neurotoxins were designated hepato- or neurotoxic irrespective of the result of the mouse bioassay. The total percentage adds up to more than 100 since several types of toxins/toxicity were detected in some samples. [b] Samples containing both microcystins and neurotoxins are included in both categories of toxicity. [c] In one third of the samples small amounts of microcystins were detected by HPLC. These have also been included as hepatotoxic

samples dominated by *Anabaena lemmermannii* contained neurotoxins, and hepatotoxins were detected in 33%. Only 13% of these samples were non-toxic, and protracted toxic response was not seen.

Aphanizomenon flos-aquae has frequently been reported toxic. However, the records of neurotoxic *A. flos-aquae* do not provide information on the varieties present. In the Danish material, *A. flos-aquae* was divided into the two varieties, *flos-aquae* and *klebahnii,* according to Komárek and Kovácik (1989). The variety *flos-aquae* was found in only a few samples, and samples dominated by the common variety *klebahnii* were often non-toxic, as were all laboratory cultures examined (17 isolated from 8 localities). The apparent discrepancy between the Danish blooms of *Aphanizomenon flos-aquae* without detectable neurotoxins and those of neurotoxic blooms from other countries may be related to the dominance of different varieties in

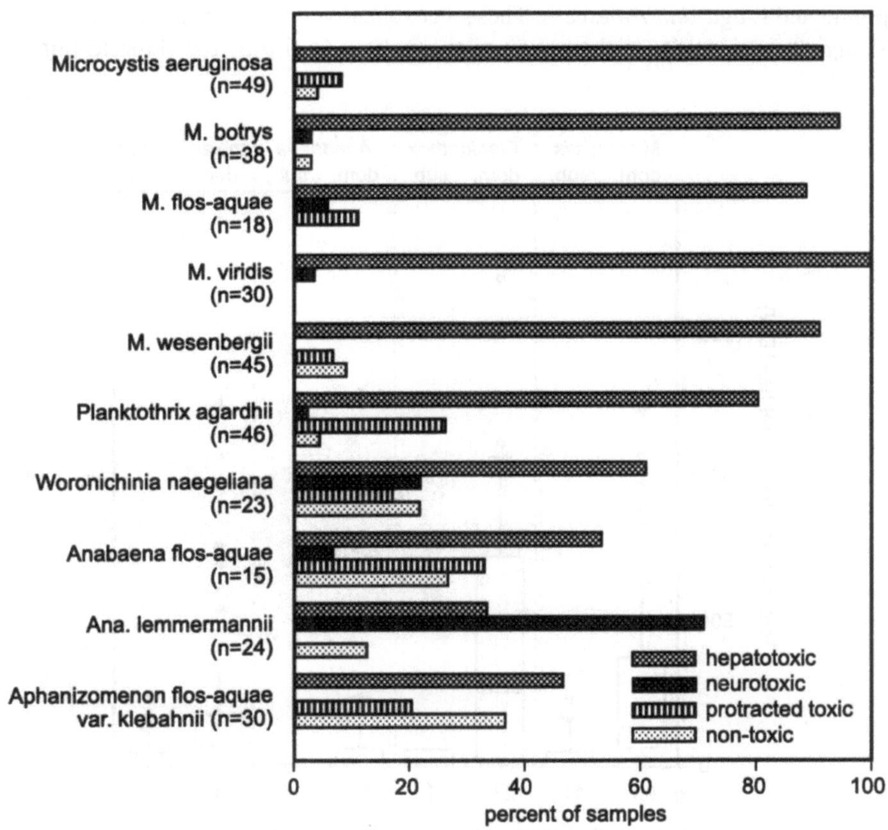

Fig. 1. Toxicity of samples dominated by the most common species.
Dominance was defined as a 50% contribution by one species (or several species combined) to the total phytoplankton biomass, based on arbitrary estimates from preserved net samples. As very few samples were dominated by a single species, all samples in which the given species was among the dominant species have been included. For each species, the total percentages may add up to more than 100% due to the presence of several different types of toxicity in some samples

different areas. The hepatoxicity associated with Danish blooms dominated by *A. flos-aquae* var. *klebahnii* was most likely due to the presence of low cell numbers of hepatoxic species (e.g. *Microcystis* spp.).

Microcystin Contents, Profiles, and Seasonal Occurrence

Qualitative and quantitative comparison of microcystin contents in samples dominated by *Planktothrix agardhii*, *Microcystis* spp., *Anabaena* spp. or multiple genera showed profound differences associated with the dominant taxa (Henriksen 1996). The highest median microcystin content (254 μg g^{-1}) was found in samples dominated by *Planktothrix agardhii* (Fig. 2). Excluding seven samples without detected microcystins, the median toxin content was 494 μg g^{-1}. Only two samples (containing 1695 and 1004 μg microcystin g^{-1}, respectively) were dominated by red-pigmented *P. agardhii/rubescens*. These were tentatively included as red-pigmented *P. agardhii*, since the trichomes were too narrow (< 6 μm) for classification as

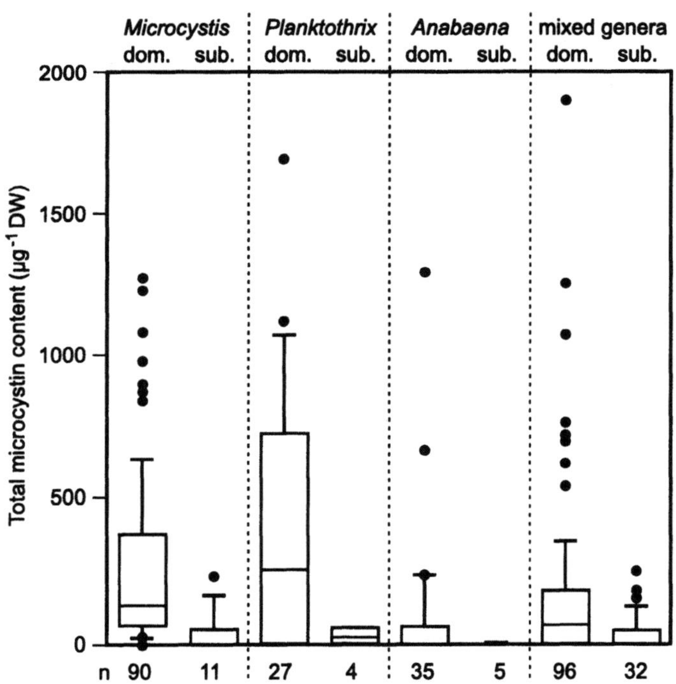

Fig. 2. Total microcystin content of samples from 1993-1995 dominated by different genera. In samples categorized as dominant (*dom.*), the given genus/species alone contributed more than 50% to the total phytoplankton biomass (arbitrary estimates from preserved net samples). Samples categorized as subdominant (*sub.*) were dominated by genera without documented microcystin-production potential. Two *Planktothrix* samples were dominated by red-pigmented *P. agardhii/rubescens*. *Boxes* show median values and the interval from the 25th to the 75th percentile, *bars* indicate the 10th and 90th percentile

P. rubescens according to Gomont (1893). Laboratory cultures of red- and green-pigmented *P. agardhii* revealed differences in toxin composition and frequency of toxicity. Eighteen red strains isolated from seven localities all produced microcystins and had almost identical toxin profiles with one major toxin ([D-Asp^3]microcystin-RR), whereas only 12 out of 20 green strains representing 9 localities produced microcystins. In addition, green strains had more variable toxin profiles, and several produced approximately equal amounts of two major toxins, some of which have not yet been chemically characterized (Fig. 3).

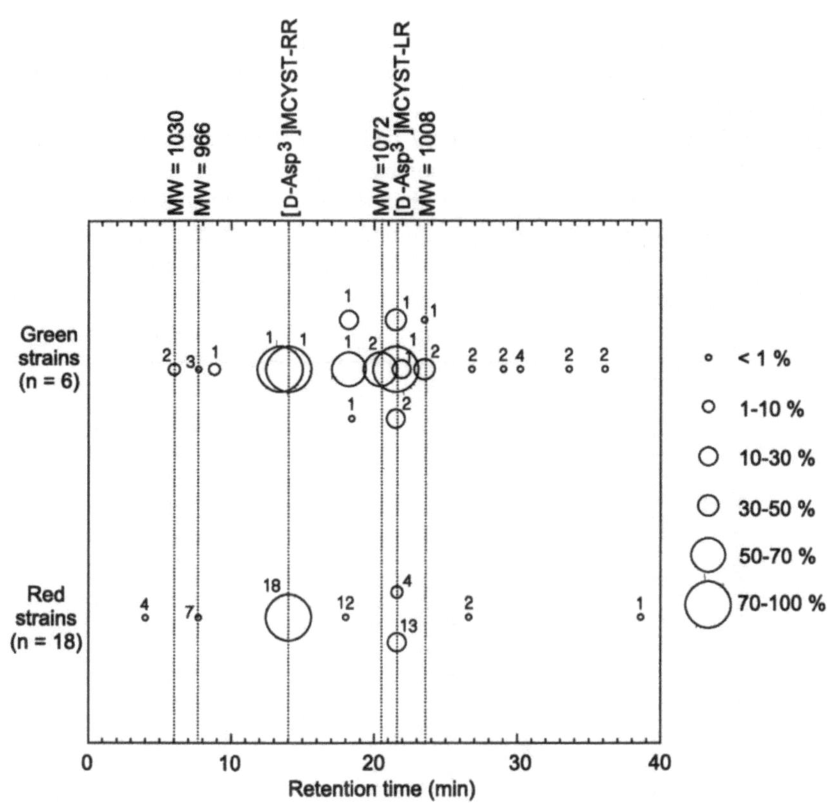

Fig. 3. Toxin profiles of green and red strains of *Planktothrix agardhii*
Aanalyzed on a Waters Symmetry C_{18} (3.9 x 150 mm) column at a flow rate of 1 ml min^{-1} using a linear gradient of 22-28% (0-25 min), 28% (25-32 min), and 28-22% (32-36 min) acetonitrile in 0.01 M ammoniumacetate (pH 5). *Markers* show percentage of total microcystin content for the toxin eluting at the given retention time. Interstrain differences in contribution by some toxins to the total microcystin content is illustrated by several markers at identical retention time. *Figures adjacent to markers* indicate the number of strains in which the toxin was found. Microcystins identified (mass spectrometry and amino acid analysis) and microcystins presently being characterized (only molecular weight given) by Jesper Laub, University of Copenhagen, are indicated by *dotted lines*. Toxin profiles of all red strains and 6 out of 12 toxic green strains are shown

The major microcystin found in all red strains was identified as [D-Asp^3]-variant of microcystin-RR byHPLC and amino acid analysis. The methods used will not separate [D-Asp^3]microcystin-RR from [D-Asp^3, Dhb^7]-microcystin-RR and thus the main microcystin in the red-pigmented Danish strains could be either of these microcystin variants.

Most samples dominated by *Microcystis* were characterized by several codominant species. The median microcystin content was 137 μg g^{-1} (162 μg g^{-1} when excluding eight samples without detectable amounts of microcystins) (Fig. 2), and the most common toxins were microcystin-LR, -RR, and -YR (Henriksen 1996). In addition, several samples dominated by *Microcystis* contained [Dha^7]-variants of microcystin-LR and -RR (Henriksen and Moestrup 1997).

Diverse toxin composition and low microcystin content (median = 0 μg microcystin g^{-1}, and 61 μg g^{-1} when excluding 19 non-toxic samples) were characteristic of samples dominated by *Anabaena* spp. (Fig. 2).

Samples dominated by several genera had a median microcystin content of 63 μg g^{-1} (111 μg g^{-1} for microcystin-containing samples only) (Fig. 2) and generally a less homogeneous toxin composition than samples dominated by *Microcystis* spp. (Henriksen 1996). This group included the highest microcystin content found in this study (1894 μg g^{-1}) in a sample co-dominated by *Anabaena* sp. and *Planktothrix agardhii*.

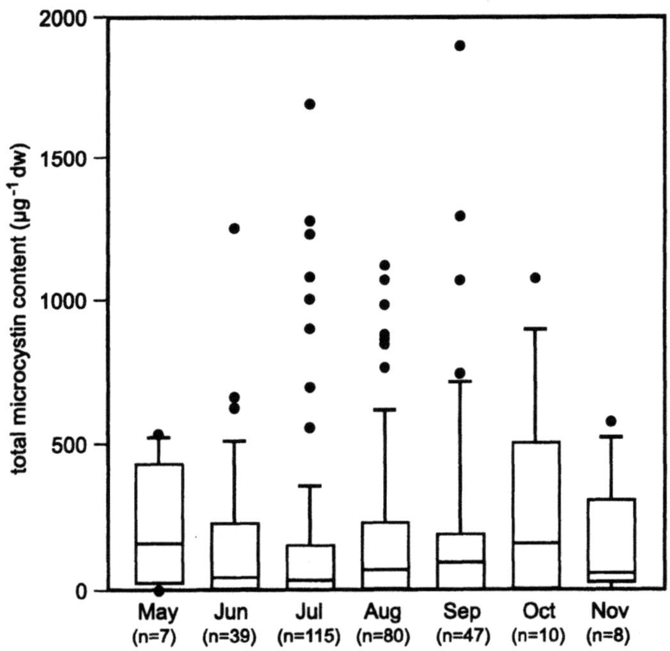

Fig. 4. Total microcystin content of samples from 1993-1995 when grouped by month of collection. *Boxes* and *bars* as in Fig. 2

The presence of toxic cyanobacteria at times other than during maximum abundance in summer/autumn was emphasized when grouping all samples by month of collection (Fig. 4). Sampling covered the season from May through November, and the highest microcystin contents of individual samples were found during high cyanobacterial abundance in July and September when *Microcystis* and *Anabaena* were the most common dominant genera, but median microcystin contents amounted to only 34 and 92 $\mu g\ g^{-1}$ (Henriksen and Moestrup 1997). In contrast, median microcystin content was high (160 and 161 $\mu g\ g^{-1}$) in samples collected in May and October where *Planktothrix agardhii* was the dominant species in 67 and 50% of the samples, respectively (Henriksen and Moestrup 1997). This species contains high microcystin concentrations, particularly under low light conditions (Sivonen 1990) and has previously formed blooms in the Nordic countries in early spring, even under ice cover (Berg et al. 1986, Lindholm et al. 1989).

Neurotoxins in Danish Cyanobacteria

The most common cyanobacterial neurotoxin, anatoxin-a, was not found in the Danish samples. The geographically most widely distributed neurotoxins were PSP-toxins, detected in 11 localities (Kaas and Henriksen 2000). The most frequent toxins were saxitoxin (STX) and gonyautoxin 4. Converting the total PSP-toxin content to STX-equivalents according to Humpage et al. (1994), the maximum toxin content was 160 μg STX-equivalents g^{-1}, and 9 out of 11 samples contained $< 40\ \mu g$ STX-equivalents g^{-1} (Kaas and Henriksen 2000). *Anabaena lemmermannii* was a dominant species in all samples collected from seven of the localities with detected PSP-toxins.

Ten Danish neurotoxic samples representing three localities gave signs of salivation and lachrymation in test mice when injected with doses around the LD_{50} and, in addition, showed inhibition of acetylcholinesterase. The toxic principle in the samples, of which nine were dominated by *Anabaena lemmermannii*, was an anatoxin-a(s)-like anticholinesterase also produced by several laboratory cultures of *A. lemmermannii* isolated from one of the blooms (Henriksen et al. 1997). The anticholinesterase from one of the natural bloom samples and a culture has subsequently been chemically characterized as anatoxin-a(s) by high-resolution FAB-MS and 1H and 13C NMR (Onodera et al. 1997).

Several neurotoxic laboratory cultures of *Anabaena* cf. *cylindrica* and *Oscillatoria* sp. were established. Neither anatoxin-a, anticholinesterase activity, nor PSP toxins was detected in these cultures (Onodera et al. 1996; own unpubl. results). They have not been analyzed for homoanatoxin-a.

In addition, neurotoxic benthic *Oscillatoria* containing yet unidentified toxin(s) were involved in intoxication of two dogs during the summer of 1996.

Episodes of Animal Poisonings

In spite of the high frequency of toxicity found in the present study (Table 1), only two samples were collected in connection with incidents of animal poisonings. They were dominated by neurotoxic *Anabaena lemmermannii* in Lake Knud Sø in June-July 1993 and 1994 in connection with registered bird-kills (Henriksen et al. 1997). These samples contained 2300 and 3300 µg anatoxin-a(s) g^{-1} DW, respectively.

Within the study period, mass mortality of fish (50000-100000) was registered during a highly hepatotoxic bloom of *Anabaena flos-aquae* in Esrum Å and Esrum Mølledam containing 2000 to 3300 µg microcystin g^{-1} DW (Christoffersen and Olrik 1996), and a cow died at Lake Torup Sø in July 1994 during a massive bloom of red-pigmented *Planktothrix agardhii*. The latter case was brought to the attention of the local county of Vejle 3 months after the episode took place and no autopsy was performed to establish the cause of death (Simon Marsbøl, pers. comm.). The possible intoxication by ingestion of water containing toxic *P. agardhii* therefore remains speculative but possible, since this bloom, when sampled on 27 July, contained approximately 1000 µg microcystins g^{-1} DW (Henriksen 1996).

In July 1996, one dog died and another showed severe convulsions but survived after drinking water containing neurotoxic benthic *Oscillatoria* sp. in a small lake situated in a bungalow area in the dunes close to the North Sea coast. The water of the lake was clear/slightly brownish coloured and without any surface cyanobacterial scum. An *Oscillatoria*-containing sample collected at the site, the lyophilized stomach content of the dead dog, and a laboratory culture of the *Oscillatoria* all proved neurotoxic in the mouse bioassay. Neither anatoxin-a, anatoxin-a(s), nor PSP-toxins was detected.

Animal intoxications by cyanobacterial toxins require a substantial biomass and high toxin production, and the very few Danish cases reported during 1993-1995 may reflect the generally rather low toxin contents of the samples examined. Thus, less than 4% of the Danish samples contained more than 1000 µg microcystins g^{-1}.

Warnings of the potential dangers of toxic cyanobacteria during times of bloom and surface scum formation have been placed at a large number of Danish lakes used for recreational purposes. The 1996 intoxication of two dogs shows, however, that highly toxic cyanobacteria may be present and pose a serious health hazard even under conditions without visible signs of colouring of the water. In addition, water management and risk assessment is hampered by yet unidentified and therefore not easily detectable cyanobacterial toxins, as illustrated by the dog intoxications and the presence of yet unidentified neurotoxin(s) in several laboratory cultures of different species.

2.2.3 Microcystin-LR and Total Microcystins in Cyanobacterial Blooms in the Czech Republic 1993-1998

Blahoslav Maršálek, Ludek Bláha, Jaroslav Turánek and Jirí Neca

Introduction

Among the cyanotoxins, microcystins are the most frequently studied and reported (Sivonen and Jones 1999). Thus, the microcystin content in buoyant scums and water blooms was widely studied in recreational reservoirs, drinking-water reservoirs, as well as in fish ponds in the Czech Republic from 1993 to 1998. Among the more than 60 different variants currently known, microcystin-LR appears to be the most often reported (Codd 1995), possibly because it has been most widely studied. To assess the relative importance of microcystin-LR, the Czech survey reported here investigated the share of microcystin-LR in relation to total microcystins.

Materials and Methods

A total of 65 environmental water bloom samples were collected from freshwater reservoirs in the Czech Republic during 1993-1998. Samples were freeze-thawed twice, lyophilized and stored at -20°C; 50 mg of the lyophylized biomass was extracted twice with deionized water (5 ml) by 5 min intensive vortexing followed by 15 min extraction in an ultrasonic bath. Pooled extracts were ultrafiltered by centrifugation through low-binding cellulose membrane in single-use filter units (Millipore Ultrafree-MC 10000 NMWL) for removal of proteins and other high molecular weight fraction molecules. Ultrafiltered extracts were analyzed with Waters 600E HPLC solvent delivery system, equipped with a Waters 717plus autosampler. Analytical column: 150x4.6 mm Supelcosil ABZ+Plus, 5 mm (Supelco). Gradient elution: mobile phase – (A) 20% acetonitrile + 0.05% tetrafluoroacetic acid (TFA); (B) acetonitrile + 0.05% TFA. Gradient composition: 0 min 95% A and 5% B, 20 min: 65% A and 35% B. Flow rate 1.0 ml min^{-1}, temperature 25°C. Photodiode array detector (Waters PDA 996, wavelength 200 - 300 nm) was used for scanning UV absorbance and evaluation of spectra. Microcystins were identified by microcystin-characteristic UV spectra.

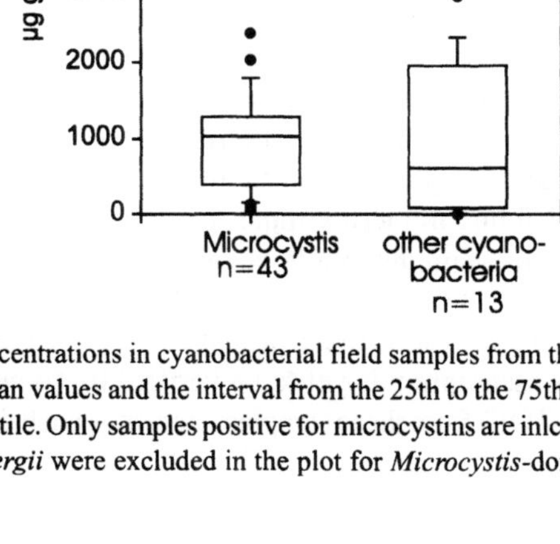

Fig. 1. Microcystin concentrations in cyanobacterial field samples from the Czech Republic 1993-1998. *Boxes* show median values and the interval from the 25th to the 75th percentile, *bars* indicate the 10th and 90th percentile. Only samples positive for microcystins are inlcuded; samples dominated by *Microcystis wesenbergii* were excluded in the plot for *Microcystis*-dominated samples

Analytically pure microcystin-LR (Calbiochem-Novabiochem Corp., La Jolla, CA) was used as an external analytical standard.

The results are expressed as micrograms of microcystin-LR g^{-1} DW of cyanobacterial biomass (µg g^{-1} DW). As only microcystin-LR was available as analytical standard, the concentration of other microcystins was determined as concentration equivalents of microcystin-LR, as suggested by Falconer et al. (1999). Taxonomic determination of dominant cyanobacterial species was performed according to Komárek (1996).

Results and Discussion

Of the total of 65 field cyanobacterial samples, 21 originated from drinking water reservoirs, 27 from reservoirs used for recreational activities, and 17 from fish ponds. Some of these waterbodies were sampled several times during a summer season. The microcystin reported most frequently – microcystin-LR – was detected in 51 samples (79.7% of samples analyzed). No microcystins were found in only 6 out of 65 samples, i.e. 90% of all samples were toxic.

The highest concentrations of microcystins detected were 4800 and 5804 µg g^{-1} DW, and median values were 820 µg g^{-1} DW for total microcystins, and 490 µg g^{-1} DW for microcystin-LR (Fig. 1); 50% of all samples contained between 250 and 1350 µg g^{-1} DW of total microcystins and between 200 and 800 of microcystin-LR (Fig. 1). Microcystin-LR amounted to roughly half of the concentration of total microcystins for all samples, but the maximal microcystin concentrations of over 4000 µg g^{-1} DW noted above consisted of microcystin-LR only.

Forty nine of the 65 samples were dominated by *Microcystis* spp. Six samples with *Microcystis wesenbergii* contained little or no microcystin (see Table 1). As indicated by the narrower percentile in Fig. 1, concentrations of total microcystins of the samples dominated by *Microcystis* spp. other than *M. wesenbergii* did not vary quite as widely as for the whole data set, and the median of 1030 is somewhat higher. Concentrations of microcystin-LR are very similar to those for the whole data set. Relatively low concentrations were also found in the samples dominated by *Anabaena* spp. and *Aphanizomenon* spp., if these were not accompanied by major populations of *Microcystis* spp. (Table 2).

Some samples dominated by *Planktothrix agardhii* showed high total microcystin concentrations of up to almost 3000 µg g^{-1} DW (Table 2). Since no data on production of microcystin-LR in this cyanobacterial species were previously reported, we can propose the following explanations of observed results: none of the samples contained a monospecific population of *P. agardhii*; *Microcystis aeruginosa* and/or *M. viridis* were also detected in these samples and constituted 5-10% of the plankton community. Thus, microcystin-LR found in these samples could be produced by these species. On the other hand, microcystin thought to be microcystin-LR in these samples (the same retention time in HPLC as microcystin-LR standard) could be a different coeluting variant of microcystin, most probably some demethylated variant of microcystin-LR. This assumption is yet to be verified by some mass spectrometry method (e.g. MALDI-TOF) from these samples; currently, such data are not available.

Table 1. Microcystin concentrations in water blooms collected in the Czech Republic 1993-1998. Samples dominated by *Microcystis* spp.

Dominant species[a]	Sample description	Main use[b]	Microcystin content ($\mu g\ g^{-1}$ DW) M-LR	Total
M. aeruginosa	Nová Ríše (1997)	dw	834	1258
M. aeruginosa	Jordán (1998)	dw	90	164
M. aeruginosa	Brno (1993)	r	686	1027
M. aeruginosa	D. Vestonice, Stredni zdrš (1993)	r	659	1096
M. aeruginosa	Nové Mlýny, Horní zdrz (1994)	r	674	1194
M. aeruginosa	D. Vestonice, Stredni zdrz (1994)	r	745	1011
M. aeruginosa	Lipno (1994)	r	941	1143
M. aeruginosa	Nové Mlýny, Dolní zdrz (1997)	r	296	1124
M. aeruginosa	Doloplazy (1994)	fp	163	810
M. aeruginosa	Pohorelice (1994)	fp	96	165
M. aeruginosa	Rímov (1995)	dw	235	394
M. aeruginosa	Vranov (1997)	r	390	1060
M. aeruginosa,	Vranov (1996)	r	565	1443
M. aeruginosa, A. spiroides	Vranov (1993)	r	325	463
M. aeruginosa, M. ichtyoblabe	Nová Ríše (1996)	dw	4800	4800
M. aeruginosa, M. novacekii	Holásky (1998)	r	Nd	163
M. aeruginosa, M. viridis	Mostište (1996)	dw	525	701
M. aeruginosa, M. viridis	Svratka river (1998)	dw	517	1298
M. aeruginosa, M. viridis	Mušov, Horní zdrz (1993)	r	484	784
M. aeruginosa, M. wesenbergii	Drnholec (1995)	r	789	1208
M. aeruginosa, M. wesenbergii	Nové Mlýny, Dolní zdrz (1995)	r	233	471
M. aeruginosa, M. wesenbergii	Fraumühln (1998)	fp	49	56,1
M. flos-aquae	Mostište (1997)	dw	29	122
M. flos-aquae	Vír (1998)	dw	58	91
M. flos-aquae, M. viridis	Vranov (1995)	r	126	386
M. flos-aquae, M. wesenbergii	Vír (1996)	dw	95	125
M. ichtyoblabe	Vír (1995)	dw	866	2036
M. ichtyoblabe	Brno (1996)	r	487	1131
M. ichtyoblabe	Jevišovice (1997)	r	561	1348
M. ichtyoblabe	Nová Ríše (1995)	dw	626	1519
M. ichtyoblabe, M. aeruginosa	Vír (1997)	dw	5804	5804
M. ichtyoblabe, M. aeruginosa	Brno (1998)	r	439	1130
M. ichtyoblabe, M. viridis	Dechtár (1997)	fp	602	1319

Table 1. (continued)

Dominant species[a]	Sample description	Main use[b]	Microcystin content (μg g-1 DW) M-LR	Total
M. viridis	Holásky (1996)	r	55	161
M. viridis	Bezdrev (1994)	fp	315	761
M. viridis	Písecný (1998)	fp	77	194
M. viridis, M. aeruginosa	Lipno (1993)	r	545	655
M. viridis, M. aeruginosa	Mušov, Horní zdrz (1995)	r	570	1030
M. viridis, M. aeruginosa	Dehtár (1994)	fp	799	1336
M. viridis, M. aeruginosa	Svct (1995)	fp	697	1739
M. viridis, M. flos-aquae	Mostište (1995)	dw	404	1232
M. viridis, M. flos-aquae	Nové Mlýny, Dolní zdrz (1998)	r	161	289
M. viridis, M. flos-aquae	Slapy (1996)	dw	852	2398
M. wesenbergii	Hosov (1995)	dw	nd c	nd
M. wesenbergii	Dalešice (1997)	dw	nd	nd
M. wesenbergii	Brno (1994)	r	nd	nd
M. wesenbergii	Stankov (1995)	fp	4.7	10.9
M. wesenbergii,	Fraumühln (1996)	fp	nd	nd
M.wesenbergii, M. novacekii	Korycany (1995)	dw	3	7

[a] Dominant species: M.- *Microcystis*, A. – *Anabaena.* [b] Main use: dw – drinking water reservoir, r – recreational, fp – fish pond. [c] nd: no microcystin detected.

The above-mentioned samples with extremely high microcystin concentrations were collected in drinking water reservoirs (Nová Ríše 1996–4800 μg g^{-1} DW; Vír 1997–5804 μg g^{-1} DW). Both samples contained mostly *M. aeruginosa* and *M. ichtyoblabe.* Microcystin-LR was the only variant of microcystin detected in both samples.

The concentrations of microcystin-LR, as well as of total microcystins, detected in the Czech Republic correspond well to the microcystin concentrations detected worldwide. In European countries, the highest amounts of microcystin-LR (2800 μg g^{-1} DW) were reported from Denmark (1992–1994) by Christoffersen (1996). Concentrations of microcystin-LR in the Finnish freshwater blooms ranged from 10 to 800 μg g^{-1} DW in 1994–1995 (Lahti et al. 1997). Fastner (1994) reported 36–360 μg g^{-1} DW concentrations of microcystin-LR in water blooms from German reservoirs in 1992. The concentrations of microcystin-LR in three Canadian lakes (Alberta, 1990–1993) ranged from 1 to 1550 μg g^{-1} DW (Kotak et al. 1995). Total microcystin concentrations ranged from 1000 to 7000 μg g^{-1} DW in Portugal (Vasconcelos et al. 1996), 70 to 3970 μg g^{-1} DW in France (Vezie et al. 1997) and 1 to 5565 μg g^{-1} DW in Germany (Fastner et al. 1999). The highest values of total microcystin content, 1900–12800 μg g^{-1} DW (based on enzyme linked immunosorbant assay, ELISA), were reported from Wisconsin, USA (McDermott et al. 1995).

The highest concentrations of microcystin-LR found in the Czech Republic (4800 and 5804 μg g^{-1} DW) are almost twice as high as the highest concentrations previously reported from European countries. Since microcystin-LR was the only variant in these samples, these are also the highest total microcystin concentrations.

Concentrations of microcystin-LR relative to total microcystins could be determined from 57 samples. In 31 of these, microcystin-LR amounted to less than 50%, and in 47 samples to less than 70% of total microcystins. These data indicate that toxicity potential in cyanobacterial samples cannot simply be expressed in microcystin-LR concentration units. Although water quality recommendations of WHO discuss only

Table 2. Microcystin concentrations in water blooms collected in the Czech Republic 1993–1998. Samples dominated by *Anabaena* spp., *Aphanizomenon* spp., *Nodularia spumigena, Planktothrix agardhii,* and *Woronichinia naegeliana*

	Dominant species[a]	Sample description	Main use[b]	Microcystin content (mg g^{-1} DW)	
				M-LR	Total
1.	*A.circinalis, Aph. gracile*	Vitek (1996)	fp	Nd[c]	nd
2.	*A. lemmermannii, M. viridis*	Lednice (1994)	fp	497	618
3.	*A. spiroides*	Jedovnice (1998)	r	Nd	85
4.	*A. spiroides, Aph. flexuosum*	Luhacovice (1996)	r	Nd	nd
5.	*Aph. flos-aquae*	Skalka (1996)	r	17	86
6.	*Aph. flos-aquae*	Jihlava (1997)	fp	nd	0.7
7.	*Aph. flos-aquae, M. ichtyoblabe*	Trnava - Preckov (1997)	fp	759	1907
8.	*Aph. flos-aquae, M. viridis*	Skalka (1997)	r	826	2204
9.	*Aph. gracile, M. viridis*	Vranov (1994)	r	180	334
10.	*Phormidium autumnale*	Moravská Nová Ves (1995)	dw	816	na[d]
11.	*Pl. agardhii*	Troubky (1997)	dw	143*	528
12.	*Pl. agardhii*	Hrušky (1995)	fp	945*	1436
13.	*Pl. agardhii*	Potišil (1997)	fp		
14.	*Pl. agardhii*	Slapy (1995)	dw		
15.	*Pl. agardhii*	Lýsky (1995)	dw		
16.	*Woronichinia naegeliana*	Štítná (1996)	fp	5.4	5.4

[a] Dominant species: A. – Anabaena, Aph. – Aphanizomenon, M.- Microcystis, Pl. – Planktothrix. [b] Main use: dw – drinking water reservoir, r – recreational, fp – fish pond. [c] nd: no microcystin detected. [d] na: data on total microcystin content not available. * possibly a demethylated variant of microcystin-LR; see text for discussion

the concentration of microcystin-LR (WHO 1998), our data indicate that other microcystins should not be neglected in analyses of cyanobacterial toxins.

The Czech survey of microcystin content in the cyanobacterial water blooms shows the need for continuous monitoring of microcystins and of toxic cyanobacteria. Future efforts should focus on establishment of an integrated program for rapid detection, management, as well as prevention of toxic cyanobacterial water blooms in the Czech Republic. Additionally, wide screening of raw and drinking waters for microcystin concentrations (especially during the summer months) is highly necessary. Since microcystins are not efficiently removed by the filtration and precipitation procedures commonly used in drinking-water treatment plants (Hrudey et al. 1999), continuous analyses of microcystins in both raw and drinking waters should be performed.

Acknowledgements

This work was supported by Ministry of Agriculture, Czech Republic. Nadatio Flos Aquae, no:12/1996, 3/1997. Veterinary Research Institute internal grant no. 331-09 / 1999. The authors thank Dr. Ingrid Chorus and Dr. Jutta Fastner for valuable comments and help with finalization of the text.

2.2.4 Freshwater Cyanobacteria and their Toxins in Portugal

Vitor Vasconcelos

Introduction

Freshwater cyanobacteria have been studied in Portugal since the early 1930s. Initially these studies focused mainly on taxonomy and diversity (Sampaio 1933), but several potentially toxic species such as *Oscillatoria formosa* and *Lyngbya majuscula* were first recorded (Frade 1951, 1954; Nauwerck 1959). In 1962, Nauwerck (1962) gave an overview of several Portuguese waterbodies taking into consideration the occurrence and dominance of cyanobacteria. Some potentially toxic cyanobacteria such as *Microcystis viridis, M. flos-aquae, Oscillatoria* (syn. *Planktothrix*) *agardhii, O. rubescens* and *Anabaena flos-aquae* were reported. During the 1980s more ecological work, mainly in southern reservoirs, added *M. aeruginosa, Aphanizomenon flos-aquae* and *Anabaena circinalis* to the list of toxic species (Santos 1973; Oliveira 1984a; Branco and Guimarães 1988).

Oliveira (1984b, 1987) relates the occurrence of toxic cyanobacteria as a possible cause of fish kills together with anoxia and high ammonium levels in the water. The same author relates the occurrence of an episode of human intoxication in 1987 to the occurrence of a bloom of *Aphanizomenon flos-aquae* (Oliveira 1991). Dead fish were also present during the decay of this bloom.

The first data on toxicity cyanobacteria originated from the analysis of a toxic bloom dominated by *Microcystis aeruginosa* that was collected in The Douro River (Crestuma reservoir) in 1989 (Vasconcelos et al. 1993a, b).

Toxicity of Cyanobacteria Blooms and Strains

A survey of the distribution of toxic cyanobacteria in Portuguese freshwaters began in 1989. Preliminary work surveyed 36 lakes, reservoirs and rivers during 1989–1992. At this time only hepatotoxic blooms were found and 60% of the blooms were toxic. The main species present in the toxic blooms were *Microcystis aeruginosa* (72%) and *Anabaena flos-aquae* (28%) (Vasconcelos 1994).

The toxicity of the bloom samples evaluated by intraperitoneal (i.p.) mouse bioassay varied from 20 to 700 mg kg^{-1} (dry weight of bloom sample kg^{-1} mouse body weight). *Microcystis* strains isolated from these blooms presented a toxicity varying from 15 to 75 mg kg^{-1} (Vasconcelos 1994) (Table 1).

The toxicity of the cyanobacteria blooms is found to vary significantly during the year: data on the LD_{50} of blooms collected in Mira lake in north Portugal show higher toxicity in early summer (Table 2). Cyanobacteria represented 96.5 to 99.8% of the total seston in all of these blooms, and changes in toxicity are attributed to shifts in strain composition rather than in species composition. The two most toxic blooms analyzed in this lake were dominated by *Microcystis aeruginosa* (June 1991) and by *Anabaena flos-aquae* (June 1992).

Further, natural populations show heterogeneity of toxic and non-toxic strains in a given lake. In a study made in three natural lakes in central Portugal, all of the *Microcystis* strains isolated from the Lakes Vela and Mira were toxic, but from Braças Lake, only 33% showed toxicity (Vasconcelos et al. 1993a).

In a study made in Crestuma reservoir, Ferreira (1994) isolated several *M. aeruginosa* strains whose the LD_{50} varied from 15–25 mg kg^{-1}. A strain of *Aphanizomenon flos-aquae* was also isolated and caused lethal toxicity on mice only with an acidic extract (Ferreira 1994). The animals showed immediate paralysis and death between 8 and 15 min.

Table 1. LD_{50} values (i.p mouse bioassay) of blooms and *Microcystis* strains collected from Portuguese freshwaters maximum, minimum, average and standard deviation (SD)

	Blooms (n=18)				Microcystis strains (n=15)			
	Max.	Min.	Average	SD	Max	Min	Average	SD
LD_{50} (mg kg^{-1})	700.0	20.0	136.0	175.0	75.0	15.0	37.5	23.5

Table 2. Toxicity and dominant species (% of total cyanobacteria density) in cyanobacteria blooms collected in Mira lake during 1991 (+ semiquantitative determination of dominance)

	1991		1992			
	June 27	Sept. 26	June 5	June 11	Oct. 7	Nov. 11
LD_{50} (mg kg^{-1})	31.0	100.0	37.5	400.0	Non-toxic	Non-toxic
Species (%)						
M. aeruginosa	+++	75.6	+	13.3	95.2	12.3
A. flos-aquae	+	7.9	+++	68.8	4.4	85.9

Oliveira (1995) registered neurotoxicity on a bloom sample collected in Mira lake composed of 98% of *Aphanizomenon flos-aquae*. Mice injected i.p. died in 2 min with characteristic neurotoxic symptoms.

Diversity of Toxins from Cyanobacteria Blooms and Strains

Microcystins

The studies that have been carried out in Portugal until now seem to point to microcystins as the dominant toxins. The major microcystins found until now are presented in Tables 3 and 4.

In the bloom samples microcystin-LR is usually the major toxin with a dominance ranging from 45.5 to 99.8% (Table 4; Vasconcelos et al. 1996). The other two dominant toxins are MCYST-YR and MCYST-RR.

Table 3. Diversity of microcystins (%) in *Microcystis* strains isolated from Portuguese freshwaters (T1 and T2 represent unidentified microcystins)

Strain Toxin	1	2	3	5	7	25	30	31	33	34	44
MCYST-LR	53.0	94.5	91.9	83.3	100	44.1	53.0	-	100	91.4	100
MCYST-LA	47.0	4.7	0.6	13.9	-	-	-	-	-	8.6	-
MCYST-[D-Asp^{3}]-LR	-	0.8	-	-	-	-	-	-	-	-	-
MCYST-AR	-	-	-	2.8	-	-	-	-	-	-	-
MCYST-YR	-	-	-	-	-	14.7	11.6	-	-	-	-
MCYST-RR	-	-	-	-	-	-	33.0	-	-	-	-
T1	-	-	7.4	-	-	5.9	2.4	100	-	-	-
T2	-	-	0.1	-	-	35.5	-	-	-	-	-

Table 4. Diversity of microcystins (%) in cyanobacterial blooms collected from Portuguese freshwaters (T1, T2 and T3 represent unidentified microcystins)

Blooms Toxins	AG	CA	ME	VB	TO	MI	CR	BM	M10	M11	M12	M13
MCYST-LR	64.2	61.1	48.8	83.0	-	45.5	99.8	-	88.6	73.1	70.9	-
MCYST-[Dha^{7}]-LR	2.3	-	-	-	-	-	-	-	-	-	-	-
MCYST-[L-MeSer^{7}]-LR	12.5	-	-	-	-	-	-	-	-	-	-	-
MCYST-HilR	2.3	-	-	-	-	-	-	-	-	-	-	-
MCYST-RR	12.7	-	-	17.0	-	-	-	-	-	-	-	-
MCYST-[D-Asp^{3}]-LR	4.2	-	-	-	-	-	0.1	-	11.4	-	-	-
MCYST-YR	-	-	48.8	-	-	-	-	-	-	11.5	23.7	-
T1	1.8	33.9	1.5	-	68.8	30.9	0.1	50.0	-	15.4	5.4	90.0
T2	-	0.6	1.0	-	23.4	11.0	-	50.0	-	-	-	10.0
T3	-	4.4	-	-	7.8	12.6	-	-	-	-	-	-

Concerning the *Microcystis* strains, the major toxin is also MCYST-LR with a dominance ranging from 43.9 to 100% (Table 3; Vasconcelos et al. 1995). In these strains the other two main toxins are MCYST-LA and MCYST-YR. Interestingly, MCYST-LA was not found in bloom samples, even though it is a major microcystin in strains 1, 5, 34, 2 and occurs in traces in strain 3. However, only strains 1 and 3 were isolated from the bloom samples in Table 4. One explanation for this rare occurrence of MCYST-LA is that the low amount of material used for toxin extraction and purification in these bloom samples was not enough to assign both microcystins that were found to any known MCYST, and T1 or T2 might be MCYST-LA. Another explanation could be that strains producing MCYST-LA might be easier to isolate and grow in culture, and selective isolation may lead to results that do not reflect strain distribution (and concominant MCYST-distribution) in the blooms.

Total content of microcystins in both the bloom and strain samples was rather high: the maximum values were 7.1 mg g^{-1} (dry weight) for blooms and 11.3 mg g^{-1} for strains (Table 5). Thus, microcystins may constitute up to 0.5–1% of total dry weight.

Neurotoxins

Neurotoxic blooms were rarely found in our monitoring programme by mouse bioassay. Since 1989 only three samples out of more than 200 gave positive results for neurotoxicity. Nevertheless, Ferreira (1994) found the occurrence of several PSP-type toxins (GTX1, GTX3 and GTX4) in an *Aphanizomenon flos-aquae* strain (see above) using reverse-phase HPLC and a fluorimetric detector.

Toxin Levels in Water Samples

Since 1994 the evaluation of total microcystins in water samples has been performed using a commercially available MCYST ELISA-kit (Ensys UK) or a monoclonal antibody ELISA (Ueno et al. 1996). The results for samples containing microcystins are summarized in Table 6. Of the total samples analyzed, around 40% did not contain microcystins.

During this period (1994–1998) the main species associated with the toxic blooms were *Microcystis aeruginosa*, followed by *Anabaena flos-aquae* and *Phormidium mucicola*. The total cyanobacterial density and that of the main hepatotoxic species are presented in Table 7. All of these waterbodies are used either for water consumption or for recreation.

In summary, from a total of 125 cyanobacteria field samples analyzed for hepatotoxicity and/or microcystin between 1989 and 1999, 60% were found positive.

Table 5. Maximum, minimum, average and standard deviation (SD) of total microcystin content (mg g^{-1} dry weight) of bloom and strains of *Microcystis aeruginosa*

	Microcystis strains (n=11)				Blooms (n=12)			
	Max.	Min.	Average	SD	Max.	Min.	Average	SD
mg mg^{-1} (dry weight)	11.3	2.1	5.5	2.5	7.1	1.0	4.1	1.9

In addition, aquatic animals such as mussels, crayfish and fish may accumulate microcystins and transfer them along food chains (Vasconcelos 1999). The accumulation of the toxins seems to be quite rapid. In an experiment with the mussel *Mytilus galloprovincialis* exposed to a toxic *Microcystis* strain at 10^5 cells ml^{-1}, the mussels accumulated 52% of the total MCYST within 24 h (Vasconcelos 1995). On the other hand, depuration is rather slow. Amorim and Vasconcelos (1999) showed that mussels contaminated with 10.7 µg MCYST g^{-1} still contained 2.5 µg MCYST g^{-1} 2 weeks after depuration had begun. One of the reasons for this slow depuration may be the reingestion of contaminated faeces, which contained up to 140 µg MCYST g^{-1} (Amorim and Vasconcelos 1999). These results pointed out the importance of monitoring programs for eutrophic rivers and estuaries where molluscs are caught for human consumption.

Table 6. Maximum, minimum, average and standard deviation (SD) of total microcystin concentration in raw water samples (µg L^{-1}) collected in Portuguese freshwaters and positive for microcystins by ELISA (1994: analyzed according to Ueno et al. 1996; 1996–1998: analyzed with the Ensys kit)

Year	1994	1996	1997	1998
No. of samples	13	6	10	23
Maximum	31.0	2.2	14.0	8.0
Minimum	0.4	0.4	0.3	0.2
Average	6.4	1.1	2.2	2.1
SD	9.2	0.7	4.2	2.3

Table 7. Maximum, minimum, average and standard deviation (SD) of total cyanobacteria density and of the main hepatotoxic species present in water samples collected in Portuguese waterbodies

Year	1994	1996	1997	1998
No. of samples	48	150	165	156
No. of waterbodies	25	45	37	33
Total cyanobacteria (cells ml^{-1})				
Maximum	516594	47965	464000	308938
Minimum	14133	11932	4609	987
Average	153329	23210	146714	59406
SD	175749	14829	385979	99290
Main hepatotoxic species (cells ml^{-1})				
Maximum	516142	47184	464000	308938
Minimum	288	11249	4604	987
Average	109466	21750	63750	48263
SD	164699	15379	141922	77636

Data obtained from wild fish species living in eutrophic waters in south and central Portuguese rivers reveal that the accumulation of microcystins in the edible parts (muscle) are not very high (Table 8) when compared with microcystin contents in mussels. Among the species of fish analyzed, carps seem to accumulate more microcystins than mullets or barbels, and this is due to their more omnivorous feeding behaviour.

The MCYST levels found during our studies (Table 9) show that very few grams of mussels and crayfish are needed to reach the TDI (tolerable daily intake) used by WHO for derivation of a guideline value for microcystins in drinkingwater (Kuiper-Goodman et al. 1999): as the TDI is 40 μg kg^{-1} body weight, an adult weighing 60 kg could consume 2400 μg of microcystins with mussels or crayfish without expecting a health risk. However, regular meals on these animals (≈100 g) contaminated with the maximum levels found in our studies (Table 9) would provide 1600 μg (mussels) and 270 μg (crayfish) of microcystins. These doses would probably be inoffensive, particularly if they are not frequently repeated. However, populations that feed frequently and in larger amounts on these products, such as fishermen, may be exposed to amounts exceeding the tolerable daily intake.

Table 8. Maximum and minimum values of MCYST in edible parts of different fish species collected in Portuguese freshwaters

Species	MCYST (ng g^{-1})
Carp (*Cyprinus* sp.)	50-280
Barbel (*Barbus* sp.)	0.7-120
Grey mullet (*Lisa* sp.)	8.5-110

Table 9. Maximum amounts of MCYST in edible parts of aquatic animals

Species	MCYST (μg g^{-1})
Fish	0.3
Crayfish (*Procambaus clarkii*)	2.7
Mussels (*Mytilus galloprovincialis*)	16.0

2.2.5
Cyanotoxins and Cyanobacterial Blooms in South Korean Lakes

Ho-Dong Park

Introduction

Progressive eutrophication in Korean waters has led to cyanobacteria becoming dominant in several man-made dams, estuary dams and reservoirs. Even though the cyanobacterial taxa found in Korea are similar to the toxic cyanobacterial species in freshwaters of the countries from which poisonings have been documented, no mortalities of wild or domestic animals have been reported in Korea. Aqueous extracts from the cyanobacteria were found to be lethal, however, upon intraperitoneal injection into mice (Kim et al. 1995). In this chapter, the distribution of dominant species of cyanobacteria and the amount of cyanotoxins in cyanobacterial blooms are detailed to assess cyanotoxin occurrence in Korean lakes which are used as drinking-water sources.

Dominant Species in Cyanobacteria Blooms

Cyanobacterial bloom samples were collected in the warm season from seven mid-stream man-made dams (Soyang, Choongju, Daechong, Jangsong, Hapchon, Andong, Imha), two estuary dams (Nakdong, Youngsan), two small reservoirs (Gunsan, Mijae) and one coastal lagoon (Younglang). With the exception of Lake Younglang and Mijae, which are used only for recreation, all of these waters are drinking-water sources. Cyanobacteria cells from natural waters were concentrated by plankton nets for the analysis of cyanotoxins and species composition of cyanobacteria. Formaldehyde (1.4%) was added to a part of the samples for identification of cyanobacteria species. The remaining bloom samples were lyophilized and stored at -30°C until the chemical analysis was performed.

Cyanobacteria were the dominant phytoplankton in all the waterbodies investigated. Three genera and 13 species of cyanobacteria were identified in 21 samples studied for species composition (Table 1): six species of *Microcystis*, five of *Anabaena,* one of *Plankothrix* and one of *Oscillatoria*. Lakes Soyang, Choongju, Daechong and Jangsong were dominated by *Anabaena, Planktothrix* or *Oscillatoria*, while *Microcystis* was predominant in the other eight lakes. The species that appeared most frequently among the 13 species was *M. aeruginosa*. Half of the cyanobacterial bloom samples were dominated by *M. aeruginosa* and *M. ichthyoblabe*. Lakes Soyang, Daechong and Jangsong were always dominated by *Anabaena* or *Oscillatoria*. In a comparison of all of the samples, *Microcystis* was the most dominant genus (dominant in 67% of the samples), followed by *Anabaena* (23%) and *Planktothrix/ Oscillatoria* (10%).

Cell-Bound Cyanotoxis

Anatoxin-a and microcystin were analyzed simultaneously with a method based on Harada et al. (1988, 1989, 1993). As summarized in Fig. 1, lyophilized cells (about 100 mg) were extracted three times with 10 ml of 0.05 M acetic acid for 30 min while stirring. The extract was centrifuged at 4000 or 10000 rpm and the supernatant was adjusted to pH 10 with 7% ammonium hydroxide. This extract was directly applied to 0.2 g of a reversed phase ODS-disposable extraction column [Bakerbond SPE 7020-03, Octadecyisilane(C18), ODS type, 3-ml; J.T. Baker, Phillipsburgh, NJ, USA] which had

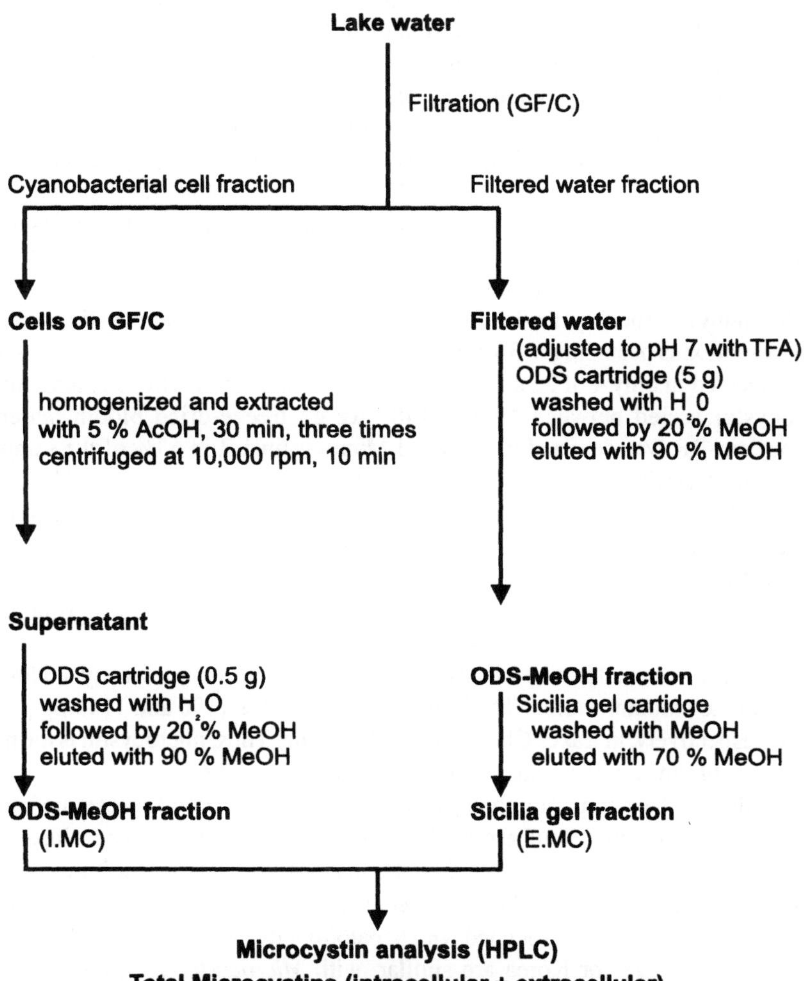

Fig. 1. System for analysis of total microcystin in lake water. *I.MC* Intracellular microcystin; *E.MC* Extracellular microcystin

been preconditioned by washing with 10 mL of methanol and 10 mL of water. The column containing toxins was washed with water (10 ml), followed by water-methanol (9:1, 10 mL). Elution from the column with methanol (20 ml) gave the fraction containing the toxins. A portion (5 μl) of the fraction was applied to the HPLC system for the analysis of microcystins. The remaining fraction, dissolved in 20 ml of water, was absorbed onto a disposable solid phase column (0.2 g) of a cation exchanger organosilan bonded to silica gel (Bond Elut 615303, CBA 1210-2038, COOH type, 3-ml; Analytichem International, Varian Associates Inc., Harbor, CA, USA) which had been preconditioned by washing with 0.1% trifluoroacetic acid-methanol (10 ml), followed by methanol (10 ml) and water (10 ml). The column was rinsed with 10 ml water and then 10 ml of methanol. The fraction containing toxins was eluted with 20 ml of 0.01% trifluoroacetic acid-methanol to give the crude toxin fraction. This fraction was dissolved with methanol and injected to an HPLC. The HPLC was carried out under the following conditions: LC: pump, Shimadzu (Kyoto, Japan) LC-9A; detector, Shimadzu SPD-M10A photodiodearray detector; Column, Cosmosil 5PH (150 x 4.6 mm; Nacalai Tesque Inc., Kyoto, Japan); mobile phase, methanol:0.1 M ammonium acetate (adjusted pH 5 with TFA) = 14:86; flow rate, 1.2 ml min^{-1}; detection, UV (227 nm).

Table 1 shows the amounts of microcystin and anatoxin-a in the 41 cyanobacterial bloom samples collected in the warm season during 1992-1996 from the 12 waterbodies in Korea. The cyanotoxins were identified as microcystins-RR, -YR, -LR; desmethyl-7-microcystin-LR (7-DMLR) and anatoxin-a. Microcystins were the main cyanotoxins occurring in 36 of the 41 cyanobacteria bloom samples. Anatoxin-a was detected in only 4 of 22 samples analyzed for this toxin. The total amounts of microcystin varied between 19 and 1489 μg g^{-1} freeze-dried bloom material. Microcystin-RR was detected in 79% and Microcystin-LR in 63% of the samples, whereas microcystin-YR was contained in only four waterbodies (Jangsong, Youngsan, Gunsan, Mijae) and 15% of the samples. 7-DMLR was detected in two lakes, Lake Jangsong and Lake Youngsan, the amounts were 80 and 39 μg g^{-1} freeze-dried bloom material (no data in Table 1), respectively. High amounts of total microcystin of more than 1 mg g^{-1} dry weight were detected in Lake Jangsong, Mijae, Younglang and Nakdong.

Anatoxin-a occurred in 4 of the 22 cyanobacteria bloom samples analyzed for this toxin, and the amounts varied. The highest amounts were 1190 and 1144 μg g^{-1} dry weight, respectively, in Lakes Choongju and Jangsong. Thus, microcystin is the main toxin in the blooms, with anatoxin-a found in only a few lakes.

According to the current morphology-based taxonomy, about 40 species are known to produce cyanotoxins (Skulberg et al. 1993). In freshwater sources in Japan, the most common bloom-forming species, *M. aeruginosa,* is also the most widely distributed (Watanabe et al. 1986b; Park et al. 1993; Park and Watanabe 1996) and thus, it is most commonly associated with hepatotoxicity (Watanabe et al. 1989). The results reported here for Korea are similar, with *Microcystis aeruginosa* and *M. ichthyoblabe* being the most abundant cyanobacteria species in natural bloom samples. In contrast, Sivonen et al. (1989) described *Anabaena* as the most frequent

Table 1. Concentrations of microcystins and anatoxin-a in cyanobacterial blooms from Korean lakes (1992–1996)

Sites	Date	MC (μg/g)			Total (intra + extra cellular) MC	Total (intra-cellular) MC	ANTX-a (intra-cellular)	Dominant cyanobact-erial species
		RR	YR	LR	(μg L^{-1})	(μg g^{-1})	(μg g^{-1})	
Soyang	09/20/92	37	N.D.[a]	2.4	-[b]	39	-	A.m, M.a
	10/26/92	133	N.D.	N.D.	-	133	-	A.m, M.a
Injae[c]	10/20/93	19	N.D.	N.D.	1.9	19	N.D.	-
	10/08/95	N.D.	N.D.	29	N.D.	29	-	-
Choongju	09/27/92	29	N.D.	5.7	-	35	1190	-
	10/05/92	24	N.D.	N.D.	-	24	-	P.a, M.w, M.i
	10/11/93	N.D.	N.D.	N.D.	N.D.	N.D.	N.D.	-
	08/12/96	-	-	-	0.4	-	N.D.	M.i, M.a, M.w
Samtan[c]	10/18/93	N.D.	N.D.	N.D.	N.D.	N.D.	N.D.	-
Danyang[c]	10/18/93	-	-	-	81	-	N.D.	-
Daechong	09/27/92	57	N.D.	39	-	96	-	-
	10/06/92	76	N.D.	N.D.	-	76	N.D.	A.s, M.i, M.a
	10/05/94	96	N.D.	45	-	141	N.D.	-
	08/30/95	140	N.D.	78	0.7	218	-	A.s, A.c, M.i
Jangsong	10/07/92	167	1062	N.D.	-	1229	1144	A.ct, M.i, M.a
	10/15/94	84	N.D.	31	12	115	N.D.	-
	08/29/95	164	N.D.	70	1.9	234	-	O.t, M.a, M.i
	09/30/95	N.D.	N.D.	N.D.	N.D.	N.D.	-	-
Hapchon	10/06/92	441	N.D.	126	-	567	N.D.	M.a, M.i, M.n, M.f
	10/06/93	514	N.D.	104	-	618	N.D.	-
	10/14/94	10	N.D.	244	-	254	N.D.	-
	08/26/95	N.D.	N.D.	207	0.9	207	-	M.i
Youngsan[d]	10/17/92	N.D.	N.D.	181	-	181	570	M.i, M.a
	09/26/94	303	N.D.	N.D.	1.6	303	N.D.	-
	08/29/95	788	N.D.	204	-	992	-	M.a, M.i
Mongtan[e]	08/28/95	595	N.D.	354	17	949	-	M.i, M.a

Table 1. (continued)

Sites	Date	MC (μg/g)			Total (intra + extra cellular) MC	Total (intra-cellular) MC	ANTX-a (intra-cellular)	Dominant cyanobact-erial species
		RR	YR	LR	(μg L⁻¹)	(μg g⁻¹)	(μg g⁻¹)	
Younglang[f]	08/17/93	793	N.D.	410	-[b]	1203	N.D.	M.i, M.a, A.f
	10/16/93	317	N.D.	91	-	408	417	-
	08/28/96	531	64	277	-	872	-	M.i, M.a
Andong	08/25/95	N.D.	N.D.	N.D.	N.D.	N.D.	-	M.a
Imha	08/26/95	N.D.	N.D.	N.D.	N.D.	N.D.	-	M.a, M.n, M.w
	08/19/96	-	-	-	1.2	-	-	M.i
Nakdong[d]	09/24/94	60	N.D.	9.0	1.0	69	N.D.	-
Mulkum[e]	07/26/94	30	N.D.	N.D.	0.6	30	N.D.	-
	07/28/94	561	N.D.	519	171	1080	N.D.	-
	08/18/94	56	N.D.	N.D.	0.9	56	N.D.	-
Myongji[d]	09/24/94	107	N.D.	N.D.	2.4	107	N.D.	-
Noksan[d]	08/27/95	433	N.D.	202	30	635	-	M.v, M.w, M.i, M.a
Gunsan R.	08/29/95	424	169	263	27	856	-	M.v, M.a
	09/30/95	146	N.D.	188	N.D.	334	-	-
Mijae R.	08/29/95	565	384	540	-	1489	-	M.i, M.a, M.w

[a] N.D. : Toxins not detected. [b] Not determined. [c] Stream inlet site. [d] Estuary dam. [e] Downstream region of the river. [f] Lagoon. R.: reservoir

genus found in freshwater blooms in Finland, where it is frequently responsible also for neurotoxic blooms. Anatoxin-a concentrations, in the Korean plankton samples that contained this toxin, varied from 417 to 1190 μg g^{-1}. This is in a range similar to anatoxin-a concentrations of Finnish bloom samples, which varied from 12 to 4360 μg g^{-1} dry weight (Sivonen et al. 1989). Anatoxin-a has also been reported in North America and northern Europe, but to our knowledge, this is the first report of detection of anatoxin-a in freshwater sources in Korea.

As codominance of several cyanobacterial taxa is a frequent phenomenon in eutrophic lakes of Korea (Kim et al. 1995), concomitant occurrence of the cyanotoxins produced by the respective taxa is to be expected. All four samples containing anatoxin-a also contained microcystins (Table 1). This may be due both to the mixture of species and to the fact that some species can simultaneously produce both microcystins and anatoxin-a, as was reported for *Anabaena flos-aquae* (NRC-44-1) from Canada by Al-Lay et al. (1988) and for *Anabaena flos-aquae* (NRC 525-17) from Canada by Harada et al. (1991). Sivonen et al. (1989) also reported simultaneous occurrence of neurotoxins, hepatotoxins as well as atypical toxic responses in some Finnish samples tested with the mouse bioassay. The method for the simultaneous determination of anatoxin-a and microcystins will contribute to the monitoring of natural blooms and strains in which these toxins often coexist.

Microcystin Concentrations in Water

Extracellular concentrations of microcystins were analyzed following the cleanup methods for trace amounts of microcystins dissolved in water given by Harada et al. (1988) and Tsuji et al. (1994) (Fig. 2). A sample (about 5 L) was filtered through a glass microfibre filter (Whatman GF/C) and adjusted to pH 7. The filter and its content were homogenized and extracted with 5% aqueous acetic acid and the supernatant was applied to an ODS cartridge (0.5 g, Bakerbond SPE 7020-03) after centrifugation. The 90% methanol-extracted eluate from the cartridge was applied to an HPLC system equipped with an ODS column (Cosmosil 5C18-AR, 4.6 x 150 mm, Nacalai, Japan). The HPLC system consisted of a Shimadzu (Kyoto, Japan) LC-9A pump coupled to a SPD-10A set at 238 nm and SPD-M10A photodiodearray detector, and C-R6A integrator. Microcystin concentrations were quantified by comparing the absorbance

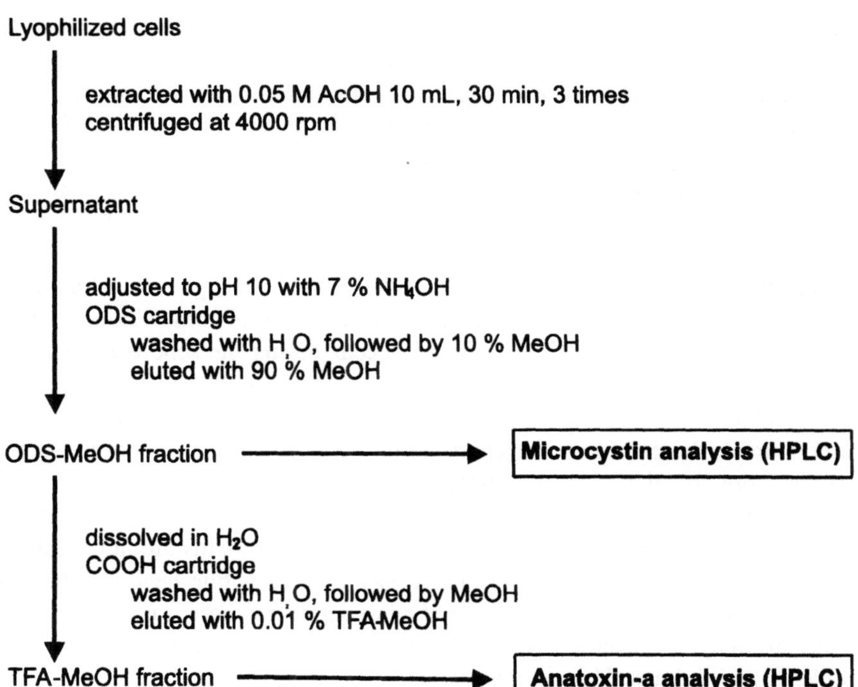

Fig. 2. Scheme for simultaneous analysis of microcystins and anatoxin-a

peak area of test samples, at 238 nm, after separation with a methanol: 0.05 M phosphate buffer (pH 3, 58:42) with those of standards. The flow rate was 1 mL min^{-1}. The filtered water was applied directly to an ODS silica gel cartridge (5 g, Chromatorex ODS, 100-200 mesh, which was packed into a polypropylene cartridge). The cartridge was rinsed with water and 20% methanol-water. The eluate from the cartridge with 90% methanol-water was evaporated to dryness, and the residue was dissolved in methanol. The methanol solution was applied to a silica gel cartridge (2 g, Sep-Pak), which was preconditioned with methanol, and the cartridge was rinsed with methanol. The eluate from the cartridge with 50% methanol-water was evaporated to dryness; the residue was then dissolved in methanol. The methanol solution was subjected to HPLC analysis for the resolution of microcystins.

Total microcystin concentrations (µg L^{-1}) calculated include intracellular microcystin and extracellular microcystin. The total microcystin concentrations of 30 samples were equal to their intracellular microcystin because extracellular microcystin was not detected in any of the samples. These concentrations varied between 0.6-171 µg L^{-1} (Table 1). High concentrations of microcystin were observed on July 28, 1994, in Mulkum and on October 18, 1993, in Danyang. In Mulkum, these were composed of 89 µg L^{-1} microcystin-RR and 82 µg L^{-1} microcystin-LR, and in Danyang of 68 µg L^{-1} microcystin-RR and 13 µg L^{-1} microcystin-LR.

Cyanotoxin concentrations have usually been expressed in units of weight per unit dry weight (e.g. µg g^{-1}). Volumetric units (e.g. µg L^{-1}) are more appropriate for the estimation of risk levels for aquatic biota or exposed humans. Few researchers, however, have reported microcystin concentrations in volumetric units (Lindholm et al. 1989; Lindholm and Meriluoto 1991; Kotak et al. 1995). Lindholm and Meriluoto (1991) reported that the highest toxin (desmethyl-microcystin-RR) levels in the metalimnion of a Finnish lake were 20-40 µg L^{-1} in the summers of 1988-90. In a Japanese hypertrophic lake, Lake Suwa, high concentrations of microcystin were found during the exponential growth phase of the bloom; the highest concentration of microcystin was 184 µg L^{-1} on October 10, 1994. However, less than 4 µg L^{-1} was found in the filtered sample, although the amount of extracellular microcystin in the filtered lake water was highest at the end of the bloom. The high concentrations of extracellular microcystin in filtered lake water at the end of some blooms suggest that release of microcystin from cells may occur during the senescence and the decomposition periods of *Microcystis* cells (Park et al. 1996, 1998).

As all of the waterbodies studied, except Lakes Younglang and Mijae, are drinking-water sources, the presence of cyanobacterial toxin can be a potential threat and requires more attention to water treatment. In Korea, rapid sand-filter systems are employed in most water-supply treatments. Slow sand-filter or carbon treatment that are more effective in removing cyanobacterial cells and toxins are not common. Although high cyanotoxin concentrations may cause a public health problem if drinking-water treatment is not capable of removing them, there is no debate about the hazard of cyanobacterial toxins in Korea. People are also suffering from odour problems, mainly caused by cyanobacteria in many reservoirs.

A comprehensive understanding of cyanotoxins in lakes cannot be obtained from only the few water samples investigated in this study. Systematic temporal and spatial measurements of cyanotoxins and toxic species composition in lakes and reservoirs are necessary to assess the risks for human health, aquatic animals, livestock and wildlife.

2.2.6 Cyanotoxin Occurrence in Freshwaters - a Summary of Survey Results from Different Countries

Ingrid Chorus

The cyanotoxin survey results reported in Chapter 2.1 and 2.2 from six different countries suggest a number of generalizations on cyanotoxin occurrence and concentration ranges. The following seven hypotheses are presented as a basis for further assessment and discussion:

1. Microcystins are found more frequently than cyanobacterial neurotoxins. Hepatotoxicity or microcystin occurrence were clearly identified as the most frequent type of acute cyanobacterial toxicity in Norway and in 60% or more of the samples investigated by the Danish, German, Portugese and Korean studies; the Czech study recorded microcystins in 90% of all samples investigated. In contrast, neurotoxic blooms were recorded in one quarter or less of the cases investigated in Norway, Denmark, Portugal, Germany and Korea, the most frequent taxon causing neurotoxicity being *Anabaena* spp., particularly *A. lemermanii* (see results from Norway and Denmark). The neurotoxin concentrations reported in these studies were usually low, with the exception of 1140 µg of anatoxin-a found g^{-1} seston dry weight in Korea during dominance of *Anabaena citrispora* and up to 3300 µg g^{-1} of anatoxin-a(S) in blooms of *Anabaena lemmermanii* in Denmark.
2. Microcystin-producing cyanobacterial taxa differ along a geographic gradient. Although some bias in favour of bloom samples cannot be excluded in the sampling programs reported in this chapter, these results do suggest significant geographic differences in dominance of microcystin-producing taxa. In southern Norway, almost half of the lakes with hepatotoxic cyanobacteria were dominated by *Anabaena* spp. whereas *Microcystis* spp. and *Planktothrix* each dominated in only about one quarter of these lakes. *Anabaena* spp. are also known as common microcystin producers from Finland (Sivonen et al. 1989). In the Danish survey, microcystins occurred most frequently in samples with several cyanobacterial taxa, followed by samples dominated by *Microcystis* spp. The German survey found *Planktothrix* spp. to be the most common dominant taxon in samples containing microcystins, closely followed by *Microcystis* spp. South of the Baltic Sea, *Microcystis* spp. were the dominant microcystin producers in more than 67-75% of the samples studied (see reports from the Czech Republic, Portugal and Korea).

3. Most populations of *Microcystis* and *Planktothrix* but only about half of the *Anabaena*-populations show hepatotoxicity or contain microcystis. More than 80-90% of all samples dominated by *Microcystis* spp. were found to contain microcystins or to show hepatotoxicity in the Danish, German, Czech and Korean surveys, the exception being the species *M. wesenbergii*, which contained little or no microcystin in six samples from the Czech Republic, but which also appears to dominate less frequently. Of the samples dominated by *Planktothrix agardhii*, 80-100% showed hepatotoxicity or contained microcystins in the Danish, German and Czech studies, and *Planktothrix* species are a major source of microcystins in Norwegian waterbodies. *Anabaena flos-aquae* does not appear to contain microcystins as consistently: in Denmark, in Germany and in the Czech Republic only half of the samples showed hepatotoxicity or contained microcystins when *Anabaena* spp. dominated (though the Czech survey includes only four samples with dominance of this genus). The Danish data indicate that differences between *Anabaena* species are important for the frequency of microcystin occurrence.

4. Characteristic ranges of microcystin concentrations within the cellular biomass can be established for common cyanobacterial taxa, with some geographic differences in absolute concentrations. For the samples dominated by *Microcystis* spp., microcystin concentrations in relation to dry weight were similar in the data sets from Germany, the Czech Republic and Korea, median values ranging from 770 to 870 and maxima from 1500 to 5800 µg g^{-1}. They were lower in the Danish data set (mean 160, max. 1280 µg g^{-1}), and higher in the data from Portugal (mean: 4100, max. 7100 µg g^{-1} DW). Samples dominated by *Planktothrix* spp. tend to have higher microcystin concentrations than those dominated by *Microcystis* spp. (see results from Denmark, Germany and the Czech Republic). The German survey found the highest microcystin concentrations in samples dominated by *Planktothrix rubescens,* the mean of 30 samples being 1890 and the maximum reaching 5600 µg g^{-1}. Less data are available for samples dominated by *Anabaena* spp. both because these did not occur as often and because they contained microcystins less frequently. Median concentrations reported from Germany and Denmark or calculated from the Czech data were only a fraction of those found with dominance of *Microcystis* or *Planktothrix,* and maxima also appear to be lower. However, in the surveys from Germany and the Czech Republic, highest concentrations are likely to have been missed in face of the low number of samples (only 4 in each of the 2 countries).

   Further, the dominant structural microcystin variants are also characteristic for different taxa. *Microcystis* spp. often contain the microcystins LR, YR and RR. In the Czech survey, microcystin-LR amounted to almost half of the total amount microcystin detected, and in many of the Portugese bloom samples, as well as in the strains isolated from them, it amounted to 90-100% of the microcystin detected, although various other microcystins also occurred in some samples and strains. *Planktothrix* species from Denmark and Germany contain demethylated microcystin variants, with red strains (or *P. rubescens*) containing predominantly only one

[D-Asp^3]microcystin-RR or dhb^7[D-Asp^3]microcystin-RR from field samples. Thus, sufficiently large regional data sets on occurrence of microcystins in relation to dominant taxa are a promising basis for predictions of toxicity ranges and for worst-case estimates.

5. Microcystins are largely found within the cellular biomass ("cell-bound" or "intracellular" toxins), and substantial extracellular concentrations are exceptional, short-lived events. Extracellular concentrations of cyanotoxins have only recently started to be included in surveys and are included in three of the six country reports presented here. Maximal concentrations detected reached 31 μg L^{-1} (sum of intra- and extracellular microcystins) in Portugal, and a few of the 183 samples investigated in the German study showed concentrations above 16 μg L^{-1}. However, most values ranged around 1 μg L^{-1} or less, and in Korea, no extracellular microcystin was found in any of the 30 samples analyzed from 5 L of water by HPLC after concentration on ODS cartridges.

6. Pelagic concentrations of cell-bound microcystins per liter of water, i.e. concentrations outside of surface scums, are usually low. In the studies presented here, they rarely exceeded several μg L^{-1} and only occasionally climb into the range of several hundreds of μg L^{-1}. In contrast, accumulation of cyanobacterial cells in surface scums may cause concentrations of several tens of mg L^{-1}. Only two of the surveys addressed concentrations of cell-bound microcystins in relation to a litre of water. In almost 80% of 270 German samples taken in the pelagic zone outside of scum areas, cell-bound microcystin concentrations amounted to less than 10 μg L^{-1}. Concentrations above 100 μg L^{-1} were detected only occasionally during dominance of *Planktothrix* spp. (see Chapter 2.1.2). In contrast, the Korean samples were bloom samples, yet the maximum concentration of 24 samples was 171 μg L^{-1}; five other samples with high concentrations ranged between 12 and 81 μg L^{-1} (see Chapter 2.2.5). In surface scums, up to 25000 μg L^{-1} were measured in a survey of bathing sites presented in Chapter 5.2. A worst case estimate discussed in this chapter shows that the maximum possible concentration, if all of the water is displaced by cyanobacterial cells, could reach 3 mg mL^{-1}, which would be 60-fold above the maximum concentration found in the above-mentioned scum of pea-soup consistency.

7. Toxicity not accounted for by the currently known cyanotoxins is observed frequently, particularly in surveys which apply both chemical analysis and bioassays. In southern Norway, non-specific toxicity was widespread, particularly in samples dominated by *Aphanizomenon flos-aquae* which showed no hepato- or neurotoxicity, but pronounced protracted toxic effects causing death of bioassay mice within 4-24 h after injection with no direct observable signs of organ damage (see Chapter 2.2.1). The Danish survey also reports this protracted toxic effect, rarely with samples dominated by *Microcystis* spp. but in 33% of those dominated by *Anabaena* spp. and in 26% of the samples dominated by *Planktothrix agardhii*. The German survey applied a bioassay using primary rat hepatocytes and found

many samples to be substantially more toxic than can be accounted for by their microcystin content alone, particularly if *Planktothrix* spp. dominated and less so during dominance of *Microcystis* spp. (see Chapter 2.1.2 and 7.2). These results indicate the existence of further, not yet identified cyanotoxins and/or of synergistic effects with other substances.

References

Al-Lay JK, Poon GK, Codd GA (1988) Isolation and purification of peptide and alkaloid toxins from Anabaena flos-aquae using high performance thin-layer chromatography, J Microbiol Methods 7:251–258

Amorim A, Vasconcelos VM (1999) Dynamics of microcystins in the mussel *Mytilus galloprovincialis*. Toxicon 37:1041–1052

Berg K, Carmichael WW, Skulberg OM, Benestad C, Underdal B (1987) Investigation of a toxic water bloom of *Microcystis aeruginosa* (Cyanophyceae) in Lake Akersvatn, Norway. Hydrobiologia 144:97–103

Berg K, Skulberg OM, Skulberg R, Underdal B, Willén T (1986) Observations of toxic blue-green algae (cyanobacteria) in some Scandinavian lakes. Acta Vet Scand 27:440–452

Branco R, Guimarães MC (1988) Características limnológicas estivais de dois lagos de barragem do rio Douro – Valeira e Régua. Actas Col. Luso-Esp. Ecol Bacias Hidrográficas e Rec Zool 35–45

Brettum P (1989) Planteplankton. Alger som indikator på vannkvalitet i norske innsjøer. Norsk institut for vannforskning, Oslo. 111 pp

Christoffersen K (1996) Ecological implications of cyanobacterial toxins in aquatic food webs. Phycologia 35:42–50

Christoffersen K, Olrik K (1996) Giftige alger i Esrum Sø og Å. Vand & Jord 3:21–24 (in Danish)

Codd G A (1995) Cyanobacterial toxins: occurrence, properties and biological significance. Wat Sci Tech 32:149–156

Cronberg G, Komarek J (1994) Planktonic cyanoprokaryotes found in South Swedish lakes during the XIIth International Symposium on Cyanophyte Research, 1992. Arch Hydrobiol/Suppl 105, Algological Studies 75:323–352

Falconer I R, Bartram J, Chorus I, Kuiper-Goodman T, Utkilen H, Burch M, Codd GA (1999) Safe levels and safe practices. In: Chorus I, Bartram J (eds) Toxic cyanobacteria in water: a guide to their public health consequences, monitoring and management. Published on the behalf of WHO by E & FN Spon, London, pp. 155–178

Fastner J (1999) Microcystins (cyanobacterial hepatotoxins) in German fresh waters: extraction, occurrence, and infuence of environmental factors on microcystin production. PhD Thesis at the Free University Berlin

Fastner J (1994) First evidence on the occurrence of microcystin-LR in Berlin and Brandenburg lakes. In Detection methods for cyanobacterial toxins In: Codd GA, Jefferies TM, Keevil CV, Potter E (eds) Special Publication No. 149. The Royal Society of Chemistry, Cambridge. pp. 149–151

Ferreira F (1994) Caracterização bioquímica de cianotoxinas produzidas por duas espécies de cianobactérias (*Microcystis aeruginosa* e *Aphanizomenon flos-aquae*) provenientes da albufeira de Crestuma-Lever (Rio Douro). MSc thesis, Porto University

Frade F (1954) Albufeira de Castelo do Bode. A água e alguns dos seus micro-organismos. Bol. da Comissão de Fiscalização das Águas de Lisboa 35:129–137

Frade F (1951) A propósito da barragem de Castelo de Bode. Bol. da Comissão de Fiscalização das Águas de Lisboa 32:95–101

Gleason FK Wood JM (1987) Secondary metabolism in the cyanobacteria, In: Fay P and Van Baalen C (eds) The Cyanobacteria. Science Publishers B.V., Amsterdam, pp. 437– 452

Gomont MM (1893) Monographie des Oscillariées (Nostocacées Homocystées). Libraire de l'académie de médcine, Paris

Harada KI, Matsuura K, Suzuki M, Oka H, Watanabe MF, Oishi S, Dahlem A, Beasely VR, Carmichael WW (1988) Analysis and purification of toxic peptides from cyanobacteria by reversed-phase high-performance liquid chromatography. J Chromatogr 448:275–283

Harada KI, Kimura Y, Ogawa K, Suzuki M, Dahlem AM, Beasley VR, and Carmichael WW (1989) A new procedure for the analysis and purification of naturally occurring anatoxin-a from the blue-green alga *Anabaena flos-aquae*. Toxicon 27:1289–1296

Harada KI, Ogawa K, Kimura Y, Murata H, Suzuki M, Thorn PM, Evans WR, Carmichael WW (1991) Microcystins from *Anabaena flos-aquae* NRC 525-17. Chem Res Toxicol 4:535–540

Harada KI, Nagai H, Kimura Y, Suzuki M, Park HD, Watanabe MF, Luukaainen R, Sivonen K, Carmichael WW (1993) Liquid chromatography/mass spectrometric detection of anatoxin-a, a neurotoxin from cyanobacteria. Tetrahedron 49:9251–9260

Henriksen P (1996) Microcystin profiles and contents in Danish natural populations of cyanobacteria/ blue-green algae as determined by HPLC. Phycologia 35 (Suppl.):102–110

Henriksen P, Carmichael WW, An J, Moestrup Ø (1997) Detection of an anatoxin-a(s)-like anticholinesterase in natural blooms and cultures of cyanobacteria/blue-green algae from Danish lakes and in the stomach contents of poisoned birds. Toxicon 35:901–913.

Henriksen P, Moestrup Ø (1997) Seasonal variations in microcystin contents of Danish cyanobacteria. Natural Toxins 5:99–106

Hrudey, S, Burch, M, Drikas, M, Gregory R (1999) Remedial measures. In: Chorus I, Bartram J (eds) Toxic cyanobacteria in water: a guide to their public health consequences, monitoring and management. Published on the behalf of WHO by E & FN Spon, London, pp. 275–312

Huber EG (1972) The crystal structure and absolute configuration of 2, 9-diacetyl-9-azabicyclo(4,2,1)non-2,3-ene. Acta Crystallographicca B28:2577–2582

Humpage AR, Rositano J, Bretag AH, Brown R, Baker PD, Nicholson BC, Steffensen DA (1994) Paralytic shellfish poisons from Australian cyanobacterial blooms. Aust J Mar Freshwater Res 45:761–771

Kaas H, Henriksen P (2000) Saxitoxins (PSP toxins) in Danish lakes. Wat Res 34:2089–2097

Kim BC, Kim EK, Pyo DJ, Park HD, Heo WM (1995) Toxic cyanobacterial blooms in Korean lakes. J KSWQ 11:231–237

Komárek J (1996) Taxonomic determination of cyanobacterial water blooms in the Czech Republic. In: Maršálek B, Marvan P, Keršner V, (eds) Cyanobacterial water blooms, Nadatio Flos Aquae, Brno, CR, pp. 22–85

Komárek J, Kovácik L (1989) Trichome structure of four Aphanizomenon taxa (cyanophyceae) from Czechoslovakia, with notes on the taxonomy and delimitation of the genus. Pl Syst Evol 164:47–64

Kotak BG, Lam AKY, Prepas EE, Kenefick SL, Hrudey, SE (1995) Variability of the hepatotoxin microcystin-LR in hypereutrophic drinking-water lakes. J Phycol 31:248-263

Kuiper-Goodman T, Falconer I, Fitzgerald J (1999) Human Health Aspects. In: Chorus I, Bartram J (eds) Toxic cyanobacteria in water: a guide to their public health consequences, monitoring and management. Published on the behalf of WHO by E & FN Spon, London, pp. 133–154

Lahti K, Rapala J, Färdig M, Niemelä SI, Sivonen K (1997) Persistence of cyanobacterial hepatotoxin, microcystin-LR in particulate material and dissolved in lake water. Wat Res 31:1005–1012

Lindholm T, Eriksson JE, and Meriluoto JAO (1989) Toxic cyanobacteria and water quality problems – examples from a eutrophic lake on Åland, south west Finland. Wat Res 23:481–486

Lindholm T, Meriluoto JAO (1991) Recurrent depth maxima of the hepatotoxic cyanobacterium Oscillatoria agardhii. Can J Fish Aquat Sci 48:629–1634

McDermott CM, Feola R, Plude J (1995) Detection of cyanobacterial toxins (microcystins) in waters of Northeastern Wisconsin by a new immunoassay technique. Toxicon 33(11): 1433–1442

Moestrup Ø (1996) Toxic blue-green algae (cyanobacteria) in 1833. Phycologia 35 (Suppl.):5

Nauwerck A (1959) Beitrag zur Kenntnis des Phytoplanktons portugiesischer Gewässer. Bol Soc Broteriana XXXIII:223–231

Nauwerck A (1962) Zur Systematik und Ökologie portugiesischer Planktonalgen. Bol Soc Broteriana XI:7–56

Nehring S (1993) Mortality of dogs associated with a mass development of Nodularia spumigena (Cyanophyceae) in a brackish lake at the German North Sea coast. J Plankton Res 13:867–872

Oliveira MR (1984a) Contribuição para o conhecimento das comunidades fitoplanctónicas das albufeiras a sul do Tejo. Bol Inst Nac Invest Pescas 11:45–58

Oliveira MR (1984b) Estrutura da comunidade fitoplanctónica e dinâmica dos "blooms" na albufeira do Maranhão. Bol Inst Nac Invest Pescas 12:37–69

Oliveira MR (1987) Estrutura de comunidades fitoplanctónicas em albufeiras portuguesas. Tese para Investigador Auxiliar. INIP, Lisboa: 307 pp

Oliveira MR (1991) Eutrofização do rio Guadiana. Blooms de Cyanophyceae e influência na ictiofauna. Relat Téc Cient INIP 42: 24 pp

Oliveira S (1995) Influência de *Microcystis aeruginosa* na biologia de *Procambarus clarkii*. MSc thesis Porto University.

Onodera H, Oshima Y, Henriksen P, Yasumoto T (1997) Confirmation of anatoxin-a(s), in the cyanobacterium *Anabaena lemmermannii*, as the cause of bird kills in Danish lakes. Toxicon 35:1645–1648

Onodera H, Oshima Y, Watanabe MF, Watanabe M, Bolch CJ, Blackburn S, Yasumoto T (1996) Screening of paralytic shellfish toxins in freshwater cyanobacteria and chemical confirmation of the toxins in cultured *Anabaena circinalis* from Australia. In: Yasumoto Y, Oshima Y, Fukuyo Y (eds) Harmful and toxic algal blooms. UNESCO

Park HD, Watanabe MF, Harada KI, Nagai H, Suzuki M, Watanabe M, Hayashi H (1993) Hepatotoxin (microcystin) and neurotoxin (anatoxin-a) contained in natural blooms and strains of cyanobacteria from Japanese freshwaters. Natural toxins 1:353–360

Park HD, Watanabe MF (1996) Toxic Microcystis in eutrophic lakes, In: Watanabe MF, Harada K-I, Carmichael WW, Fujiki H (eds), Toxic Microcystis. CRC Press, Boca Raton, Florida, pp. 57–77

Park HD, Iwami C, Watanabe MF, Harada KI, Okino T, and Hayashi H (1996) Seasonal changes of toxic Microcystis and amount of microcystin in Lake Suwa, Japan. In T. Yasumoto, Y. Oshima and Y. Fukuyo (eds) Harmful and toxic algal blooms, Intergovermental Oceanographic commission of UNESCO, Paris, pp 555-558

Park HD, Iwami C, Watanabe MF, Harada KI, Okino T, Hayashi H (1998) Temporal variabilities of the concentrations of intra- and extracellular microcystin and toxic *Microcystis* species in a hypertrophic lake, Lake Suwa, Japan (1991–1994). Environ Toxicol Water Qual 13:61–72

Sampaio J (1933) Apontamentos para o estudo das cianófitas portuguesas. Anais da Fac Ciências Porto XVIII:49–59

Santos MF (1973) Contribuição para o conhecimento das algas de água doce de Portugal. III. Bol Soc Brot XLVII:105–124

Sivonen K (1990) Effects of light, temperature, nitrate, orthophosphate, and bacteria on growth of and hepatotoxin production by *Oscillatoria agardhii* strains. Appl Environ Microbiol 56: 2658–2666

Sivonen K, Himberg K, Luukkainen R, Niemela SI, Poon GK, Codd GA (1989) Preliminary characterization of neurotoxic cyanobacteria blooms and strains from Finland. Toxic Assess 4:339–352

Sivonen K, Jones G (1999) Cyanobacterial toxins. In: Chorus I, Bartram J (eds) Toxic cyanobacteria in water: a guide to their public health consequences, monitoring and management. Published on the behalf of WHO by E & FN Spon, London, pp. 41–111

Skulberg OM (1978) Sestonobservasjoner ved vassdragsundersøkelser, (Observations of seston supporting investigations of watercourses.) Fauna 31:48–54

Skulberg OM and Skulberg R (1985) Planktic species of Oscillatoria (Cyanophyceae) from Norway – characterization and classification, Arch. Hydrobiol/Suppl 71 Algological Studies 38/39:157–174

Skulberg R, Skulberg OM (1990) Forskning med algekulturer – NIVAs kultursamling av alger. Research with algal cultures – NIVAs Culture Collection of Algae. Norsk institutt for vannforskning, Oslo. 32pp

Skulberg OM, Carmichael WW, Codd GA, Skulberg R (1993) Taxonomy of toxic cyanophyceae (cyanobacteria). In: Falconer IR (ed) Algal toxins in seafood and drinking water, Academic Press, London, pp. 145–164

Skulberg OM (1996) Toxins produced by cyanophytes in Norwegian inland waters – health and environment. In: Låg J (ed) Chemical data as a basis of geomedical investigations. The Norwegian Academy of Science and Letters, Oslo, pp.197–216

Tsuji K, Naito S, Kondo F, Watanabe MF, Suzuki S, Nakazawa H, Suzuki M, Shimada, Harada KI (1994) A clean-up method for analysis of trace amounts of microcystins in lake water. Toxicon 32:1251–1259

Ueno Y, Nagata S, Tsutsumi T, Hasegawa A, Yoshida F, Sutajjit M, Mebs D, Vasconcelos V (1996) Survey of microcystins in environmental water by a highly sensitive immunoassay based on monoclonal antibody. Natural Toxins 4:271–276

Underdal B, Nordstoga K, Skulberg OM (1999) Protracted toxic effects caused by saline extracts of *Aphanizomenon flos-aquae* (Cyanophyceae/Cyanobacteria). Aquatic Toxicol 46:269–278

Utermöhl H (1958) Zur Vervollkommnung der quantitativen Phytoplankton-Methodik. Mitt internat Verein Limnol 9:1–38

Vasconcelos VM (1994) Toxic cyanobacteria (blue-green algae) in Portuguese freshwaters. Arch Hydrobiol 130: 439–451

Vasconcelos VM (1995) Uptake and depuration of the heptapeptide toxin microcystin-LR in *Mytilus galloprovincialis*. Aquatic Toxicol 32:227–237

Vasconcelos VM (1999) Cyanobacterial toxins in Portugal: effects on aquatic animals and risk for human health. Braz J Med Biol Res 32:249–254

Vasconcelos VM, Campos T, Amorim A, Soares AMVM (1993a) Toxicidade de estirpes de cianobactérias isoladas a partir das lagoas das Braças, Vela e Mira. Bol UCA U Algarve UCTRA 1:193–201

Vasconcelos VM, Evans W, Carmichael WW, Namikoshi M (1993b) Isolation of microcystin-LR from a *Microcystis* (Cyanobacteria) bloom collected in the drinking water reservoir for Porto, Portugal. J Env Sci Health 28A:2081–2094

Vasconcelos VM, Sivonen K, Evans WR, Carmichael WW, Namikoshi M (1995) Isolation and characterization of microcystins (heptapeptide hepatotoxins) from Portuguese strains of *Microcystis aeruginosa*. Arch Hydrobiol 134:295–305

Vasconcelos VM, Sivonen K, Evans WR, Carmichael WW, Namikoshi M (1996) Hepatotoxic microcystin diversity in cyanobacteria blooms collected in Portuguese freshwaters. Wat Res 30:2377–2384

Vezie C, Brient L, Sivonen K, Betru G, Lefeuver JC, Salkinoja-Salonen M (1997) Occurrence of microcystins containing cyanobacterial blooms in freshwaters of Brittany (France). Arch Hydrobiol 139:401–413

Watanabe Y, Watanabe MF, Watanabe M (1986) The distribution and relative abundance of bloom forming *Microcystis* species in several eutrophic waters. Jpn J Limnol 47:87–93

Watanabe MF, Harada KI, Matsuura K, Oishi S, Watanabe Y, Suzuki M (1989) Heptapeptide toxins contained in natural samples of *Microcystis* species. Toxicity Assessment 4: 487–497

WHO (1998) Guidlines for drinking-water quality, Second edition, Addendum to Volume 2 Health criteria and other supporting information. World Health Organisation, Geneva

2.3
Release and Persistence of Microcystins in Natural Waters

Martin Welker, Christian Steinberg and Gary Jones

Introduction

Since most incidences involving human exposure to cyanotoxins have been linked to ingestion or contact with cyanobacterial bloom material, dissolved cyanotoxins have, until recently, drawn much less attention. More recent concerns about chronic, low-level exposure of humans to microcystin in drinking water have led to a recommendation by the World Health Organization that microcystin concentration in drinking water should not exceed 1 μg L^{-1} (WHO 1998). Exposure of humans to hazardous microcystin levels through drinking water can be expected mainly in regions with little choice of drinking-water resources and where treatment is rudimentary or absent. In these regions, monitoring microcystin concentrations is crucial for assessment of exposure risks. The Caruaru incident in Brazil, which resulted in the deaths of more than 50 renal dialysis patients, was a fateful combination of human error, lack of knowledge, and cyanotoxin occurrence (Jochimsen et al. 1998).The inadequate maintenance of water-processing facilities coincided with blooms of toxic *Microcystis* in the reservoir used for the treatment of the dialysis patients.

As with other contaminants, the potential for persistence of microcystins is crucial for the evaluation of their effect on aquatic organisms, livestock and humans, and the understanding of their persistence and degradation is a prerequisite for thorough risk assessment. This requires concomitant study of the dynamics of microcystins and their persistence in natural waters. Understanding of persistence under natural conditions will further be useful for investigation of the behaviour of microcystins during water-treatment processes.

One reason for the scarcity of reports on concentrations of dissolved microcystins in surface waters is the complexity of sample preparation and analytical procedures required for measurement of dissolved microcystins in natural waters, particularly at low concentrations. The ease with which dissolved microcystins can be detected has been increased markedly with the advent of routine analytical tests based on immuno- or protein phosphatase inhibition assays.

Release of Microcystins from Water Blooms

A detailed study determining the ratio of dissolved microcystins to total microcystins in Japanese waterbodies applying a monoclonal ELISA revealed, however, that only in a few cases was more than 50% of the total amount of microcystin found dissolved in the water and, in many cases, no dissolved microcystins were detected, although the total concentration reached values as high as several hundred μg L^{-1} (Ueno et al. 1996). In the Bautzen Reservoir in three subsequent years, Jungmann et al. (1996)

detected microcystins in the cells of *Microcystis* sp. blooms from May to October but never detected dissolved microcystins at a detection limit of 1 µg L^{-1}.

In the study presented here, the concentration of microcystins in Müggelsee (Berlin, Germany) were monitored over weekly intervals in the open water (pelagic zone) and at different sites near the shore. The microcystins were quantified using an HPLC method modified from Lawton et al. (1994), and particle free lake water was analyzed for dissolved microcystins by solid phase extraction at a detection limit of 0.3 µg L^{-1}.

In both years microcystins (mainly the RR and the LR variants) were detectable in samples from June to October, and the concentrations were positively correlated to the biovolume of *Microcystis* sp. cells (Welker et al., 1999). The concentrations of cell-bound microcystins ranged from 0.2 (detection limit) to 3 µg microcystin L^{-1} in the pelagic zone.

At sites near the shore, concentrations of cell-bound and extracellular microcystins were highly variable. At one site concentrations of up to 30 µg L^{-1} cell-bound microcystins were obtained, but no dissolved microcystin was detected. At other near-shore sites in Müggelsee considerable amounts of extracellular, dissolved microcystins could be detected. In an artificial cove protected by a small peninsula, a

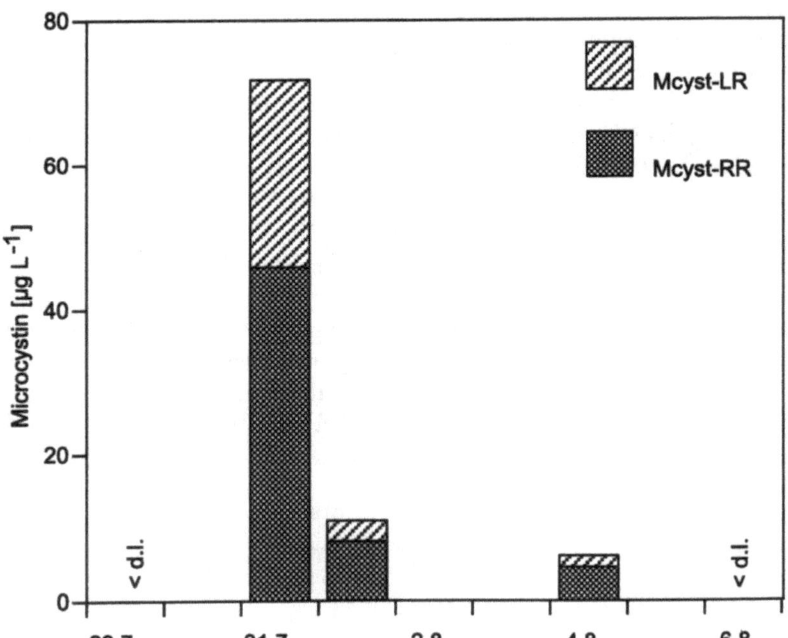

Fig. 1. Dissolved microcystins in an artificial bight of Lake Müggelsee (Berlin, Germany) during a water bloom in 1997. Samples were prepared by applying 0.5 L of filtrated lake water to a SPE cartridge (Waters, SepPak C_{18}, 1 g). Samples were analyzed by HPLC. <*d.l.* Concentration below the detection limit of 0.3 µg L^{-1}. All samples were prepared and analyzed in duplicate

thick surface scum had accumulated for several days before lysis and rotting became conspicuous. The concentration of dissolved microcystins increased from non-detectable to over 70 μg L^{-1} and dropped to non-detectable values again within a few days (Fig. 1). The observed decline is suspected to be caused by hydrological effects rather than by degradation of the toxins (see below).

The highest microcystin concentrations were found at a site within the fringing aquatic reed belt on the downwind side of the lake. There, scums of cyanobacteria had accumulated 10 to 15 cm thick and were already showing visible signs of rotting when sampled. Samples yielded a concentration of 120 μg L^{-1} total dissolved microcystins which was, however, only a minor part (about 10%) of the total microcystins (intracellular and dissolved) found at concentrations of up to 1200 μg L^{-1} (Fig. 2). In contrast, samples taken on several occasions along the shore of Müggelsee during lysis of a bloom (mainly *Microcystis* sp.) – as evidenced by the

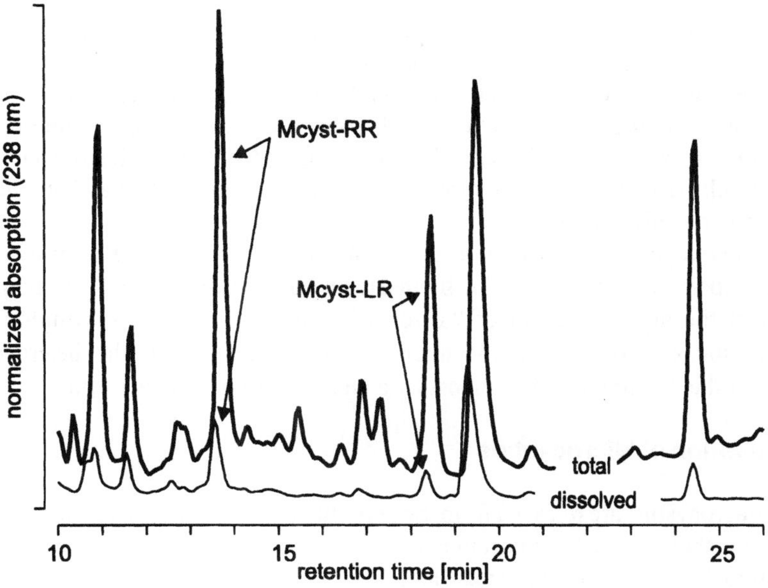

Fig. 2. HPLC chromatograms of two fractions of lake water on 4.8.1997. Samples were collected at the reed belt of lake Müggelsee (Berlin). The chromatograms are normalized allowing direct comparison. *Total* Raw water; *Dissolved* Filtrate (GF/C). Samples were freeze-dried and the residuals extracted with 10% methanol and sonicated

blueish colour of the filtrates or thick layers of proteinic foam – were negative for dissolved microcystins, underlining the ephemeral and unpredictable concentration of the toxins in natural waters.

The release of microcystins from cyanobacterial cells is considered to occur when the cells are decomposing (Berg et al. 1987; Sivonen et al. 1992; Watanabe et al. 1992; Sivonen and Jones 1999). Release of the toxins from healthy cells seems to be negligible, and no evidence for toxin exudation has been published to date. Algicide treatments of lakes with populations of microcystin containing cyanobacteria will therefore cause a rapid increase in the concentration of dissolved microcystins (e.g. Kenefick et al. 1993; Jones and Orr 1994) and may enhance rather than reduce the risk of human exposure through drinking water. If algicide treatment is to be applied to drinking-water reservoirs, careful monitoring and management are essential, and application should follow the state of the art (Lam et al. 1995a; Hrudey et al. 1999).

Generally, microcystin concentrations well in excess of the WHO guideline of 1 $\mu g\ L^{-1}$ would be expected at sites where toxic cyanobacteria accumulate, rather than in the open water. The accumulation of cyanobacteria is the result of their positive buoyancy and wind-driven hydrodynamics, with downwind (leeward) shorelines and embayments being the likely sites for dense scums or accumulations (Hutchinson and Webster 1994). Once accumulated, several natural mechanisms may lead to their lysis, such as photodamage and heat injury (Ibelings and Maberly 1998; Abeliovich and Shilo 1972) or infection by cyanophages (Padan and Shilo 1972).

Rapid lysis of dense cyanobacterial blooms may lead to very high concentrations of dissolved microcystins. Jones and Orr (1994) reported up to 1800 ug L^{-1} from a lake which was treated with an organic copper algicide. Under natural field conditions, however, toxins most likely are released from lysing cells rather gradually over a longer period of time, although this has not been well studied. Ecologically, it is of importance that high concentrations of dissolved microcystins may preferentially occur in sheltered areas which are often preferred by fish as breeding habitats, such as reed belts (for effects of cyanotoxins on early life stages of fish and amphibians see Chapter 6, this Volum).

The heterogeneous distribution of dissolved microcystins is also relevant for human exposure. Routine sampling limited to one or two (often central) locations may result in the announcement of "all-clear", whereas at other unsampled sites concentrations can be very high. Monitoring programmes must take this heterogenous nature of the distribution of cyanobacteria and their toxins into account.

Degradation of Microcystins

Once microcystins are released from the cells into the water several mechanisms can determine their fate in the environment.

- dilution with non-contaminated water,
- absorption to sediments and suspended solids or dissolved organic matter,
- incorporation into aquatic organisms such as plants and fish,
- microbial degradation by natural aquatic bacteria,

- direct chemical breakdown, e.g. by hydolysis and oxidation,
- degradation by indirect photolysis in the presence of humic substances and/or pigments.

In natural waters, dilution has been observed to decrease the concentration of microcystins (Jones and Orr 1994). Thus, a decrease in concentration at one site does not automatically allow the conclusion that a breakdown of the toxins has occurred.

Microcystins are chemically very stable. Complete chemical hydrolysis is achieved only by harsh treatments, for example with 6N hydrochloric acid or with trifluoroactetic acid under reflux (Harada 1996). However, for detoxification, complete or even partial hydrolysis may not be necessary. The Adda side-chain of microcystin, in particular its conjugated diene, is essential for toxic activity, but is also reasonably susceptible to the action of strong oxidants such as ozone or chlorine (Hrudey et al. 1999). The degradation of microcystins can also be achieved by semiconductor photocatalysis under UV irradiation in the presence of titanium dioxide (Lawton et al. 1999; Feitz et al. 1999). These findings, however promising, are of more technical interest and will not be applicable to natural waterbodies.

Adsorption of microcystins to particle surfaces has mostly been investigated under aspects of water treatment such as removal by activated carbon (e.g. Lambert et al. 1996). Some microcystin removal has been observed through adsorption onto suspended solids in water, but this seems to be only a low-affinity pathway for microcystin removal. Only about 10% of the initial 5 $\mu g\ L^{-1}$ microcystins were found adsorbed on natural seston particles after three days of contact (Rivasseau et al. 1998). Lahti et al. (1998) have shown that significant microcystin removal can occur during soil filtration of microcystin-contaminated water. However, this study did not differentiate between adsorption onto clay particles and the effect of soil biota potentially involved in degradation.

Complexation of microcystins by dissolved or particle-associated humic acids, as has been observed for other contaminants (e.g. Laor et al. 1998), has not been investigated, but could be of significance. Complexation with humic material could lessen the toxicity of contaminated waters to aquatic organisms by reducing the bioconcentration of the toxins, as it is known from other xenobiotic compounds (Haitzer et al. 1998).

The uptake of microcystins by aquatic biota is of importance both for these organisms and for determining the partitioning and fate of microcystin in aquatic environments. Once introduced into organisms, the toxins may induce the synthesis of detoxification enzymes (see Chapter 6, this Volume). In other cases, microcystins may simply bioaccumulate. Because they are water-soluble, microcystins are less likely to accumulate in organisms of higher trophic levels (Kotak et al. 1996; Laurén et al. 1995) to the degree known for other more lipophilic contaminants (e.g. Morrison et al. 1997).

All four mechanisms for microcystin breakdown or removal from the waterbody discussed so far are likely to be of only minor importance on a lake-wide scale. Because of their substantial degradation potential, microbial as well as photolytic mechanisms will be subject to the following, more detailed discussion.

Microbial Degradation of Microcystins

As oligopeptides, microcystins may have a high nutritional value either as source of amino acids, or of energy to microorganisms. However, their cyclic structure renders them resistant to many common bacterial proteases. Common peptidases such as trypsine proved unable to hydrolyze any peptide bond of the cyclic molecule (Harada 1996), indicating that the ability to degrade microcystins requires rather specific enzymes. Therefore quick and complete degradation by bacterial consortia of surface waters may not necessarily be expected, especially if specific bacteria and/or unusual enzyme systems must be induced by exposure to microcystin.

Several studies showed a prolonged lag phase until degradation of the toxins commenced, as is frequently observed with bacterial degradation processes in natural environments. When the degradability of microcystins was tested experimentally using natural microbial communities, persistence of toxins ranged from a period of a few days to several weeks, probably mainly dependent on the origin of the heterotrophic inocula. In an experiment subjecting a nonaxenic culture of *Microcystis aeruginosa* to decomposition, dissolved microcystins were recorded in detectable amounts for 70 days despite a constantly high density of heterotrophic bacteria (Watanabe et al. 1992). Interestingly, the degradation of the microcystins began when the density of heterotrophic bacteria reached its lowest value on day 42 of the experiment. Lam et al. (1995b) reported a lag phase of 10 days for degradation with sewage effluent as inoculum, Cousins et al. (1996) found total degradation within less than 1 week in reservoir water, and Jones et al. (1994) found lag-phases ranging from 3 to 20 days in different surface waters and at different initial concentrations of microcystins. In surface waters from 14 lakes near Berlin, microcystin persisted for 13 to more than 53 days (when the experiment had to be terminated) in batch cultures (Fig. 3). A biphasic degradation pattern was most distinct in samples with short persistence for microcystin. The longer the toxin persisted, the less marked became the shift from the lag phase to the degrading phase. Indeed, in two samples in which about half of the initial concentration of microcystin could still be detected after 53 days, degradation was rather gradual and continued for several weeks at a low rate. The degradation kinetics in individual subcultures were in close agreement to each other, but differences between spatially and hydrologically linked samples were marked: in the sample Löcknitz/Kienbaum microcystin-LR persisted for 35 days whereas a complete degradation was achieved in 13 days in the sample Löcknitz/Fangschleuse. Both water samples were taken in a small stream at a distance of some few kilometers from each other on the same sampling date. A similarly high variability in the persistence was found in cultures from the Schlaube system, a series of lakes and fishponds connected to each other by the River Schlaube. However, degrading activity was observed in all water samples. In summary, it can be generalized that in most of the laboratory studies degradation proceeded rapidly after the lag phase and was completed within a few days, mostly in less than 1 week, but the lag phase could last for several weeks.

Extended observations of microcystin persistence on a lake-wide scale were conducted by Jones and Orr (1994) following an algicide treatment. The concentration declined from more than 1600 µg L^{-1} to about 120 µg L^{-1} after a lag phase of 9 days days and further to 40 µg L^{-1} on day 21 until a flooding event diluted the toxins to undetectable concentrations. The observed temporal dynamics do not reflect the processes of release and degradation under natural conditions, because the toxins were released lake-wide in one pulse by the algicide. Further, heterotrophic bacteria may have been affected by the treatment also, and temporal dynamics may include the time required for recovery of a microcystin-degrading bacterial consortium.

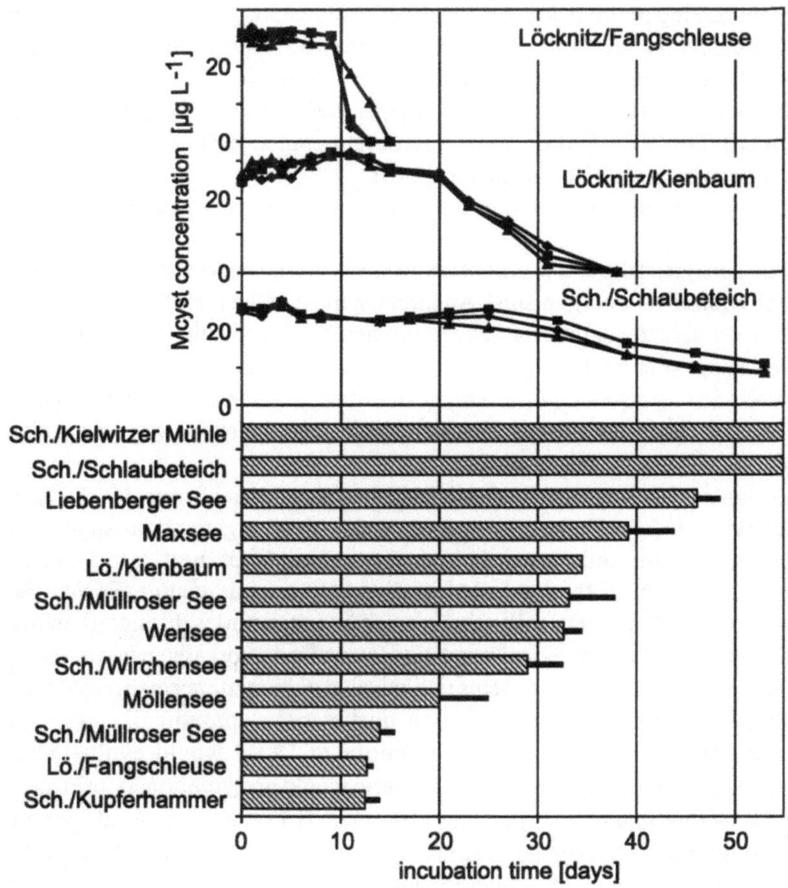

Fig. 3. Kinetics of microbial degradation of microcystin-LR (examples, *upper panel*) and mean persistence ± SE of microcystin-LR in triplicate cultures derived from water bodies near Berlin, Germany (*lower panel*). Raw waters were derived from eutrophic water bodies and filtered (Whatman GF/C) to remove most bacterivores. Purified microcystin-LR was added to the cultures which were incubated at 25°C in Erlenmeyer flasks. Subsamples were taken daily, centrifuged, and the supernatant injected directly in the HPLC (detection limit: 0.5 ng on column). Persistence in individual cultures was calculated as the mean between the last positive sampling and the first sample in which no microcystin-LR was detected. *Lö* River Löcknitz; *Sch.* River Schlaube, both Brandenburg/ Germany

For estimation of the persistence of cyanobacterial toxins it is crucial to know in which waters and sediments degrading bacteria may be expected. Current understanding of the microorganisms able to degrade microcystin is limited. Seventeen strains of bacteria with the ability to degrade microcystins isolated from Finnish lakes and sediments by Lahti et al. (1998) were all Gram-negative and oxidase positive, possessing weak catalase activity. Since the degradation rates of these strains for microcystins differ substantially, the degradation pathways may also differ.

One enzymatic pathway of microcystin degradation by a strain of *Sphingomonas* sp. was studied by Bourne et al. (1996). The initial step was found to be the opening of the peptide ring at the Adda-Arg-peptide bond resulting in a linear oligopeptide which subsequently was further digested. The enzyme responsible for the initial step was named microcystinase and is a rather rare metallo-protein. The enzyme is expressed constitutively, i.e. in the absence of microcystin, and this raises the possibility that its function is not only the hydrolysis of microcystin, but also the breakdown of other naturally occurring cyclic or unusual peptides.

Regarding the specificity of enzymes for microcystin degradation the questions are:

1. Are such enzymes the equipment of some few bacterial taxa or are they widespread among aquatic microorganisms? Apparently, most of the microbial inocula tested so far were able to degrade microcystins when incubated long enough. Degrading bacteria appear to be distributed all over the world and several strains have been isolated and characterized from a range of differing waterbodies (Lahti et al. 1998; Bourne et al. 1996). Whether or not microcystin degradation is always a constitutive function in these bacteria is unknown.
2. Are these enzymes synthesized only when microcystins are present or can their synthesis be induced by a wider group of dissolved organic compounds? Even when microcystins are present in high concentrations of mg L^{-1}, they are still a minor component of the total pool of dissolved organic matter (DOM). DOM concentrations are generally three to four magnitudes higher than likely concentrations of dissolved microcystins (expressed as organic carbon). From an evolutionary point of view, it seems unlikely that an enzyme specialized in one step of metabolizing a compound which occurs only sporadically and in low concentrations compared to other fractions of DOM would evolve. Only if microcystins were generally toxic to bacteria would an evolutionary pressure that could select for microcystin degraders be given. However, cyanobacteria contain numerous other cyclic peptides (Namikoshi and Rinehart 1996), and microcystinase could prove to be capable of hydrolyzing other such molecules.

Degradation of microcystins was studied in experiments conducted with a microbial community derived from a sewage plant oxidation pond and at different concentrations of microcystin-LR (Fig. 4). The toxins were added as cell free extract of a toxic *Microcystis* bloom to simulate the suite of compounds dissolved in water following natural lysis. The extract contained microcystin-LR as the only microcystin variant. Growth of the bacterial population was monitored photometrically by measuring turbidity at λ=540 nm. In all treatments microcystin was decomposed after a lag phase of 5 to 12 days. During the first days the bacterial community grew well in all concentration treatments, with rates positively correlated to the amount of cyanobacterial extract added to the cultures (R^2=0.995, linear regression). In the low concentration treatments bacterial density reached a maximum after 2 days. For the highest extract concentration net bacterial growth ended after 4 days but the turbidity remained at high levels until it began to decline after 12 days. Microcystin-LR was not degraded during the initial phase of intensive bacterial growth and when degradation set on, the toxin declined to undetectable values within 2 to 5 days.

The lag phases in the particular treatments (here and below expressed as the time until 50% of the initial toxin was degraded, graphically obtained from plots of residual microcystin versus time for each particular culture) were dependent on the initial concentration of cell-free extract in the cultures (expressed in concentration of microcystins) with lag-phases increasing with increased initial concentration of extract in the cultures (R^2=0.945, exponential regression).

In all treatments, dense populations of small flagellates developed after several days, and the grazing activity of these flagellates may have influenced the degradation kinetics either by suppressing bacterial growth in general or by species-specific

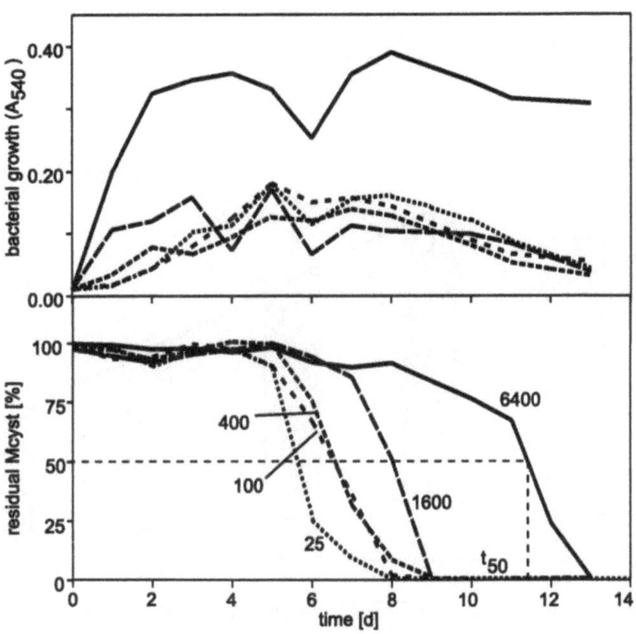

Fig. 4. Growth of a microbial community derived from an oxidation pond near Griffith, Australia (*upper panel*) and degradation of microcystin-LR (*lower panel*). Microcystin-LR was added as aqueous cell-free extract of dried bloom material in concentrations indicated in the lower plot (µg L^{-1}). Incubation was done as described above. Bacterial growth was monitored as absorbance at 540 nm, the patterns of the lines refer to the different microcystin concentrations. Data represent means of triplicates. *Thin dotted lines* visualize the determination of the time until 50% of initial microcystin was degraded (t_{50}) which is used as a measure of the lag phase

predation. This raises the possibility that trophic interactions among microorganisms may influence the overall kinetics of microcystin degradation under field conditions. To test this hypothesis, we incubated different fractions of pond-water filtered to sequentially exclude different-sized organsisms from the raw water sample (<1.2, <2.7, <10, <30 µm, and raw pond water). In all treatments degradation was completed within a few days. However, a Spearman rank correlation revealed a significant relationship between the size fraction of the particles and organisms in the culture and the lag phase of toxin degradation, with slightly shorter lag phase in the larger size fractions (R^2=0.681, p<0.0005). Overall, the difference in effect was rather small (lag phase differences of less than 1 day between treatments), but the results highlight the potential for underlying food web processes to impact on what might otherwise be seen as simply a single-component bacterial process.

As mentioned above, microcystin degrading enzymes, or bacteria which contain them, are possibly induced by other cyclic peptides rather than by microcystin alone. The *Sphingomonas* sp. isolated by Jones et al. (1994) maintained its ability to degrade microcystin when grown on peptone or yeast extract – both of which are sources of unusual cyclic and linear peptides – in the absence of microcystin. Furthermore, the *Sphingomonas* sp. studied by Jones et al. (1994) was isolated from a waterbody with

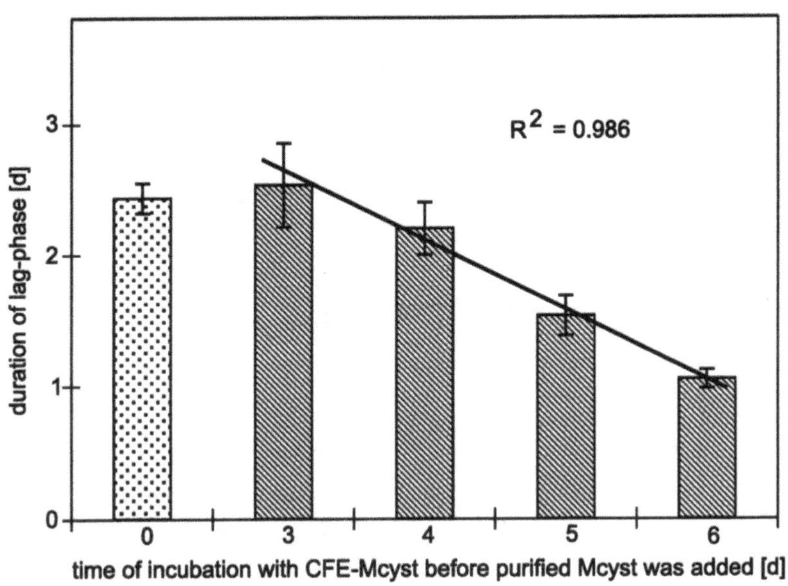

Fig. 5. Effect of the incubation of a bacterial community derived from an oxidation pond near Griffith, Australia, with a cyanobacterial cell-free extract without microcystin (CFE-microcystin) on the duration of the lag phase in degradation of microcystin-LR. Microcystin-LR was added to the cultures on different days after preincubation with microcystin-free CFE as indicated by the categories. The duration of the lag phase is determined as described in the text and Fig. 4. *Columns* represent means of triplicates ± SE. The linear regression was calculated for data represented by the *dark columns*

no history of *Microcystis* blooms. In order to test the efficacy of the induction of microcystin-degradation by other compounds, the following experimental approach separated the induction and growth of a microbial community from the time of exposure to microcystins, and then observed microcystin degradation kinetics. For this purpose, pond water was incubated with a cell-free extract of *Microcystis* from which all microcystins had been removed by passage through an open chromatographic column packed with C_{18} gel (Waters). The aqueous eluant fraction from the C_{18} column, which was presumed to contain a range of proteins, polysaccharides and other polar compounds, was added to the pond-water at the beginning of the experiment. After incubation of the cultures with microcystin-free extract for a defined period of time, purified microcystin was added at a concentration of 250 µg L^{-1}. This was done at days 0, 3, 4, 5 and 6 of the experiment, with each treatment consisting of three independent triplicates. Microcystin concentrations were then monitored daily by HPLC analysis until the concentration fell below the detection limit and average lag-phases (i.e. t_{50}; see above and Fig. 4) were calculated for each treatment (Fig. 5).

For the first 3 days with growth on microcystin-free cell-extract, we observed an increased turbidity (at 540 nm) in the cultures following a pattern comparable to Fig. 4 (upper panel). The mean lag phase in the cultures to which microcystin was added on day 3 was not significantly different from the mean lag phase in the cultures to which the toxin was added in the beginning ($P>0.05$). However, in the cultures to which microcystin was added on the following days, the lag phase was significantly reduced ($P<0.001$, $R^2=0.986$). Overall, the longer the pond water was incubated with cell extract free of microcystin, the more rapidly microcystin was degraded after its addition. In fact, for the later days of incubation no lag phase in degradation of microcystin was observed at all.

From these observations we conclude that the induction of the ability to degrade microcystins by endemic bacteria may be independent of the presence of microcystin itself. We suggest that organic compounds other than microcystin present in cyanobacterial exudates, or perhaps present as background DOM in surface waters, induce the growth of a diverse bacterial population, some species of which are constitutively able to degrade microcystin. This supports the hypothesis that microcystin-degrading bacteria are rather common in surface waters, whether or not these waters have ever been exposed to blooms of *Microcystis* or other microcystin producing species.

Photosensitized Degradation of Microcystins

Although the peptide bonds of microcystins are chemically and physically very stable (see above), direct degradation under field conditions, for example by photooxidation, may be feasible through absorption of UV energy by the conjugated diene of the Adda side chain (Kaya and Sano 1998), which is the chromophore widely recognized and utilized for HPLC detection of microcystins (Jones et al. 1994). Direct photodegradation, however, of microcystins by sunlight is negligible since the absorbance spectrum of microcystin does not overlap with the solar spectrum and only absorbance of photons can cause transformations in a molecule. Therefore, photodegradation can only take place indirectly, via sunlight-absorbing substances occurring in natural waters such as pigments or humic substances. Tsuji et al. (1994) showed that microcystins are degraded by natural sunlight rather slowly, but that this rate is greatly enhanced by cyanobacterial pigments which act as photosensitizers. In their experiments, however, the concentration of pigments required for effective degradation within a week or two was quite high (1-5 mg L^{-1}) and generally higher than would be observed in natural waters, except following the lysis of a dense cyanobacterial bloom.

In addition to algal pigments, other photosensitizers present in high concentrations in every waterbody are humic substances (Steinberg and Münster 1985). Humic substances can act as photosensitizers by absorption of sunlight in a range upwards from UV-B to the wavelengths of photosynthetically active radiation (PAR). Absorbed light energy can than be transferred from the excited humic molecule to other molecules while the humic molecule returns to its ground state. This process generates highly reactive molecules such as the hydroxyl radical, singlet oxygen or H_2O_2 (Cooper et al. 1994) which can react with otherwise photochemically inert molecules (Zepp et al. 1985), possibly microcystin.

This hypothesis was tested with a degradation experiment exposing microcystins to natural sunlight in the presence of humic substances. Sterile filtered solutions of purified humic substances (fulvic acids derived from a drinking-water plant in Fuhrberg, Germany) and microcystins in water were exposed in UV-transparent quartz tubes. In a first experiment no difference in degradation efficiency among three variants of microcystins was found (Fig. 6). All variants had residual concentrations of 50% after 8 h of irradiation in the presence of fulvic acids, whereas no degradation was observed in pure water (Welker and Steinberg 1999). In all non-irradiated treatments – with and without humic substances – the concentration of microcystins did not decline significantly. This means that adsorption to humic molecules was not evident and declines in microcystin concentrations were therefore the result of indirect photolysis. Interestingly, we did not observe a difference in degradation efficiency when the concentration of humic substances was varied above 2 mg L^{-1} (Fig. 6). This was unexpected, because the photochemical formation of reactive species like H_2O_2 via photoexcitation of humic substances depends functionally on the concentration of the latter (Scully et al. 1996). Since irradiation in water alone did not result in degradation this suggests that the lowest concentration of fulvic acids (2 mg DOC L^{-1}) was already higher than the saturation concentration.

A further degradation experiment with refined experimental setup showed a dependence of the degradation rate on the concentration of humic substances which can be described by a saturation function in a similar range of FFA-concentrations (0-12 mg DOC L^{-1}; Welker and Steinberg 2000). This was expected, since with increasing concentrations of humic substances the optical density also increases and at a saturation concentration incident active solar radiation is completely absorbed by the solution.

Although the actual molecular mechanisms involved in the indirect photolysis are unclear, there is evidence for involvement of the reactive Adda moiety, especially the conjugated double bond, in the degradation process (Welker and Steinberg 1999). In all experiments performed so far, no new peaks were detected in HPLC chromatograms with a microcystin-like absorption spectrum. A derivatized microcystin molecule with an intact Adda moiety, however, would still have a similar absorption spectrum

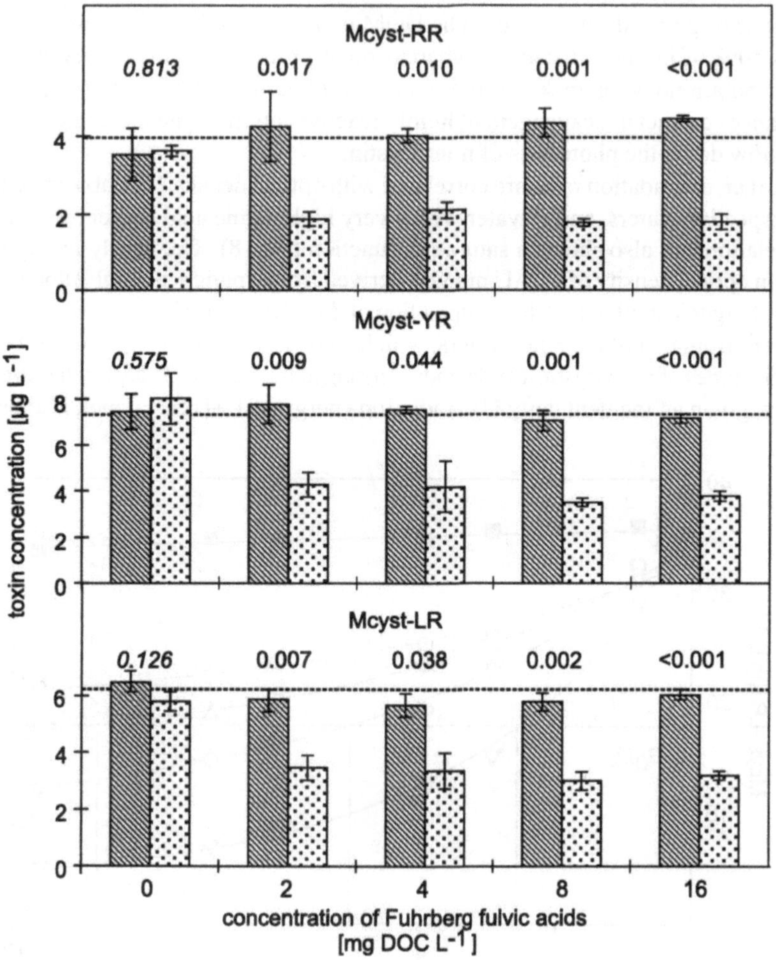

Fig. 6. Photolytical degradation of three microcystin variants in the presence of humic substances under irradiation with natural sunlight. *Dark columns* control, not irradiated; *light columns* irradiated; *numbers above each pair of columns* are probabilities for control treatment = irradiated treatment, $t_{a(2),n}$. Pure microcystins were incubated in Duran or quartz tubes containing aqueous solutions of Fuhrberg fulvic acids of different concentrations. After 8 h of irradiation samples were analyzed by HPLC. *Columns* represent means of triplicates. *Dotted horizontal lines* indicate initial concentrations of microcystins. (From Welker and Steinberg 1999, with permission from Elsevier Science)

(see Bourne et al. 1996). Further, conjugated dienes like in the Adda side chain are highly susceptible to oxidation by reactive oxygen species generated by the photosensitation of humic substances (Zepp et al. 1985; Hoigné et al. 1989).

In particle-free natural waters, from different types of waterbodies like rivers, lakes and ponds sampled near Berlin, Germany, degradation of microcystins due to indirect photolysis could be demonstrated (Fig. 7; Welker and Steinberg 2000). Degradation was, however, much slower when compared to the degradation in solutions of purified humic substances: first-order degradation rates were about threefold higher in solutions of purified fulvic acids. The DOM pool of natural waters consists of only about 50% of humic substances, whereas other groups like polysaccharides, fatty acids and amino acids make up the remainder (Thurman 1985). These non-humic substances can act as scavengers of highly reactive intermediate molecules and thus may slow down the photolysis of microcystin.

Further, degradation rates are correlated with optical density (UV absorptivity) of the respective waters, and if water with a very high humic acid content is included, this relationship also shows a saturation function (Fig. 8). The highly humic water with an optical density of 11 AU nm was derived from a pond with high allochtonous input of organic matter, and the composition of dissolved organic matter is likely to be different from that of the other waters, which were derived from larger water bodies. Half-life times of microcystin-LR degradation ranged from 2.5 to 20 days in the particular waters, given an incident daily UV-radiation energy of 1 MJ m^{-2}, which is about the

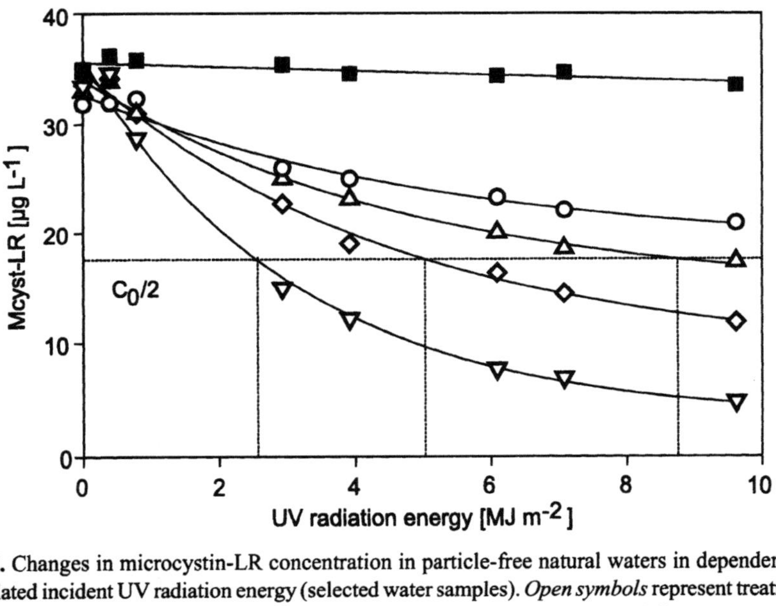

Fig. 7. Changes in microcystin-LR concentration in particle-free natural waters in dependence of cumulated incident UV radiation energy (selected water samples). *Open symbols* represent treatments with natural waters derived from different waterbodies near Berlin, Germany, *full squares* represent a control treatment with MilliQ-water. *Lines* represent fitted first order decay functions. Daily incident UV radiation energy was about 1 MJ m^{-2} during the experiment. (Data from Welker and Steinberg 2000, with permission from the American Chemical Society)

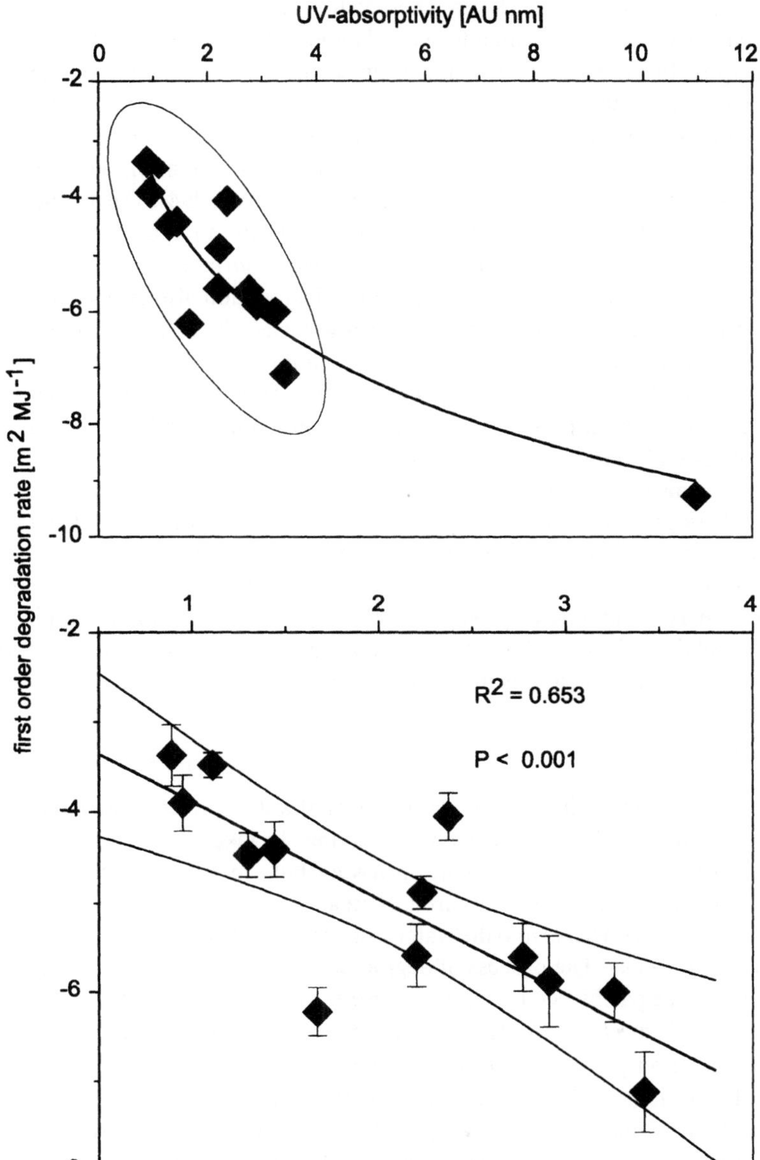

Fig. 8. Dependence on UV absorptivity of first-order rates of photosensitized degradation of microcystin-LR in natural waters from Berlin and Brandenburg. Rates were calculated from time series of concentrations of microcystin-LR in sunlight-irradiated samples (n=11). Radiation energy was recorded with a spectroradiometer during the experiment. UV absorptivity of water samples was calculated from absorption spectra by integrating A=f(l) from 300 to 400 nm. *Upper plot* Results for all waters; *the fitted curve* is a logistic growth function; lower plot *data points encircled* in the upper plot enlarged; statistics are related to a linear regression model; *error bars* indicate standard errors. (Modified from Welker and Steinberg 2000)

average midsummer value at Berlin. These values are, however, strictly related to the actual dimensions of experimental vessels and cannot be extrapolated to whole waterbodies. In the range of UV absorptivity up to 4 AU nm, the relationship of degradation rate to absorptivity was linear (R^2=0.65). This means that the degradation rate of microcystins in natural waters by indirect photolysis was proportional to the absorbed UV radiation. UV radiation is generally absorbed in full within the upper part of the water column of inland waters (Scully and Lean 1994). The degradation per surface area is, therefore, not dependent on the concentration of humic substances but rather on the solar radiation penetrating the water surface. Consequently, within areas of similar sunlight intensity the per-area degradation also is similar. In highly humic waters degradation is limited to a narrow surface layer, but is highly efficient within this layer, whereas in waters with low humic content degradation activity reaches deeper layers but with lower efficiencies. Therefore, indirect photolysis is more relevant for the elimination of microcystins in shallow, well-mixed lakes than in deeper lakes with stable water columns.

A first estimate of the half-life time of microcystins with humic-sensitized photolysis as the only degradation pathway results in about 70 days per meter of water column depth depending on the initial concentration of microcystins, solar radiation, and on the solar elevation. The degradation rates on a lake-wide scale certainly cannot reach rates that may be attained by bacterial activity during the degrading phase, but photolytical degradation will set on without a lag phase and therefore can contribute to the elimination of the toxins in surface waters.

Conclusions

Microcystins occur chiefly intracellularly in natural waters. Release into the waterbody appears when cyanobacterial cells are decaying. In lakes, high concentrations of microcystins rarely are reported for the open water but concentrations of some hundred µg L^{-1} are occasionally encountered at near-shore sites where cyanobacteria accumulate. Once released into the water from cells, the toxins can persist for several days or a few weeks. The process of degradation in large waterbodies is difficult to follow due to the patchy occurrence of release and barely quantifiable dilution effects.

The two main mechanisms of toxin elimination in natural waters are microbial activity and degradation photosensitized by humic substances. Microcystins have been shown to be degraded by bacterial consortia derived from many waterbodies worldwide. It is not yet known whether the capability to degrade microcystin is a unique feature of a widely distributed bacterial group or a common feature among many bacteria. Typically, bacterial breakdown of the toxins sets on after a lag phase of several days and is then performed at high rates. The duration of the lag phase is mediated by the rate of growth and establishment of the bacteria population and by the concentration of the toxins. The induction of toxin degrading activity is not dependent on the presence of microcystins, and other soluble components of cyanobacteria appear to be involved.

The otherwise chemically stable toxins are degraded in the presence of humic substances when irradiated with natural sunlight. The three variants microcystin-RR, -YR, and -LR are degraded at similar rates. On a lake-wide scale photolytical degradation rates depend mainly on the surface to volume ratio of the waterbody and the actual radiation energy reaching and penetrating the surface of the respective waters. Though the rates of photochemical degradation are lower than those that can be achieved by bacteria, they can significantly contribute to the dissipation of microcystins in natural waters because the process is not delayed by a lag phase.

Acknowledgements

We wish to thank Philip and Cheryl Orr and Denise de Paoli for technical and personal assistance during MW's stay at CSIRO in Griffith, and Linda Tidwell and Bernhard Rübe for technical assistance at the IGB in Berlin.

References

Abeliovich A, Shilo M (1972) Photooxidative death in blue green algae. J Bact 111:682–689

Berg K, Skulberg OM, Skulberg R. (1987) Effects of decaying blue-green algae on water quality – a laboratory study. Arch. Hydrobiol. 108:549–563

Bourne DG, Jones GJ, Blakeley RL, Jones A, Negri AP, Riddles P (1996) Enzymatic pathway for the bacterial degradation of the cyanobacterial cyclic peptide toxin microcystin-LR. Appl Environ Microbiol 62:4086–4094

Cooper WJ, Shao C, Lean DRS, Gordon AS, Scully FE (1994) Factors affecting the distribution of H_2O_2 in surface waters. In: Environmental chemistry of lakes and reservoirs. Advan Chem Ser 237:393–422

Cousins IT, Bealing DJ, James HA, Sutton A (1996) Biodegradation of microcystin-LR by indigenous mixed bacterial populations. Wat Res 30:481–485

Feitz AJ, Waite TD, Jones GJ, Boyden BH, Orr PT(1999) Photocatalytic degradation of the blue green algal toxin microcystin-LR in a natural organic-aqueous matrix. Environ Sci Technol 33:243–249

Haitzer M, Höss S, Traunspurger W, Steinberg C (1998): Effects of dissolved organic matter (DOM) on the bioconcentration of organic chemicals in aquatic organisms – A review. Chemosphere 37:1335–1362

Harada K (1996) Chemistry and detection of microcystins. In: Watanabe MF, Harada K-I, Carmichael WW, Fujiki H (eds) Toxic *Microcystis*. CRC, Boca Raton, USA

Hoigné J, Faust BC, Haag WR, Scully FE, Zepp RG (1989) Aquatic humic substances as sources and sinks of photochemically produced transient reactants. In: Suffet IH, MacCarthy P (eds) Aquatic humic substances: influence on fate and treatments of pollutants. Adv Chem Ser 219:363–381

Hrudey S, Burch M, Drikas M, Gregory R (1999) Remedial measures. In: Chorus I, Bartram J (eds) Toxic cyanobacteria in water: a guide to their public health consequences, monitoring and management. Published on the behalf of WHO by E & FN Spon, London, pp. 275-312

Hutchinson PA, Webster IT (1994) On the distribution of blue-green algae in lakes: wind-tunnel tank experiments. Limnol Oceanogr 39:374–382

Ibelings BW, Maberly SC (1998) Photoinhibition and the availability of inorganic carbon restrict photosynthesis by surface blooms of cyanobacteria. Limnol Oceanogr 43:408–419

Jochimsen EM, Carmichael WW, An JS, et al. (1998) Liver failure and death after exposure to microcystin at a hemodialysis center in Brazil. New England J Med 338:873–878

Jones GJ, Bourne DG, Blakeley RL, Doelle H (1994) Degradation of cyanobacterial hepatotoxin microcystin by aquatic bacteria. Natural Toxins 2:228–235

Jones GJ, Orr PT (1994) Release and degradation of microcystin following algicide treatment of a *Microcystis aeruginosa* bloom in a recreational lake, as determined by HPLC and protein phosphatase inhibition assay. Wat Res 28:871–876

Jungmann D, Ludwichowski KU, Faltin V, Benndorf J (1996) A field study to investigate factors that could effect microcystin synthesis of a *Microcystis* population in the Bautzen reservoir. Int Rev Hydrobiol 81:493–501

Kaya K, Sano T (1998) A photodetoxification mechanism of the cyanobacterial hepatotoxin microcystin-LR by ultraviolet irradiation. Chem Res Toxicol 11:159–163

Kenefick SL, Hrudey SE, Peterson HG, Prepas EE (1993) Toxin release from *Microcystis aeruginosa* after chemical treatment. Wat Sci Technol 27:433–440

Kotak BG, Zurawell RW, Prepas EE, Holmes CFB (1996) Microcystin-LR concentration in aquatic food web compartments from lakes of varying trophic status. Can J Fish Aquat Sci 53:1974–1985

Lahti K, Vaitomaa J, Kivimäki AL and Sivonen K (1998) Fate of cyanobacterial hepatotoxins in artificial recharge of groundwater and in bank filtration. In: Peters et al. (ed) Artificial recharge of groundwater. Balkema, Rotterdam, pp. 211-216

Lahti K, Niemi MR, Rapala J, Sivonen K (1998) Biodegradation of cyanobacterial hepatotoxins - characterisation of toxin degrading bacteria. In: Reguera B, Blanco J, Fernández ML, Wyatt T (eds) Harmful algae. Xunta de Galicia, Int Oceanogr Comm UNESCO

Lam AKY, Prepas EE, Spink D, Hrudey SE (1995a) Chemical control of hepatotoxic phytoplankton blooms: implications for human health. Wat Res 29:1845–1854

Lam AKY, Fedorak PM, Prepas EE. (1995b) Biotransformation of the cyanobacterial hepatotoxin microcystin-LR, as determined by HPLC and protein phosphatase bioassay. Environ Technol 29:242–246

Lambert TW, Holmes CFB, Hrudey SE (1996) Adsorption of microcystin-LR by activated carbon and removal in full scale water treatment. Wat Res 30:1411–1422

Laor Y, Farmer WJ, Aochi Y, Strom PF (1998) Phenanthrene binding and sorption to dissolved and to mineral-associated humic acid. Wat Res 32:1923–1931

Laurén MC, Hietala J, Reinikainen M, Walls M. (1995) Do *Microcystis aeruginosa* toxins accumulate in the food web: a laboratory study. Hydrobiologia 304:23–27

Lawton LA, Edwards C, Codd GA (1994) Extraction and high-performance liquid chromatography method for the determination of microcystins in raw and treated waters. Analyst 119:525–1530

Lawton LA, Robertson PK, Cornish BJPA, Jaspars M (1999) Detoxification of microcystins (cyanobacterial hepatotoxins) using TiO_2 photocatalytic oxidation. Environ Sci Technol 33:771–775

Morrison HA, Gobas FAPC, Lazar R, Whittle DM, Haffner GD (1997): Development and verification of a benthic/pelagic food web bioaccumulation model for PCB congeners in western Lake Erie. Environ Sci Technol 31:3267–3273

Namikoshi M, Rinehart KL (1996) Bioactive compounds produced by cyanobacteria. J Ind Microbiol 17:373–384

Padan E, Shilo M (1972) Cyanophages – viruses attacking blue-green algae. Bacteriol Rev 37:343–370

Rivasseau C, Martins S, Hennion MC (1998) Determination of some physicochemical parameters of microcystins (cyanobacterial toxins) and trace level analysis in environmental samples using liquid chromatography. J Chromatogr 799:155–169

Scully NM, Lean DRS (1994) The attenuation of ultraviolet radiation in temperate lakes. Arch Hydrobiol Beih Ergebn Limnol 43:135–144

Scully NM, McQueen DJ, Lean DRS, Cooper WJ (1996) Hydrogen peroxide formation: the interaction of ultraviolet radiation and dissolved organic carbon in lake waters along a 43-75°N gradient. Limnol Oceanogr 41:540–548

Sivonen K, Jones GJ (1999) Cyanobacterial toxins. In: Chorus I, Bartram J (eds) Toxic cyanobacteria in water: a guide to their public health consequences, monitoring and management. Published on the behalf of WHO by E & FN Spon, London, pp. 41–111

Sivonen K, Namikoshi M, Evans WR, Carmichael WW, Sun F, Rouhiainen L, Luukkainen R, Rinehart KL (1992) Isolation and characterisation of a variety of microcystins from seven strains of the cyanobacterial genus *Anabaena*. Appl Environ Microbiol 58:2495–2500

Steinberg C, Münster U (1985) Geochemistry and ecological role of humic substances in lake water. In: Aiken RG, et al. (eds) Humic substances in soil, sediment, and water. Wiley, New York

Thurman EM (1985) Organic geochemistry of natural waters. Nijhoff/Jonk Publishers

Tsuji K, Naito S, Kondo F, Ishikawa N, Watanabe MF, Suzuki M, Harada K (1994) Stability of microcystins from cyanobacteria: effect of light on decomposition and isomerization. Environ Sci Technol 28:173–177

Ueno Y, Nagata S, Tsutsumi T, Hasewaga A, Yoshida F, Suttajit M, Mebs D, Pütsch M, Vasconcelos VM (1996) Survey of microcystins in environmental water by a highly sensitive immunoassay based on monoclonal antibody. Natural Toxins 4:271–276

Watanabe MF, Tsuji K, Watanabe Y, Harada K, Suzuki M (1992) Release of heptapeptide toxin (Microcystin) during the decomposition process of *Microcystis aeruginosa*. Natural Toxins 1:48–53

Welker M, Steinberg C (1999) Indirect photolysis of cyanotoxins: one possible mechanism for their low persistence. Wat Res 33:1159–1164

Welker M, Steinberg C (2000) Rates of humic substance photosensitized degradation of microcystin-LR in natural waters. Environ Sci Technol 34:3415–3419

Welker M, Hoeg S, Steinberg C (1999) Hepatotoxic cyanobacteria in lake Müggelsee. Hydrobiologia 408/409:263–268

WHO (1998) Guidelines for drinking-water quality. Second edition, addendum to volume 2: Health criteria and other supporting information. World Health Organization, Geneva

Zepp RG, Schlotzhauer PF, Sink RM (1985) Photosensitized transformations involving electronic energy transfer in natural waters: role of humic substances. Environ Sci Technol 19:74–81

3 Factors Controlling Cellular Microcystin Content

It is still unclear why cells contain cyaontoxins i.e. what the benefit is of producing cyanotoxins in competition against cells without them? One approach to addressing this question is the study of their toxin content (or their toxicity) in relation to environmental factors. If toxin content or toxicity increased substantially under certain growth conditions, this would indicate a benefit for toxin production under those conditions. From such results the biochemical function of these substances in the cells could perhaps be inferred. Earlier laboratory experiments and field studies, chiefly addressing microcystin concentrations or heptatotoxicity, indicated some gradients in relation to environmental factors relevant for growth, such as temperature, light, nutrient concentrations, salinity and pH. A comprehensive overview of experiments published up to 1998 is provided by Sivonen and Jones (1999).

Two levels of regulation of the toxin content of natural cyanobacterial populations by environmental factors are conceivable:

1. Limiting resources could directly influence toxin contents of the cells. Toxin production has been postulated to be a general response to stress. Conversely, if toxins are secondary metabolites not essential to the cells, their production could decline if limiting resources are more immediately needed for maintenance metabolism or synthesis of vital compounds.
2. Field populations are a mixture of morphologically often indiscernible, but genetically different cells, i.e. of strains or genotypes. These may vary in their toxin content, as shown in Chapter 4.2. If some can utilize certain resources slightly better than others, they will be able to grow faster when these resources are limiting. As these strains increasingly dominate the population in the sense of winning competition, their toxin concentration would increasingly determine that of the field population.

Chapters 3.1 to 3.4 address both levels of regulation by measuring microcystin content and/or growth rates of different strains under different resource limitations (Chapter 3.1 to 3.3) and by investigating genetic differences between strains with and without microcystins (Chapter 3.4). In contrast to much of the earlier work in this field, the experiments on growth rates and microcystin content used continuous or semicontinuous culture systems in order to ensure defined experimental conditions.

Chapter 3.1 investigates the effects of both light and nutrient limitation on growth rates, nutrient storage capacity and microcystin contents for six different strains of *Microcystis aeruginosa*. Coping with the large number of cultures and contitions was made possible by using a semicontinuous culture system. Chapter 3.2 investigates the impact of light limitation on growth and microcystin content of *Planktothrix agardhii* in comparison to *Microcystis aeruginosa* while assessing the applicability of a new turbidostat system for cyanobacterial cultures. Both studies demonstrate that neither light limitation nor phosphorus or nitrogen limitation substantially influences microcystin contents in the cells of *Microcystis aeruginosa*, and for light limitation this also applies for *Planktothrix agardhii*. Cell quotas appear to show this most clearly, whereas usage the ratio of microcystins to biomass indicators such as dry weight or biovolume may cloud the picture somewhat by reflecting the response of these biomass indicators to resource limitation, rather than the response of microcystin production as such. Furthermore, both studies show that differences in growth rate between different strains of a species are not very pronounced, and Chapter 3.1 demonstrates that they are not related to microcystin production.

Together with other recent studies (Orr and Jones 1998; Long et al. 2001) the observations reported in Chapter 3.1 and 3.2 consolidate the emerging picture that resource limitation affects microcystin content in the cells only slightly or not at all. This hypothesis is supported by the results reported in Chapter 3.4 showing that microcystin production is closely linked to the presence or absence of one crucial gene.

However, the results reported in Chapter 3.3 imply that some growth conditions may induce substantial changes in microcystin content: this project followed a different continuous culture approach of "stepping down" in the concentrations of phosphate or iron in the medium. As was to be expected, *Microcystis aeruginosa* responded with an according decrease of culture density until a new steady state was reached. This was accompanied by substantial carbohydrate accumulation, and also by pronounced changes in the ratio of microcystins to proteins as well as in energy charge (determined from the ratio of the adenylate nucleotides). Microcystins and energy charge were determined both during phases of transition and when cultures had reached a stable optical density. Remarkably, microcystin to protein ratios correlate well with energy charge. Although further elucidation of microcystin biosynthesis will be necessary to understand the implications of this observation, the results are exciting in the light of new work on genetic regulation of microcystin production: Kaebernick et al. (2000) show that short-term exposure to some light quantities and qualities may enhance transcription of the microcystin gene cluster. Changes in energy charge might reflect the changes in cell physiology under non-steady state conditions that also lead to enhanced microcystin production.

3.1 Effects of Light and Nutrient Supply on Growth and Microcystin Content of Different Strains of *Microcystis aeruginosa*

Karina Hesse and Johannes-Günter Kohl

Introduction

The question whether differences between growth rates of cyanobacterial strains with and without microcystins are significantly related to microcystin production can only be answered by comparing a number of strains of each group. This study investigates six strains of *Microcystis aeruginosa* which differ in morphology, microcystin content and pigment patterns. It addresses the effects of light and nutrient limitation on both the growth rates and microcystin concentrations. In contrast to previously published studies which used single cells, this study included three strains growing in their natural form as colonies.

Experimental Design

The organisms used in the experiments were non-axenic strains of *Microcystis aeruginosa* growing as single cells (strains HUB 5-3, HUB 5-2-4 and HUB 063) as well as strains newly isolated during this project from Wannsee which still grew in colonies (W75, W334 and W386; culture collection of our laboratory, cf. Rohrlack Chapter 4.2, this Volume).

Following the two-compartment-model of Droop (cf. Kohl and Nicklisch 1988) growth is a function of nutrient supply, which is characterized by two independent processes – first the rapid uptake of nutrients by highly efficient carrier systems (e.g. PO_4^{-3}) and secondly the conversion of the internal nutrients into biomass. Growth rates depend directly upon internal nutrient concentrations. Due to the high affinity of the cellular uptake systems, the external nutrient concentrations in the surrounding medium often decrease below the detection limits. For determination of the strength of nutrient limitation it is therefore necessary to investigate the intracellular contents, i.e. the cell quotas (sensu Droop).

In the semicontinuous culture system applied (Kohl and Nicklisch 1988), physiological conditions were maintained in a quasi-steady state by regular dilution of the cultures. Experiments addressing light limitation were conducted according to the turbidostat principle, which maintains constant cell densities, i.e. constant turbidity, and thus ensures defined photon flux densities. At steady state, growth rates are directly proportional to the dilution rates needed to maintain constant turbidity. Here, the cultures were diluted back to their original density at intervals of 2-3 days. Within this time span, culture density had maximally increased by a factor of 2.

Nutrient limitation experiments were performed according to the chemostat principle by semi-continuous dilutions with a defined daily dilution rate, thus obtaining a quasi-steady state. As for the turbidostat principle, at steady state the dilution rate is directly related to the growth rate of the culture, and resource-limited growth may be expressed as a quotient of maximal growth rates ($\mu_{limited} = \mu/\mu_m$). Further, as shown by Droop (1968, cf. Kohl and Nicklisch 1988), nutrient-limited growth rates are directly related to the cell quotas, and vice versa the growth rates determine cell quotas. Thus, in chemostats, nutrient limitation is determined by the dilution rate chosen. Here, it was adjusted to half of the maximal growth rates.

Nutrient concentrations in the medium determine cell density in the culture (rather than growth rates). In order to avoid light limitation in the experiments addressing nutrient limitation, cultures must be sufficiently dilute. This was established by running cultures with different nutrient concentrations and testing for linearity between nutrient concentration applied and culture density.

According to the classical theory of continuous cultures, turbidostat control allows maintaining growth near the maximal specific growth rate in the steep section of these curves. Chemostat control is applicable for nutrient-limitation experiments at relatively low growth rates, where modifications of the dilution rate (growth rate) induce only a small change in the related biomass concentration.

Culture growth was followed with optical density at 436 nm (Eppendorf-Photometer 1101 M, double-frame technique, cf. Kohl and Nicklisch 1988). In order to relate nutrient limitation and microcystin to biovolume, a conversion factor was determined by counting cell numbers ml^{-1}, measuring the mean cell volume, calculating biovolume and relating it to optical density. Colony-forming cultures of strains W75, W334 and W368 were treated with short-pulsed ultrasonication to disintegrate colonies into aggregates small enough to allow cell counting.

Erlenmeyer flasks (500 ml) were used as culture vessels. Two parallels were run for each parameter tested, for the irradiance experiments subsequently and for the nutrient limitation experiments simultaneously. Temperature was kept constant at 20°C by a water bath. Light was provided by fluorescent tubes (Lampi warm white) in a light-dark cycle of 12:12 h. Experiments for determination of light dependence of growth and microcystin content were run under nutrient-saturated conditions with M IV/2 solution (Woitke et al. 1997). Photon flux densities were varied between 5 and 150 $\mu mol\ m^{-2}s^{-1}$. Nutrient-limited cultures were held under saturated light supply at 110 $\mu mol\ m^{-2}s^{-1}$. Cleaning of the culture vessels with 1 N HCl for 24 h and triplicate washing with deionized water proved to be important for avoiding nutrient exchange between glass vessel walls and medium.

Cellular contents of microcystins, phosphorus and nitrogen were determined on 3 successive days of steady state in parallel samples. Microcystins were analyzed according to the method of Lawton et al. (1994; see also Chapter 9.1). Frozen aliquots of total cell suspension with predetermined cell content (biovolume per sample volume) were freeze-dried at -20°C. Subsequently, the cell material was disintegrated and extracted with methanol. The toxins were detected and quantified using HPLC with standards (microcystins LR, RR, YR – Calbiochem, La Jolla).

Cells for phosphorus assays were taken from a defined culture volume by filtration through P-free polycarbonate filter with a poresize of 1 µm (MEMFIL PC-PCTE, Membrapure) thoroughly washed with deionized water. Filters with algal cells were combusted for 1 h at 550°C. Ashes were diluted in 3 ml 1 N HCl and polyphosphates were hydrolyzed over 1 h at 125°C under pressure (autoclave).

Cells for nitrogen analysis were isolated from the cultures with defined biovolume content by filtration over PC-PCTE filters. After washing, cells were resuspended in deionized water and digested (peroxidisulfate) as organic nitrogen was transformed to nitrate. The inorganic nitrogen and phosphorus samples were measured with a flow-injection analyzer (EVA-System, Eppendorf): total nitrogen content was determined as NO_2^- after reduction of nitrate on a cadmium reduction column and phosphorus concentration as inorganic phosphate according to the method described by Vogler (1976).

As the mucilaginous envelope itself contains amino-nitrogen as well as nitrogen and phosphorus in the associated bacteria (Rhee 1972; Vadstein and Olson 1989) this extracellular nutrient pool was also investigated. The mucilage of algal colonies was removed by dissolution with deionized water and centrifugation of the algal suspension at about 4500 rpm, washing the cells with deionized water followed by second centrifugation (4500 rpm). The dissolved mucilage in the combined supernatants was analyzed for nitrogen content after digestion, or freeze-dried and prepared for phosphorus determination as described above.

Results and Discussion

Effect of Irradiance on Growth Rates

The dependence of growth rates on photon flux density was described by fitting a saturation curve according to the modified model of Mitscherlich (1909, cf. Kohl and Nicklisch 1988) to the experimental data by non-linear regression (Table 1, Fig. 1).

Table 1. Parameters estimated for light-limited growth curves from non-linear regressions with a modified Mitscherlich model: $\mu=\mu_m\{1-\exp[-\ln 2(I-i_0)/(K_I-i_0)]\}$[a]

	Single cells			Colonies		
	HUB 063	HUB 5-3	HUB 5-2-4 (MC)	W75	W334 (MC)	W368 (MC)
μm day^{-1}	0.225	0.288	0.272	0.324	0.251	0.307
	a	b	b	a b	a b	a b
I_0 µmol m^{-2}s^{-1}	3.68	3.43	3.48	-31.9	1.35	3.63
	a	a	a	a	a	a
K_I µmol m^{-2}s^{-1}	15.02	15.16	16.42	18.07	19.13	25.87
	a	a	a	a	a	a
R^2	0.975	0.979	0.982	0.957	0.980	0.992

[a] μ_m maximum specific growth rate, i_0 minimum light demand for growth, K_I half-saturation constant for light, R^2 regression coefficient

Significant differences between strains are marked with a and b; i.e. differences between strains were significant only if marked with a different letter

The high accuracy of fitting proved this model to be well suited for describing growth kinetics of single-cell cultures (cf. Kohl and Nicklisch 1988). For the colony-forming strains accurate measurements of the growth rates were more difficult because photometric biomass determination was less accurate, due to the uneven dispersion of biomass in the colony suspension; this resulted in less accurately determined dilution rates. Therefore, the growth rates and the estimated parameters of the model show rather wide confidence limits for the strains growing in colonies and allow only an approximate comparison with the unicellular strains (Table 1).

None of the strains investigated showed any suppression of growth by high irradiance up to the photon flux density of 150 µmol $m^{-2}s^{-1}$ (Fig. 1). HUB 5-3, W75, HUB 5-2-4 and W368 exhibit similar maximum specific growth rates of about

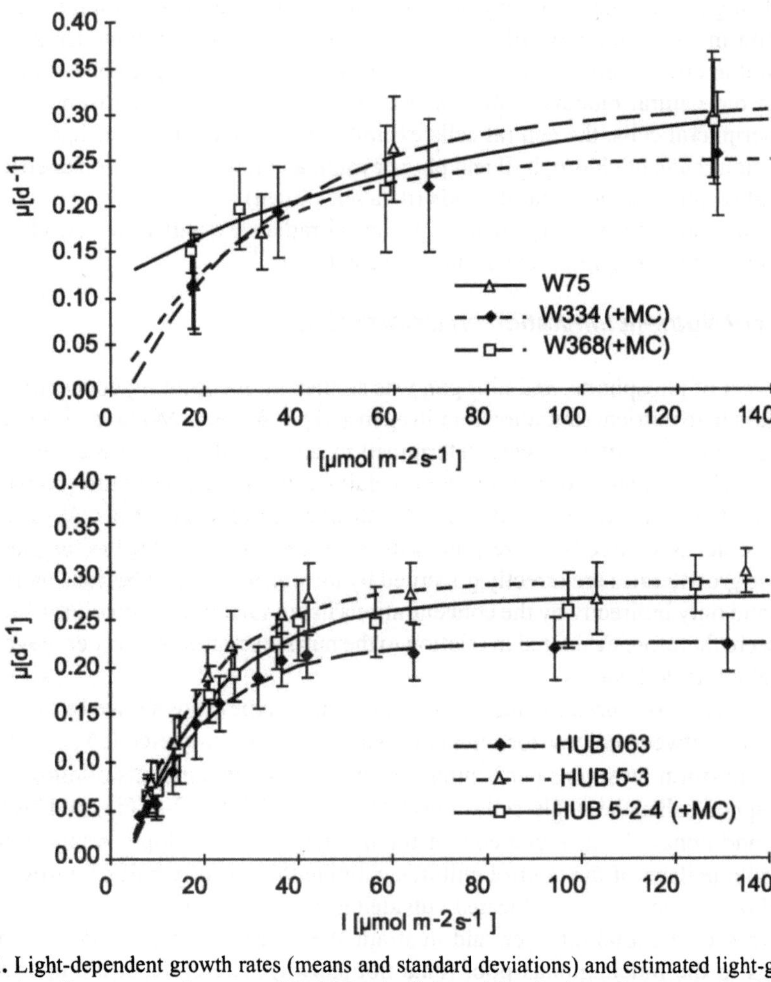

Fig. 1. Light-dependent growth rates (means and standard deviations) and estimated light-growth curves (non-linear regression to modified Mitscherlich model) of colony-forming and unicellular *Microcystis aeruginosa* strains. *+MC* microcystin-containing strains

0.3 day^{-1}, the μ_m value of strains W334 and HUB 063 are lower – significantly for HUB 063 in comparison with the other single-cell strains (Table 1).

Light demand for growth is described by the minimum photon flux density required for growth, i_0 (X-axis intersection of light vs. growth curves; growth is only possible above this maintenance energy requirement) and by the half-saturation constant (K_I). The single-cell strains (HUB 063, HUB 5-3, HUB 5-2-4) show no differences in i_0 and K_I. Comparison with the results for the colony-forming strains is hampered by the above-mentioned problems of biomass evaluation of colonies. However, as compared to the colony-forming strains, the single-celled strains tend to use light more efficiently: they attain light saturation (calculated as $\mu > 0.95\ \mu_m$) already at 50 μmol m^{-2}s^{-1}, whereas W368 requires 100 μmol m^{-2}s^{-1} and W334 needs 80 μmol m^{-2}s^{-1} to attain light saturation.

This higher light demand may be caused by the light gradient within the colony, resulting in light limitation of the central cells. This could be demonstrated with cytophotometric determination of the absorption spectra of single cells of colonies taken from natural blooms of *Microcystis aeruginosa*. In a comparison of central with peripheral cells, the central cells exhibit higher attenuation coefficients in the absorption range of chlorophyll and phycocyanin and lower attenuation coefficients in the absorption range of carotenoids (data not shown).

Notably, no difference in growth response to irradiance could be detected between the strains containing microcystin and those without.

Effect of Nutrient Limitation on Growth Rates

The effect of phosphorus and nitrogen was studied at one limiting concentration in comparison to nutrient saturation for nitrogen and phosphorus. Maximal growth rates without nutrient limitation were determined in semicontinuous cultures run at 110 μmol m^{-2}s^{-1}. To induce nutrient limitation, daily dilution rates were then adjusted to nearly half of the maximal specific growth rate of the saturated controls. As discussed above, at steady state cells will respond with lower cell quotas for the limiting nutrient, because growth rates are directly governed by the nutrient concentrations within the cells (and only indirectly by the concentrations in the surrounding medium). Growth rates were therefore evaluated in relation to the nutrient concentration per biovolume (in analogy to cell quotas).

In the nutrient-saturated controls, nutrient to biovolume ratios showed a substantial difference between colony-forming and two of the three single-celled strains, with the colony-forming strains containing substantially higher amounts of nitrogen and of phosphorus than the single-cell strains HUB 5-3 and HUB 5-2-4 (Fig. 2). This is due to the additional N and P contents of the mucilaginous envelope. Analysis of the isolated mucilage of the control cultures exhibited that nearly half of nitrogen and phosphorus, respectively, is located outside the cells in the mucilage. In strain W75 even 75% of the total nitrogen and in strain W368, approximately 65% of total P belongs to the mucilaginous pool (data not shown). Therefore, it is important to know whether this pool can be mobilized during nutrient shortage or not.

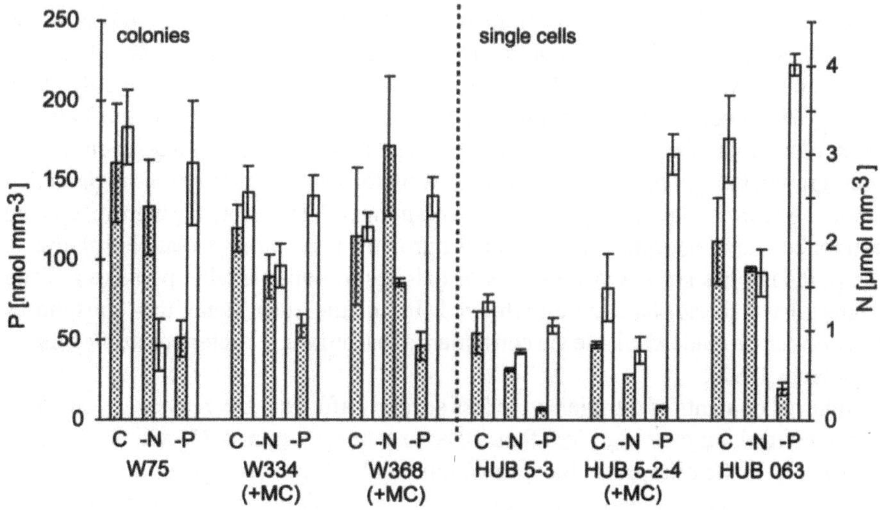

Fig. 2. Cellular contents of phosphorus (*shaded columns*) and nitrogen (*empty columns*) – means and standard deviations – for strains forming colonies (W strains) and for strains growing as single cells (HUB-strains). *C* saturated control; *-N* nitrogen-limited cultures; *-P* phosphorus-limited cultures; *(+ MC)* microcystin-containing strains

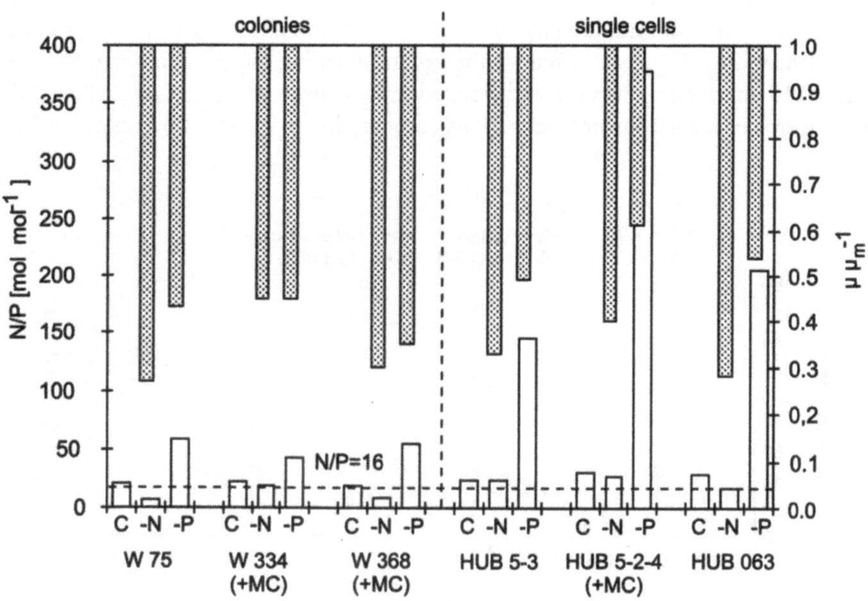

Fig. 3. N/P ratios (*empty columns*), Redfield ratio of N/P=16 (*dotted line*) and reduction of relative growth rates (*shaded columns*) under nutrient limitation and in nutrient-saturated controls. *C* saturated control; *-N* nitrogen-limited cultures; *-P* phosphorus-limited cultures; *(+ MC)* microcystin-containing strains

In all the experiments nutrient-limited growth led to lower internal contents of that nutrient as compared to the control (Fig. 2). This is especially pronounced for phosphorus, which is reduced by up to 70% whereas growth rates were only about 0.5 μ_m. However, phosphorus-limited colonial strains still contained substantially higher amounts of phosphorus per biovolume than unicellular strains, even though their growth rates in relation to μ_m (Fig. 3) were lower than those of the single-celled strains. As lower relative growth rates reflect stronger P limitation, the higher P content cannot be due to phosphorus available for growth. This suggests that phosphorus in the mucilage has structural functions or belongs to non-available pools (e.g. from bacteria) which can hardly be mobilized. In aquatic ecosystems this phosphorus fraction can be made available via consumers of bacteria, e.g. bacterivorous flagellates (Rothaupt 1992).

Internal contents of nitrogen were less severely affected. Decreases, as compared to the control, amounted to less than 50%, except for strain W75, which showed a sharp drop in the nitrogen to biovolume ratio under N limitation (Fig. 2).

Strains differed in their internal contents of the non-limiting nutrient. In most cases the concentrations of the non-limiting nutrient was unchanged compared with the saturated control. Only strain W368 showed some additional P-accumulation under nitrogen-limited growth; all other strains exhibit an even slightly lower P content. Under P limitation strains HUB 5-2-4 and HUB 063 showed a significant increase in N content in relation to the control.

The internal molar ratio of nitrogen to phosphorus (N/P) is optimal at 16 mol mol^{-1} for most phytoplankton taxa (Redfield ratio, cf. Kohl and Nicklisch 1988). Under conditions of nutrient saturation all six *Microcystis* strains show somewhat higher N/P ratios (mean 24). However, because the colony-forming strains contained more P (mean 0.13 µmol mm^{-3}) than the single-celled strains (mean 0.07 µmol mm^{-3}) their N/P ratios were somewhat lower (mean 20 as compared with mean 28 for single-celled

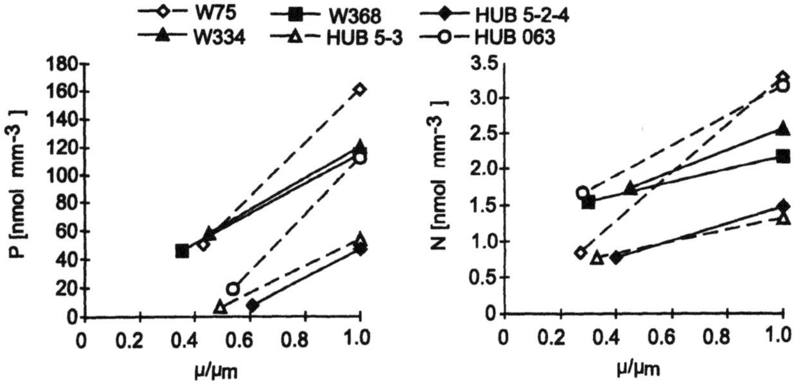

Fig. 4. Internal phosphorus and nitrogen concentrations in relation to strength of limitation (relative growth rates). *Solid symbols* Toxic strains; *open symbols* non-toxic strains; *broken lines* single-cell strains; *solid lines* colony-forming strains

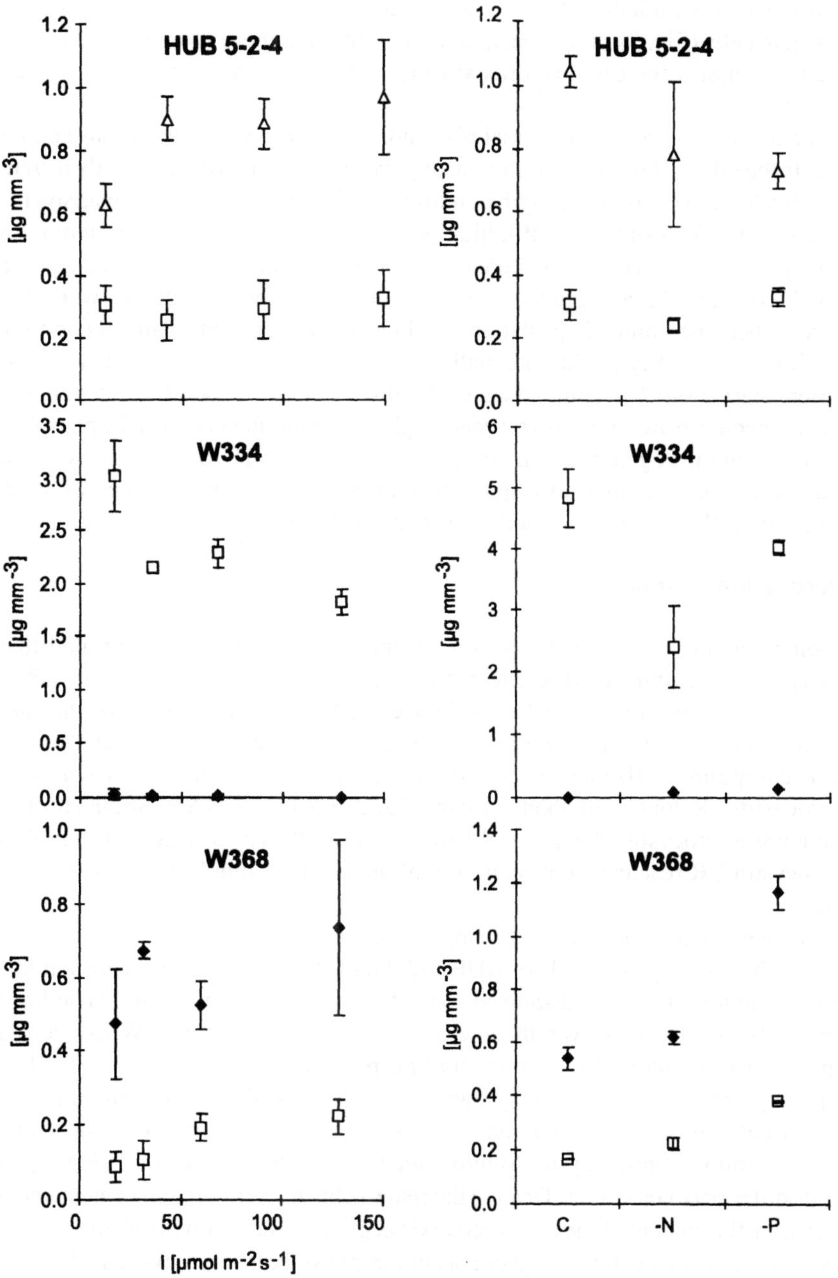

Fig. 5. Microcystin concentration of *Microcystis aeruginosa* strains HUB 5-2-4, W334 and W368 at different photon flux densities (left-hand *figures*) and under nutrient limitation (*right-hand figures*) – means and standard deviations. *C* control; *-N* nitrogen limitation; *-P* phosphorus limitation with relative growth rates. Note different scales of the axis for MC mm^{-3}

strains). Under P limitation, the N/P ratio increases drastically to 50-380, particularly for single cells (Fig. 3). In contrast, under N limitation there is only a slight change because N quotas decrease less dramatically, and P quotas were at least not higher in most strains.

The relationship between internal phosphorus or nitrogen concentrations and μ/μ_m reflects the effect of nutrient limitation on growth of strains with inherently different growth rates as well as the strength of nutrient limitation (Fig. 4) In comparison to all other strains, W75 (under N, P limitation) and HUB 063 (under P limitation) are characterized by a smaller reduction in growth rate at comparable decreasing internal N or P contents. In consequence, these strains may have advantages when these nutrients become limited (e.g. stationary bloom under constant weather conditions). This implies that they could potentially outcompete the others under conditions of nutrient starvation. The comparison of the limiting nutrient pools in relation to the relative specific growth rate also reveals a highly dynamic adaptation of the phosphorus as well as the nitrogen pools in strains W75 and strain HUB 063 (Fig. 4). These strains are especially able to respond to nutrient shortage by a change of storage pools. Strain W75 seems to be able to use nutrients from its mucilage.

Microcystin Content

Chromatographic toxin determinations in the investigated strains revealed that a variety of microcystins could be detected in HUB 5-2-4, W334 and W368 with different strain-specific distribution patterns. However, LR, YR and RR were the major microcystins found in these strains; therefore, data evaluation concentrated on these three compounds. Highest toxin concentrations were found in W334 with a microcystin-LR content of about 5 $\mu g\ mm^{-3}$ and only traces of YR. Strain HUB 5-2-4 contains microcystins-LR (up to 0.4 $\mu g\ mm^{-3}$) and RR (up to 1 $\mu g\ mm^{-3}$). In W368 microcystin YR reached a concentration of up to 1.1 $\mu g\ mm^{-3}$ and LR up to 0.3 $\mu g\ mm^{-3}$.

Changes in photon flux density changed microcystin content maximally by a factor of 2-3. The microcystin levels of HUB 5-2-4 and W368 were nearly constant in the range of light saturation and showed some decrease under limiting light conditions, maximally by 50%. However, the microcystin-LR concentration of W334 increases significantly by about 50% under limiting photon flux density (Fig. 5). This is surprising in face of its very low growth rate (μ/μ_m=0.4) due to light limitation.

Nutrient limitation also has only a moderate impact on microcystin content, changing ratios of microcystin to biovolume at most by a factor of 2 (Fig. 5), and trends differ between strains. Despite the remarkably higher nitrogen content under P limitation, the content of the measured toxic peptides is substantially changed only in strain W368, which exhibits higher concentrations of both microcystins YR and LR under phosphorus limitation as compared to unlimited controls. Nitrogen limitation reduced microcystin contents in relation to controls in strains HUB 5-2-4 and W334 (but not in W368); phosphorus limitation had a similar, but less pronounced, effect on the contents of the predominant toxins RR and YR in these two strains. Strain W334 also shows higher contents during phosphorus-limited growth than under nitrogen limitation, but, in contrast to strain W368, levels are highest if non-limited.

Discussion

These results show that some physiological traits show a wide range of variability between strains. The six investigated strains differ in many aspects: microcystin content, pigment pattern, mucilage composition (see Chapter 4.2) and nutrient storage. For field populations, this suggests that the responses of these traits to changing environmental conditions may mainly be due to shifts in genotypic composition, i.e. shifts in dominance of different strains.

Further, it became evident that microcystin contents of the various strains are not sharply regulated by either energy (light) or nutrient availability. Although two strains showed slightly lower microcystin to biovolume ratios under energy limitation of growth (which would be in accordance to expectations from the results of Bickel et al. (Chapter 3.3, this Volume), one strain (W334) shows a different response, with the content of microcystin-LR increasing under light limitation, for reasons unknown.

Thus, the results provide no general evidence to support the hypothesis that peptide synthesis (and specifically that of microcystins) may be favoured under excess nitrogen supply, particularly under P limitation, as suggested by Sivonen (1990). Only the results for one of the three strains (W368) show a slight response in accordance with this assumption.

Previous reports on the effects of nutrient and light availability on microcystin concentration in cyanobacterial cells are controversial. Utkilen and Gjølme (1992), as well as Watanabe and Oishi (1985), found a tendency to lower microcystin content at very low light supply. Codd and Poon (1988) observed no change in microcystin content at light intensities between 5 and 50 μmol $m^{-2}s^{-1}$. Van der Westhuizen and Eloff (1985) and Lehtimäki et al. (1994) also described only a small effect of fluence rate on toxicity in *Microcystis aeruginosa* and *Nodularia spumigena*.

Former investigations of nutrient influence carried out by Watanabe and Oishi (1985) show slightly lower toxicity of cells grown under phosphorus deficiency in comparison to saturated controls. Further, reduced nitrogen concentrations in the growth medium caused a decline in toxicity but not in relation to the supplied N concentrations. In contrast, Sivonen (1990) reports a continuous increase in microcystin concentration with rising nitrate application for various strains of *Oscillatoria agardhii*. However, these examinations take into consideration only supply of the nutrient in the delivered medium rather than intracellular content, thus the limitation status of the cells was not clearly defined.

Not only was microcystin content scarcely affected by growth-limiting factors, but, vice versa, no substantial differences between strains with and without microcystin were found for light-dependent growth. In theory, additional energy demand for microcystin synthesis might be negligible if this compound amounts to less than 1% of total dry weight. However, the investment costs for synthesis of the large multifunctional enzyme complexes which produce microcystins are also required (Dittmann et al. 1997). This suggests that the strains without microcystin may be using a similar amount of energy for the synthesis of alternative products.

Remarkably, no distinctly higher light demand for growth of colony-forming strains was measurable, in spite of their substantial mucilage production. Also, microcystin-producing strains do not differ substantially from non-toxic strains in light-use efficiency, but slight differences may be concealed by high variances in growth measurements with colony-forming strains.

Cyanobacteria are able to survive periods with shortages of essential nutrients, probably by accumulation of reserve compounds while the relevant nutrients are still available. Cyanophycin, a non-ribosomally synthesized polypeptide of aspartate and arginine, acts as dynamic nitrogen reservoir in *Microcystis* (Mackerras et al. 1990). Phycocyanin, a structural compound of light-harvesting complexes involved in transmission of light energy to photosystem II, plays an additional role as nitrogen reserve (Carr 1988) because of its proteinaceous character. Nevertheless, the light-harvesting complex (chlorophyll, phycobiliproteins), as well as the energy-dissipation machinery (carotenoids), is changed in response to phosphorus limitation in a similar way (cf. Kohl and Nicklisch 1988; Rücker et al. 1995). Therefore, the change in phycocyanin content may be preferentially a response to avoid the absorption of excess energy in relation to the demand due to nutrient shortage rather than the mobilization of stored nitrogen. Polyphosphate granulae clearly serve as phosphorus storage compounds (Shively 1988).

The results show the different nutrient storage and mobilization capacities of the strains. Only strain W368 accumulates N and P if the alternative nutrient is limiting. Two other strains (HUB 5-2-4, HUB 063) increase only their nitrogen content under phosphorus deficiency. The other strains show no remarkable change in their nutrient content under limitation of the alternative nutrient. They are apparently not well adapted to frequent changes in nutrient limitation. Comparing strains with and without microcystins, no consistent differences are visible. Two non-microcystin containing strains (W75, HUB 063) show very strong changes in nutrient content under limiting as compared to saturating conditions, but the third non-microcystin containing strain HUB 5-3 demonstrates only low storage ability under saturating supply but low nutrient concentrations also under limiting conditions. Thus, no relationship between microcystin content and nutrient management could be detected.

Future research should examine the complete storage dynamics of nitrogen and phosphorus (q_0, Q_{max}) with axenic (bacteria-free) strains in order to assess the potential biomass formation of toxic and non-toxic strains after exhaustion of the external nutrient pools at expense of stored nutrients. Interactive relations with bacteria in the mucilage may be important for understanding competition between different strains for limiting nutrients and their storage ability.

In summary, all our results demonstrate how diversely the strains respond to changes of environmental conditions. Every strain has genotypically determined morphological and physiological characteristics and specific adaptation to changing environmental conditions. None of these traits showed a relationship to microcystin production. Therefore, the prognosis of the composition of *Microcystis* blooms with respect to the different infraspecific genotypes in relation to resource utilization is rather complicated. Additionally, strain-specific interactions with the herbivorous zooplankton have to be taken into consideration.

Experiments concerning these interactions, also carried out in this project (Rohrlack et al. 1999; Henning Chapter 6.4, this Volume) suggest that *Microcystis* can influence the zooplankton community by suppressing species which feed on the microcystin-containing strains. On the other hand, zooplankton species consuming only competitors of *Microcystis* for nutrients and light may be supported. Under these conditions, *Microcystis aeruginosa* is able to grow and to form massive blooms in spite of its relatively low growth rate.

Thus, bloom toxicity appears to be preferentially determined by its composition of toxic and non-toxic genotypes. Light and nutrient supply influence the dominance of genotypes with most advantageous physiological strategies to exploit these resources under specific circumstances. The selection of strains with microcystins might be regulated mainly by interaction with the herbivorous zooplankton, rather than by growth-limiting resources.

Acknowledgements

We are grateful to Mrs. M. Dewender, Mrs. C. Martin and Mrs. G. König for the excellent technical support at HPLC- and Flow-Injection-Analysis.

3.2
Light-Limited Growth and Microcystin Content of *Planktothrix agardhii* and *Microcystis aeruginosa* in Turbidostats

Gabriele Böttcher, Ingrid Chorus, Silvia Ewald, Thomas Hintze and Norbert Walz

Introduction

This Chapter compares the effect of irradiance on two strains of *Microcystis aeruginosa* and on three strains of *Planktothrix agardhii*, one strain of each species containing microcystins, in order to address two key questions:

1. Does irradiance influence microcystin concentrations in cells of *Microcystis aeruginosa* and *Planktothrix agardhii*? Do results obtained from a continuous culture system and with a further taxon confirm the findings of Hesse and Kohl presented in Chapter 3.1 from semi-continuous cultures?
2. How pronounced are differences in the response of growth rates to light limitation between the two strains of *Microcystis aeruginosa* and between the three strains of *Planktothrix agardhii?*

Cyanotoxin concentrations in cyanobacterial cells have been related to various parameters indicative of cell biomass, i.e. to dry weight, biovolume, protein and chlorophyll-a. However, the relative share of various cellular components and even cell size may vary substantially as a reaction to limiting conditions and growth rates. Protein has been proposed as the least variable cellular component by Utkilen and Gjolme (1995) and is used by Bickel and Lyck in Chapter 3.3 as biomass parameter. Orr and Jones (1998), as well as Long et al. (2001), emphasize that for physiological studies addressing microcystin production rates the absolute amount of microcystin in the cells is the only meaningful point of reference. In the study by Long et al. (2001), cells actually became smaller with increasing growth rate, yet their microcystin content per cell increased. These authors therefore suggest using the concept of cell quotas, i.e. microcystin per cell, in analogy to the concept originally introduced by Droop (1973) for relating nutrient-limited growth to cellular nutrient contents. A further aim of our study therefore is to explore the response of ratios between microcystins and the biomass indicators dry weight, biovolume (i.e. cell numbers multiplied with mean cell volume) and chlorophyll-a to different light limitations.

Extracellular microcystin concentrations have scarcely been addressed in studies with continuous cultures. They have proven to be low in most of the few field situations where they were studied (Chapter 2.2; Sivonen and Jones 1999). Therefore, our study includes extracellular microcystin concentrations in relation to irradiance.

For these investigations, a new development in turbidostat systems (Walz et al. 1997), not yet employed for experiments with cyanobacteria, was combined with features of a continuous culture system previously optimized for cyanobacteria (van Liere et al. 1975). Our study assesses the suitability of this new continuous culture system for work with two species belonging to different genera of cyanobacteria.

The Turbidostat System

The turbidostat system for continuous cultures (Fig. 1) sets photon flux density in the culture at defined levels and adjusts medium flow to maintain constant turbidity. This is optimal for experiments addressing the influence of irradiance on growth and cellular components. The automatic, computer-controlled system used here follows the principles of Walz (1993) and Walz et al. (1997), modified to calculate dilution rate from overflow, and allows tight control of turbidity.

Turbidity is measured with an improved sensor according to the principle of Boraas and Bennett (1988) by the attenuation of a light beam at 880 nm. At this wavelength, turbidity was shown to relate linearly to algal biovolume measured by a Coulter counter (Walz et al. 1997) and cell constituents such as pigments do not affect turbidity. In order to avoid distortion of turbidity readings through wall growth in the culture vessel - commonly a substantial problem in turbidostat designs - the reading is taken outside the vessel through the wall of a glass syringe which draws 5 ml of culture from the vessel. Turbidity is measured after 10 s for another 10 s, then the culture is pushed back into the vessel by the plunger, whose regular activity prevents wall growth in the syringe. Turbidity is calculated according to ISO 7027 (1990) and compared to a predetermined threshold value, which is set at the rather low level of 0.04532 to avoid light gradients in the culture. Van Liere et al. (1978) showed that irradiance varies only little throughout cylindrical culture vessels at low culture densities in the range of 30 mg L^{-1} DW, as used here. Turbidity readings above the threshold value activate the peristaltic medium pump, which then proceeds to introduce fresh medium until a subsequent turbidity reading is below the threshold and deactivates the pump. Time intervals for turbidity control, pump activation/deactivation and all other recordings were set at 2 min.

The volume of the culture within the vessel is held constant, because as fresh medium is introduced, excess culture volume flows out of the vessel through the nozzle into an output vessel located on an electronic balance which is connected to the computer. Output volume is determined by weight increment which is divided by the culture volume to obtain the current dilution rate.

Mineral medium M III (modified according to Kohl and Nicklisch 1988) is pumped into the vessel by a peristaltic pump from a 20-L reservoir through double glass bulbs to avoid back-flow. Part of the air stream is diverted to these bulbs to blow the medium droplets out of the pipe for rapid mixing, and to assist in keeping the glass tube clean.

Light was provided by neon lamps (OSRAM L 36 W/20, cool white; Kohl and Nicklisch 1981) with a light:dark rhythm of 16:8 h. This was chosen because growth rates close to maximal rates at continuous light may be expected, as shown for *Planktothrix agardhii* by Zevenboom and Mur (1984), and also because this pattern reflects diurnal patterns characteristic in the field during the cyanobacterial season in Germany. Irradiance is adjustable by the number of lamps and by a venetian blind, and is controlled both at the wall of the culture vessel and within the vessel by measurements with a quantum scalar irradiance meter. At the turbidity threshold chosen, no reduction or maximally 3 µmol $m^{-2}s^{-1}$ (with wall growth) of photon flux density in relation to readings taken directly at the wall closest to the lamp was detected throughout the vessel, thus confirming the observations of van Liere et al. (1978).

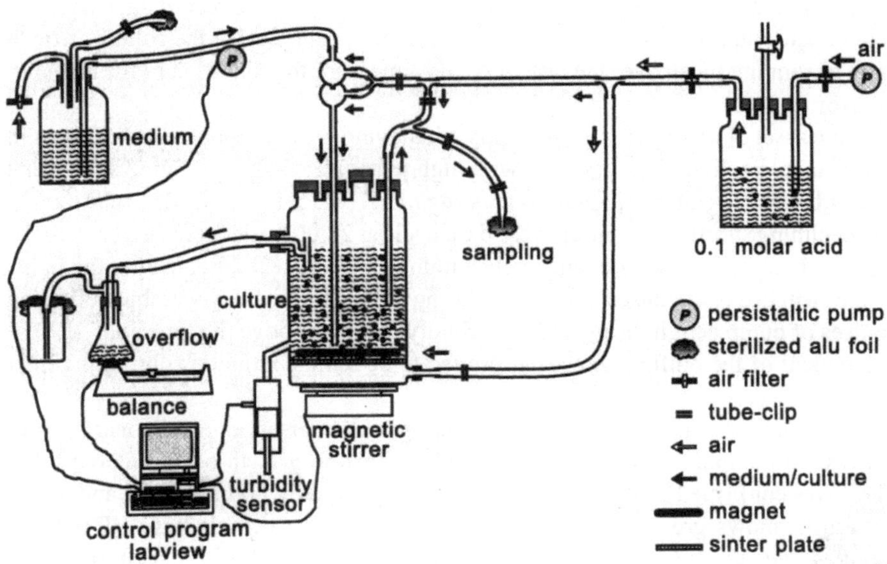

Fig. 1. Design of the turbidostat system

Culture vessels were constructed largely following the design of van Liere et al. (1975) with cylindrical vessels containing a volume of approximately 2.5 L, fitted with a sinter plate through which fine air bubbles can be evenly distributed throughout the culture at a constant flow of approximately 60 L h^{-1}. The air is sterilized through a PTFE Filter (Sartorius Midisart 2000), moistened and cleaned by passing it through a bottle filled with hydrochloric acid (0.1 mol L^{-1}).

As already observed by van Liere and Mur (1979) for *Planktothrix agardhii*, bubbling the culture through the bottom sinter plate alone proved insufficient to prevent sedimentation of *Microcystis* and *Planktothrix*, which then adhere to the sinter plate and grow into the pores. Daily cleaning with a stirrer visibly resuspended cells, elevated turbidity and thus triggered growth, so that growth rates proved dependent upon such regular cleaning actions. Therefore, a magnetic stirrer was introduced which gently cleans the sinter plate for 10 min once per hour at a slow rotation rate of 10-30 min^{-1}. This ensured suspension and stable growth rates, while no evidence of cell rupture through stirring was detected by checking for increased turbidity in the filtrate immediately after 10 min of stirring. All the results presented for *Microcystis aeruginosa* HUB 5-2-4 and HUB 5-3, as well as for *Planktothrix agardhii* NIVA-CYA 12 and HUB 076 were obtained after introduction of the stirrer; the results for *P. agardhii* NIVA-CYA 34 were obtained without stirring. Formation of flocs and wall growth (growth of filaments at the inner wall of the culture vessel), a further potential problem in *Planktothrix* cultures, scarcely occurred after regular automatically timed stirring had been introduced.

The control program was designed with the graphic Labview program (National Instruments Inc.). On the monitor it shows the active controlling instruments of the turbidostat system (turbidity sensor, medium pump, balance and magnetic stirrer), the data most recently recorded and a situation diagram of turbidity with the running mean of dilution rates over a time span of the last 48 h. For each recording, data saved cover date and time, current turbidity, the threshold value set for the culture, growth rate as running mean over the last 24 h, and a measure for condition of the turbidity sensor.

One asset of this system is the option of studying growth responses to changes in irradiance, i.e. light:dark intervals with high temporal resolution. Fig. 2 shows that the turbidity of a culture of *Microcystis aeruginosa* HUB 5-2-4 responded quite rapidly by declining shortly after onset of the dark cycle at 22:00 h and increased back to the levels characteristic for the light cycle within approximately 2 hours after the lights were switched on (06:00 h). Flow rates during light hours are very variable, reflecting pulses of pump activity triggered by turbidity readings above the threshold. Weight increment of the outflow, however, proved to be quite continuous during the light hours.

The medium pump was switched off during dark hours as a precautionary matter in case of computer failure and in assumption that no growth occur without light. This was confirmed by declining turbidity during the dark hours (Figs. 2 and 3).

Fig. 3 shows growth rate and turbidity of *Microcystis aeruginosa* HUB 5-2-4 studied over about 3 weeks, beginning with steady-state conditions at an irradiance of 19 µmol $m^{-2}s^{-1}$. Irregular growth rate peaks reflect the effect of harvesting the

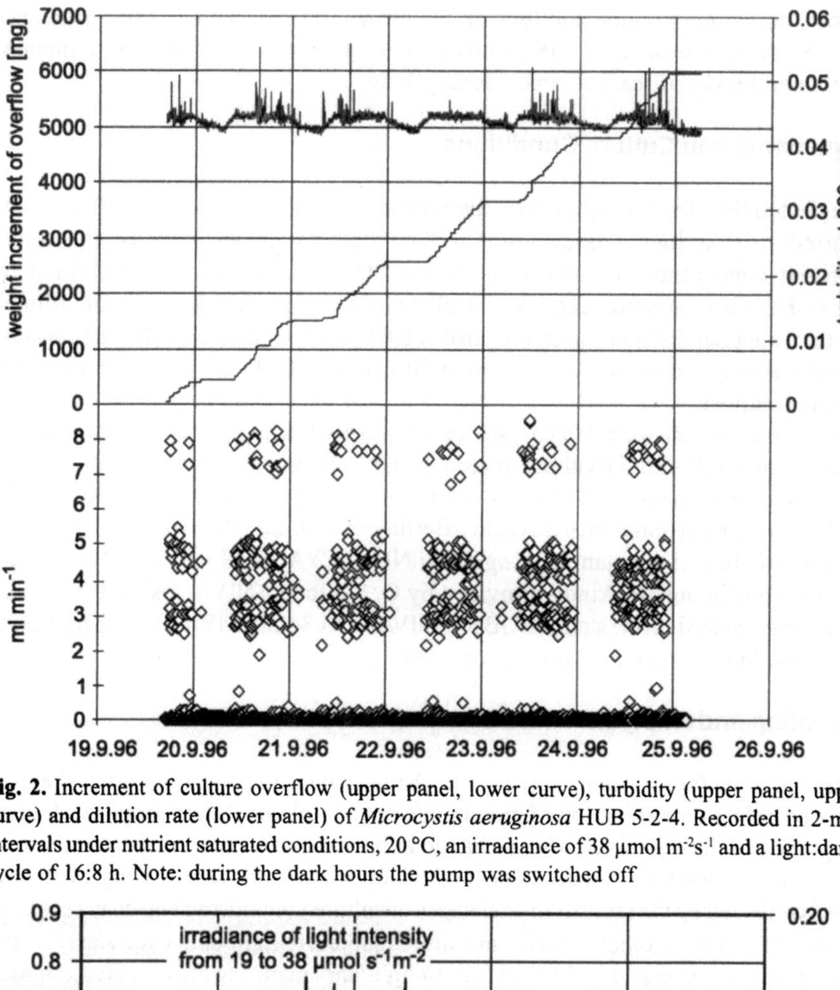

Fig. 2. Increment of culture overflow (upper panel, lower curve), turbidity (upper panel, upper curve) and dilution rate (lower panel) of *Microcystis aeruginosa* HUB 5-2-4. Recorded in 2-min intervals under nutrient saturated conditions, 20 °C, an irradiance of 38 µmol $m^{-2}s^{-1}$ and a light:dark-cycle of 16:8 h. Note: during the dark hours the pump was switched off

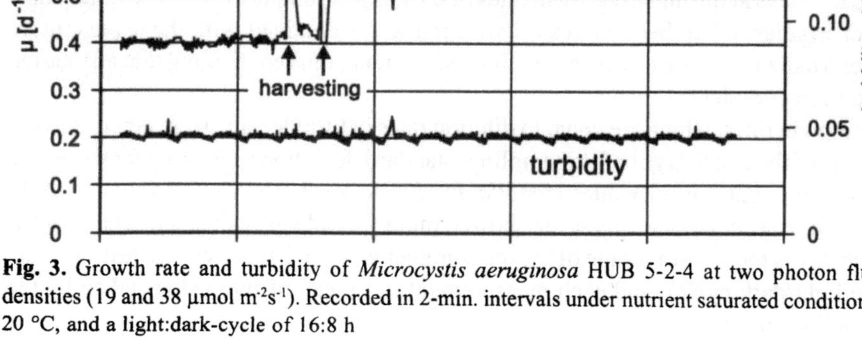

Fig. 3. Growth rate and turbidity of *Microcystis aeruginosa* HUB 5-2-4 at two photon flux densities (19 and 38 µmol $m^{-2}s^{-1}$). Recorded in 2-min. intervals under nutrient saturated conditions, 20 °C, and a light:dark-cycle of 16:8 h

culture on culture volume and thus on the growth rate registered. After irradiance was set up to 38 μmol $m^{-2}s^{-1}$, the growth rate of *M. aeruginosa* adapted to the new light conditions and reached a new steady state.

Organisms and Culture Conditions

The photon flux densities applied for the experiments were 2.4, 4.8, 9.5, 19.1, 38.2 and 73 μmol $m^{-2}s^{-1}$ at light:dark cycles of 16:8 h. Other parameters were kept constant. Nutrient concentrations were in excess of demand, which was confirmed by measurements of soluble reactive phosphate, nitrate and ammonium in the filtered medium after harvest. pH was maintained at 8.0 (± 0.2) through the buffering capacity of the medium. Temperature was kept at 20°C (± 0.5°C) by air conditioning of the culture chamber.

The strains used were *Microcystis aeruginosa* HUB 5-2-4 (with microcystins), *M. aeruginosa* HUB 5-3 (without microcystins), both growing as single cells, as well as *Planktothrix agardhii* HUB 076 (with microcystins), all three kindly provided by Prof. J.-G. Kohl (Humboldt-University Berlin), *P. agardhii* NIVA-CYA 12 (without microcystins), and a mutant of *P. agardhii* NIVA-CYA 34 which ceased to contain microcystins in our lab, kindly provided by O. Skulberg (NIVA, Oslo). Occasional microcystin analysis in the strains HUB 5-3, NIVA-CYA 34 and NIVA-CYA 12 confirmed that these indeed contained no microcystins.

Sampling and Analysis

A crucial issue for assessing responses to changes in resource availability is the time allowed before determination of growth rate or physiological state for physiological adaptation of the organisms after a change in limitation. A criterion frequently used is the time needed for a certain number of cell divisions at the respective growth rate. This is reflected by the number of exchanges of culture volume, each taken to represent one cell division. Another criterion in cultures subject to light/dark cycles may be the number of these cycles, e.g. 14 days (i.e. 14 light/dark cycles) were used by Loogman (1982) for various light/dark cycles. Our cultures were sampled after 3 total exchanges of the culture volume (including only those after growth rates had attained a new steady state), which amounted to 100 days at the lowest growth rates of 0.01 day^{-1} and to 5 days at the highest growth rates of 0.67 day^{-1}. Comparison of results obtained after irradiance had been increased with replicates obtained with the same culture after irradiance had been decreased showed no trend, thus confirming that adaptation had been complete.

Growth rates, which are equal to dilution rates in turbidostats, were determined as means of the last 5 days before sampling; standard deviations are given for these five successive daily mean values (for *Planktothrix agardhii* NIVA-CYA 12, up to 40 days were used because this strain showed more variable growth rates). The value of each day is the mean average of the measurements every 2 min over the last 24 h (i.e. one light/dark cycle), and each measurement is a mean of five values taken by the turbidity sensor.

Sampling of the cultures was performed in the mornings, 3-4 h after lights were switched on. Sampling at a fixed time of day is necessary when using light/dark cycles, because most cellular components show diurnal cycles even if cell density and cell size does not, as demonstrated for *Planktothrix agardhii* by van Liere et al. (1979) for carbohydrate content, dry weight, chlorophyll-a and C-phycocyanin at a light/dark cycle of 16:8 h.

Parameters determined in the culture samples were dry weight, cell numbers or total filament length ml^{-1}, biovolume, chlorophyll-a concentration and intracellular as well as extracellular microcystin concentrations. Results are given as means of two to six samples; the error bars show minimum and maximum values of all samples measured.

Dry weight was determined on preweighed membrane filters (pore diameter: 0.45 µm) after a fixed time of drying in a chamber held at a constant humidity of 50%.

Cell numbers of *Microcystis aeruginosa* were determined by counting using an inverted microscope at 400-fold magnification and differentiating single cells and cells in division (see Chapter 9.2). For *Planktothrix agardii*, the length of filaments within the counting grid was cumulatively assessed rather than counting filaments, as filament length may be extremely variable. Cell and filament diameters were measured at 1000-fold magnification concentration. For *Microcystis aeruginosa*, mean cell diameters were 3-6 µm and mean cell volume amounted to 27 μm^3 ($\pm$ 10 μm^3) for single cells and 32 μm^3 ($\pm$ 8 μm^3) for cells undergoing division. Biovolume was calculated by multiplication of these mean cell volumes with the cell numbers ml^{-1}. For *Planktothrix agardhii*, mean filament diameter was 3 µm, and biovolumes were calculated from mean filament cross-section areas multiplied by total filament length ml^{-1}. Biovolumes allow comparison between these two morphologically very different taxa, whereas cells cannot be readily distinguished for *Planktothrix agardhii.*

Chlorophyll-a was determined according to ISO 10260 (1992). Microcystin concentrations in the cells were analyzed by HPLC, as described in Chapter 9.1. Extracellular microcystin concentrations in the filtrate of the culture were analyzed by the immunoassay as described by Flury et al. in Chapter 8.3.

For comparison with results of other authors, absorption was also determined photometrically at 436, 740, 860 and 880 nm immediately after harvesting. Data are given as mean of three values. Absorption at 436 nm reflects the chlorophyll-a content. 740 nm is another wavelength frequently used to determine optical density. Readings at 860 and 880 nm served to check the functioning of the turbidity sensor.

Results and Discussion

Growth Rates

Growth rates relate to irradiance as described by the saturation model as used by Hesse and Kohl in Chapter 3.1 and derived from the original Mitscherlich model by Kohl and Nicklisch (1988) (Fig. 4). Maintenance energy requirements obtained by extrapolating the curves for growth over light to their intersection with the x-axis

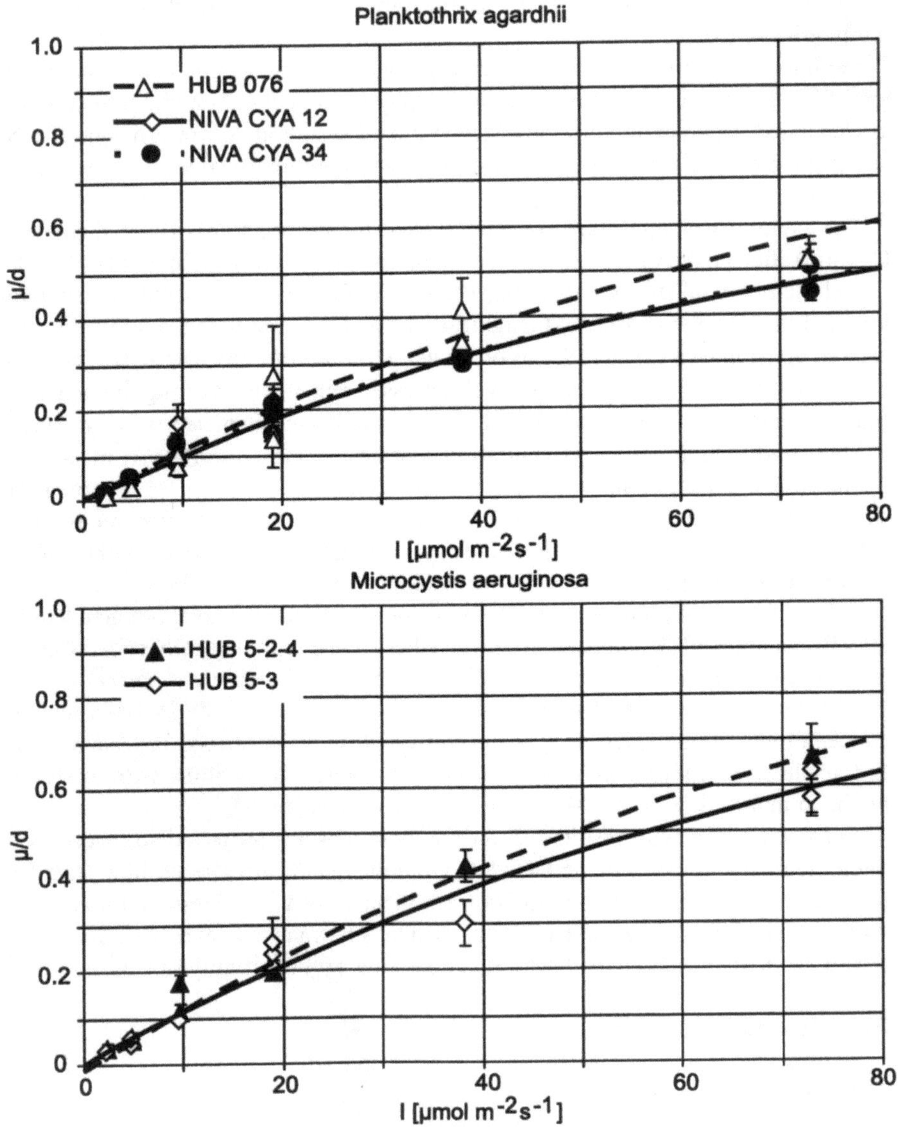

Fig. 4. Growth rates in relation to photon flux density for 3 strains of *Planktothrix agardhii* and 2 strains of *Microcystis aeruginosa*. *Solid curves* Strains with microcystins; *dotted or dashed curves* strains without microcystins; *bars* show standard deviation of 5 successive days before harvesting; *symbols* show means. Note: growth curves for *Planktothrix agardhii* strains NIVA-CYA 34 and NIVA-CYA 12 are identical

Strain	Culture system	µm and L:D	Irradiance for µm	Source
Planktothrix agardhii Gomont HUB 076 Isolated from Lake Müggelsee in Germany (1986)	Semicontinuous. Turbidity readjusted to a constant value daily or every 2nd day; irradiance was chosen to be saturating or near-saturating	0.62 day^{-1} At L:D of 12:12 0.66 day^{-1} Under continuous light	80 µmol $m^{-2}s^{-1}$ 64 µmol $m^{-2}s^{-1}$	Nicklisch (1998)
Planktothrix agardhii Gomont Isolated from Veluwemeer in The Netherlands	Continuous. At constant turbidity, varying irradiance up to 3300 µmol $m^{-2}s^{-1}$	0.86 day^{-1} Under continuous light	≅ 47-94 µmol $m^{-2}s^{-1}$ (converted from 10 - 20 W m^{-2}, applying the factor of 4.7 for warm white fluorescent lamps)	van Liere and Mur (1979)
Planktothrix agardhii Gomont Mixture of strains from Lake Norrviken in Sweden	Continuous. Under constant irradiance, varied dilution rates and concentrations of N and P	0.46 day^{-1} Under continuous light	37 µmol $m^{-2}s^{-1}$; Higher irradiance not measured	Ahlgren (1977)
Planktothrix agardhii Gomont Isolated from Veluwemeer in the Netherlands	Continuous. At constant turbidity, varying photoperiod as well as irradiance	0.74 day^{-1} At L:D of 16:8	≅ 140 µmol $m^{-2}s^{-1}$ (Converted from 10-20 W m^{-2}, applying the factor of 4.7 for warm white fluorescent lamps)	Loogman (1982)
Microcystis aeruginosa HUB 5-2-4	Semicontinuous. At constant turbidity, varying irradiance	0.27 day^{-1} At L:D of 12:12	≅ 70 µmol $m^{-2}s^{-1}$ and higher	Hesse and Kohl (Chapter 3.1)
Microcystis aeruginosa HUB 018	Semicontinuous. At constant turbidity, varying irradiance	0.33 day^{-1} Under continuous light	≅ 100 µmol $m^{-2}s^{-1}$ and almost as high at ≅ 70 µmol $m^{-2}s^{-1}$	Nicklisch and Kohl (1983)
Microcystis aeruginosa	Not given	1.11 day^{-1} At 23°C Under continuous light	"Saturating"	Reynolds (1984)

Table 1. Maximal growth rates reported for different strains of *Planktothrix agardhii* and *Microcystis aeruginosa*. At 20°C (unless stated otherwise); *L:D* h light to h dark

(Loogman et al. 1980) result in extremely low photon flux densities of 0 to 0.36 µmol $m^{-2}s^{-1}$ for *Microystis aeruginosa* and of 0.2 to 0.77 µmol $m^{-2}s^{-1}$ for *Planktothrix agardhii* (Fig. 4).

Differences in growth rates between the two strains of *Microcystis aeruginosa* and between the three strains of *Planktothrix agardhii* are not pronounced. For both species slightly higher growth rates were observed for the two microcystin-producing strains, but at the low number of strains tested, coincidence cannot be excluded. This result is coherent with the observations of Hesse and Kohl for six strains of *M. aeruginosa* reported in Chapter 3.1: their semi-continuous cultures showed some variation in growth rates between species, independently of whether they contained microcystins. Thus, it may be concluded that microcystin production does not markedly influence growth rates, either for *Microcystis aeruginosa* or for *Planktothrix agardhii.*

Standard deviations in growth rates averaged from 5 successive days before harvesting were low for *Microcystis aeruginosa,* usually substantially less than 20% of the mean. For the *Planktothrix agardhii* strains they were sometimes higher, occasionally ranging up to 30% of the mean. This may be due to different light-scattering properties of filaments as compared to spherical cells, but also to a stronger tendency of *Planktothrix agardhii* to form flocs, particularly strains NIVA-CYA 12 and HUB 076.

Growth rates were similar for all five strains investigated at photon flux densities up to 38 µmol $m^{-2}s^{-1}$, but at higher irradiance they were slightly lower for all three *Planktothrix* strains as compared to the two strains of *Microcystis* (Fig. 4). Generally, they correspond well to results previously reported (Table 1), if differences in light/dark cycles or temperature are taken into account. For both species, irradiance approaches light saturation in the range of 70 µmol $m^{-2}s^{-1}$, as reported by other authors: for *Planktothrix agardhii* van Liere and Mur (1979) show a maximal growth rate of 0.86 day^{-1} observed in continuous cultures at an irradiance between 10 and 20 W m^{-2}, which corresponds to 47-94 µmol $m^{-2}s^{-1}$ (applying the factor of 4.7 for conversion of warm white fluorescent lamps); growth rate rapidly declined in their experiments as irradiance exceeded 40 W m^{-2} (i.e. 190 µmol $m^{-2}s^{-1}$). For *Microcystis aeruginosa,* Nicklisch and Kohl (1983) report only a slight increase in growth rate between 15 and 20 W m^{-2} (70-94 µmol $m^{-2}s^{-1}$), but a temperature increase from 20 to 25°C almost doubled growth rates.

Microcystin Contents

The amount of microcystins (sum of all variants) per cell is given as cell quotas in Fig. 5A for *Microcystis aeruginosa.* As individual cells cannot readily be distinguished for the filamentous *Planktothrix agardhii,* trichome quotas are suggested as an adequate analogy. Further, concentrations of microcystins within the cells are given in terms µg of microcystins mm^{-3} of biovolume. As no changes were detected in mean filament diameters of *P. agardhii* and changes in mean cell volume of *M. aeruginosa* were only minor, trends for microcystin per biovolume are very similar to cell quotas and identical to the trichome quotas.

Microcystis aeruginosa HUB 5-2-4 contained between 20 and 45 femtograms of microcystins (fg) cell^{-1}, and *Planktothrix agardhii* HUB 076 contained between 7 and 10 fg µm^{-1} trichome length. In relation to biovolume, microcystin concentrations ranged from 0.8 to 2.2 µg mm^{-3} for *M. aeruginosa* and for *P. agardhii* from 0.9 to 1.4 µg mm^{-3} (Fig. 5A). The most important result is that for both species, variations of cell quotas or trichome quotas in relation to photon flux density were only in the range of a factor of 2. Even photon flux densities in the range of maintenance energy requirements did not induce lower microcystin quotas. A slight decrease can be detected at 9.5 µmol m^{-2}s^{-1}, particularly for *Microcystis aeruginosa* HUB 5-2-4. However, this is still within the margin of error indicated by variation between replicates, and a larger database would be necessary to ascertain a statistical significance of these differences.

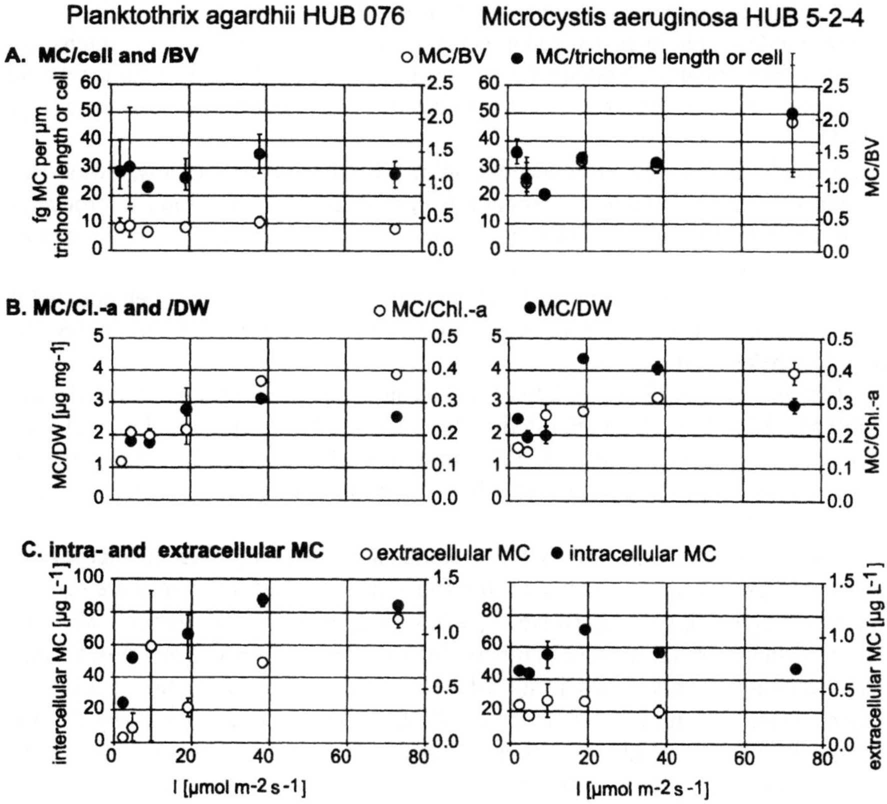

Fig. 5. Microcystin content of *Planktothrix agardhii* HUB 076 and *Microcystis aeruginosa* HUB 5-2-4. **A** Amounts of microcystin cell^{-1} for *M. aeruginosa* and mm^{-1} trichome for *P. agardhii* – i.e. cell quotas and trichome quotas – and concentrations per biovolume. **B** Ratios microcystins/dry weight and microcystins/chlorophyll-a. **C** Concentrations of intra- and extracellular microcystins L^{-1} culture volume. All values are means of two to six samples; vertical bars show the corresponding minimum and maximum values

Ratios of microcystins to other biomass indicators and to culture volume do show trends:

- Microcystin per DW increased for both species up to irradiances of 38 µmol $m^{-2}s^{-1}$ by more than two-fold: from 1.6 to 4.0 µg mg^{-1} for *M. aeruginosa*, and from 1.5 to 4.9 µg mg^{-1} for *P. agardhii* (Fig. 5B). The values as such are in a range similar to those found for N-limited growth of *Microcystis aeruginosa* (MASH-01A19) by Long et al. (2000), who report concentrations increasing with growth rate from values around 1 µg mg^{-1} at 0.1 day^{-1} up to more than 6 µg mg^{-1} at a growth rate of 1.1 day^{-1}.
- For both species, values of microcystins per chlorophyll-a amounted to 0.1 - 0.4 (Fig. 5B) and tended to increase with irradiance (Fig. 5C). Long et al. (2000) report somewhat higher mean values of microcystins per chlorophyll-a values of 0.5-0.7 for N-limited growth of *Microcystis aeruginosa* (MASH-01A19).
- Trends are also evident for of microcystin concentrations L^{-1} culture volume (Fig. 5C), both *Microcystis aeruginosa* and *Planktothrix agardhii* show the highest intracellular microcystin concentrations at 10-20 µmol $m^{-2}s^{-1}$. Concentrations were lower at low light intensity, and they tended to decline slightly at light intensities above 20 µmol $m^{-2}s^{-1}$. Notably, for *M. aeruginosa*, trends of microcystins in relation to irradiance (Fig. 5C) were inverse to trends of cell quotas over irradiance (Fig. 5A).
- Extracellular microcystin concentrations in the culture were very low, always less than 0.5 µg L^{-1}, which corresponds to less than 1% of the total microcystin concentration (Fig. 5C). For both species, they followed the concentrations of intracellular microcystins L^{-1} culture volume.

For both species, all these trends reflect trends for biovolume and cell counts L^{-1} culture volume, rather than reflecting a true increase in microcystin content of the cells (see below and Fig. 7).

Planktothrix agardhii HUB 076 contains chiefly one microcystin variant ([Asp^3, Dhb^7] MC-RR) (Fig. 6). At photon flux densities of 19 µmol $m^{-2}s^{-1}$ and more, traces of microcystin-LR and [Asp^3] MC-RR were found. These were not present at photon flux densities below 19 µmol $m^{-2}s^{-1}$, which might be an effect of the detection limit (see Chapter 9.1).

Microcystis aeruginosa HUB 5-2-4 contains up to nine different microcystin variants (Figs. 6A, C). At higher photon flux densities, their relative share of total microcystins scarcely varies. However, below 19 µmol $m^{-2}s^{-1}$ three variants were not detected, and the relative share of others shifted in favour of MC-RR. Thus, though very low photon flux densities do not reduce total microcystin production in the cells, they may affect microcystin composition. Hesse and Kohl (Chapter 3.1) also show a stronger effect of low light intensities on some microcystin variants as compared to others; however, in their experiments the concentration of MC-RR declined at irradiances of 5 µmol $m^{-2}s^{-1}$. As for total microcytins (Figs. 5A, C), in relation to culture volume, the individual microcystins show a trend over irradiance which is different from and for some even inverse to that of cell quotas (Fig. 6B).

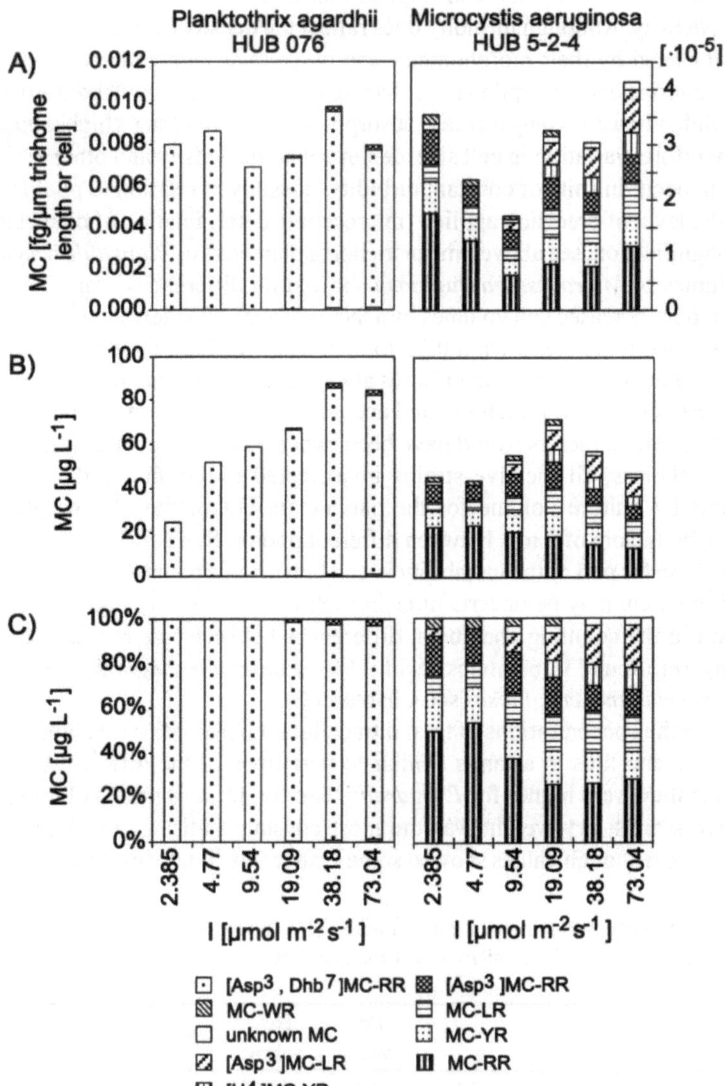

Fig. 6. Microcystin variants in the culture of *Planktothrix agardhii* HUB 076 and *Microcystis aeruginosa* HUB 5-2-4. **A** fg microcystin variants μm^{-1} trichome length for *P. agardhii* and $cell^{-1}$ for *M. aeruginosa*. **B** µg microcystin variants L^{-1} culture volume. **C** Relative share of each variant. All values are means of two to six samples. *MC* microcystins

Other Biomass Indicators in Relation to Light-Dependent Growth

The experimental design of turbidostats not only intends to ensure a defined photon flux density in the cultures, but also implies that this is achieved by maintaining constant cell density. Although turbidity does reflect the density of cells or trichomes, it is also influenced by their morphometry and by cellular components which may affect light scattering and absorption (e.g. changes in gas vesicle or thylakoid structure at the lower light intensities may increase absorption, thus mimicking a higher number of cells). Therefore, variation in cell size, cell or trichome density and other biomass indicators can occur in spite of constant turbidity, possibly in relation to photon flux density. At the level of precision applied (microscopic determination of diameters at 1000-fold magnification; see above), mean trichome diameters of *Planktothrix agardii* and cell volumes of *Microcystis aeruginosa* showed no differences. Van Liere and Mur (1979) report increased cell volumes by a factor of approximately 2 for *P. agardhii* as growth rates were increased from 0.25 to 0.60 day^{-1} for both light-saturated and light-limited cultures. However, our cultures always grew at the maximal growth rate (μ_m) for the respective light limitation, and according to van Liere and Mur's results (1979), smaller cell diameters would have been expected only at growth rates lower than μ_m. Nonetheless, all the five strains investigated did exhibit differences in concentrations L^{-1} culture volume for the parameters biovolume, dry weight and chlorophyll-a by factors of 2 to 3 between different photon flux densities (Table 2).

A further cause for part of the variability in cell or trichome counts and biovolumes calculated from them may be uncertainties in trichome counts, e.g. due to irregular sedimentation in the counting chambers: differences between five replicate counts (max to min) were found to be almost twofold for *Planktothrix agardhii* and up to 1.5-fold for *Microcysis aeruginosa* (see Chapter 9.2).

Variations in the concentrations of total intracellular microcystins L^{-1} culture were two- to fourfold and thus in a range similar to variations of the different biomass indicators, and they were higher for *P. agardhii* than for *M. aeruginosa* (Table 2).

For all five strains, dry weight was the most constant indicator of biomass L^{-1} culture (Fig. 7C). Although values showed some scatter (maximally by a factor of 2.7,

Table 2. Ratio of maximum to minimum concentrations (L^{-1} culture) of biovolume, dry weight, chlorophyll-a and total intracellular microcystins for each strain

Strain	Dry-weight	Bio-volume	Chlorophyll-a	Total MC (intracellular)
Planktothrix agardhii NIVA-CYA 34 (MC -)	2.74	2.01	2.03	n.d.
Planktothrix agardhii NIVA-CYA 12 (MC -)	1.69	1.50	1.92	n.d.
Panktothrix agardhii HUB 076 (MC +)	1.62	3.45	1.58	3.88
Microcystis aeruginosa HUB 5-3 (MC -)	2.21	3.14	2.34	n.d.
Microcystis aeruginosa HUB 5-2-4 (MC +)	1.95	3.31	2.46	1.65

n.d. not detected; *MC -* strains without microcystins; *MC +* strains with microcystins

for *Planktothrix agardhii* NIVA-CYA 34; Table 2), variability was not related to photon flux density. This corresponds to van Liere and Mur's (1979) observation that "dry weight is a linear function of optical density over a wide range" (p. 31). However, Loogman (1982) points out that growth rates obtained using optical density showed less fluctuation than those based on dry weight. Consequently, as turbidity is held constant in our culture system, some fluctuation in dry weight is to be expected.

The relationship of cell numbers L^{-1} culture volume and mm trichome length L^{-1} culture volume to photon flux density is almost identical to that of the biovolumes (Figs. 7A, B), because the mean cell or trichome diameters from which biovolumes were calculated showed little or no differences in relation to irradiance. Generally,

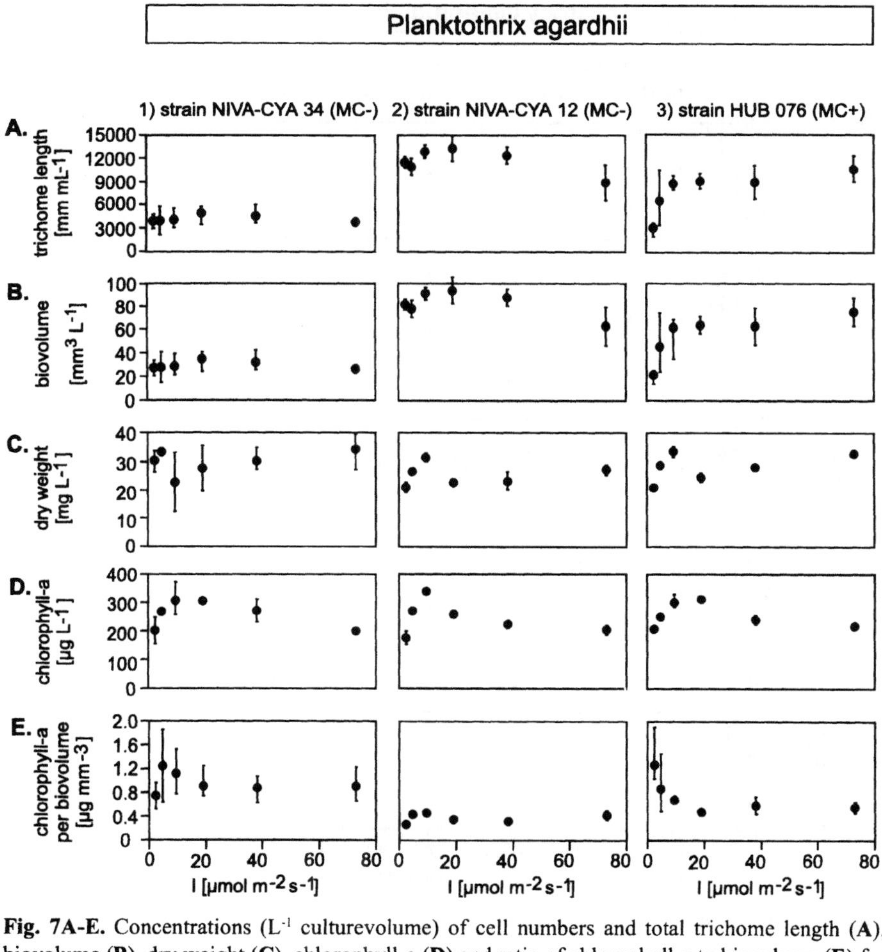

Fig. 7A-E. Concentrations (L^{-1} culturevolume) of cell numbers and total trichome length (**A**), biovolume (**B**), dry weight (**C**), chlorophyll-a (**D**) and ratio of chlorophyll-a to biovolume (**E**) for each strain of *Planktothrix agardhii* and *Microcystis aeruginosa*. All values are means of two to six samples; *vertical bars* show the corresponding minimum and maximum values. *MC-* strains without microcystins; *MC+* strains with microcystin

values were lower at the lowest irradiance of 2.4 µmol $m^{-2}s^{-1}$, which was close to the maintenance energy requirement. Both strains of *Microcystis aeruginosa* attained highest cell densities and biovolumes at 10 µmol $m^{-2}s^{-1}$ and showed a substantial decline at irradiances above 20 µmol $m^{-2}s^{-1}$ (Figs. 7A, B). The three strains of *Planktothrix agardhii* exhibited a similar trend, however, it was less pronounced for the two strains without microcystins (NIVA-CYA 34 and NIVA-CYA 12), and the strain containing microcystin (HUB 076) showed more scatter and no decline at higher irradiance.

Concentrations of chlorophyll-a L^{-1} culture volume decreased with increasing light intensity for all five strains (Fig. 7D). This may be expected, as less pigment is needed for light harvesting at higher irradiances (see Post et al. 1985 for adaptation of

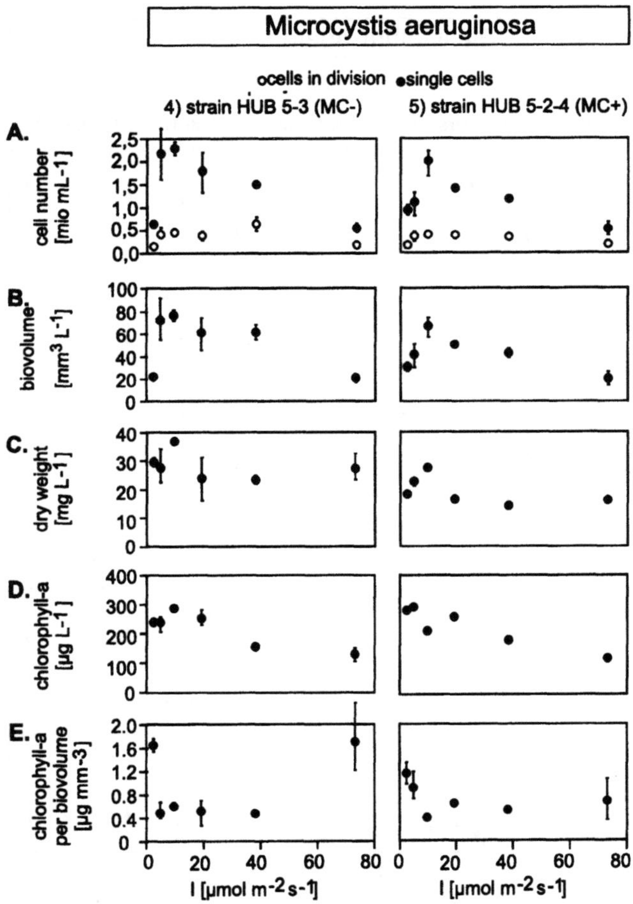

Fig. 7A-E. (continued)

Planktothrix agardhii to low irradiance). However, in relation to biovolume, this trend is confirmed for only two strains – *Microcystis aeruginosa* HUB 5-2-4 and *Planktothrix agardhii* HUB 076. For the others, lower concentrations of chlorophyll-a paralleled lower biovolume concentrations, and the ratio between both parameters did not decline (Fig. 7E), thus implying that cellular concentrations of chlorophyll-a remained unchanged. At the lowest photon flux density of 2.4 µmol $m^{-2}s^{-1}$ most strains also showed reduced chlorophyll-a concentrations L^{-1} culture volume. For two strains (*P. agardhii* NIVA-CYA 34 and *P. agardhii* NIVA-CYA 12) similar reductions of biovolume concentrations caused chlorophyll-a to biovolume ratios to remain unchanged, but *P. agardhii* HUB 076 and *M. aeruginosa* HUB 5-2-4 showed higher ratios of chlorophyll-a to biovolume at the lowest irradiances. Interestingly, both *Microcystis* strains show a slight increase of the ratios in the highest light intensity.

Optical density is frequently measured at 740 nm rather than at 880 nm. An influence of pigments is possible at 740 nm, and assessment of comparability between both measures is relevant for comparison between projects. Fig. 8 shows that optical density at 740 nm is indeed more variable, and for some strains it appears to show a response to irradiance. For all strains, it declined slightly below 4.8 µmol $m^{-2}s^{-1}$. In terms of optical density at 740 nm, cell density varied by a factor of almost 2 between culture experiments with different strains, reflecting the variability of other biomass parameters shown in Fig. 7 and Table 2.

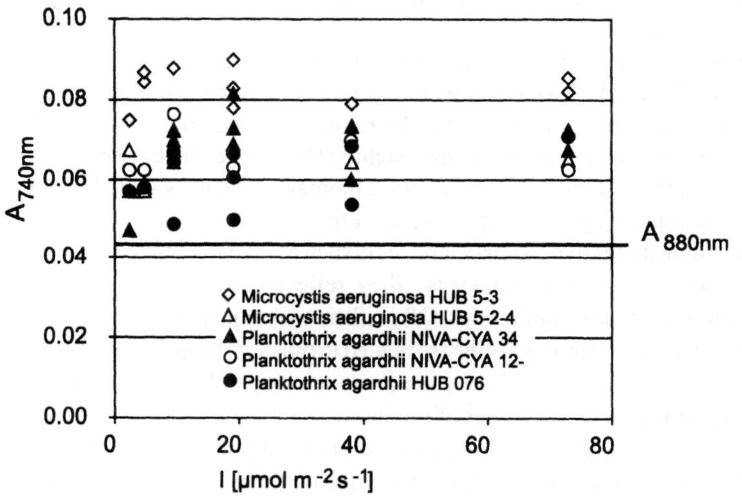

Fig. 8. Ratio of turbidity at 880 nm to optical density (adsorption) at 740 nm for the 5 strains at the photon flux densities investigated

Summary and Conclusions

The computer-controlled turbidostat proved to be an effective continuous culture design for assessing growth rates and cellular composition of these cyanobacterial taxa. Growth rates were very stable after introduction of automatic gentle cleaning of the sinter plate at regular intervals. Advantages over continuous culture systems without automatic turbidity control are the very constant growth conditions even at light/dark intervals, as well as continuous monitoring of culture growth. This tool is likely to be particularly useful for more detailed studies of various light/dark intervals because it can record the instantaneous response of turbidity to changes in conditions. Even if used as chemostat with a constant dilution rate directly controlled by the medium pump, continuous registration of turbidity and calculation of current growth rates at 2-min intervals may be of advantage as compared to manual daily sampling and turbidity determinations. The only drawback is dependence on proper functioning of the computer, but a safeguard against potential computer crashes and subsequent washout of the culture through uncontrolled continuous pumping of medium is possible by adjusting the medium pump to transport medium at a rate only somewhat above that determined by the program in response to turbidity increments.

Although Walz et al. (1997) observed a linear relationship between turbidity and biovolume, our results show that they may differ by a factor of up to 3, particularly at very low light intersities (see above and Fig. 7), probably due to changes in cell structure which affect the cell's light absorption. However, this does not impair growth rate determination, because the growth rate is calculated as relative change of weight in the output vessel.

The results obtained for the two cyanobacterial species show that growth rates differ only slightly between strains. Further, they show that total microcystin cell or trichome quotas are scarcely affected by photon flux densities over a wide range of irradiances down to levels almost approaching the maintenance energy requirements. In contrast, ratios of microcystins to other biomass indicators varied by a factor of 2 to 3, as has frequently been reported from other studies.

This demonstrates that biomass indicators such as dry weight, biovolume or pigment content may be misleading because they reflect the responses of other cellular components to resource limitations. Often, their variation is not sufficiently addressed in culture studies, and microcystins are related to only one biomass indicator. Variations in microcystin to biomass ratios by factors in the range of 2 or 3 may then reflect the variations of the cellular content of the biomass indicator chosen, rather than a true impact of resource limitation on microcystin production. Cell or trichome quotas thus prove to be a useful concept for studies of toxin production, even if the (yet unclear) function of microcystins in the cells is certainly different from that of nutrients, for which the cell quota concept was originally developed.

Nonetheless, biovolume has an important position in studies on microcystin production because it is the only reference for comparing microcystin content between species of different cell size and morphometry.

Against this background, and in conjunction with the results presented by Hesse and Kohl in Chapter 3.1, it may be concluded that light as the growth-limiting resource has little or no impact on microcystin content of the cells. Variations in microcystin content of field populations then must be due to the – probably often rather subtle – differences in growth rates of strains with different microcystin content.

Acknowledgements

Thanks are due to especially to Ingeborg Flieger for microcystin analysis, to Katrina Laskus for cell counts and to Elke Pawlitzky for analysis of chlorophyll-a. We also thank Luuc Mur and Andreas Nicklisch for valuable support in the planning phase for these experiments and discussion of the results, as well as Gary Jones for helpful discussions on interpretation of the data. The *Planktothrix* strains NIVA-CYA 12 and NIVA-CYA 34 were kindly supplied by Olav Skulberg.

3.3
Importance of Energy Charge for Microcystin Production

Heike Bickel and Susanne Lyck

Introduction

The literature discussing the effects of nitrogen, phosphorus and iron concentration on the cellular microcystin content is contradictory (Watanabe and Oishi 1985; Codd and Poon 1988; Sivonen 1990; Lukac and Aegerter 1993; Utkilen and Gjølme 1995; Lyck et al. 1996; Rapala et al. 1997). All these previous studies have focused on "direct" effects of nutrients on the production of microcystin.

Recent studies on the mechanism controlling the synthesis of microcystin have shown that microcystin is synthesized by a non-ribosomal thio-template mechanism that requires energy in the form of ATP (Arment and Carmichael 1996; Dittmann et al. 1996). The information at hand led us to suppose that the effects of nutrient limitation on microcystin production reported in some studies may have been caused by changes in the energetic state of the cyanobacteria.

The hypothesis behind our experimental work was that the energetic state of the cyanobacterial cells would change under stress conditions. Energy charge could then be the controlling factor for the production of microcystin. The purpose of our study was to examine the relationship between energy charge and the cellular content of microcystin in *Microcystis aeruginosa* NIVA-CYA 228/1 at three levels of phosphorus and iron concentrations in continuous cultures.

Experimental Design

The microcystin-producing clone *Microcystis aeruginosa* NIVA-CYA 228/1 was obtained from the culture collection of the Norwegian Institute for Water Research. The continuous cultures were grown as described in Bickel et al. (2000). The dilution rates of the experimental systems were kept at 0.01 h^{-1} (corresponding to 0.24 day^{-1}).

An incident photon flux density of 25 µmol $m^{-2}s^{-1}$ was applied because previous investigations had shown that the microcystin content in laboratory cultures of cyanobacteria increased with irradiance up to about 50 µmol $m^{-2}s^{-1}$, which is comparable with integrated light intensities in cyanobacterial blooms (Whitton 1973; van der Westhuizen and Eloff 1985; Watanabe and Oishi 1985; Sivonen 1990; Utkilen and Gjølme 1992; Annadotter et al. 1993; Wicks and Thiel 1990; Rapala et al. 1997).

In the phosphorus experiments two sets of *Microcystis aeruginosa* NIVA-CYA 228/1 cultures were initially fed with O2 medium (van Liere 1979, modified as given in Bickel et al. 2000) containing 150 µM P. The medium was subsequently shifted to O2 medium containing 15 and finally 5 µM P. In analogy, for the iron experiments the O2 medium containing initially 10 µM Fe was shifted to O2 medium containing 1.5 and finally 0.5 µM Fe, respectively.

In both the phosphorus and iron experiments the optical densities at 740 nm were followed daily in order to ascertain when the culture became stable. At stable optical density as well as during phases of shiftdown, chlorophyll-a, protein, carbohydrate, dry weight, microcystin, ATP, ADP and AMP contents were measured for 5 successive days. The 5th day of sampling at stable optical density with complete O2 medium was defined as starting point of an experiment.

All samples for the analysis were taken as triplicates. The analytical methods were performed as described in Bickel et al. (2000).

The energy charge (EC, Chapman et al. 1971) is the fractional degree to which the AMP-ADP-ATP system is filled with high-energy phosphates, expressed as: EC = ([ATP] + ½ [ADP]) : ([ATP] + [ADP] + [AMP]). The adenylate nucleotides were assayed by means of bioluminescence applying the luciferine-luciferase system in firefly (*Photinus pyralis*) lantern extract according to a modified method of Karl and Holm-Hansen (1978).

Results and Discussion

Responses of the Continuous Cultures to Phosphorus and Iron Limitation

The density and protein contents of the continuous cultures decreased tenfold in response to decreasing phosphorus and iron concentrations from complete O2 media with 150 µM P and 10 M Fe to the lowest P concentration of 5 µM P or 0.5 µM Fe. The periods required for establishment of a new stable optical density were nearly half as long in the phosphorus- as in the iron-limited cultures: after the first step down from 150 to 15 µM P in the phosphorus experiment the cultures took an average of 12 turnovers to reach a new stable optical density (Fig. 1), and after the second step from 15 down to 5 µM P the new stable optical density was established after about 10 turnovers. In the iron experiment, stable optical densites were attained after an average of 24 turnovers after the first step from 10 to 1.5 µM and after about 12 turnovers following the second step down from 1.5 to 0.5 µM Fe.

These differences between the time needed for adaptation of the cultures could be explained by different grades of nutrient reduction: the reduction of phosphorus in the medium covered the range of a factor of 30 and thus was greater than the reduction of iron, which covered only the range of a factor of 20. Further, growth can be maintained over a wide range of nutrient concentrations as the nutrient quotas in the cells gradually decrease. The ranges of cellular iron quota are typically ten times higher than for phosphorus and therefore it will take longer to draw cellular reserves down to growth-limiting quotas (Morel 1987). Finally, the iron-limited cells might have had a more effective alternative to compensate the nutritional deficiency than the phosphorus-limited cells (reviewed by Straus 1994).

Phosphorus and iron limitation proved to modulate the cellular composition of *M. aeruginosa* NIVA-CYA 228/1: carbohydrates constituted 6-10% and protein constituted on average 65% of the dry weight when the cultures were grown in complete O2 medium. During prolonged phosphorus and iron limitation the carbohydrate contents increased as shown for two examples in Fig. 2.

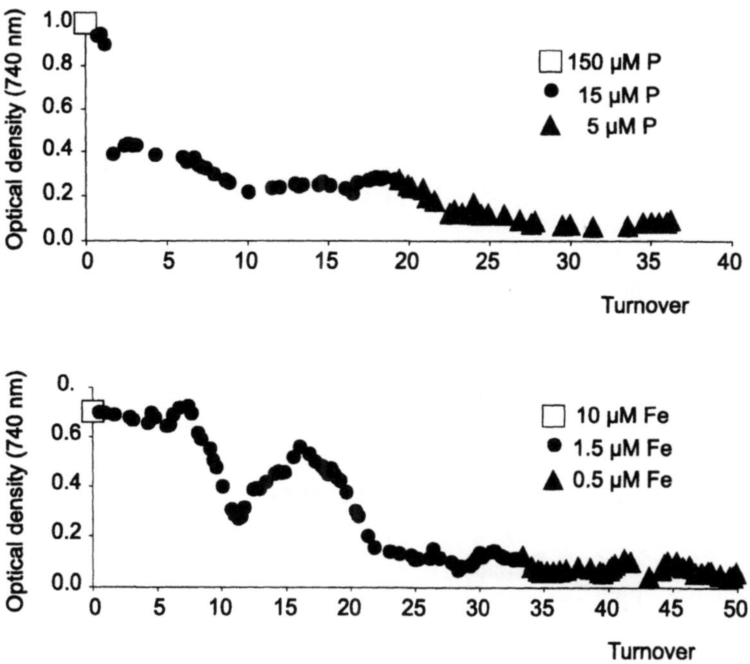

Fig. 1. Optical density at 740 nm in continuous cultures of *M. aeruginosa* NIVA-CYA 228/1. The change in phosphorus concentration in the medium from 150 to 15 μM P is shown as starting point of the experiment. After 19 turnovers the phosphorus concentration was changed from 15 to 5 μM P. The iron concentration was changed from 10 μM to 1.5 μM Fe at the starting point and from 1.5 to 0.5 μM Fe after 34 turnovers. Dilution rate: 0.01 h^{-1}. Incident continuous photon flux density: 25 μmol $m^{-2}s^{-1}$

The iron-limited cultures accumulated much more carbohydrate (on average up to 70% DW) than the phosphorus-limited cultures (up to 40% DW). This is in accordance with investigations reported in the literature: Hardie et al. (1983a) obtained a ninefold increase in carbohydrate content in iron-limited cultures of *Agmenellum quadruplicatum*. In comparison, as a response of cyanobacteria to phosphorus deficiency, carbohydrate content increased only 2.5-fold (reviewed by Grossman et al. 1994). Iron as a part of ferredoxin is important for the activities of nitrate and nitrite reductase. Therefore, iron limitation is likely to lead to cellular nitrogen limitation and then to an accumulation of glucose (Stevens et al. 1981; Hardie et al. 1983a, b; Sherman and Sherman 1983; Stanier et al. 1987; Grossman et al. 1994). Our results show more marked changes in carbohydrate than in protein content of the cells. Furthermore, to a certain extent, the decreases in protein relative to cell weight could be due to the increases in cell weight. Since protein was a more stable factor during different growth conditions compared with cell weight, the cell toxin content was expressed as the ratio of microcystin per protein.

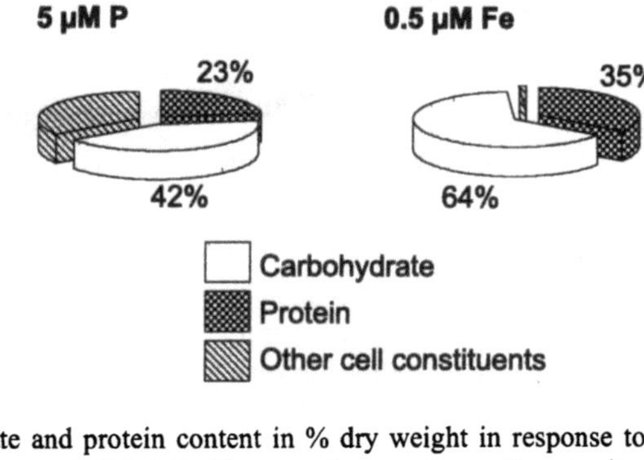

Fig. 2. Carbohydrate and protein content in % dry weight in response to phosphorus and iron limitation. Dilution rate: 0.01 h^{-1}. Incident continuous photon flux density: 25 µmol $m^{-2}s^{-1}$

Microcystin Content

Microcystis aeruginosa NIVA-CYA 228/1 chiefly contained microcystin-RR and minor concentrations of microcystin-LR. Their relative amounts showed no changes between different nutrient concentrations in the medium. Therefore, the total amount of microcystins (MC-RR plus MC-LR) was used for calculations.

The microcystin content of the organisms was examined as ratio of microcystin both to protein and to dry weight. The ratio of protein to dry weight varied by a factor of only 2, substantially less than the ratio of carbohydrates to dry weight, which varied by a factor of 6 (Table 1; Fig. 2). Variation of microcystin to protein ratios appears to be largely due to pronounced carbohydrate accumulation in the more dilute media rather than to true decrease in protein content. The apparent decreases

Table 1. Microcystin content (± SD) in relation to protein and to dry weight at three different phosphorus and iron concentrations and ratio between highest and lowest values

P in culture medium (μM)	150	15	5	max/min
ng microcystin μg^{-1} protein (± SD)	8.03 (± 2.58)	3.74 (± 0.83)	22.85 (± 4.56)	6.1
ng microcystin μg^{-1} dry weight (± SD)	3.47 (± 1.28)	1.44 (± 0.18)	3.24 (± 0.68)	2.4
Fe in culture medium (μM)	10	1.5	0.5	
ng microcystin μg^{-1} protein (± SD)	9.64 (± 0.23)	1.97 (± 0.56)	7.83 (± 2.06)	4.9
ng microcystin μg^{-1} dry weight (± SD)	7.4 (± 1.3)	1.43 (± 0.1)	1.18 (± 0.74)	6.3

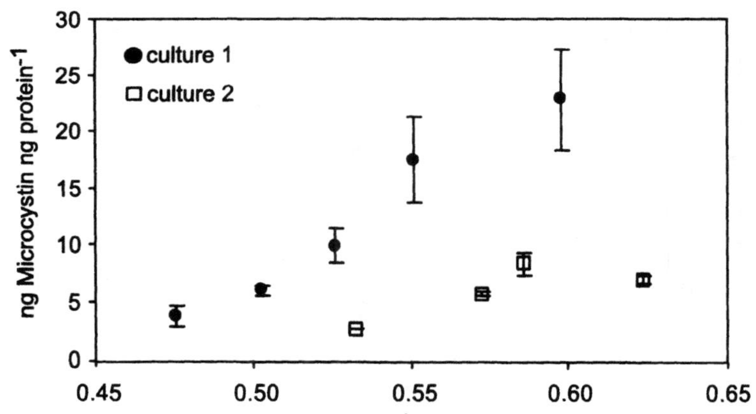

Fig. 3. Ratio of microcystin to protein (± SD) and adenylate energy charge at three different phosphorus concentrations (150, 15 and 5 μM). Dilution rate: 0.01 h^{-1}. Incident continuous photon flux density: 25μmol $m^{-2}s^{-1}$. (After Bickel et al. 2000, with permission from Phycologia)

in the ratio of microcystin to dry weight could be a result of carbohydrate accumulation and thus increased cell weight (Grossman et al. 1994; Utkilen and Gjølme 1995). Utkilen and Gjølme (1995) showed that the ratio of microcystin per protein of this strain of *Microcystis aeruginosa* remained constant (variation maximally by a factor of 1.1) when the phosphorus concentration was reduced from 150 to 15 μM, whereas the ratio of microcystin per dry weight decreased by a factor of 2.3; further, the ratio of microcystin per dry weight minus carbohydrate content remained constant. Therefore, in the following the ratio of microcystin to protein is used as indicator of microcystin content.

The changes in microcystin per protein were found to be more pronounced in the phosphorus- than in the iron-limited cultures, as shown in Figs. 3 and 4 and Table 1. They vary by a factor of 6 between different phosphate concentrations and thus are substantially higher than the variation of protein content (protein to dry weight varied by a factor of 2).

The range of microcystin content in the iron-limited cells is in accordance with results reported earlier (Utkilen and Gjølme 1995; Lyck et al. 1996). In contrast, the results of the phosphorus limitation experiments show a disagreement with the results obtained by Utkilen and Gjølme (1995), who found no changes in microcystin related to protein in phosphorus-limited cultures (Table 2). One reason for this disagreement may be the lower phosphorus concentration used in our experiments. However, our results for microcystin related to dry weight are consistent with the results obtained by Utkilen and Gjølme (1995).

Comparison of results concerning effects of iron or phosphorus on microcystin content (examples are summarized in Table 2) is hampered by several differences in

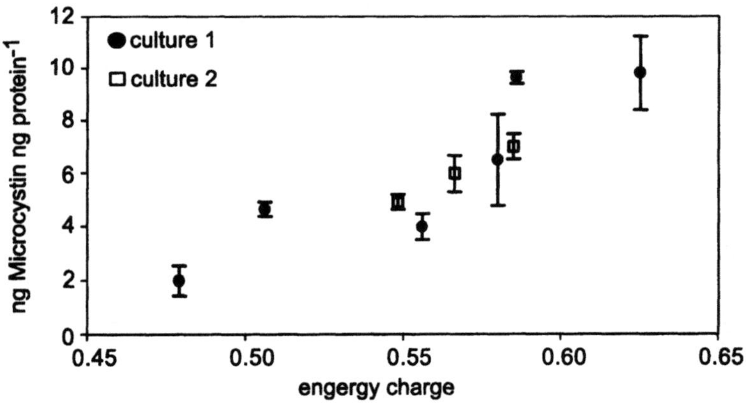

Fig. 4. Ratio of microcystin to protein (± SD) and adenylate energy charge at three different iron concentrations (10, 1.5 and 0.5 μM). Dilution rate: 0.01 h^{-1}. Incident photon flux density: 25 μmol m^{-2} s^{-1}. (After Bickel et al. 2000)

Table 2. Laboratory studies on the effect of P and Fe limitation on microcystin content

Test range (µM unless stated otherwise)	Culture conditions	Methods	Nutrient limitation effect	Range of effect (max/min)	Reference
Phosphorus					
P-biovolume ratios in the cells reduced to 1/5 of those required for maximal growth rates	Turbidostat, Semicontinuous	HPLC, Microcystin per biovolume	Strain-dependent Increase or decrease of microcystin	2-fold	Hesse and Kohl (Chapter 3.1)
452, 45, 23 µM	Batch cultures	Mouse bioassay, toxicity per dry weight	Lower toxicity	3.8-fold	Watanabe and Oishi (1985)
226 µM, deplete	Batch cultures	Mouse bioassay, toxicity per dry weight	No change of toxicity	1.7-fold	Codd and Poon (1988)
178-3.2 µM	Batch cultures	HPLC, microcystin per dry weight	Lowest microcystin content at 3.2 µM P	1.8 - 2.5-fold	Sivonen (1990)
178-1.6 µM	Batch cultures	HPLC, microcystin per dry weight	Decrease of microcystin content	5-fold	Rapala et al. (1997)
150, 15 µM	Chemostat, continuous	HPLC, microcystin per dry weight; per protein	Decreased microcystin per dry weight, no change in microcystin/protein	2.3-fold for dry weight	Utkilen and Gjølme (1995)
Iron (µM Fe)					
28 µM, deplete	Batch cultures	HPLC, microcystin per dry weight	Increased microcystin at ≤ 2.5 µM Fe	1.7-fold	Lukac and Aegerter (1993)
10-0.3 µM	Chemostat, continuous	HPLC, microcystin per dry weight; per protein	Decreased microcystin per dry weight and per protein	2-fold	Utkilen and Gjølme (1995)
10 µM, deplete	Chemostat, continuous	HPLC, microcystin per dry weight; per protein	Increased microcystin per protein, decreased microcystin per dry weight	3-fold for protein	Lyck et al. (1996)

experimental approaches: (1) Different culturing methods affect conditions of limitation. In chemostat and turbidostat cultures in steady state, nutrient uptake, cell quotas and growth rates are balanced, whereas in batch cultures light limitation increases as they become denser so that no steady state is possible and determinants of the cells' physiological status may be ambiguous. (2) Various strains have been used. Variations in biosynthetic activity might be influenced by the intrinsic biosynthetic capacity of the selected clones (e.g. differences in microcystin content or microcystin variants). (3) Nutrient limitation has been assessed at different light intensities in different studies, and changes of cell density caused by reduction of substrate nutrient concentrations may affect light availability. Thus even if the same culture systems were used (e.g. continuous cultures, Utkilen and Gjølme 1995; Lyck et al. 1996) the results could not simply be compared without regarding the changed light intensities. (4) The toxin content has been expressed as the ratio of microcystin to the dry weight or to the protein or to the cell numbers, but relative shares of these different cellular components may vary, as may cell size.

The results summarized in Table 2 suggest some effect of nutrient limitation on the ratio of microcystins to other cellular components or to biovolume, the maximal range of variation (four- and fivefold) was observed in batch cultures, whereas in continuous cultures variations were in the range of a factor of 2. The increased microcystin/protein ratios of chemostat cultures in response to iron limitation (Lyck et al. 1996) might be explained with the increased light availability caused by lower nutrient concentrations.

The most striking result is a positive correlation between the energy charge (EC) and the ratio of microcystin to protein in both the phosphorus- and iron-reduction experiments (Figs. 3 and 4). The Spearman rank correlation coefficients were 1.0 and 1.0 for both phosphorus reduction experiments and 0.943 for one iron reduction experiment. Combining the data of both iron experiments results in a correlation coefficient of 0.933 (the correlation coefficient of 1.0 for the second iron reduction experiment was not considered because of the low number of data). The significance of all of these correlations was at most $p=0.01$.

The energy charge values of the phosphorus- and iron-reduced cultures ranged from 0.47 to 0.63 and indicate that the organisms were physiologically stressed. Laboratory and field studies indicate that energy charge in exponentially growing cells may range between 0.7 and 0.9 depending on the species (Chapman et al. 1971; Falkowski 1977; Witzel 1979). Chapman et al. (1971) demonstrated energy charge values ranging between 0.5 and 0.7 for stressed cultures of *E. coli* that maintained viability but not growth.

In our cultures, there was no correlation between the optical density at 740 nm and energy charge or between the optical density and the ratio of microcystin to protein. Decreases in the energy charge and the ratio of microcystin to protein during decreasing optical density may have been a response to nutrient limitation. Increases in energy charge and the ratio of microcystin to protein, which were also observed during decreasing optical density could be explained by the increased light availability: the cell density decreased in response to reduced nutrient concentrations in the medium (Fig. 1). These decreases in cell density resulted in an increased transparency of the culture. This is also reflected by a decrease of chlorophyll-a in relation to protein content in average from 2.6 to 0.8% of protein content.

Conclusions

The results obtained suggest a direct relationship between energy charge and the ratio of microcystin to protein (Figs. 3 and 4). This is in agreement with the fact that ATP is essential for peptide synthesis by the thio-template mechanism – the mechanism that synthesizes microcystin (Arment and Carmichael 1996; Dittmann et al. 1996). The relationship between microcystin, protein and energy charge also suggests the hypothesis that at lower levels of energy charge the energy was used for protein rather than for microcystin synthesis. The observation of a direct relationship between energy charge and the ratio of microcystin to protein are supported by Kaebernick et al. (2000), who observed increased microcystin synthetase transcript levels after short-term increases in irradiance, thus also indicating a direct influence of energy supply on mircocystin production.

Factors affecting temporal variation – increase or decrease in the biosynthetic activity – are connected with stimuli from the physical environment (e.g. light) or from the components of the culture medium (e.g. phosphorus and iron concentration). Both appear to have indirect effects on yield and accumulation of microcystin in the cells by their effects on the energy state of the organisms. Phosphorus, iron and light are essential environmental factors for the promotion of metabolic energy. Phosphorus is an important component in DNA and RNA and in energy-transferring compounds such as ATP. Iron is essential for photosynthesis, nitrogen metabolism and respiration. Iron is of importance in chlorophyll-a synthesis, and as a component in iron-sulphur proteins and in heme enzymes as electron carrier (Hardie et al. 1983a, b; Straus 1994). Light is the major energy source for phytoplankton. Thus, changes in the intensity and concentration of these environmental factors may affect the adenylate energy charge, and this will give rise to variations in the microcystin content.

Acknowledgements

We gratefully acknowledge the input of H. Utkilen, K. Christoffersen an O.M. Skulberg in numerous helpful discussions.

3.4
Characterization of Microcystin Synthetase Genes in *Microcystis aeruginosa*

Elke Dittmann, Marcel Erhard, Daniel Tillett, Brett A Neilan, Hans von Döhren and Thomas Börner

Microcystins are cyclic heptapeptides and presumed to be synthesized non-ribosomally by peptide synthetases. Peptide synthetases are multifunctional proteins composed of homologous units. Each of these synthetase units contains specific functional domains for recognition, aminoacyl adenylation, thioesterification and modification of its amino acid substrate, as well as for elongation of the growing peptide product (Kleinkauf and von Döhren 1996). Adenylate-forming domains of peptide synthetases share a high degree of sequence conservation.

Using degenerated PCR techniques and screening of genomic libraries, we could identify putative peptide synthetase genes in the microcystin-producing strains *Microcystis aeruginosa* PCC 7806 and HUB 5.2.4 obviously encoding homologous enzymes. Using natural transformation and homologous recombination, we could insertionally inactivate this gene in strain PCC 7806 (2). HPLC analyses of wild type

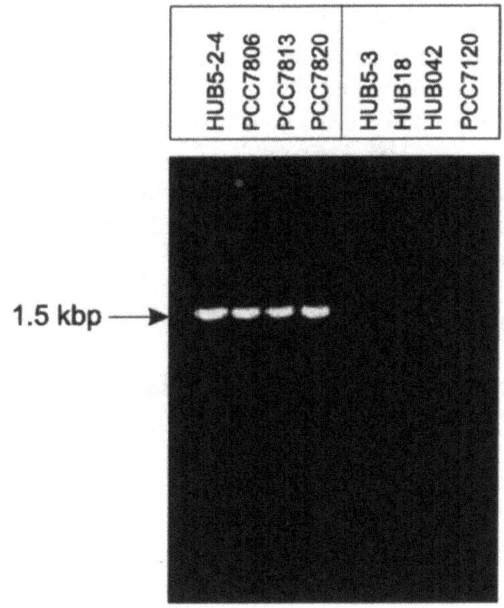

Fig. 1. Amplification of *mcy*B-specific PCR products from DNA of the microcystin-producing strains *Microcystis aeruginosa* HUB 5-2-4, PCC 7806, PCC 7813 and PCC 7820. In contrast, the non-toxic strains HUB 5-3, HUB 18, HUB 042 and PCC 7120 lack *mcy*B

and mutant extracts could demonstrate the loss of all microcystin forms in the mutant cells. Additional MALDI-TOF analyses confirmed these data and they could moreover show that the amount of the other cyclic peptide compound in strain PCC 7806, cyanopeptolin (Martin et al. 1993), was the same in both wild type and mutant extract. A protein phosphatase assay using a colorimetric substrate revealed the lack of the whole protein phosphatase inhibition activity in the microcystin-deficient mutant cells.

According to these results, the peptide synthetase gene cluster was named *mcy* gene cluster (Dittmann et al. 1997). Further sequencing of the *mcy* region of strain PCC 7806 revealed the existence of several genes involved in microcystin and ADDA biosynthesis including peptide synthetase genes and genes encoding polyketide synthases, together comprising more than 60 kb (unpubl. data). The MCY synthetases show all features known from various peptide synthetase sequences but, in addition, they contain sequences without homology to known enzymes. The mutants thus obtained may be useful as tools for elucidating the precise effects of microcystin on cyanobacterial physiology and studying the regulation of toxin biosynthesis. They clearly demonstrate that all microcystin variants produced by a single strain are produced by one and the same enzyme. Moreover we can conclude that microcystin is the sole protein phosphatase inhibitor produced by strain PCC 7806 (Dittmann et al. 1997).

These results provided the means to answer several questions. Do cyanobacteria in general possess peptide synthetase genes? Is the occurrence of the *mcy* gene cluster limited to *Microcystis aeruginosa*? Is the *mcy* gene cluster conserved enough to develop molecular techniques for the detection of microcystin-producing cyanobacteria?

In order to answer these questions, fragments of the *mcyB* gene were used as probes in Southern hybridization and dot blot experiments and primers derived from this sequence were provided for PCR experiments. Interestingly, first Southern experiments could demonstrate that the basic difference between toxic and non-toxic *Microcystis* strains is the presence and absence, respectively, of the *mcyB* gene (Meißner et al. 1996) and most probably of the other *mcy* genes. All hepatotoxic strains of *Microcystis* were found to contain the *mcyB* gene. Among many tested non-hepatotoxic *Microcystis* strains, we found only a few which seem to possess the *mcyB* gene. It remains to be shown whether the *mcyB* gene is down-regulated or mutated in these strains. Another possibility is that these non-hepatotoxic strains contain a gene very similar to *mcyB* that might be involved in the biosynthesis of another peptide. In agreement with the hybridization data, we could amplify *mcy* gene-specific fragments by PCR using DNA from toxic *Microcystis* strains, but not from non-toxic strains (Fig. 1).

This type of experiment has been extended to other cyanobacterial strains, belonging to the species *Anabaena, Nostoc, Oscillatoria, Nodularia* and *Aphanizomenon*, which are all known to include neuro- or hepatotoxin producers. Probing of dot blot filters revealed a good correlation between microcystin content of the strains and hybridization signals with the *mcyB* probe. Nevertheless, the intensity of these signals was quite different and three of the non-toxic strains showed significant cross-reactions (Neilan et al. 1999). As the *mcyB* gene is obviously not highly conserved among the microcystin-producing species, we are currently checking additional sequences of the *mcy* region of strain PCC 7806 for their utility to serve as probes or primers in DNA-based techniques for the discrimination of strains with and without microcystins.

References

Ahlgren G (1977) Growth of *Oscillatoria agardhii* in chemostat culture. 1. Nitrogen and phosphorus requirements. Oikos 29:209–224

Annadotter H, Cronberg G, Jönsson L (1993) The toxicity of blue-green algae (cyanobacteria) in Lake Finjasjön, Sweden in relation to environmental factors - a 5-year study. In: 5th International Conference on the Conservation and Management of Lakes. Strategies for Lake Ecosystems Beyond 2000, Stresa, Italy, pp. 239-244. Instituto Italiano di Idrobiologia, Verbania Pallanza

Arment AR, Carmichael WW (1996) Evidence that microcystin is a thio-template product. Journal of Phycology 32:591–597

Bennett WN, Boraas ME (1988) Isolation of a fast-growing strain of the rotifer *Brachionus calyciflorus* Pallas using turbidostat culture. Aquaculture 73 (1-4):27–36

Bickel H, Lyck S, Utkilen H (2000) Energy state and toxin content – experiments on *Microcystis aeruginosa* (Chroococcales, Cyanophyta). Phycologia 39:212–218

Boraas ME, Bennett WN (1988) Steady-state rotifer growth in a two-stage, computer-controlled turbidostat. In: IRL Press Ltd., Oxford, England, pp. 1023–1038

Carmichael WW (1994) The toxins of cyanobacteria. Scientific American 270/1:64–72

Carr NG (1988) Nitrogen reserves and dynamic reservoirs in cyanobacteria. In: Rogers LJ, Gallon JR (eds) Biochemistry of the algae and cyanobacteria. Proceedings of the Phytochemistry Society of Europe, vol. 28, Oxford University Press, Oxford, pp. 12–21

Chapman AG, Fall L, Atkinson DA (1971) Adenylate energy charge in *E. coli* during growth and starvation. Journal of Bacteriology 108:1072–1086

Codd GA, Poon GK (1988) Cyanobacterial toxins. In: Rogers LJ, Gallon JR (eds) Biochemistry of the algae and cyanobacteria. Proceedings of the Phytochemistry Society of Europe, vol. 28, Oxford University Press, Oxford, pp. 283–296

Dittmann E, Meissner K, Börner T (1996) Conserved sequences of peptide synthetase genes in the cyanobacterium *Microcystis aeruginosa*. Phycologia 35:62–67

Dittmann E, Neilan BA, Erhard M, von Döhren H, Börner T (1997) Insertional mutagenesis of a peptide synthetase gene that is responsible for hepatoxin production in the cyanobacterium *Microcystis aeruginosa* PCC 7806. Mol Microbiol 26:779–787

Droop MR (1973) Some thoughts on nutrient limitation in algae. J Phycol 9:264–272

Falkowski PG (1977) The adenylate energy charge in marine phytoplankton: the effect of temperature on the physiological state of *Skeletonema costatum* (Grev.) Cleve. Journal of Experimental Marine Biology and Ecology 27:37–45

Grossman AR, Schaefer MR, Chiang GG, Collier JL (1994) The response of cyanobacteria to environmental conditions: light and nutrients. In: DA Bryant (ed) The molecular biology of cyanobacteria. Kluwer Academic Publishers, Amsterdam, pp. 641–675

Hardie LP, Balkwill DL, Stevens JSE (1983a) Effects of starvation on the physiology of the cyanobacterium *Agmenellum quadruplicatum*. Appl Environm Microbiol 45:999–1006

Hardie LP, Balkwill DL, Stevens JSE (1983b) Effects of iron starvation on the ultrastructure of the cyanobacteria *Agmenellum quadruplicatum*. Appl Environm Microbiol 45:1007–1017

ISO 10260 (1992) Water quality – measurement of biochemical prameters – spectrometric determination of the chlorophyll a concentration

ISO 7027 (1990) Water quality – determination of turbidity

Jespersen AM, Christoffersen K (1987) Measurement of chlorophyll a from phytoplankton using ethanol as extraction solvent. Archiv für Hydrobiologie 3:445–454

Kaebernick M, Neilan B, Börner T, Dittmann E (2000) Light and the transcriptional response of the microcystin biosynthetic gene cluster. Appl Environm Microbiol 66:3387–3392

Karl DM, Holm-Hansen O (1978) Methodology and measurements of adenylate energy charge ratios in environmental samples. Marine Biology 48:185–197

Kleinkauf H, von Döhren H (1996) A nonribosomal system of peptide biosynthesis. Eur J Biochem 236:335–351

Kohl JG, Nicklisch A (1981) Chromatic adaptation of the planktonic blue-green alga *Oscillatoria redekei* van Goor and its ecological significance. Int Revue ges Hydrobiol 66 (1):83–94

Kohl JG, Nicklisch A (1988) Ökophysiologie der Algen. Akademie-Verlag, Berlin, 253 p

Lawton LA, Edwards C, Codd GA (1994) Extraction and high-performance liquid chromatographic method for the determination of microcystins in raw and treated waters. Analysts 119:1525–1530

Lehtimäki J, Sivonen K, Luukkainen R, Niemelä SI (1994) The effects of incubation time, temperature, light, salinity and phosphorus on growth and hepatotoxin production by *Nodularia* strains. Arch Hydrobiol 130/3:269–282

Long B, Jones G, Orr P (2001) Cellular microcystin content in N-limited *Microcystis aeruginosa* can be predicted from growth rate. Appl Environm Microbiol 67:278–283

Loogman JG (1982) Influence of photoperiodicity on algal growth kinetics. Academisch Proefschrift, Amsterdam, 112 p

Loogman JG, Post AF, Mur LR (1980) The influence of periodicity in light conditions, as determined by the trophic state of the water, on the growth of the green alga *Scenedesmus protuberans* and the cyanobacterium *Oscillatoria agardhii*. In: Barica J, Mur LR (eds) Developments in hydrobiology 2:79–82

Lukac M, Aegerter R (1993) Influence of trace metals on growth and toxin production of *Microcystis aeruginosa*. Toxicon 31:293–305

Lyck S, Gjølme N, Utkilen H (1996) Iron starvation increases toxicity of *Microcystis aeruginosa* CYA 228/1 (Chroococcales, Cyanophyceae). Phycologia 35:120–124

Mackerras AH, De Chazal NM, Smith GD (1990) Transient accumulations of cyanophycin in *Anabaena cylindrica* and *Synechocystis* 6308. J Gen Microbiol 136:2057–2065

Martin C, Oberer L, König WA, Busch M Weckesser J (1993) Cyanopeptolins, new depsipeptides from the cyanobacterium *Microcystis aeruginosa* PCC 7806. J Antibiot 46:1550–1556

Meißner K, Dittmann E, Börner T (1996) Toxic and non-toxic strains of the cyanobacterium *Microcystis aeruginosa* contain sequences homologous to peptide synthetase genes. FEMS Microbiol Lett 135:295–303

Morel FMM 1987. Kinetics of nutrient uptake and growth in phytoplankton. J Phycol 23:137–150

Neilan BA, Dittmann E, Rouhiainen L, Bass RL, Schaub V, Sivonen K, Börner T (1999) Nonribosomal peptide synthesis and toxigenicity of cyanobacteria. J Bacteriol 181:4089–4097

Nicklisch A (1998) Growth and light absorption of some planktonic cyanobacteria, diatoms and Chlorophyceae under simulated natural light fluctuations. J Plankton Res 20:105–119

Nicklisch A, Kohl JG (1983) Growth kinetics of *Microcystis aeruginosa* (Kütz) Kütz as a basis for modelling its population dynamics. Int Revue ges Hydrobiol 68 (3):317–326

Orr PT, Jones GJ (1998) Relationship between microcystin production and cell division rates in nitrogen-limited *Microcystis aeruginosa* cultures. Limnol Oceangr 43 (1):1604–1614

Post AF, de Wit R, Mur LR (1985) Interactions between temperature and light intensity on growth and photosynthesis of the cyanobacterium *Oscillatoria agardhii*. J Plankton Res 7 (4):487–495

Rapala J, Sivonen K, Lyra C, Niemelä SI (1997) Variation of microcystins, cyanobacterial hepatotoxins, in *Anabaena* spp. as function of growth stimuli. Appl Environm Microbiol 63:2206–2212

Reynolds C (1984) The ecology of freshwater phytoplankton. Cambridge University Press, Cambridge, 384 p

Rhee GY (1972) Competition between an alga and an aquatic bacterium for phosphate. Limnol Oceanogr 17:505–514

Rohrlack T, Dittmann E, Henning M, Börner T, Kohl JG (1999) Role of microcystins in poisoning and food ingestion inhibition of *Daphnia galeata* caused by the cyanobacterium *Microcystis aeruginosa*. Appl Environm Microbiol 65:737–739

Rothhaupt KO (1992) Stimulation of phosphorus-limited phytoplankton by bacterivorous flagellates in laboratory experiments. Limnol Oceanogr 37:750–759

Rücker J, Kohl JG, Kaiser K (1995) Responses of carotenoids and chlorophylls to variations of growth-limiting factors in three filamentous blue-green algae. Algological Studies 77:51–65

Sherman DB, Sherman LA (1983) Effect of iron deficiency and iron restoration on ultrastructure of *Anacystis nidulans*. J Bacteriol 156:393–401

Shively JM (1988) Inclusions: granules of polyglucose, polyphosphate and poly-ß-hydroxybutyrate. In: Methods in enzymology, vol. 167, Academic press, pp. 195–203

Sivonen K (1990) Effects of light, temperature, nitrate, orthophosphate and bacteria on growth of and hepatotoxin production by *Oscillatoria agardhii* strains. Appl Environm Microbiol 56:2658–2666

Sivonen K, Jones G (1999) Cyanobacterial toxins. In: Chorus I, Bartram J (eds) Toxic cyanobacteria in water: a guide to their public health consequences, monitoring and management. Published on the behalf of WHO by E & FN Spon, London, pp. 41–112

Stanier RY, Ingraham JL, Wheelis ML, Painter PR (1987) The photosynthetic eubacteria. In Stanier RY, Ingraham JL, Wheelis ML, Painter PR (eds) General microbiology. Prentice-Hall, Englewood Cliffs, New Jersey, pp. 355–371

Stevens SE Jr, Paaone DAM, Balkwill DL (1981) Accumulation of cyanophycin granules as a result of phosphate limitation in *Agmenellum quadruplicatum*. Plant Physiology 67:716–719

Straus NA (1994) Iron deprivation: physiology and gene regulation. In: Bryant DA (ed) The molecular biology of cyanobacteria, Kluwer Academic Publishers, Amsterdam, pp. 731–750

Utkilen H, Gjølme N (1992) Toxin production by *Microcystis aeruginosa* as a function of light in continuous cultures and its ecological significance. Appl Environm Microbiol 58:1321–1325

Utkilen H, Gjølme N (1995) Iron-stimulated toxin production in *Microcystis aeruginosa*. Appl Environm Microbiology 61:797–800

Vadstein O, Olsen Y (1989) Chemical composition and phosphate uptake kinetics of limnetic bacterial communities cultured in chemostats under phosphorus limitation. Limnol Oceanogr 34:939–946

van der Westhuizen AJ, Eloff JN (1985) Effect of temperature and light on the toxicity and growth of the blue-green alga *Microcystis aeruginosa* (UV-006). Planta 163:55–59

van Liere L (1979) On *Oscillatoria agardhii* (Gomont), experimental ecology and physiology of a nuisance bloom-forming cyanobacterium. Unpublished Ph.D Thesis, University of Amsterdam, Amsterdam

van Liere L, Zevenboom W, Mur LR (1975) Growth of *Oscillatoria agardhii* Gom. Hydrobiol Bull 9:62–70 van Liere L, Mur LR (1978) Light-limited cultures of the blue-green alga *Oscillatoria agardhii*. Mitt Internat Verein Limnol 21:158–167

van Liere L, Loogman JG, Mur LR (1978) Measuring light irradiance in cultures of phototrophic microorganisms. FEMS Microbiol Lett 3:161–164

van Liere L, Mur LR (1979) Growth kinetics of *Oscillatoria agardhii* Gomont in continuous culture, limited in its growth by the light energy supply. J Gen Microbiol 115:153–160

van Liere L, Mur LR (1979) Materials and methods, cultivation of *Oscillatoria agardhii* Gomont, and some properties of the strain. In: van Liere L (ed) On *Oscillatoria agardhii* Gomont – experimental ecology and physiology of a nuisance bloom-forming cyanobacterium. Academisch Proefschrift, Amsterdam, pp. 19–40

van Liere L, Mur LR, Gibson CE, Herdman M (1979) Growth and physiology of *Oscillatoria agardhii* GOMONT cultivated in continuous culture with a light/dark cycle. In: van Liere L (ed) On *Oscillatoria agardhii* Gomont – experimental ecology and physiology of a nuisance bloom-forming cyanobacterium. Academisch Proefschrift, Amsterdam, pp. 68-74

Vogler P (1976) Analysenautomation in Wasserlaboratorien mit flow-stream-Automaten Teil 4. Die teilautomatische Bestimmung von organisch gebundenem Stickstoff und Gesamtphosphat. Acta Hydrochim Hydrobiol 4/1:43–58

Walz N (1993) Methods of single- and two-stage chemostat experiments. In: Walz N (ed) Plankton regulation dynamics. Experiments and models in rotifer continuous cultures. Ecological Studies 98:21–29

Walz N, Hintze T, Rusche R (1997) Algae and rotifer turbidostats: studies on stability of live feed cultures. Hydrobiologia 358:127–132

Watanabe MF, Oishi S (1985) Effects of environmental factors on toxicity of a cyanobacterium (*Microcystis aeruginosa*) under culture conditions. Appl Environm Microbiol 49:1342–1344

Whitton BA (1973) Freshwater plankton. In: Carr NG, Whitton BA (eds) The biology of blue-green algae. Blackwell Scientific Publications, Oxford, pp. 353–367

Wicks RJ, Thiel PG (1990) Environmental factors affecting the production of peptide toxins in floating scums of the cyanobacterium *Microcystis aeruginosa* in a hypertrophic African reservoir. Environmental Scientific Technology 24:1413–1418

Witzel K-P (1979) The adenylate energy charge as a measure of microbial activities in aquatic habitats. Archiv für Hydrobiologie. Beiheft Ergebnisse der Limnologie 12:146–165

Woitke P, Hesse K, Kohl JG (1997) Changes in the lipophilic photosynthetic pigment contents of different *Microcystis aeruginosa* strains in response to growth irradiance. Photosynthetica 33/3-4:443–453

Zevenboom W, Mur LR (1984) Growth and photosynthetic response of the cyanobacterium *Microcystis aeruginosa* in relation to photoperiodicity and irradiance. Arch Microbiol 139:232–239

4 Factors Affecting Cyanotoxin Concentrations in Natural Populations

As discussed in the introduction to Chapter 3, understanding how environmental factors influence cyanotoxin concentrations in waterbodies requires differentiation between two potential mechanisms: a direct impact on cellular toxin content, and/or an impact on competition between genotypes which would lead to dominance of strains or species with or without microcystin. The field results reported in Chapter 2 reinforce published observations that some taxa produce microcystins, others contain certain neurotoxins, and some may contain both or neither. However, further differentiation is possible: Chapter 4.1 shows that *Microcystis* spp., *Planktothrix agardhii* and *Planktothrix rubescens* each typically contain specific microcystin variants, though relative shares of variants as well as total microcystin content of the field populations may vary. Section 4.2 also addresses the level below that of species – i.e. the level of strains or genotypes – by investigating microcystin content in different strains of *Microcystis aeruginosa* isolated from two lakes. These results demonstrate a substantial variability of microcystin content between different strains of *M. aeruginosa*. In conjunction with the culture study results reported in Chapter 3 (which showed little impact of environmental factors on cellular toxin content), this result suggests competition between strains of a species to be a decisive determinant of microcystin concentrations in natural populations.

To assess the role of environmental factors in determining microcystin concentrations in field populations, Chapter 4.3 analyses the data-base compiled by the German cyanotoxin survey both for direct relationships of microcystin concentrations to environmental factors as well as for indirect relationships which act through an impact of environmental factors on the population density of the cyanobacterial taxa and strains containing these toxins.

4.1 Microcystin Variants in *Microcystis* and *Planktothrix* Dominated Field Samples

Jutta Fastner, Marcel Erhard and Uwe Neumann

Currently, more than 60 structural variants of microcystins have been isolated and characterized from cyanobacterial blooms and cultures (for an overview see Sivonen and Jones 1999). However, most of these variants were found only in isolated strains or in low amounts in field samples. Only some microcystins are reported regularly as quantitatively abundant in isolated strains and field samples, although exceptions occur. Microcystin-LR, microcystin-RR and -YR are, in varying proportions, often the dominant microcystins in blooms and strains of *Microcystis* sp. (Sivonen and Jones 1999). A similar pattern of microcystin specificity is found within the genus *Planktothrix*. Strains and field samples of this genus seem to produce exclusively demethylated microcystin variants (Lindholm and Meriluoto 1991; Luukkainen et al. 1993). In contrast, the microcystin composition in *Anabaena* strains and populations was found to be more diverse (Sivonen et al. 1994; Henriksen 1996).

This Chapter addresses the question which microcystin variants were found within different cyanobacterial taxa in Germany. For this purpose samples were selected from the waterbodies investigated in this study containing almost uni-cyanobacterial populations of the genera *Microcystis* and *Planktothrix*. A brief characterization of the waterbodies is given in Table 2.1, Chapter 2. The chemical structure of microcystins in these samples was elucidated using amino acid analysis in combination with FABMS, and/or MALDI TOF-MS (Fastner et al. 1999, see Chapter 9.2 and 9.3).

Results and Discussion

All major and some of the minor microcystins were isolated and characterized from 22 field samples dominated by either *Microcystis* sp., *Planktothrix agardhii, Planktothrix rubescens* or *Planktothrix* sp. by amino acid analysis followed by FABMS, and/or MALDI-TOF-MS. From these, the identity of 16 different microcystins was determined (Fastner et al. 1999; Neumann, unpubl. results).

The results revealed an obvious relationship between the dominant cyanobacterial taxa and the major microcystins detected. Usually four to ten different microcystins were detected in the *Microcystis* spp.-dominated field samples. Among these, microcystin-RR, microcystin-YR and microcystin-LR were the main toxin constituents in all samples (Fig. 1). These microcystins were also frequently observed in Japanese, Korean and Danish *Microcystis* blooms and strains, although they do not always concur and relative shares were variable (Watanabe et al. 1991; Park et al. 1993; Henriksen 1996; Park et al. 1998). In other parts of the world, MC-LR is more often reported as the main constituent microcystin in blooms and strains of *Microcystis*, sometimes accompanied by minor amounts of MC-RR and/or MC-YR, or other microcystins [e.g. Vasconcelos et al. 1995; 1996 (Portugal); Vezie et al. 1997 (France); Stotts et al. 1993 (USA)].

In contrast, the major microcystins in Finnish *Microcystis* strains and blooms were demethylated variants of MC-RR, MC-YR and MC-LR (Luukkainen et al. 1994), and some of these variants were also occasionally detected in substantial amounts in Danish and Japanese populations of this genus (Park et al. 1993; Henriksen 1996). Various demethylated variants such as [Dha^7] MC-RR, [D-Asp^3] MC-RR, [Dha^7] MC-YR, and [D-Asp^3] MC-LR were also found together with *Microcystis* spp. within the investigated field samples; however, they rarely amounted to more than 10% of the total microcystin content (Fastner et al. 1999). Other minor microcystins identified with *Microcystis* sp. were MC-WR and [H_4] MC-YR (Fastner et al. 1999).

Unicyanobacterial populations of the genus *Planktothrix* exclusively contained demethylated microcystin variants. The samples contained between two and eight different microcystins, and those dominated by *Planktothrix agardhii* showed a more diverse microcystin profile than those dominated by *Planktothrix rubescens* (Fig. 1). The major toxins in *Planktothrix agardhii* were [D-Asp^3] MC-RR, [D-Asp^3] MC-HtyR or a variant of [D-Asp^3] MC-HtyR not conclusively identified and [D-Asp^3] MC-LR (Fig. 1). Further, minor microcystins were [D-Asp^3, Dhb^7] MC-RR, [D-Asp^3] MC-YR and [D-Asp^3, Dhb^7] MC-YR (Fig.1; Fastner et al. 1999; Neumann, unpubl. results). In 13 Finnish strains of *Planktothrix agardhii* Luukkainen and coworkers (1993) usually identified either [D-Asp^3] MC-RR or [Dha^7] MC-RR as major microcystin. [D-Asp^3] MC-RR was also the major microcystin in strains of the red-pigmented variant of *P. agardhii* as well as in field samples of *P. agardhii* from Scandinavia (Fig. 1; Lindholm and Meriluoto 1991; Henriksen 1996, this Volume). Similarly to our results, Henriksen (this Volume) showed green-pigmented strains of *P. agardhii* from Danish waterbodies having a more complex microcystin pattern, as did Danish field samples dominated by this species (Henriksen 1996).

[D-Asp^3, Dhb^7] MC-RR was the main component in all samples dominated by *Planktothrix rubescens*, beside small proportions of [D-Asp^3] MC-RR, [D-Asp^3] MC-LR and [D-Asp^3] MC-YR (Fig.1). [D-Asp^3, Dhb^7] MC-RR was previously isolated and characterized from *Oscillatoria* (syn. *Planktothrix*) *agardhii* NIES-610 (Sano and Kaya 1995). We could show that this microcystin variant is widespread in field samples of the genus *Planktothrix*.

Two samples contained filaments from the *P. agardhii/rubescens* group, which were tentatively denoted as *Planktothrix* sp. since they could not be clearly assigned to either species. Their microcystin pattern was similar to that of *Planktothrix rubescens* with [D-Asp^3, Dhb^7] MC-RR and [D-Asp^3] MC-RR being the main microcystins (Fig. 1). Further examinations will reveal the actual species present at that time.

Our data revealed that species-specific patterns of microcystins showed high similarity both between the populations in the waterbodies examined as well as between the two years investigated (Fig.1). In addition, the microcystin composition scarcely changed over time as long as a species persisted (Fastner et al. 1999). This is in accordance with all studies investigating waterbodies over a longer period: usually microcystins were found in all successive years of investigation and also within years as long as a particular species persisted (Kotak et al. 1995; Jungmann et al. 1996; Park et al. 1998; Welker et al. 1999). Furthermore, all authors, although some examined only the occurrence of MC-LR (Kotak et al. 1995; Jungmann et al. 1996), noted a similar qualitative microcystin composition during the entire period of investigation. For *Microcystis* spp. blooms Park et al. (1998) demonstrated persisting patterns of MC-RR and MC-LR during 4 years in Lake Suwa, Japan and Welker et al. (1999) also reported almost constant ratios of MC-RR to MC-LR in Müggelsee, Germany, within a 2-year period. Changes in the microcystin composition over time were also observed to be quantitative rather than qualitative by Henriksen and Moestrup (1997) in *Microcystis* spp. and *Planktothrix agardhii* blooms from Danish

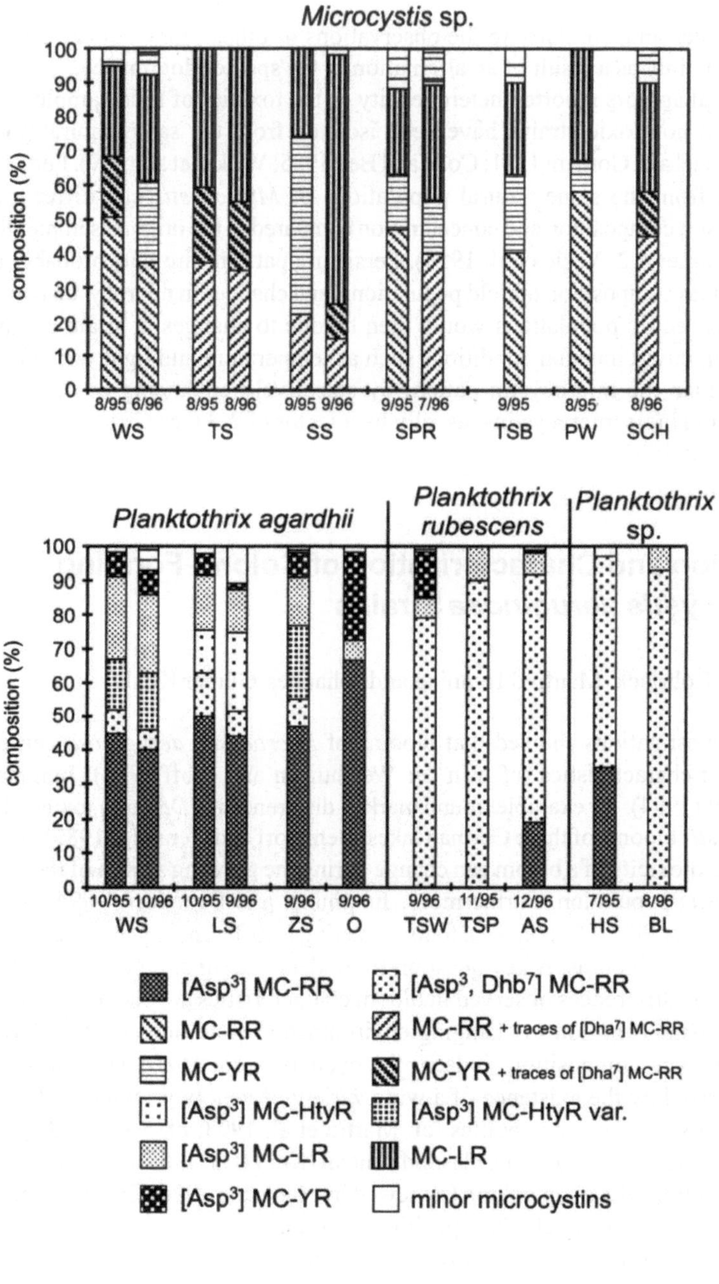

Fig. 1. Composition of microcystins (%) in samples dominated by *Microcystis* sp., *Planktothrix agardhii*, *Planktothrix rubescens* or *Planktothrix* sp. from Wannsee (*WS*), Tegeler See (*TS*), Schlachtensee (*SS*), Speicher Radeburg (*SPR*), Talsperre Bautzen (*TSB*), Postweiher (*PW*), Schwielowsee (*SCH*), Langer See (*LS*), Zeuthener See (*ZS*), Oestertalsperre (*O*), Talsperre Weida (*TSW*), Talsperre Pöhl (*TSP*), Arendsee (*AS*), Haussee (*HS*), Breiter Luzin (*BL*)

waterbodies and, similarly to the observations of others, they noticed a qualitative change mainly as a result of an alternation in the species dominance.

Several authors reported heterogeneity of the toxicity of field samples, and both toxic and non-toxic strains have been isolated from the same natural population (Carmichael and Gorham 1981; Codd and Bell 1985; Welker et al. 1999). Further, strains isolated from the same natural populations of *Microcystis* can differ greatly in microcystin composition and concentration compared to the original sample (Rohrlack et al. Chapter 4.2; Vezie et al. 1998). Persisting patterns are thus probably due to a stable strain composition of field populations, and changes in microcystin composition of monospecific populations would then be due to changes in strain composition. However, environmental conditions such as temperature and light were also shown to influence the microcystin pattern by selectively enhancing or decreasing the amount of single microcystins, usually by a factor of 2-3 (see Chapter 3.1 to 3.3 and Rapala et al. 1997).

4.2 Isolation and Characterization of Colony-Forming *Microcystis aeruginosa* Strains

Thomas Rohrlack, Manfred Henning and Johannes-Günter Kohl

Many investigations showed that blooms of *Microcystis aeruginosa* can differ in important characteristics (cf. van der Westhuizen and Eloff 1983). Jungmann and Benndorf (1994), for example, found marked differences in *Daphnia* toxicity between *Microcystis* blooms of three German lakes. Benndorf and Henning (1989) pointed out that hepatotoxicity of a bloom can change during the growing season of the respective *Microcystis* population. Furthermore, Jungmann and Benndorf (1994), as well as Fastner et al. (1995, Chapter 4.2), found remarkable differences in microcystin content of *Microcystis* blooms. A key question for understanding microcystin production is whether the differences observed in bloom characteristics are due to the reactions of *Microcystis aeruginosa* to changing environmental conditions or to differences in their genotype composition. Indeed, the cyanobacteria *Microcystis aeruginosa* is characterized by the existence of a wide variety of genotypes which differ in their content of secondary metabolites (cf. Martin et al. 1993; cf. Moore 1996), growth strategies (Hesse, Chapter 3.1), plasmid content (Kohl et al. 1988; Schwabe et al. 1988) and interaction with zooplankton (reviewed by Hanazato 1996). Due to some of these differences, it seems likely that specific environmental conditions (e.g. nutrient and light supply, composition of grazers and competitors) select some well-adapted genotypes, whose characteristics would determine the properties of the respective *Microcystis* water bloom to a great extent. This hypothesis is supported by findings of Hesse (Chapter 3.1) on the influence of light and nutrient supply on microcystin production of some *Microcystis aeruginosa* strains isolated from the same water bloom and of Henning et al. (Chapter 6.4) on their interactions with zooplankton.

These authors concluded that strain-specific differences in the features assessed are much more important than changes induced by variations in growth resources. These results suggest that strain composition is of high significance in determination of *Microcystis aeruginosa* bloom characteristics.

The study presented in this Chapter aimed to test the hypothesis that *Microcystis* water blooms contain several strains differing strongly in their ecological and physiological properties.

Origin of *Microcystis* Samples and Isolation of Strains

Two lakes were chosen for the isolation of new *Microcystis aeruginosa* strains. The eutrophic shallow Lake Pehlitzsee (max. depth 2 m, an area of 2 ha), situated in the northern part of state Brandenburg (Germany), is mainly influenced by agricultural sewage and fertilizer. Due to its nutrient status, *Microcystis* species are the most common phytoplankton in summer. The other lake area was the eutrophic Wannsee, located at the western border of Berlin. The characteristics of this shallow lake and the frequent formation of toxic *Microcystis* water blooms were already described by Fastner et al. (1995) and Danowski (1996).

The isolation procedures started with plankton net samples of *Microcystis* blooms taken from Lake Pehlitzsee and Wannsee in July and August 1995, respectively. In a first step samples were centrifuged at 3000 *g*. After this procedure most algae sedimented, whereas cyanobacteria with gas vesicles (mainly *Microcystis* and *Aphanizomenon*) floated. This floating material was subdivided and used for isolation of new *Microcystis aeruginosa* strains in different ways:

- One aliquot was washed with sterile culture medium (according to Zehnder and Gorham 1960) and transferred onto agar plates (1% agarose mixed with culture medium). After incubation for 1 month in daylight and 20-22°C *Microcystis aeruginosa* colonies were transferred separately into sterile culture medium. In this way the Wannsee strains HUB-W15, HUB-W238, HUB-W309, HUB-W332, HUB-W333, HUB-W334, HUB-W335 and HUB-W336 were obtained (the characters W or P in strain names describe their origin, W-Wannsee and P-Pehlitzsee).
- A second aliquot was transferred into 100-ml vessels and treated with overpressure of 0.2 MPa for 2 min. In consequence, gas vesicles and, in some cases, cells of most cyanobacteria (eg. *Aphanizomenon*) collapsed, whereas *Microcystis* cells were still intact and buoyant. After the pressure treatment, the suspension was centrifuged (3000 *g*) to separate *Microcystis* colonies from the destroyed cells of other cyanobacteria. The *Microcystis aeruginosa* colonies were washed with sterile culture medium and transferred separately into culture vessels filled with liquid medium described by Zehnder and Gorham (1960). This produced the strains HUB-W61, HUB-W75, HUB-W77, HUB-W368, HUB-W396 and all Pehlitzsee strains.

More than 500 different cultures were inoculated with *Microcystis aeruginosa* colonies and this resulted in isolation of 22 new strains. However, the success of isolation procedures was strongly affected by the presence of epiphytic algae in or on the mucilaginous envelope of *Microcystis* colonies (such as several *Nitzschia* species, and *Chlorella* cf. *minutissima*). These algae are able to overgrow and suppress the *Microcystis* cells very rapidly (Täuscher 1980). In contrast, other planktonic algae or cyanobacteria did not disturb the isolation process due to their easier removal by the cleaning procedures applied. The pressure treatment described proved to be very useful in separation of *Microcystis* colonies from other accompanying planktonic cyanobacteria, especially *Aphanizomenon* trichoms.

Chemical Analyses of the *Microcystis* Strains and Water Blooms

Chemical analyses of strain properties was performed on the net samples and on the non-axenic batch cultures isolated from them. The quantitative analyses of most important microcystins (LR, demethyl LR, YR, RR) were carried out by HPLC (Lawton et al. 1994) and qualitative assessment was performed by MALDI-TOF-MS (analyses by M. Erhard, Institut for Molecular Biology, TU Berlin, Germany; see Chapter 8.4). This revealed no further microcystins beyond these four. In most cases it was difficult to separate microcystin LR from its demethyl variants by the HPLC method applied. Therefore, these two microcystins were combined for the quantitative evaluation. The carotenoid composition in the *Microcystis aeruginosa* cultures and lake samples was analyzed using a non-end capped C18-RP-column (Woitke et al. 1994). The cell diameter was measured using an interference microscopic method (accuracy $\leq 0.1\ \mu m$).

Characterization of Water Samples Used for Isolation of *Microcystis* Strains

In August 1995, when the samples used for isolation of new strains were taken, a heavy *Microcystis* water bloom had developed in Wannsee with seven species of *Microcystis* and dominance of *M. aeruginosa* (Chorus and Komárek, pers. comm.). Comparable conditions were observed in Lake Pehlitzsee (own unpubl. data). Quantitative microcystin analyses showed the Wannsee population to contain the variants LR, RR and YR and the Pehlitzsee population to contain only the microcystin variants LR and YR (Fig. 1). Differences between both lakes were also found in the carotenoid composition. The samples from both lakes contained ß-carotene, echinenon, zeaxanthin, myxoxanthophyll and coloxanthin, while nostoxanthin was only found in Lake Pehlitzsee (data not shown).

Morphological Characterization of the Strains

All of the 22 strains newly isolated from the two lakes continued to grow in colonies with voluminous mucilaginous envelope and without showing any tendency to disaggregate. The dimensions and the cell arrangement of colonies appeared to depend on culture age. Freshly inoculated cultures of new strains usually started with small compact colonies. The dimensions of the colonies increased with culture age and density, while the colony habitus changed from a compact cell arrangement to a typical *Microcystis aeruginosa* shape. According to their typical colony habitus, all strains were classified as *Microcystis aeruginosa*.

Based on their cell diameter the new isolated *Microcystis* strains can be subdivided into two significantly different clusters (T-test at 95% level of significance; Fig. 2).

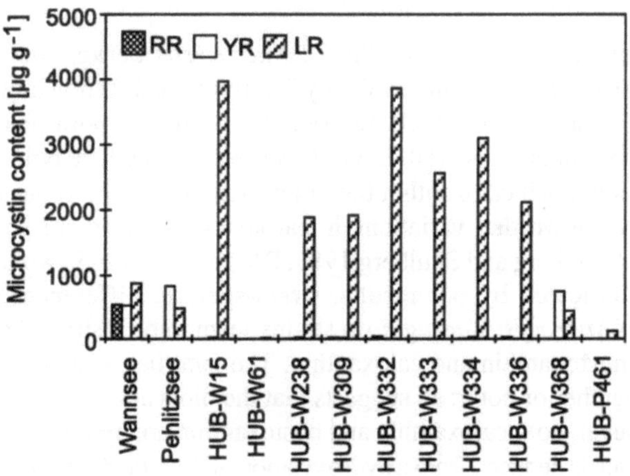

Fig. 1. Microcystin content analyzed by HPLC. *Bars* represent means of two replicates, strains without microcystins are not shown

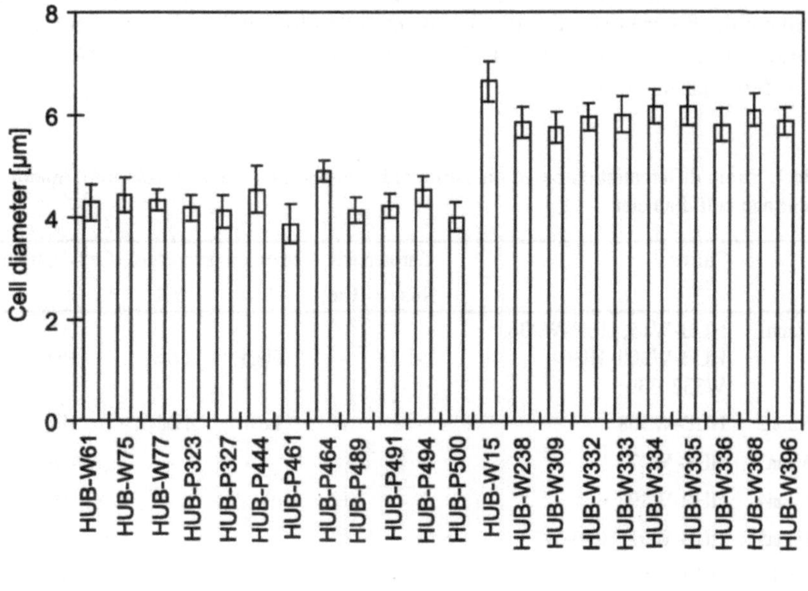

Fig. 2. Cell diameter of the strains investigated. *Bars* represent mean values ± SD, n=20

The first one included all strains from Lake Pehlitzsee and some strains isolated from Lake Wannsee and was characterised by relatively low cell diameters (3.8-4.9 µm) compared to the strains of cluster 2 (5.7-6.6 µm). Both dimensions were within the range of the cell size of *Microcystis aeruginosa* (Komárek, pers. comm.), supporting the species determination of the described strains.

Composition of Carotenoids and Microcystins in New Isolated Strains

Based on carotenoid composition, the 22 *Microcystis aeruginosa* strains can be subdivided into those which contained only ß-carotene, echinenon, myxoxanthophyll and zeaxanthin, and those which additionally produced both nostoxanthin and caloxanthin. Only in one case (HUB-W77) was an intermediate type of both groups found, which contained caloxanthin but not nostoxanthin (see Table 1).

It is already known that variations in carotenoid patterns are very common in cyanobacteria (Skulberg and Skulberg 1985; Rücker et al. 1995; Woitke et al. 1997). However, as indicated by our results, strain-specific differences in carotenoid composition of *Microcystis aeruginosa* strains seem to be restricted to the presence or absence of nostoxanthin and caloxanthin. The fact that both carotenoids usually occur either together or not at all suggests that the biosynthesis of these carotenoids is linked. However, both caloxanthin and nostoxanthin are hydroxylated derivates of zeaxanthin, which is deduced from a hydroxylation of ß-carotene (Böger and Sandmann 1993). Thus, it is conceivable that these compounds are synthesized by an additional enzyme, or that the ß-carotene hydroxylases of the respective strains exhibited a decreased positional specificity (Woitke et al. 1997). In contrast to caloxanthin and nostoxanthin, ß-carotene, zeaxanthin, myxoxanthophyll and echinenon were found in all strains investigated in this study. Therefore, they are unsuitable for characterizing and differentiating strains.

Table 1. Strain differentiation based on carotinoids (nostoxanthin and caloxanthin), microcystin content and cell diameter

	Strains	Carotenoids		Microcystin content		Cell diameter
		Nos	Cal	LR	YR	
Wann1	HUB-W15, HUB-W238, HUB-W309, HUB-W332..336	-	-	High	Low	High
Wann2	HUB-W368	-	-	Low	High	High
Wann3	HUB-W77	-	+	-	-	Low
Wann4	HUB-W396	+	+	-	-	High
Wann5	HUB-W61, HUB-W75	+	+	-	-	Low
Pehl1	HUB-P461	+	+	Low	-	Low
Pehl2	Remaining Pehlitzsee strains	+	+	-	-	Low

Most of the strains isolated from Lake Pehlitzsee did not contain microcystins (data not shown); in only one of them (HUB-P461) were small amounts of microcystin LR detected (Fig. 1). Due to marked differences in microcystin composition, the strains isolated from Lake Wannsee can be subdivided into three distinct clusters. No microcystins were found in the strains HUB-W61, HUB-W75, HUB-W77 and HUB-W396. Another group of strains contained almost exclusively microcystin LR, in concentrations several-fold above those of the field population. One strain (HUB-W368) contained two microcystins, YR and LR.

Differentiation of the Strains

It cannot be ruled out that some strains were repeatedly isolated by the procedures applied and are actually genetically identical. Therefore, the characteristics described were used to subdivide the strains into distinct groups. This subdivides the 13 strains isolated from Wannsee into 5 groups, sometimes represented by only one strain (Table 1). Group Wann1 was mainly characterized by a very high content of microcystin-LR and by the absence of both carotenoids, caloxanthin and nostoxanthin. The strains of groups Wann 3, 4 and 5 (see Tab. 1) did not synthesize any microcystin variant, but they differ in their carotenoid composition and cell diameter. As demonstrated in Fig. 1, the microcystin-LR content of the Wann1 strains varied by about a factor of 2. However, these variations may have arisen from differences in culture conditions (see Chapter 3). Therefore, the differences found in the microcystin LR content were not used for a further subdivision of group Wann1.

The strains isolated from Lake Pehlitzsee probably descended from two groups. Typical for most of them was the absence of microcystins and the presence of all of the carotenoids assessed. However, the strain HUB-P461 contained comparably low amounts of microcystin LR.

Discussion

The main concern of the present study was to characterize and group the strains by general properties. As these varied strain-specifically and largely independently of variations in environmental conditions, they allowed a comparison with the original *Microcystis* lake samples.

In Wannsee, for example, some strains which did not synthesize any microcystin variant (groups Wann 3, 4 and 5, see Tab. 1) were accompanied by some microcystin-containing genotypes (Wann 1 and 2, Tab. 1). Hesse (Chapter 3.1) pointed out that the strains derived from Wannsee also differ in their reactions to nitrogen or phosphorus limitation. Henning et al. (Chapter 6.4) and Rohrlack et al. (1999) found significant differences in effects of some Wannsee strains on filter-feeding zooplankton. Thus, *Microcystis aeruginosa* blooms may consist of strains which differ not only morphologically and biochemically, but consequently also in their physiological and ecological characteristics. These are probably related to different genotype compositions which can be understood as intraspecific biocoenotic adaptation resulting from interactions with abiotic and biotic factors.

The comparison of strain characteristics with properties of the respective bloom allow some conclusions on the dominance structure of strains in the bloom. For example, the sample of Lake Wannsee did not contain any nostoxanthin, which was, nevertheless, found in cells of strains comprising groups Wann4 and Wann5. Therefore, it can be concluded that these strains were present in the Wannsee bloom, but only in very minor cell density. The very high cellular microcystin-LR content of the Wann1 strain is in striking contrast to the comparable low amounts in the natural bloom, which contained only1/4 to 1/2 of the concentrations found in the individual strains. This suggests that the strains containing microcystin were also subdominant. Further, the isolation procedures probably did not include all strains of the bloom investigated, as the original samples of both lakes contained microcystin variants (RR in Lake Wannsee, YR in Lake Pehlitzsee) which did not occur in detectable amounts in the strain cultures. This additional microcystin indicates the presence of other *Microcystis* species or strains (or cyanobacteria) in the original bloom which were not obtained by the isolation procedures.

In conclusion, our results suggest that *Microcystis aeruginosa* water blooms contained several strains which differ in chemical characteristics as well as in their physiology and ecology. On account of these differences, changes in environmental conditions (nutrient supply, grazing pressure) may result in a shift in strain composition and consequently in bloom characteristics.

Acknowledgements

The authors would like to express their gratitude to Mrs. Marion Dewender for her analyses of microcystins as well as for her excellent laboratory assistance. We also wish to thank Mr. M. Erhard (Institut for Molecular Biology, TU Berlin, Germany) for providing results of MALDI-TOF-MS analyses of microcystin composition. In addition to the grant of the Federal Ministry of Education and Research (BMBF) this study was supported by the Fazit Foundation.

4.3 Environmental Factors an Microcystin Levels in Waterbodies

Ingrid Chorus, Verena Niesel, Jutta Fastner, Claudia Wiedner, Brigitte Nixdorf and Karl-Erich Lindenschmidt

Introduction

All Environmental factors can influence microcystin levels in waterbodies through two mechanisms, which can act either individually or simultaneously:

1. By determination of the microcystin content in cyanobacterial cells; i.e. differences in the availability of resources such as light or nutrients can influence the microcystin production process within the cells.
2. By determination of the biomass of microcystin-producing cyanobacteria, through limiting the carrying capacity for phytoplankton in general and through influencing the outcome of competition between cyanobacterial species and strains of different inherent toxin content.

Microcystin content in field populations was shown to differ between cyanobacterial taxa (see Chapter 2.1.2). In part, this is due to the specific current strain composition; within one and the same species, strains with no, low or high microcystin content may be found in a given population (see Chapter 4.2), and genetic work has shown that microcystin is constituent to some strains, but not to all (see Chapter 3.4). This would imply the second mechanism to be the most important determinant of microcystin concentrations in field samples. However, it does not exclude the relevance of mechanism 1 as well. Indeed some impact of resource limitation on microcystin production has been demonstrated by laboratory experiments, e.g. addressing nitrogen limitation (Long et al. 2001) and the role of irradiance for transcription (Kaebernick et al. 2000).

A number of publications have used field data to address correlations between environmental factors and microcystin content of seston. Although some correlations have been found, results are in part contradictory or interpretations are inconclusive. The key problem is that most studies investigated microcystins in relation to total seston dry weight, but in field samples, changes in this relationship may reflect changes both in the toxin content of cyanobacterial cells as well as in the relative contribution of toxic cyanobacteria to total seston, i.e. the study of microcystin per dry weight cannot differentiate between the two mechanisms described above. Thus, correlations found may merely reflect an increasing level of cyanobacterial biomass and/or an increasing share of cyanobacteria in total seston (Wicks and Thiel 1990; Vezie et al. 1998).

Kotak et al. (1995, 2000) are the only authors to follow a different approach by relating microcystins not only to seston dry weight, but also to cyanobacterial biomass and water volume. Their data set indicates an effect of total phosphorus on microcystin concentrations, but data evaluation did not differentiate whether this is due to a direct effect on cellular toxin content or only to the indirect effect through the influence they show for total P on the biomass of *Microcystis* spp.

Approach

The cyanotoxin survey in Germany produced a large data set for microcystin concentrations in relation to dry weight, chlorophyll-a and L^{-1} of water together with data on environmental factors potentially impacting on toxin concentrations (see Chapter 2.1 for an overview). They cover a large span of conditions from a wide array of lakes from different regions with different ecological conditions. For a subset of these samples, data are also available for the biovolume of microcystin-producing cyanobacterial taxa, which in most cases belonged to either the genera *Microcystis* or *Planktothrix*.

Data evaluation followed two approaches: one was to evaluate scatterplots of microcystin concentrations against environmental parameters for envelope functions in order to show the maximal microcystin concentrations attained at a given level of the environmental parameter. The other was a statistical evaluation of the potential influence of environmental factors on microcystin concentrations using multivariate analysis in order to differentiate between direct effects on microcystin concentrations in the cyanobacterial biomass (Fig. 1A), indirect effects acting through effects on population sizes of the microcystin producers (Fig. 1B), and the combination of both (Fig. 1C).

Among the range of potentially relevant environmental factors, we addressed nutrients (phosphorus and nitrogen) and light availability. As data on photosynthetically available radiation within the waterbody were available only from a small number of the waterbodies studied, two indicators calculated from transparency

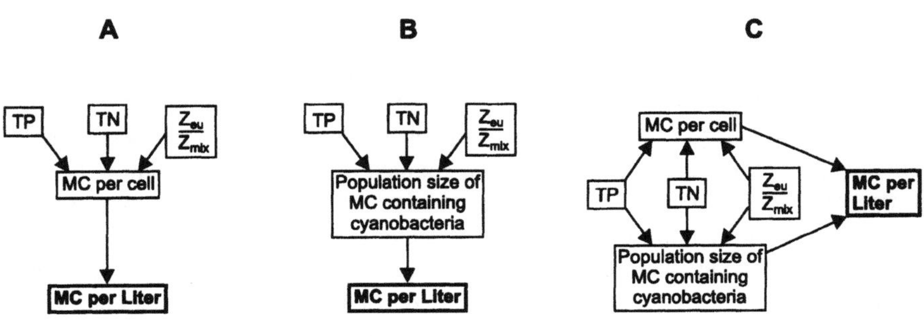

Fig. 1. Potential impact of environmental factors on microcystin concentration through (**A**) effects on the microcytin content of the cells, (**B**) effects on the population size of microcystin-producing taxa, or (**C**) through both mechanisms. *MC* microcystin; *TP* total phosphorus; *TN* total nitrogen; Z_{eu}/Z_m ratio of euphotic to mixing depth as indicator of average light exposure of cyanobacterial cells

(measured as Secchi depth) were used: the depth of the euphotic zone and the relationship of euphotic depth to mixing depth. The latter was expected to best reflect the average cell's exposure to light during entrainment by epilimnetic mixing.

Limnological Methods and Data Used

Data on microcystin concentrations together with at least one of the environmental parameters studied were available from 27–35 waterbodies, depending on the respective parameter. Data on concentrations of nutrients (phosporus and nitrogen), Secchi depths, mixing depths, chlorophyll-a and phytoplankton biovolume were provided by a range of partner laboratories (see Chapter 2.1.1) and determined with their respective in-house methods. For cyanobacterial biomass determination, standard counting techniques using an inverted microscope according to Utermöhl (1958) were applied in the partner laboratories. Depths of the euphotic zone (Z_{eu}) were calculated from Secchi depth data by multiplication with 2.5. Mixing depths (Z_m) were determined by the authors from the temperature profiles supplied by the partner laboratories; for non-stratifying waterbodies mean depths were used.

Microcystins (MC) were determined by three laboratories within the project using the same analytical method, in most cases from seston samples collected with a plankton net. Both microcystins and chlorophyll-a were determined from aliquots of these. Microcystin concentrations L^{-1} were then calculated from (1) the microcystin/chlorophyll-a ratios determined for the plankton net samples and from (2) the field data for chlorophyll-a L^{-1}. Selective sampling of larger phytoplankton forms (particularly colonies) by nets may introduce a bias that would lead to an overestimation of microcystin concentrations if cyanobacteria are selectively enriched. For a small number of samples, i.e. those collected from Mueggelsee in 1997, microcystin concentrations L^{-1} were determined directly from filters. (See Chapter 2.1.2 and 9 for more detail on methods.)

For envelope functions showing the maximal microcystin concentration found in relation to total phosphorus (TP), total nitrogen (TN) and the ratio of euphotic to

Table 1. Descriptive statistics of variables used for envelope functions

	Valid N	Mean	Median	Minimum	Maximum	Std. dev.
Microcystin concentration (L^{-1})	573	15.8	0.40	0.0	465	163
Total nitrogen ($\mu g\ L^{-1}$)	190	1372	1080	0.7	7443	1274
Total phosphorus ($\mu g\ L^{-1}$)	330	122	776	11	719	127
Ratio of euphotic to mixing depth	338	1.0	0.7	0.1	13	1.0

mixed depth (Z_{eu}/Z_{mix}), all of the available data were used, regardless of the number of samples available from each waterbody, which ranged from 1 to 43. Depending on the variable investigated, the number of available samples ranged from 190 to 573, and values covered a wide range (Table 1).

For regression analyses, we sought a more even representation of waterbodies. Among those waterbodies sampled regularly, data were available for the key months of the summer cyanobacterial season, from June to September. In analogy to Shreurs' (1992) approach of "lake years", we treated each summer as a separate "event", regardless of whether a given waterbody is represented with data from one or from three summers. However, instead of using summer mean values, we used monthly means in order to balance different sampling frequencies without averaging out the effects to be investigated. Thus, a given waterbody may be represented with 4 (in a few cases only 3) to 8 values in the data set (Table 2). Whereas for all of these months, data are available for microcystin concentrations L^{-1} and for the biovolume of microcystin-producing taxa (i.e. species of *Microcystis*, *Planktothrix*or *Anabaena*), in some cases data for nutrients, euphotic or mixing depth are lacking, so the number of monthly means included in some regressions may be smaller.

Dissolved nutrients (soluble reactive phosphorus, SRP, and total inorganic nitrogen, TIN) were not addressed for two reasons: for SRP, no meaningful results are to be expected in face of detection limits well above the concentrations limiting uptake rates: where phytoplankton "leaves over" more than 3–6 μg L^{-1} P (as soluble reactive phosphate, SRP), this nutrient is not likely to limit uptake rates and thus growth (Reynolds 1997). In such situations, cellular nutrient contents (cell quotas) are saturated, and no impact on processes such as microcystin production is to be expected.

Table 2. Number of monthly means (June to September) included in regression analyses

	1995	1996	1997	Total
Grosser Wannsee	4	4		8
Langer See	3	4		7
Melangsee		4		4
Mueggelsee		4	4	8
Radeburg		3		3
Schlachtensee	3			3
Storkower See	4	3		7
Tegeler See	4	4		8
TS Weida		3		3
Zeuthener See		4		4
Total	18	33	4	55

Table 3. Descriptive statistics of parameters used for regression analysis (monthly means, June to September) for all data sets with microcystin concentrations above 0

Parameter	Valid N	Mean	Median	Minimum	Maximum	Std. Dev.
Microcystin concentration (g L^{-1})	47	20	1.9	0.01	276	52
Biovolume of microcystin-producing taxa (mm^3 L^{-1})	47	5.6	2.0	0.01	33	7.3
Total inorganic N (µg L^{-1})	20	1714	125	10.4	9337	3369
Total N (µg L-1)	21	2772	1313	677	7443	2389
Total P (µg L^{-1})	25	76	86	24	141	42
Euphotic depth (m)	45	3.4	2.4	0.75	10.1	2.4
Ratio of euphotic to mixing depth	42	0.79	0.62	0.28	3.3	0.54

Further, though negative exponential relationships between microcystin concentrations and TIN as well as SRP can be shown from the data, they are likely to reflect primarily the depletion of these nutrients at high levels of cyanobacterial biomass rather than to indicate any impact of nutrient limitation on microcystin content of the cells, as postulated by Kotak et al. (2000) for similar results with respect to inorganic nitrogen.

Furthermore, total phosphorus data included in regression analysis were limited to samples with values lower than 150 µg L^{-1}, as the envelope functions (Fig. 2) indicated no further impact of total phosphorus on microcystin concentrations above this threshold. Table 3 shows the parameters studied by regression analysis and their respective range of values.

Statistical Methods for Regression Analysis

For regression analysis, data sets were tested for normal distribution with the Kolmogoroff-Smirnofftest. Data which were non-normally distributed data were ln-transformed and where the transformation resulted in negative values, ln (10) or ln (100) were added to all of the values of the respective data set. Subsequently, transformed data sets were again tested for normal distribution.

Data sets transformed into a normal distribution were analyzed with linear regression analysis using the Pearson correlation coefficient as an estimator with a significance level of 0.05. In the case of total nitrogen, where ln-transformation did not result in normal distribution, the data were analyzed by Spearman rank correlation with a significance level of 0.05.

Multiple regression analysis was used to test the simultaneous influence of two parameters on microcystin concentrations, again using the Pearson correlation coefficient as estimator, after testing for normal distribution. Analysis of variance with estimations of the coefficient of determination B and levels of significance was first performed for the pairwise parameters individually and then compared to the results of multiple regression analysis. The influence of a further parameter was estimated by comparing squared coefficients between simple linear (R^2) and multiple linear regressions (R^2) analysis.

Results and Discussion

Scatterplots and Envelope Functions

Total Phosphorus An envelope function may be drawn which suggests that microcystin concentrations can maximally attain levels amounting to approximately half of the total phosphorus concentrations at TP levels below 120–170 µg L^{-1} TP (Fig. 2, 3 out of 330 samples disregarded). At higher TP concentrations, maximal microcystin concentrations level off. Microcystin concentrations well below the maximum level of the envelope function occurred frequently. The three samples with microcystin concentrations above the envelope function are probably the result of overestimations due to methodological problems, as indicated by the microcystin/chlorophyll-a ratios determined in the net samples: for the samples from Langer See and Gloertalsperre they amounted to 1.2–2.2, whereas for all other samples they were below 1, usually below 0.7.

Similarly, thresholds have been observed for the dependency of phytoplankton biomass on total phosphorus concentrations. Their cause has been shown to be a limitation of phytoplankton through other factors such as light and nitrogen, when phosphorus is amply available (Reynolds 1997; Chorus and Mur 1999). Up to such threshold concentrations, total phosphorus determines the carrying capacity for phytoplankton – and thus also for cyanobacteria; above them, a further increase in total phosphorus will not lead to a further increase in phytoplankton biomass because another factor is limiting. Depending upon the depth of the waterbody or mixed layer (i.e. the epilimnion) of a given waterbody, threshold TP concentrations for phytoplankton biomass are found to lie in the range of 20–150 µg L^{-1}. As many of the waterbodies included here are shallow, a dependency of phytoplankton biomass on TP is to be expected up to TP concentrations of approximately 150 µg L^{-1}.

Following the Redfield ratio, 150 µg of phosphorus could sustain 6300 µg of carbon, which would correspond to 16 mg of phytoplankton dry weight. According to the microcystin/dry weight relationships of up to 6 µg mg^{-1} dry weight described in Chapter 2.1.2, these could contain up to 100 µg of microcystins in situations where phytoplankton is dominated by microcystin-producing cyanobacteria. Thus, the maximal microcystin concentrations observed here at TP levels in the range of 150 µg L^{-1} correspond well to the expectations derived from the carrying capacity for biomass by total phosphorus. Thus, the threshold pattern for microcystin concentrations in relation to total P concentrations may reflect the strong dependency of microcystin concentrations on cyanobacterial biomass.

This envelope function can be used for practical purposes of water quality management, as it indicates a relationship between concentrations of total phosphorus

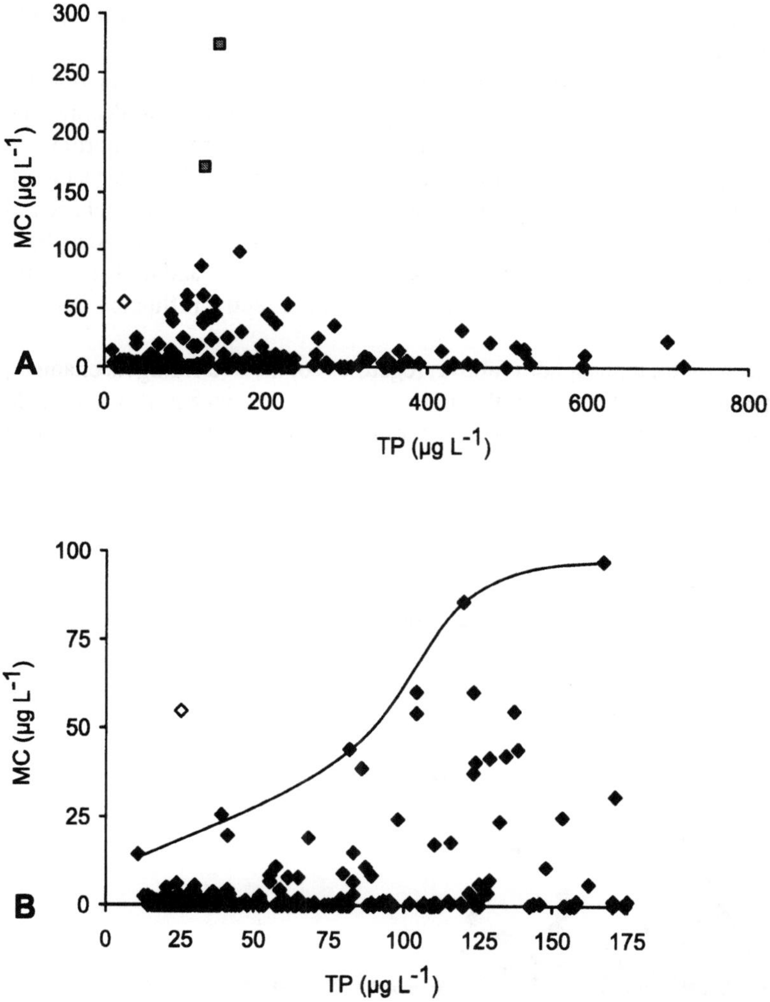

Fig. 2. Scatterplots of microcystin concentrations over total phosphorus (330 samples from 35 waterbodies). **A** Total data set. **B** Magnification of lower range. *Open diamond* Sample from Gloertal Reservoir; *shaded squares* Samples from Langer See; all three with exceptionally high ratios of MC to chlorophyll-a

and maximum microcystin concentrations to be expected outside of scum areas. This is particularly relevant for waterbodies inhabited by non-scum-forming cyanobacterial taxa such as *Planktothrix agardhii.*

Total Nitrogen As for total phosphorus, scatterplots for microcystin concentrations over total nitrogen show a threshold pattern: the envelope function indicates that microcystin concentrations are N-limited at concentrations below approximately 1500 µg L^{-1} TN (Fig. 3). Again, the key mechanism may be N limitation of cyanobacterial biomass: following the Redfield ratio, 1500 µg of nitrogen could sustain 9000 µg of carbon, which would correspond to 22 mg of dry weight and could contain 100–130 µg of microcystins.

Light Availability, Indicated by the Ratio Z_{eu}/Z_{mix} The envelope function for microcystin concentrations in relation to Z_{eu}/Z_{mix} shows an exponential decline: the clearer the water in relation to mixing depths, the lower are the microcystin concentrations (Fig. 4). In this relationship, causality may comprise three mechanisms: the water is likely to be clearer if phytoplankton biomass – and thus also cyanobacterial biomass – is lower, which would impact upon microcystin concentrations. Also, light availability may influence phytoplankton species composition – and thus indirectly affect microcystin concentrations. Further, it could directly influence cellular microcystin content.

As discussed for total phosphorus above, regardless of the overriding mechanism, this envelope function indicates that the highest microcystin concentrations (outside of scums) are to be expected in turbid waters, often with a euphotic depth lower than the mixing depth.

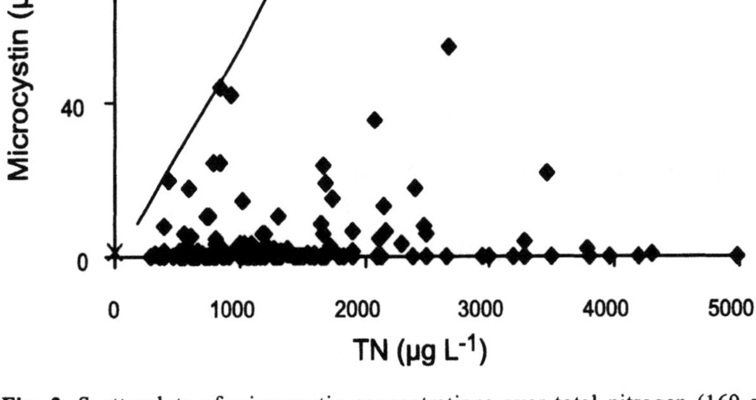

Fig. 3. Scatterplots of microcystin concentrations over total nitrogen (169 samples from 28 waterbodies)

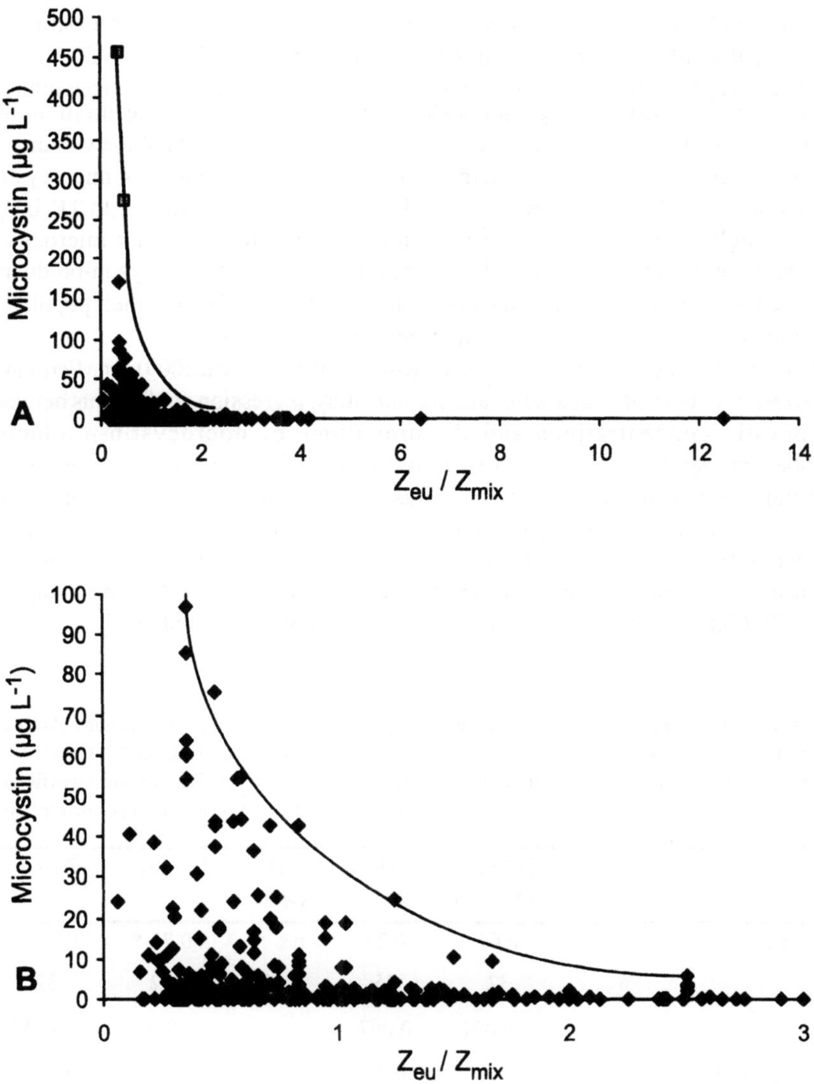

Fig. 4. Scatterplots of microcystin concentrations over the ratio of euphotic to mixed depth (338 samples from 27 waterbodies) **A** Total data set. **B** Magnification of lower range. *Shaded squares* Samples from Langer See with exceptionally high ratios of MC to chlorophyll-a

Simple Linear Regression Analysis

Among the 55 monthly means available for simple regression analysis, 8 showed microcystin concentrations L^{-1} of water of 0 (i.e. below the detection limit) in spite of the presence of a potentially microcystin-producing taxon. Seven of these originated from Springsee and one from Lake Tegel, and in all cases, the only potentially microcystin-producing taxon was *Anabaena* spp., found at low biovolumes of less than 1 $mm^3\ L^{-1}$. These low biovolumes imply microcystin concentrations well under the detection limit. Further, *Anabaena* spp. was found to contain microcystins less frequently than *Microcystis* spp. and *Planktothrix* spp. (see Chapter 2.1.2).

Pairwise regression analysis for the monthly means with microcystin concentrations larger than zero shows a strong relationship between microcystin concentrations and the biovolume of microcystin-producing taxa (Table 4), with 67% of the variance of microcystin concentrations (MC) being explained by the biovolume of microcystin-producing taxa (BV_{mcp}). Both BV_{mcp} and MC correlate significantly with TP, BV_{mcp} more strongly than MC. These results support the hypothesis that microcystin concentrations in water depend largely on the population size of microcystin-producing cyanobacteria. They further indicate that total phosphorus influences their population size more strongly than it influences microcystin concentrations.

This result is coherent with that found by Kotak et al. (1995 and 2000) for *Microcystis aeruginosa*: for both of their studies and for our study, regression coefficients between microcystin concentrations and the biovolume of microcystin-producing cyanobacteria were all higher or at least as high as those for microcystin concentrations and total phosphorus concentrations. Thus, the field data from all these studies provide no indication of an impact of this environmental variable on toxin content of the cells. Interestingly, values for r^2 for the correlation between concentrations of microcystins and total phosphorus were similar in both studies, 0.28 in this study and 0.21–0.29 (depending on the data set used) in the studies by Kotak et al.

Table 4. Results of pairwise simple regression analysis for microcystin concentrations (*MC*) and biovolume of microcystin producers (BV_{mcp}) against total phosphorus (*TP*), total nitrogen (*TN*), euphotic depth (Z_{eu}) and ratio of euphotic to mixing depth (Z_{eu}/Z_{mix}). Results are significantly different from 0 at $p < 0.05$; * negative correlations; ** results of Spearman rank correlation

		BVmcp ($mm^3\ L^{-1}$)	TP ($\mu g\ L^{-1}$)	TN ($\mu g\ L^{-1}$)	Zeu (m)	Z_{eu}/Z_{mix}
MC ($\mu g\ L^{-1}$)	R^2	0.67	0.28	n.s. **	0.73 *	0.19 *
	n	32	25	21	32	32
	p	< 0.001	0.007		< 0.001	< 0.013
BVmcp ($mm^3\ L^{-1}$)	R^2	-	0.33	n.s. **	0.58 *	0.22 *
	n		25	21	32	32
	p		0.003		< 0.001	< 0.006

No correlation was found for either BV_{mcp} or MC against total nitrogen (TN), although the envelope function shown in Fig. 3 for MC against TN indicates a relationship between maximal MC concentrations and those of TN. However, the data set is dominated by a high number of cases with low values for both MC and BV_{mcp} throughout the range of TN concentrations observed. This indicates that for the majority of samples in this database, TN was not the parameter limiting cyanobacterial biomass. In these cases, TN would not be expected to influence microcystin concentrations.

The strongest correlations were found between MC and BV_{mcp} against the depth of the euphotic zone, Z_{eu}. (R^2= 0.73, Table 4). In contrast to the results for TP, MC correlates more strongly with Z_{eu} than it does with BV_{mcp}. This may indicate a potential direct influence of Z_{eu} on microcystin content of the cells, in addition to the indirect effect through the relationship between Z_{eu} and BV_{mcp}, a hypothesis to be further tested by multiple regression analysis.

Surprisingly, the correlation of both MC and BV_{mcp} to the ratio between euphotic and mixing depth (Z_{eu}/Z_{mix}) was much weaker than the correlation to Z_{eu} (Table 4). In theory, this ratio would be expected to reflect the influence of average light availability much better than Z_{eu} alone, because the depth of cell entrainment within the mixed water layer (Z_{mix}) influences the amount of time they are exposed to light within the euphotic layer (Z_{eu}). Two causes are conceivable for the poor relationship observed. One is the large margin of error involved in determining Z_{mix} with sufficient accuracy from temperature gradients that do not show a clear lower boundary of the mixed layer, or for shallow, non-stratified lakes for which Z_{mix} was taken to be equivalent to mean depth. The other is the strong causal relationship between BV_{mcp} and Z_{eu} based on more than one mechanism: low light availability has been shown to favour some cyanobacterial taxa (particularly *Planktothrix agardhii*) in competition against other phytoplankton (Mur et al. 1993; Schreurs 1992), and high phytoplankton biovolumes are often the key cause for low euphotic depths. Thus, BV_{mcp} will be not only the result of low values for Z_{eu}, but also their cause, and R^2 is therefore likely to be high.

The two studies by Kotak et al. (1995, 2000) also showed a correlation between microcystin concentrations and Secchi depth transparency. However, their values reported for R (-0.35 and -0.38; i.e. R^2= 0.12 and 0.14) were substantially lower than in our data set; possibly our stronger correlation reflects the wider range of transparencies covered in the data set. As for total phosphorus, their results provide no evidence of a direct impact of light availability on microcystin content of the cells.

Multiple Linear Regression Analysis

Multiple regression analysis was employed to address two questions:

1. Is more variability of MC explicable if TP is included in addition to BV_{mcp}?
2. Is more variability of MC explicable if Z_{eu} is included in addition to BV_{mcp}?

For this evaluation, only monthly means could be used for which all of these data are available; the subset of data is thus smaller than that used for the simple regression analyses. Thus, to produce a basis for comparison of the results of pairwise regression

Table 5. Results of pairwise simple regression analysis for microcystin concentrations (*MC*) and biovolume of microcystin producers (BV_{mcp}) against total phosphorus (*TP*) and euphotic depth (Z_{eu}) for the subset of 24 monthly means, for which all of these data are available. Results are significantly different from 0 at p < 0.05; negative correlations are indicated with *

		BVmcp (mm^3 L^{-1})	TP (μg L^{-1})	Zeu (m)
MC (μg L^{-1})	R^2	0.71	0.31	0.64 *
	n	24	24	24
	p	< 0.001	0.005	< 0.001
BVmcp (mm^3 L^{-1})	R^2	-	0.31	0.53 *
	n		24	24
	p		0.004	< 0.001

Table 6. Comparison of coefficients for simple and multiple analysis and levels of significance of the partial correlations as test for the additional influence of total phosphorus (*TP*) on the correlation between microcystin concentrations and the biovolume of microcystin-producing cyanobacteria (BV_{mcp}); $n = 24$

R^2 from simple regression analysis (see Table 5)	R^2_{adj} from multiple regression analysis	p of partial correlation for BV_{mcp}	p of partial correlation for TP
0.71	0.69	< 0.001	0.391

Table 7. Comparison of coefficients for simple and multiple analysis and levels of significance of the partial correlations as a test for the additional influence of eupthotic depth (Z_{eu}) on the correlation between microcystin concentrations and BV_{mcp}; $n = 24$

R^2 from simple regression analysis (see Table 5)	R^2_{adj} from multiple regression analysis	p of partial correlation for BV_{mcp}	p of partial correlation for TP
0.71	0.79	0.008	0.03

analysis to those of multiple regression analysis, pairwise regression analysis was first repeated for the subset of data, for which monthly means of all the necessary data are available. The results for this subset of the data (Table 5) are very similar to those shown in Table 4. Total nitrogen and Z_{eu}/Z_{mix} were not included in multiple regression analysis, as the simple analysis shown above proved these to be of no or little relevance.

Multiple regression analysis for the correlation between microcystin concentrations and total phosphorus including BV_{mcp} results in an adjusted R^2 (R^2_{adj}) of 0.69, which is not stronger than the R^2 of 0.71 from the simple regression analysis (Table 6). Further, whereas the R^2 for MC and BV_{mcp} is highly significantly different from zero ($p < 0.001$), the partial correlation between MC and TP ($p = 0.391$) is not. Both results indicate that TP has no additional direct influence on microcystin concentrations beyond the indirect influence of TP on MC through its influence on BV_{mcp}.

In contrast to these results for total phosphorus, multiple regression analysis for testing the additional direct influence of euphotic depth on microcystin concentrations results in a substantially higher correlation coefficient if Z_{eu} is included in the regression between MC and BV_{mcp} (Table 7). Furthermore, the coefficient for MC and Z_{eu} is also significantly different from zero, though at a lower level than for MC and BV_{mcp}. This result suggests a direct influence of Z_{eu} on microcystin concentrations in addition to the indirect influence through the biovolume of microcystin-producing cyanobacteria.

In summary, regression analysis of the field data obtained in the German survey supports the hypothesis derived from the physiological experiments (see Chapter 3) that microcystin concentrations are largely governed by the biomass of microcystin-producing cyanobacteria, although some direct influence of light, indicated by euphotic depth, cannot be ruled out. The envelope functions for microcystin concentrations over nutrient concentrations also support this, as maximal microcystin concentrations at a given nutrient concentration correspond to expectations derived from the carrying capacity for biomass (as estimated from Redfield ratios) and microcystin/biomass ratios.

Assessment of Study Designs

Field data would be expected to reflect an impact of environmental factors, i.e. growth conditions of cyanobacteria, on toxin content only if this impact is more substantial than the background noise in the data. One potential source of noise in our data was the differences in laboratory methods used by the different contributors of the data. In particular, both our study and those by Kotak et al. (1995 and 2000) are affected by the indirect calculation of microcystin concentrations from seston net samples and ratios to another parameter: we used microcystin/ chlorophyll-a ratios determined from aliquots of the same samples, and Kotak et al. (1995 and 2000) used a factor from the literature ratio between dry weight and chlorophyll-a, supported by their own data. An improved design for future studies would be the determination of microcystins L^{-1} by filtering samples of defined water volume.

The assessment of the potential impact of light availability on toxin content could be improved by using mean photosynthetically available radiation in the mixed upper water layer rather than euphotic depths estimated from Secchi depth readings, as the former parameter better includes global irradiation.

The most critical aspect of the design of field studies on the influence of environmental factors on toxin concentrations, however, is the question of what the results can tell us about the mechanisms in action. Our statistical approach differentiated between indirect effects through effects on the biomass of microcystin-producing taxa, and further direct effects that would be indicated by multiple regression analysis. This approach showed indirect effects through the biomass of microcystin-producing taxa to be overriding, with some indication of a direct effect of euphotic depth by the higher value for R^2_{adj} (see Table 7). For the latter, it could not separate physiological responses from population responses, i.e. distinguish between the impact on toxin production and cellular content versus the impact on strain composition. Thus, the potential direct influence of euphotic depth on microcystin content indicated could be due to both shifts in strain composition and changes in cellular microcystin content.

The approach necessary for addressing the direct physiological impact of environmental factors on cellular microcystin production and content would be the determination of both cell quotas for microcystins. If the environmental factors to be studied are nutrients, cell quotas for nutrients would also need to be assessed, as limitation of growth rates, which consequently causes changes in other cellular processes, depends on the amount of nutrients incorporated within the cells (i.e. the cell quotas). Both would need to be done on the level of strains rather than species.

On the level of species, average microcystin cell quotas for a given population can be determined if only one potentially microcystin-producing taxon is present, or if the microcystins present can be clearly allocated to individual dominant taxa (as may be the case in mixed populations of *Planktothrix* and *Microcystis* which contain different microcystin variants). Microcystin cell quotas may then be obtained by counting cells from aliquots of the samples used for microcystin determination, be they unconcentrated grab samples for filtration, or plankton net samples (see Fastner 1999 and Wiedner 2000).

For nutrients, cell quotas are usually more difficult to determine from field populations because they are not specific to individual species. Methods of TP and TN analysis would require separation of the particle mixture into fractions which represent species. Even for larger taxonomic groups of phytoplankton, such data are rarely available. Only in situations where the toxin-producing cyanobacterial taxon heavily dominates total seston, may the particulate nutrient fractions (i.e. total nutrient minus the dissolved fraction) roughly reflect cell quotas and thus serve as an approximate surrogate. Unfortunately for the cell-quota approach to field samples, such situations often occur particularly in hypertrophic conditions when light is limiting rather than nutrients.

On the level of strains, however, it is currently not possible to determine cell quotas for toxins or nutrients from field populations with mixed strains and species.

Thus, we may conclude that field data might be useful for answering the question of how environmental factors affect toxin production and cellular content once methods are developed that can differentiate between shifts in strain composition on the one hand, and changes in toxin quotas as well as nutrient quotas of a given strain on the other.

Conclusions

While improvements in study design are conceivable, current evidence suggests strain composition to be a key mechanism that determines microcystin concentrations in cyanobacterial field populations. Vezie et al. (1998) found microcystins in field samples of a shallow eutrophic lake on all occasions between the end of June and early October 1994, but total microcystin concentrations in the seston as well as patterns of microcystin variants varied between sites from not detectable at some sites to up to 2300 $\mu g\ g^{-1}$ dw at another. Interestingly, only 16 of 98 strains of *Microcystis aeruginosa* isolated from the lake during the 5-month study period contained microcystins, and some of these showed concentrations of up to 5060 $\mu g\ g^{-1}$ dw. The authors attribute the pronounced variability of microcystin/dry weight ratios over time to a "mosaic structure" of the populations, which consist of varying shares of strains with and without microcystins.

In conjunction with similar observations in our study (see Chapter 4.2), this suggests that strain composition and mechanisms of competition between strains are the overriding mechanism governing microcystin concentrations. This is further supported by the results of the studies on the effect of light and nutrients in continuous culture reported in Chapters 3.1 and 3.2 of this book, as well as by the data from physiological experiments reviewed by Sivonen and Jones (1999), all of which indicate that potentially growth-limiting resources influence the microcystin content of cyanobacterial cells by a factor of 2–3. Even if improved study designs, as outlined above, reduce the ratio of the resulting noise, it is unlikely that they will be able to pick up physiological responses of microcystin content to environmental variables in this range. This may be of minor importance in relation to the impact of strain composition, unless methods are found for assessing microcystin quotas in individual strains.

For future studies aiming at acquiring an understanding of the mechanisms which determine the microcystin content of field populations, the key issue to be investigated will therefore be: What governs the outcome of competition between strains with different microcystin content?

For practical purposes of management, problem prediction and worst-case estimation, however, the recognition of the biomass of microcystin-producing cyanobacteria as a key determinant of microcystin concentrations is of major importance. This implies that the currently available knowledge of phytoplankton ecology is of direct relevance for estimating the occurrence of microcystin producers. As long as factors leading to the dominance of strains containing more or less microcystin are not known, assessments of exposure risks will need to be based on worst-case estimates from the biomass of potentially microcystin-producing taxa, unless direct monitoring of microcystin concentrations is carried out.

The envelope functions shown here for microcystin concentrations in relation to Secchi disc transparency and to nutrients determining the carrying capacity for pelagic biomass, i.e. P and N, may be useful. While their further substantiation with additional data would be generally desirable, locally or regionally established statistical evaluations of microcystin concentrations in relation either to the biomass of toxic cyanobacteria or to environmental variables, based on experience over several years, may provide a valuable tool for the estimation of human health and environmental risks from these toxins.

Acknowledgements

We thank in particular our project partners who provided the data and patiently helped scrutinize the database for their correct transfer into it. Further, we are very grateful to Ursula Gaedke and Hannelore Liero for advice and support with the statistical approach.

References

Benndorf J, Henning M (1989) *Daphnia* and toxic blooms of *Microcystis aeruginosa* in Bautzen reservoir (GDR). Int Revue ges Hydrobiol 74:233–248

Böger P, Sandmann G (1993) Pigment biosynthesis and herbicide interaction. Photosynthetica 28:481–493

Carmichael WW, Gorham PR (1981) The mosaic nature of toxic blooms of cyanobacteria. In: Carmichael WW (ed) The water environment, algal toxins and health., Plenum Press, New York. pp. 161–172

Chorus I. and Mur L (1999) Preventive measures. In: Chorus I, Bartram J (eds) Toxic cyanobacteria in water: a guide to their public health consequences, monitoring and management. Published on the behalf of WHO by E & FN Spon, London, pp. 235–273

Codd GA, Bell SG (1985) Eutrophication and toxic cyanobacteria in freshwaters. Wat Pollut Control 84:21–30

Danowski A (1996) Beiträge zur Abundanzdynamik potentiell toxischer Cyanophyceen im Großen Wannsee und Tegler See. Dissertation at the Humboldt-University Berlin

Fastner J (1999) Microcystins (cyanobacterial hepatotoxins) in German fresh waters: Extraction, occurrence, and influence of environmental factors on microcystin production. Doctoral thesis at the Free University Berlin, Germany

Fastner J Erhard M, Carmichael WW, Sun F, Rinehart KL, Rönicke H, Chorus I (1999) Characterization and diversity of microcystins in natural blooms and strains of the genera *Microcystis* and *Planktothrix* from German freshwaters. Arch Hydrobiol 2:147–163

Fastner J, Heinze R, Chorus I (1995) Microcystin content, hepatotoxicity and cytotoxicity of cyanobacteria in some German waterbodies. Wat Sci Tech 32:165–170

Hanazato T (1996) Toxic cyanobacteria and the zooplankton community. In: Watanabe MF, Harada K, Carmicheal WW, Fujiki H (eds) Toxic microcystis. CRC Press, New York, pp. 79–102

Henriksen P (1996) Microcystin profiles and contents in Danish populations of cyanobacteria/blue-green algae as determined by HPLC. Phycologia 35:102–110

Henriksen P, Moestrup Ø (1997) Seasonal variations in microcystin contents of Danish cyanobacteria. Natural Toxins 5:99–106

Jungmann D, Benndorf J (1994) Toxicity to *Daphnia* of a compound extracted from laboratory and natural *Microcystis* spp., and the role of microcystins. Freshwat Biol 32:13–20

Jungmann D, Ludwichowski K-U, Faltin V, Benndorf J (1996) A field study to investigate environmental factors that could affect microcystin synthesis of a *Microcystis* population in the Bautzen Reservoir. Int Revue ges Hydrobiol 81:493–501

Kaebernick M, Neilan B, Börner T, Dittmann E (2000) Light and the transcriptional response of the microcystin biosynthetic gene cluster. Appl Enviornm Microbiol 66:3387–3392

Kohl JG, Börner T, Henning M, Schwabe W, Weihe A (1988) Plasmid content and differences in ecologically important characteristics of different strains of *Microcystis aeruginosa*. Arch Hydrobiol Suppl 80 (Algological Studies 50–53):195–201

Kotak BG, Lam AK-Y, Prepas EE, Kenefick SL, Hrudey SE (1995) Variability of the hepatotoxin microcystin-LR in hypertrophic drinking water lakes. J Phycol 31:248–263

Kotak BG, Lam AK-Y, Prepas EE, Kenefick SL, Hrudey SE (2000) Role of chemical and physical variables in regulating microcystin-LR concentrations in phytoplankton of eutrophic lakes. Can J Fish Aquat Sci 57:1584–1593

Lawton LA, Edwards C, Codd GA (1994) Extraction and high-performance liquid chromatographic method for the determination of microcystins in raw and treated waters. Analyst 119: 1525–1530

Lindholm T, Meriluaoto JAO (1991) Recurrent depth maxima of the hepatotoxic cyanobacterium *Oscillatoria agardhii*. Can J Fish Aquat Sci 48:1629–1634

Long BM, Jones GJ, Orr PT (2001) Cellular microcystin content in N-limited *Microcystis aeruginosa* can be predicted from growth rate. Appl Environ Microbiol 67: 278–283

Luukkainen R, Sivonen K, Namikoshi M, Färdig M, Rinehart KL, Niemelä SI (1993) Isolation and identification of eight microcystins from thirteen *Oscillatoria agardhii* strains and structure of a new microcystin. Appl Env Microcbiol 59:2204–2209

Luukkainen R, Namikoshi M, Sivonen K, Rinehart KL, Niemelä SI (1994) Isolation and identification of 12 microcystins from four strains and two bloom samples of *Microcystis* spp.: structure of a new hepatotoxin. Toxicon 32:133–139

Martin C, Oberer L, Ino T, König WA, Busch M, Weckesser J (1993) Cyanopeptolins, new depsipeptides derived from the cyanobacterium *Microcystis aeruginosa* PCC 7806. J Antibiot 46:1550–1556

Moore RE (1996) Cyclic peptides and depsipeptides from cyanobacteria: a review. Journal of Industrial Microbiology 16:134–143

Mur Lr, Schreurs H, Visser P (1993) How to control undesirable cyanobacterial dominance. In: Guissani G, Callieri C (eds) Strategies for Lake Ecosystems Beyond 2000. 5th International Conference on the Conservation and Management of Lakes, Stresa, Italy, C.N.R. Ist Ital Idrobiol 565–569

Park HD, Iwami C, Watanabe MF, Harada KI, Okino T, Hayashi H (1998) Temporal variabilities of the concentration of intra- and extracellular microcystin and toxic *Microcystis* species in a hypertrophic lake, Lake Suwa, Japan (1991–1994). Env Tox Wat Qual 13:61–72

Park HD, Watanabe MF, Harada KI, Nagai H, Suzuki M, Watanabe M, Hayashi H (1993) Hepatotoxin (microcystin) and neurotoxin (anatoxin-a) contained in natural blooms and strains of cyanobacteria from japanese freshwaters. Natural Toxins 1:353–360

Rapala J, Sivonen K, Luukkainen R, Niemelä SI (1997) Variation of microcystins, cyanobacterial hepatotoxins, in *Anabaena* spp. as a function of growth stimuli. App Environ Microbiol 64:2206–2212

Reynolds CS (1997) Vegetation Processes in the Pelagic: a Model for Ecosystem Theory. Excellence in Ecology. Ecology Institute, Oldendrof-Luhe, 371 pp.

Rohrlack T, Henning M, Kohl JG (1999) Does the toxic effect of *Microcystis aeruginosa* on *Daphnia galeata* depend on microcystin ingestion rate? Archiv für Hydrobiologie 146: 385–395

Rücker J, Kohl JG, Kaiser K (1995) Responses of carotenoids and chlorophylls to variations of growth-limiting factors in three filamentous blue-green algae. Algological Studies 77:51–65

Sano T, Kaya K (1995) A 2-amino-2-butenoic acid (Dhb)-containing microcystin isolated from *Oscillatoria agardhii*. Tetrahedron Letters 36:8603–8606

Schreurs H (1992): Cyanobacterial dominance - relations to eutrophication and lake morphology. Thesis at the Universtiy of Amsterdam, 198 pp.

Schwabe W, Weihe A, Börner T, Henning M (1988) Plasmids in toxic and nontoxic strains of the cyanobacterium *Microcystis aeruginosa*. Current Microbiology 17:133–137

Sivonen K, Namikoshi M, Luukkainen R, Färdig M, Rouhiainen L, Evans WR, Carmichael WW, Rinehart KL, Niemelä S (1994) Variation of cyanobacterial hepatotoxins in Finland. In: Codd GA, Jefferies TM, Keevil CW, Potter E (eds) Detection methods for cyanobacterial toxins. The Royal Society of Cemistry, Cambridge, pp. 152–154

Sivonen K, Jones J (1999) Cyanobacterial toxins. In: Chorus I, Bartram J (eds) Toxic cyanobacteria in water: a guide to their public health consequences, monitoring and management. Published on the behalf of WHO by E & FN Spon, London, pp. 41–112

Skulberg OM, Skulberg R (1985) Planktic species of *Oscillatoria* (Cyanophyceae) from Norway. Characterization and classification. Arch Hydrobiol Suppl 71 (Algolocigal Studies 38/39): 157–174

Stotts RR, Namikoshi M, Haschek WM, Rinehart KL, Carmichael WW, Dahlem AM, Beasely VR (1993) Structural mofifications imparting reduced toxicity in microcystins from *Microcystis* spp. Toxicon 31:783–789

Täuscher L (1980) Bemerkungen zu endo- und epiphytischen Algen von *Microcystis* Kütz. im Großen Müggelsee. Limnologica 12:313–314

Utermöhl H (1958) Zur Vervollkommnung der quantitative Phytoplankton-Methodik. Mitt int Verein theor angew Limnol 5:567–596

van der Westhuizen AJ, Eloff JN (1983) Effect of culture age and pH of culture medium on the growth and toxicity of the blue-green alga *Microcystis aeruginosa*. Z Pflanzenphysiol 110:157–163

Vasconcelos VM, Sivonen K, Evans WR, Carmichael WW, Namikoshi M (1995) Isolation and characterization of microcystins (heptapeptide hepatotoxins) from Portugese strains of *Microcystis aeruginosa* Kutz. emend Elekin. Arch Hydrobiol 134:295–305

Vasconcelos VM, Sivonen K, Evans WR, Carmichael WW, Namikoshi M (1996) Hepatotoxic microcystin diversity in cyanobacterial blooms collected in Portuguese freshwaters. Wat Res 30:2377–2384

Vezie C, Brient L, Sivonen K, Bertru G, Lefeuvre JC, Salkinoja-Salonen M (1997) Occurrence of microcystin-containing cyanobacteria in freshwaters of Brittany (France). Arch Hydrobiol 139:401–413

Vezie C, Brient L, Sivonen K, Bertru G, Lefeuvre J-C, Salkinoja-Salonen M (1998) Variation of microcystin content of cyanobacterial blooms and isolated strains in Lake Grand-Lieu (France). Microb Ecol 35:126–135

Watanabe MF, Watanabe M, Kato T, Harada KI, Suzuki M (1991) Composition of cyclic peptide toxins among strains of *Microcystis aeruginosa* (blue-green algae, canobacteria). Bot Mag Tokyo 104:49–57

Wicks RJ, Thiel PG (1990) Environmental factgors affecting the production of peptide toxins in floating scums of the cyanobacterium *Microcystis aeruginosa* in hypertrophic African reservoir. Environ Sci Technol 24:1413–1418

Wiedner C (1999) Toxische und nicht-toxische Cyanobakterien in Gewässern der Scharmützelseeregion: Ihr Vorkommen in Gewässern unterschiedlicher Trophie und Morphometrie und Steuermechanismen ihrer Dynamik in polymiktischen Flachseen. Doctoral thesis, Brandenburg Technical University, Germany

Welker M, Hoeg S, Steinberg C (1999) Hepatotoxic cyanobacteria in the shallow lake Müggelsee. Hydrobiologia 408/409:263–268

Woitke P, Martin CD, Nicklisch A, Kohl JG (1994) HPLC determination of liophilic photosynthetic pigments in algal cultures and lake water samples using a non-endocapped C18-RP-column. Fresenius J Anal Chem 348:762–768

Woitke P, Hesse K, Kohl JG (1997) Change in liophylic photosynthetic pigment contents of different *Microcystis aeruginosa* strains in response to growth irradiance. Photosynthetica 33:443–453

Zehnder A, Gorham PR (1960) Factors influencing the growth of *Microcystis aeruginosa* Kütz. emend. Elenkin. Canadian Journal of Microbiology 6:645–660

5 Cyanobacterial Toxicity and Human Exposure

In spite of numerous documented cases of animal deaths after drinking water with cyanobacteria, data on exposure of humans to cyanotoxins are scarce worldwide, usually because incidents of illness are not related to cyanotoxins as potential cause quickly enough in order to investigate whether they occur, and at which concentrations. Toxicological knowledge of the effects of cyanobacterial toxins on mammals is, however, sufficiently developed to indicate diverse health hazards qualitatively, and for some cyanotoxins, quantitative assessments are also possible. Chapter 5.1 provides an overview of the information currently available for hazard assessment. Chapter 5.2 assesses ambient microcystin concentrations found in recreational waters in Germany and relates them to concentrations likely to have a health impact.

Until recently, concern for human exposure to cyanotoxins has focused on microcystins, in part because more toxicological information is available for microcystin-LR than for any other cyanotoxin, but also because of its mechanism of inducing liver damage that may not readily be detected, but potentially may cause long-term harm (see Chapter 5.1). Chapter 5.3 compiles six studies addressing microcystin concentrations in drinking water and their removal through methods of treatment and abstraction. The bottom line of these studies is that removing cyanobacterial cells is the most effective method of avoiding microcystins in drinking water, whereas chlorination alone is not likely to ensure sufficient oxidative degradation. However, these studies scarcely address dissolved microcystins, and none of them addressed the valid concerns of toxin liberation through some treatment proceedures.

The case of a substantial outbreak of gastroenteritis in southern Sweden reported in Chapter 5.3 is one of the few incidents where – coincidentally – cyanobacterial population densities and microcystin concentrations had been studied shortly before and after an accidental false connection of river water severely polluted a village drinking-water supply. Thus, it is known that microcystin concentrations were probably only in the range of 1 $\mu g\ L^{-1}$, and the symptoms described for those who fell ill appear to be associated with other cyanobacterial components, possibly lipopolysaccharides. Together with the symptoms typically observed after recreational exposure (Pilotto et al. 1997), this result demonstrates that while concern about long-term liver damage through continued exposure to microcystins still holds, responses to cyanobacterial exposure with short-term illness are due to other substances they contain. Protection of human health therefore should not exclusively target microcystins, but also encompass the irritative and inflamatory effects of whole cyanobacterial cells.

5.1 Cyanotoxins and Human Health – Overview

D. James Fitzgerald

Introduction

Humans can be exposed to a range of cyanotoxins either contained in cyanobacterial cells or released into the water. In addition, the cyanobacterial cellular material has the potential to initiate a toxic response. This chapter will discuss the relevant toxicology of cyanobacteria and cyanotoxins from the viewpoint of the different modes of exposure that humans experience. Where human data are lacking, experimental and/or field animal information will be outlined in an effort to flag the possibility for toxicity to humans.

Reports of human cyanobacterial intoxication often do not meet all generally accepted epidemiological criteria. One major difficulty in assigning observed health effects to specific species or toxins is that often several cyanobacterial species coexist in water resources used by humans.

Ingestion of Cyanotoxins with Drinking Water

Exposure to toxic cyanobacteria via drinking water is the key concern for human health regarding both acute and chronic effects. In some countries cyanobacteria may dominate in water resources over extended periods of time and result in continued exposure to subacute concentrations (Ressom et al. 1994; Hitzfeld et al. 2000). Episodes of acute sickness have been reported after treatment of cyanobacterial blooms with copper sulphate to kill a cyanobacterial bloom, which then resulted in release of cyanotoxins into the water and breakthrough of dissolved toxins into drinking-water supplies. Exposure to cell material of any cyanobacterial taxon can cause illness such as fever, probably evoked by the lipopolysaccharides contained in the cell wall of cyanobacteria (Lippy and Erb 1976; Keleti et al. 1979). Other toxins are more specific to certain taxa (see also Chapter 2.1.2 and 4.1), and quite specific in their mechanism of toxicity. Drinking-water guideline values exist for one cyanotoxin, and guideline derivation for further cyanotoxins is progressing.

Microcystins

Human health effects associated with ingestion of water containing toxic *Microcystis* and/or other microcystin-producing genera are known from several reports. These effects range from elevated plasma liver enzyme levels suggesting liver damage (Falconer et al. 1983; de Olivera Araújo 1995), to gastroenteritis (Zilberg 1966; Annadotter et al., this Volume) and even death (Teixera et al. 1993). This range of effects is not unexpected in view of the numerous documented cases of toxicity and fatality from *Microcystis* in field and domestic animals (Ressom et al. 1994). In addition, studies in experimental animals and in vitro cell systems using *Microcystis* extracts and pure microcystins have provided much pathological and mechanistic detail (Carmichael 1994; Codd et al. 1997).

Microcystins enter the body from the intestine probably via the bile acid transporter system (Falconer et al. 1992; Ito et al. 2000). From the blood, microcystins then bind quite specifically to liver tissue where they form adducts with protein phosphatases 1 and 2A (Craig et al. 1996; Yoshida et al. 1998), resulting in inhibition of these important enzymes (Honkanen et al. 1990). In experimental animals, this occurs in the cytoplasm and nuclei of hepatocytes in the centrilobular regions (Yoshida et al. 1998), and is followed by intrahepatic haemorrhage, apoptosis, progressive liver necrosis, hepatomegaly and death if the dose is sufficient (Ressom et al. 1994). Apoptosis seen after oral dosing is considered one of the distinguishing features in microcystin-induced hepatotoxicity (Yoshida et al. 1997).

Oral administration in mice of single high doses of microcystin-LR revealed a 24 h LD_{50} between 5 mg kg^{-1} (Fawell et al. 1999) and 10.9 mg kg^{-1} (Yoshida et al. 1997). The latter study showed liver pathology changes similar to those known for intraperitoneal dosing. It was also noted that there were no abnormalities in the stomach, intestine, skin or other organs. In another oral microcystin-LR study in mice, it was seen that older animals were more susceptible to liver injury; this was attributed to age-related changes in the gastric and small-intestinal mucosa (Ito et al. 1997a, 2000).

Of considerable concern are the several lines of evidence indicating that microcystins (and nodularin) may pose a cancer risk to humans. In epidemiological studies in China, an elevated incidence of hepatocellular carcinoma in certain areas has been associated with the regular ingestion of *Microcystis*-contaminated surface water (Yu 1989, 1995). Although it is possible that these data are confounded by other liver cancer aetiological factors such as hepatitis B virus and dietary aflatoxin, they may also reflect promotion of tumour growth by microcystins after tumour induction by aflatoxins or hepatitis B virus..

Animal studies have indicated that, although microcystins appear not to be genotoxic, they may have carcinogenic potential (Ressom et al. 1994; Fujiki et al. 1996). The results of some of these studies, summarized in Table 1, have been interpreted as demonstrating a tumour-promoting activity. In vitro work with extracts of *Microcystis aeruginosa* has shown promotion of neoplastic cell transformation (Wang and Zhu 1996) which is an effect, along with experimental tumour promotion (Fitzgerald and Yamasaki 1990), elicited also by other protein phosphatase inhibitors (Katoh et al. 1990; Fujiki and Suganuma 1993).

Of particular relevance to evaluating carcinogenicity potential is chronic oral dosing of cyanotoxins in animal studies. To date, the few such reports have not shown tumour induction. Mice given microcystin-LR 100 times over 28 weeks did not develop liver neoplasms or show evidence of chronic injury (Ito et al. 1997b).

In mice initiated with N-methyl-N-nitrosourea, no primary liver tumours or promotion of lymphoid or duodenal tumours occurred after oral administration of *Microcystis* extract over 22 weeks (Falconer and Humpage 1996). An 18-month study in mice given microcystin-LR ad libitum in drinking-water at 20 μg L^{-1} (a total of 35 μg $mouse^{-1}$) showed no evidence of chronic toxicity or accumulation of the toxin in the liver (Ueno et al. 1999). While the search for signs of liver cancer has clearly been the focus of most research in this area, the low incidence of this tumour type and the high incidence of gastrointestinal cancer in many human populations has led to the suggestion that this latter site should be examined for susceptibility to microcystin-induced cancer (Falconer 1998). A recent initiation/promotion study in mice has shown that *M. aeruginosa* extract provided in drinking water increased the area of aberrant crypt foci in the colon, suggestive that microcystins promote preneoplastic colonic lesions (Humpage et al. 2000).

Drinking-Water Guidelines for Microcystins

In developing drinking-water guidelines for the protection of human health, WHO expert committees have evaluated the literature on microcystin toxicology. Comprehensive dose-response oral administration studies of liver toxicity in pigs and in mice revealed a lowest-observed-adverse-effect level (LOAEL; Falconer et al. 1994) and a no-observed-adverse-effect level (NOAEL; Fawell et al. 1994, 1999) to which various uncertainty factors have been applied, leading to a consensus life time consumption safe level of 1.0 μg microcystin-LR equivalents L^{-1} (WHO 1998; Falconer et al. 1999a; Kuiper-Goodman et al. 1999). Health Canada has proposed a value of 1.5 μg microcystin-LR L^{-1} on the basis of daily consumption over a full year (Health Canada 1998). For considerations of toxicity in the short term, a health alert of 10 μg microcystin-LR equivalents L^{-1} has been proposed (Fitzgerald et al. 1999).

Table 1. Experimental evidence suggesting that microcystins have carcinogenic activity

Species	Dosing	Finding	Reference
Rat	i.p., 6 weeks, 2 times per week, 1-50 μg MC-LR kg^{-1} after DEN initiation	Increased no. and size of liver GST-P foci	Nishiwaki-Matsushima et al. (1992)
Mouse	i.p., 28 weeks; MC-LR, 100 injections of 20 μg kg^{-1}	Persistent neoplastic liver nodules	Ito et al. (1997b)
Rat	i.p., 6 weeks, 2 times per week, 1 or 10 μg MC-LR kg^{-1} ± aflatoxin B1 initiation	Increased no. and size of liver GST-P foci; synergism	Sekijima et al. (1999)
Mouse	Oral, 7 weeks; *M. aeruginosa* extract after topical DMBA initiation	Increased no. and size of skin papillomas	Falconer (1991) Falconer and Buckley (1989)
Rat	Oral, 8 months; pond-ditch water residuals containing microcystins, ± aflatoxin B1 initiation	Increased no. of liver GGT foci; synergism	Yu (1995)
Mouse	Oral, 7 months; *M. aeruginosa* extract after i.p. azoxymethane initiation; 0.382 and 693 μg kg^{-1} day^{-1} of microcystins	Increased area of aberrant colon crypt foci	Humpage et al. (2000)

Nodularin

There is no unequivocal information on the toxicity of ingested nodularin and its parent cyanobacterium *Nodularia spumigena* to humans, though health problems through recreational exposure have been documented (see Chapter 5.1.4). However, toxicity to animals from ingestion of water containing *N. spumigena* is well known (Ressom et al. 1994), with the necrosis, haemorrhage and other effects in the liver being similar to those seen with microcystins (Runnegar et al. 1988). Nodularin itself, in intraperitoneal exposure studies in experimental animals, is at least as hepatotoxic as microcystin-LR (Carmichael et al. 1988). Mechanistically, nodularin binds to and inhibits the same protein phosphatases as microcystin-LR (Yoshizawa et al. 1990).

Carcinogenicity Considerations for Nodularin

A study involving the induction of liver preneoplastic foci by i.p. administration of nodularin after diethylnitrosamine initiation in rats indicates that this toxin may have liver tumour-promoting activity of at least the same level as that of microcystin-LR (Ohta et al. 1994). In a more recent rat study using a similar protocol, nodularin was shown to also induce proliferating cell nuclear antigen in hepatocytes and to promote preneoplastic enzyme-altered liver foci which enlarged and persisted for several weeks after cessation of nodularin exposure (Song et al. 1999).

Drinking-Water Guidelines for Nodularin

Generally, *N. spumigena* resides in brackish water but it is known to occur in freshwaters used for drinking-water abstraction (Baker and Humpage 1994; Heresztyn and Nicholson 1997). In view of the absence of oral studies using nodularin, and in view of nodularin's similar potency to microcystin-LR in other studies, it would be reasonable to adopt the same life-time drinking-water guideline value for nodularin, i.e. 1.0 $\mu g\ L^{-1}$. A short-term exposure health alert for nodularin of 10 $\mu g\ L^{-1}$ has also been proposed (Fitzgerald et al. 1999).

Cylindrospermopsin

In the only reported incident of sickness after ingestion of water which probably contained *Cylindrospermopsis raciborskii*, 139 children and 10 adults required hospital treatment (Byth 1980). This incident occurred in an Aboriginal community on Palm Island in North Queensland in 1979, and was later linked to the drinking of water from the island's Solomon Dam which had a history of algal bloom formation and which had been treated with copper sulphate to destroy a dense bloom just prior to the outbreak of illness (Bourke et al. 1983). The reported symptoms (headache, vomiting, painful liver enlargement, bloody diarrhoea, anorexia) and the results of blood and urine analyses indicated acute hepatoenteritis together with marked renal damage. While everyone eventually recovered, more than half of those affected had required intravenous treatment and several individuals suffered hypovolaemic/acidotic shock (Byth 1980).

Subsequently, *C. raciborskii* was obtained from the Solomon Dam and grown in the laboratory. Intraperitoneal administration of cell material to mice resulted in a toxic response characterized by diarrhoea, anorexia and irregular respiration, and death at certain doses (Hawkins et al. 1985). This and more recent work with an extract from another *C. raciborskii* isolate has further shown liver, kidney and small intestine haemorrhage and lung congestion, with histopathology revealing adrenal cortex necrosis, liver centrilobular necrosis and renal tubular epithelium damage (Hawkins et al. 1985, 1997).

Cylindrospermopsin, one of the toxicologically active substances in *C. raciborskii*, has been purified (Ohtani et al. 1992) and characterised in various ways. Acute mouse toxicity studies have shown a 5-6 day LD_{50} (i.p.) of 200 $\mu g\ kg^{-1}$ (0.51 $\mu mol\ kg^{-1}$) (Ohtani et al. 1992; Moore et al. 1998) which, compared on a molar basis to the mouse i.p.-LD_{50} of microcystin-LR (0.05 $\mu mol\ kg^{-1}$), is an order of magnitude less potent. Several groups have dosed mice intraperitoneally with purified cylindrospermopsin or *C. raciborskii* extract and reported on the progressive necrosis in liver and kidney (Harada et al. 1994; Tereo et al. 1994; Falconer et al. 1999b) though in one report liver damage only was noted (Chiswell et al. 1997). Moore et al. (1998) showed that mice which survive a near-lethal i.p. dose fully recover and display normal livers after 14 days.

Studies on the mechanism of action of cylindrospermopsin have shown that in mouse hepatocytes in vivo the toxin disrupts protein synthesis (Tereo et al. 1994), and in rat hepatocytes in vitro the toxin is metabolized by cytochrome P-450 whereupon a metabolite acts to inhibit glutathione synthesis; pharmacological inhibition of cytochrome P-450 protected against the toxicity of cylindrospermopsin (Runnegar et al. 1995). This is then a mechanism very different to the action of microcystins and nodularin.

Of particular relevance to human exposure, laboratory studies in mice have been recently conducted using oral administration of a *C. raciborskii* extract containing 0.2% cylindrospermopsin; the toxin was shown to be bioavailable by this route (Chiswell et al. 1997; Moore et al. 1998; Seawright et al. 1999). Animal lethality could be effected with a single gavage dosing of the aqueous extract, the median lethal dose being in the range of 4.4 to 6.9 mg toxin-equivalent kg^{-1} (Seawright et al. 1999). This oral administration resulted in the same liver pathology changes as those caused by i.p. dosing, with a NOAEL for repeated oral dosing of about 0.1 mg kg^{-1} (Moore et al. 1998). At minimally toxic doses the liver was the primary target organ with some evidence of spleen lymphophagocytosis, while at higher doses effects included thymic atrophy, renal ischaemic acute tubular necrosis, subepicardial and myocardial haemorrhage, adrenal cortex sinusoid congestion, and microscopic ulcerations of the oesophageal part of the gastric mucosa with moderate stomach wall oedema (Moore et al. 1998; Seawright et al. 1999); no pathological alterations were seen in the ileum and duodenum (Seawright et al. 1999).

This knowledge of the toxicity of cylindrospermopsin, together with the discovery that the toxin is produced also by other cyanobacterial species, namely *Umezakia natans* in Japan (Harada et al. 1994) and *Aphanizomenon ovalisporum* in Queensland and Israel (Moore et al. 1998), and the fact that toxic *Cylindrospermopsis* is widely distributed (Chiswell et al. 1997; Falconer 1998), are sufficient reasons for heightened public health concern. Though experimental evidence indicates that in acute tests cylindrospermopsin is slower acting and less toxic than microcystins, chronic testing is required to more fully appraise its toxicity and to assist in developing a drinking-water guideline value.

Cyanobacterial Neurotoxins – Anatoxins and Saxitoxins

The principal neurotoxins produced by cyanobacteria are anatoxin-a, anatoxin-a(s) and the saxitoxins (or PSP-toxins) which include saxitoxin, neosaxitoxin, the gonyautoxins and the C toxins. Anatoxin-a, anatoxin-a(s) and saxitoxin only will be considered here.

Anatoxin-a is a neurotoxin which acts by binding to and activating neuronal nicotinic acetylcholine receptors, and is about 100 times more potent than nicotine (Fawell and James 1994). Subsequent depolarization at neuromuscular junctions can lead to paralysis, asphyxiation and death in experimental animals. Numerous cases of fish, bird and domestic animal deaths have also been recorded in areas of neurotoxic cyanobacterial blooms (Ressom et al. 1994; Falconer 1998). In studies in rats, anatoxin-a was shown to stimulate the sympathetic nervous system via the release of catecholamines from nerve endings (Dube et al. 1996). The toxicity of anatoxin-a on a molar basis appears to be about 3 to 5% that of microcystin-LR, at least for acute toxicity as measured by i.p.-LD_{50} in mice (Fitzgeorge et al. 1994); the oral-LD_{50} is at least an order of magnitude greater than the i.p.-LD_{50} (Carmichael 1992; Fitzgeorge et al. 1994).

Of different molecular structure and mode of action to anatoxin-a is anatoxin-a(s), an organophosphate which inhibits acetylcholinesterase activity. There are currently no oral studies, but anatoxin-a(s) is considerably more toxic than anatoxin-a and has an i.p.-LD_{50} of 20 μg kg^{-1}, which is slightly lower than that of microcystin-LR (Matsunaga et al. 1989; Kuiper-Goodman et al. 1999).

Apart from anatoxin-a, the main cyanobacterial neurotoxin of concern is saxitoxin. This toxin and its analogue neosaxitoxin are extremely toxic, acting as blockers of sodium ion channels of the excitable membranes of nerves resulting in altered impulse generation in peripheral nerves and skeletal muscles (Catterall 1980). In experimental animals, the consequent nerve dysfunction can lead to paralysis, respiratory depression, violent spasms and convulsions, respiratory failure and death in 1 to 2 min; the i.p.-LD_{50} for saxitoxin and neosaxitoxin in mice is 5 μg kg^{-1} (Codd 1992; Kao 1993).

Saxitoxins are also produced by various dinoflagellate algae and are known to accumulate in certain shellfish which filter-feed on these algae; consumption of such shellfish can result in a range of health effects including muscular and respiratory paralysis and even death; hence the toxins have been termed paralytic shellfish poisons or PSP (Falconer 1993; Carmichael 1994). Though there are no reports of human health effects from ingestion of drinking water containing saxitoxin-producing cyanobacteria or PSP-producing dinoflagellates, many documented and often fatal PSP episodes as well as animal deaths from water containing saxitoxins have alerted to the potential for saxitoxin toxicity in humans (Ressom et al. 1994; Negri et al. 1995; Fitzgerald et al. 1999). Reversible symptoms can occur within 5 min and can involve nausea, vomiting, paraesthesia of the mouth and throat, dizziness and weakness, while paralysis and death can occur within 2 to 12 h (McFarren et al. 1960; Rodrigue et al. 1990).

Drinking-Water Guidelines for Cyanobacterial Neurotoxins

Presently, there are no appropriate experimental data on anatoxin-a or saxitoxin for use in preparing drinking-water guidelines. However, information on exposure and human health effects from reported cases of paralytic shellfish poisoning has been employed to develop a health alert for saxitoxin equivalents of 3 $\mu g\ L^{-1}$ (Fitzgerald et al. 1999). In Australia at least, where saxitoxins are produced by *Anabaena circinalis* (Humpage et al. 1994), this concentration is likely to be associated with cell densities greater than 20000 cells ml^{-1} (Fitzgerald et al. 1999).

Drinking-Water Guidelines for Cyanobacterial Cells

In addition to the attempts to develop drinking-water guidelines for individual cyanotoxins, guidelines have also been generated for cyanobacterial cell numbers. Based principally on taste and odour perceptions, an acceptable cell concentration of 2000 cells ml^{-1} (or 0.2 $mm^3\ L^{-1}$ biovolume or 1 $\mu g\ L^{-1}$ chlorophyll-a) had been used in Australia (Jones et al. 1993; NHMRC/ARMCANZ 1996). Based on the Australian approach, WHO recommends a comprehensive alert levels framework involving a vigilance level, two alert levels, and a range of monitoring and health risk assessment phases (Bartram et al. 1999).

Haemodialysis Exposure

Only one incident involving parenteral exposure of humans to cyanotoxins is well documented. This involved a group of renal dialysis patients in a haemodialysis centre in Caruaru, Brazil, who, in early 1996, were inadvertently provided with dialysate improperly prepared from water taken from a reservoir which was contaminated with toxic cyanobacteria (Pouria et al. 1998; Jochimsen et al. 1998). The reservoir has been found to harbour *Microcystis*, *Anabaena*, *Anabenopsis*, *Aphanizomenon* and *Oscillatoria* species (Pouria et al. 1998), each of which produces microcystins and/or neurotoxins.

All 126 patients became ill to various degrees and 60 of them eventually died, principally of liver failure; 6 had died by 2 weeks after exposure, 30 by 6 weeks, 44 by 10 weeks, and 55 by 27 weeks (Pouria et al. 1998). The details of the epidemiological, clinical, diagnostic and pathology study of the patients are shown in Table 2. It was

Table 2. Profile of cyanotoxin-exposed dialysis patients in the Caruaru haemodialysis poisoning incident. (After Pouria et al. 1998 and Jochimsen et al. 1998)

Epidemiological details:	
Total no. of dialysis patients	126 male
No. developing symptoms	126 (68 male, 58 female)
Evidence of liver injury	108 (at least)
No. of deaths	60
Median age of fatalities[a]	47 years
Median age of survivors[a]	35 years

Clinical, diagnostic and pathological details:
Initially, within hours - malaise, weakness and gastrointestinal disturbance (vomiting, nausea), upper abdominal discomfort and pain
Later - dizziness, confusion, fever, headaches, vertigo, tinnitus, mild deafness, massive tender hepatomegaly, jaundice; in severe cases - visual disturbance, blindness and grand mal convulsions
Laboratory tests - elevated serum hepatic enzymes, alkaline phosphatase, bilirubin, triglycerides
Biopsy/necropsy - acute toxic hepatitis involving necrosis, canalicular cholestasis, apoptosis, enlarged cell nuclei and multinucleation, neutrophil infiltration; no evidence of intrahepatic haemorrhage

[a] Determined at the time of the 50th fatality ($p < 0.001$ for age comparison, Jochimsen et al. 1998)

evident that these patients had suffered serious neurological and enterohepatic trauma, with the hepatotoxicity displaying many similarities to experimental animals injected with microcystin-LR (Ressom et al. 1994). However, it was considered that not all deaths of the dialysis patients could be directly attributable to microcystin toxicity (Pouria et al. 1998). It is possible that the pyrogenic reactions were caused by lipopolysaccharide endotoxin which is common to cyanobacteria and other algae. Such endotoxin from an unspecified algal species was the likely inducer of pyrogenic reactions reported in a group of haemodialysis patients exposed to dialysis water-contaminated with cyanobacteria (Hindman et al. 1975).

The detection of microcystin-LR, microcystin-YR and microcystin-AR in liver tissue and serum samples at concentrations which in experimental animals cause liver injury further implicated microcystins in the hepatotoxicosis (Pouria et al. 1998; Jochimsen et al. 1998). In a trial in one patient whose serum microcystin level was approximately 100 ng ml^{-1}, haemoperfusion was effective in markedly reducing this level (Pouria et al. 1998). Currently, the survivors are being routinely monitored for any evidence of long-term effects, such as hepatocellular carcinoma. A concern for DNA damage at high microcystin doses comes from a study in mice where lethal and near-lethal intraperitoneal administration of microcystin-LR-induced DNA unwinding and random DNA fragmentation in liver tissue (Rao and Bhattachary 1996).

A more detailed account of this human tragedy is provided by Pouria et al. (1998) and Jochimsen et al. (1998). This incident demonstrates clearly the toxicity of cyanotoxins to humans after intravenous exposure, and highlights the need for special surveillance of water sources which are utilized by haemodialysis clinics. This exposure route involves contact with over 100 L of water several times per week and thus may provide a dramatically higher dose than oral uptake through water ingestion or recreational activities.

Recreational Exposure

Recreational activities in or on water containing cyanobacteria result in exposure to the skin and can also involve the facial region which may lead to accidental ingestion, inhalation, intranasal and ocular contact. Exposure by these routes is also possible where contaminated domestic supplies are used for washing and bathing.

Many case reports exist of illness associated with recreational exposure to cyanobacteria, and some of these are summarised in Table 3. These instances involved exposure to predominantly hepatotoxic blooms, including *Microcystis* species and *Nodularia spumigena*. Similar symptoms have been reported for recreational exposure involving neurotoxic species (Dillenberg and Dehnel 1960; El Saadi and Cameron 1993; Ressom et al. 1994).

It is evident that recreational exposure to cyanobacteria can result in a range of symptoms including severe illness. Some of these symptoms are reminiscent of allergic reactions, and studies have suggested an immunological basis for such responses to cyanobacteria (Ressom et al. 1994).

Interestingly, the symptoms reported after recreational exposure may be interpreted as responses of inner or outer body surfaces to irritation of varying degree, possibly in part due to lipopolysaccharides contained in the cell walls of all cyanobacteria. The study by Pilotto et al. (1997; see also summary in Chapter 5.2) demonstrated that symptoms were related to exposure to cyanobacteria as such, but showed no relationship to the concentrations of microcystins or neurotoxins. This would be expected because – as described above – these substances have completely different

Table 3. Examples of incidents of recreational exposure to cyanobacteria and related health effects

Activity	Effects	Reference
Swimming	Headache, nausea, muscular pains, painful diarrhoea	Dillenberg and Dehnel (1960)
Canoe training and rolling, swimming	Vomiting, diarrhoea, central abdominal pain, blistering of lips, sore throat; severe pneumonia in 2/20 cases	Turner et al. (1990)
Showering, bathing	Skin rashes, sore red eyes, sore throat, hay fever, asthma	Soong et al. (1992)
Swimming	Diarrhoea, vomiting, flu symptoms, skin rashes, mouth ulcers, fevers, skin and eye irritation	Pilotto et al. (1997)

mechanisms of toxicity and target organs. Thus, the assessment of health effects of recreational exposure to cyanobacteria must differentiate between these general irritative and/or immunological effects of cyanobacterial cells and the systemic effects of toxins ingested, resorbed by body fluids and transported to target organs such as the liver (for microcystins) or the nervous system (for the neurotoxins). Recreational water users need to be protected from both.

Of interest also, since cyanobacterial blooms commonly comprise both hepatotoxic and neurotoxic species, is a study of intranasal exposure in mice showing that microcystin-LR and anatoxin-a can act synergistically to induce lethality (Fitzgeorge et al. 1994).

WHO Recreational Water Guidelines

Based principally on the results of the epidemiological survey by Pilotto et al. (1997) of health effects associated with cyanobacterial exposure during recreational water-related activities, a guideline value of 20000 cyanobacterial cells ml^{-1} has been developed by a WHO expert group (Falconer et al. 1999a). At this cell concentration

Table 4. WHO guidelines for safe practice in managing recreational waters containing cyanobacterial cells. (Falconer et al. 1999a)

Guidance level or situation	How guidance level derived	Health risks	Recommended action
Cyanobacterial scum formation in bathing areas	Inference from oral animal lethal poisonings Actual human illness case histories	Potential for acute poisoning Potential for long-term illness with some cyanobacterial species Short-term adverse health outcomes, e.g. skin irritations, gastrointestinal illness	Immediate action to prevent contact with scums; possible prohibition of swimming and other water-contact activities Public health follow-up investigation Inform relevant authorities
100000 cells cyanobacteria ml^{-1} or 50 μg chlorophyll-a L^{-1} with dominance of cyanobacteria	From provisional drinking water guideline for microcystin-LR, and data concerning other cyanotoxins	Potential for long-term illness with some cyanobacterial species Short-term adverse health outcomes, e.g. skin irritations, gastrointestinal illness	Watch for scums Restrict bathing and further investigate hazard Post on-site risk advisory signs Inform relevant health authorities
20000 cells cyanobacteria ml^{-1} or 10 μg chlorophyll-a L^{-1} with dominance of cyanobacteria	From human bathing epidemiological study	Short-term adverse health outcomes, e.g. skin irritations, gastrointestinal illness, probably at low frequency	Post on-site risk advisory signs Inform relevant authorities

only relatively mild adverse health effects would be expected. For microcystin-producing species, this cell concentration could equate with microcystin levels of about 2 to 10 μg L^{-1}, i.e. up to 1 order of magnitude greater than the life-time drinking-water guideline value. In instances of cyanobacterial cell concentrations exceeding 100000 cells ml^{-1} or where scum formation is evident in bathing areas, moderate to severe health risks are considered likely, potentially also through microcystins, and access to such localities should be restricted to avoid scum contact (Falconer et al. 1999a). The WHO guidelines for safe practice in managing recreational waters containing cyanobacteria are summarized in Table 4.

Sensitive Subpopulations

From the preceding discussion, there are indications that certain groups of people may have a particular sensitivity to some of the cyanotoxins. Since microcystins specifically target the liver, those populations already at risk of liver disease may be at even higher risk if microcystins exist regularly in their drinking water. Infection with hepatitis B virus and consumption of aflatoxin-contaminated food has already been mentioned (and possible synergism), while liver dysfunction or damage incurred by alcohol is another precondition likely to enhance sensitivity to cyanotoxins. Previously, mention was made of allergic-type reactions to cyanobacteria, thus people with general hypersensitivity may be predisposed to an allergic response to cyanobacterial dermal or inhalational exposure. Finally, from allometric principles, children may be more prone to cyanobacterial toxicity because their water intake kg per body weight is greater than that of adults. On the other hand, there is the interesting observation from experimental studies in mice that absorption of microcystin-LR from the gut appears to occur to a greater extent in older animals (Ito et al. 1997a, 2000); the relevance of such data to humans is not known.

Treatment of Intoxication from Cyanotoxins

Various approaches exist, though mostly experimental, for alleviating the effects of cyanotoxin intoxication. For the microcystins at the level of absorption from the gut, inhibitors of the bile acid transporter system are able to antagonize toxicity. The immunosuppressant cyclosporin-A and the antibiotic rifampicin are the most effective such inhibitors, and have been shown to protect rats and mice against microcystin-induced lethality (Hermansky et al. 1990a, b; Dawson 1998). Monoclonal antibodies to microcystin-LR have been developed and shown to protect mice against microcystin-LR-induced toxicity when coadministered intraperitoneally (Nagata et al. 1995). A potential therapeutic approach is based on the finding that microystins enhance the intracellular generation of reactive oxygen species in rat hepatocytes in vitro, leading to oxidative damage; desferrioxamine significantly decreased reactive oxygen species generation caused by exposure to cyanobacterial extract (Ding et al. 1998). Other antidote approaches have been researched (Ressom et al. 1994), but of those showing promise the majority have required prior or coexposure to microcystin-LR. However, some studies with rifampicin in mice have shown a measure of protection against microcystin-LR-induced lethality and liver damage for rifampicin applied after microcystin-LR (Hermansky et al. 1990b). In the event of an actual human intoxication circumstance, the use of haemoperfusion led to a dramatic and rapid reduction in the serum level of microcystins. However, the subsequent death of the patient suggests that this technique may be more successful if used at an early stage of toxin exposure.

In regard to the neurotoxic cyanotoxins, their extremely rapid action would render ineffective any postintoxication remedy. However, recovery from sublethal dosing may be assisted by artificial respiration, lavage and use of activated carbon to reduce absorption (Beasley et al. 1989).

5.2 Recreational Exposure to Cyanotoxins

Ingrid Chorus and Jutta Fastner

A feature of cyanobacterial toxicity important for health risk assessments is the observation that cyanotoxins are usually contained within the cells, and that concentrations dissolved in water are rarely reported above a few μg L^{-1} (see Chapter 2 and Sivonen and Jones 1999). This has substantial implications for the assessment of exposure through swimming and other water sports: people are most likely to be exposed to hazardously high cyanotoxin concentrations where cell densities are high, particularly in surface scums.

Cyanotoxins such as microcystins, neurotoxins or cylindrospermopsin contained in the cells may reach target organs in the human body (such as the liver or the nervous system) if the cells are accidentally ingested or aspired. This is relevant particularly for sports with intensive water contact such as sailboarding, water-skiing and competitive sailing in stormy weather. Swimmers can scarcely prevent nasal and oral uptake of water if they immerse their head, and exposure estimates in recreational water toxicology are often based on the assumption of an average uptake of 100 ml per swim. Children are particularly at a risk of ingesting major amounts of cyanobacterial cells when playing in the shallow zones along the shore where scums accumulate.

Dermal exposure, particularly if cells are accumulated under swimsuits and wet suits, may lead to skin irritations and allergic reactions. Symptoms involving irritation of internal and external mucous membranes, i.e. gastro-intenstinal or respiratory organs, eyes, ears, mouth and throat are also reported. In spite of ample anecdotal evidence from many countries, scientific substantiation of these reports is poor. This is due to the high costs of prospective epidemiological studies with sufficiently large groups of people and sufficiently well-controlled exposure situations to allow association of cause and effect. The design for such prospective studies of recreational exposure is piloted by the epidemiological work of Kay et al. (1994) on risk of infectious disease through exposure during swimming at marine beaches. For recreational exposure to cyanobacteria, one prospective epidemiological investigation is available. It involves 852 persons exposed during swimming and recorded symptoms such as diarrhoea, vomiting, flu symptoms, skin rashes, mouth ulcers, fevers, eye or ear irritations occurring within 7 days after exposure (Pilotto et al. 1997; see Box 1 for a summary). Of the 295 persons who had contact with cyanobacteria in the water on the day of the study and had neither engaged in water activities nor shown symptoms in the week before exposure, the odds ratio (i.e. likelihood) for illness after exposure was threefold higher than in the control group (43 persons), and it was statistically significantly related to duration of exposure and cyanobacterial cell density. Although the rate of symptoms reported in this study correlated with cyanobacteral cell density, it showed no relationship to the concentration of microcystins. Further, the above-mentioned symptoms reported after exposure to cyanobacteria are not typical for any of the known cyanotoxins other than – possibly – the lipopolysaccharides. However, the toxicology of cyanobacterial lipopolysaccharides has scarcely been studied (see Chapter 1).

Thus, recreational exposure to cyanobacteria should be assessed with respect to the possibility of health hazards both through dermal exposure to cyanobacterial cells and their poorly understood irritative and/or allergenic components as well as through systemic exposure due to accidental ingestion and aspiration of cell material.

Box 1. Health effects of recreational exposure to cyanobacteria

Louis Pilotto[1], Mike Burch, Robert Douglas, Scott Cameron, Graham Rouch, Christine Cowie, Mary Beers, Priscilla Robinson, Martyn Kirk, Sean Hardiman, Clare Moore, Robyn Attewell
[1] National Centre for Epidemiology and Population Health, Australian National University

The objectives of this study were to investigate health effects of exposure to cyanobacteria as a result of recreational water activities, and to examine dose-response relationships. We conducted the study in Australia during the summer of 1995 on Lake Alexandrina in South Australia, on the Hawkesbury River in New South Wales, and on two lakes in Victoria. Based on previous experience, algal blooms were expected to occur at these sites during summer. Interviewers attended these sites on various Sundays and invited all people in attendance, aged over 6 years old, to participate. A structured questionnaire was used by the interviewers to determine the health status and recreational water-related activities, including the duration of water contact, of each participant both for the day of interview and in the previous week. At the time of interview, appointments were made for telephone follow-up 7 days later. At follow-up, any diarrhoea, vomiting, flu symptoms, skin rashes, mouth ulcers, fevers, eye or ear irritations since the initial contact were recorded. Swimming and water activities subsequent to the initial interview were also recorded. On the Sunday of the initial interview, water samples were collected from each of the sites, according to a predetermined sampling protocol, and sent to a nominated State laboratory for cyanobacterial counts and toxin analysis.

Out of a total of 1029 people approached by interviewers, 921 (90%) agreed to participate. Approximately two thirds of participants were male, a pattern that was fairly consistent across all age groups. The majority of subjects were aged between 20 and 29. The higher male to female ratio was also consistent across all of the study sites. Ninety-three % of people took part in the follow-up.

A subset of 338 of these participants did not experience any illness or engage in any water activities in the week prior to interview. Of these, 43 did not have water contact on the day of interview and were considered unexposed. The 295 who had water contact were exposed to caynobacteria.

There was a significant trend to increasing symptom occurrence between the unexposed, those exposed for up to 60 minutes and those exposed for more than 60 min ($p = 0.03$). A significant trend to increasing symptom occurrence was also found between unexposed subjects, and those exposed to water with cell counts of less than 5000, 5000-20000, 20000-80000 and >80000 cells ml^{-1} ($p = 0.04$). Participants exposed to more than 5000 cells ml^{-1} for more than 1 h experienced a statistically significantly higher symptom occurrence rate than the unexposed, and a significant trend ($p = 0.004$) was found for duration combined with cell densities above and below 5000 cells ml^{-1} (Fig. 1). Symptoms were not correlated with the presence of microcystins. These results suggest that increasing symptom occurrence was associated with increasing duration of contact with water containing cyanobacteria, and with increasing cyanobacterial cell density. The findings suggest that the currently accepted threshold in Australia for exposure of 20000 cells ml^{-1} may be too high.

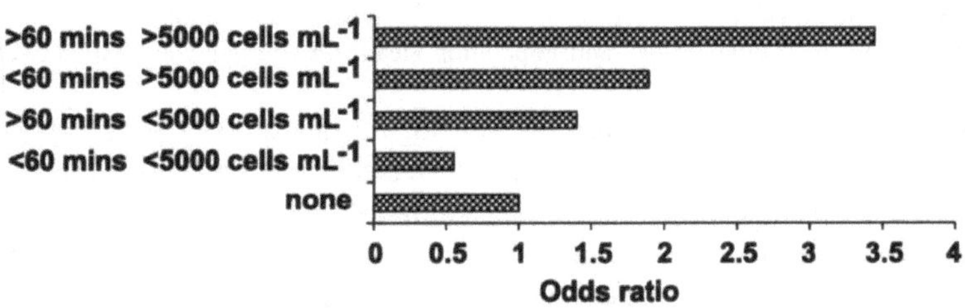

Fig. 1. Increase in odds ratio for illness after exposure to cyanobacteria in relation to duration of exposure and cell density

Potential for Recreational Exposure

A key question for recreational exposure is whether cyanobacteria are sufficiently toxic to threaten human life. This appears reasonable on the basis of the numerous animal deaths through ingestion of scum material. Nonetheless, cases of human fatalities through recreational exposure have not been documented. One approach to estimating the risk is calculation of a worst-case scenario from toxicological data and data on cyanotoxin concentrations, as given in Table 1 using the example of microcystins. It applies equally to other cyanotoxins with similar values for oral LD_{50} and similar cell-internal concentrations. This calculation is based on extreme assumptions, each of which, however, is in itself not unreasonable:

- *Microcystis* populations containing chiefly microcystin-LR have been found (Chapter 4.1), and some other microcystins frequently encountered are equally as toxic as microcystin-LR in intraperitoneal mouse assays (e.g. microcystin-YR found in *Microcystis*).
- Though it is possible that humans are less sensitive than mice, the LD_{50} gives only the mean death rate of the population, and sensitive individuals are likely to show lethal responses at lower concentrations.
- The most unlikely of the assumptions is the total displacement of water by cyanobacterial cells. Volumetric ratios of cyanobacteria to surrounding water in scums have not been investigated. Very dense scums are reported, and the author's observations include two waterbodies in which scum material showed gelatinous consistency.

Finally, the question remains whether recreational activities are likely to occur in such situations. One health authority in northern Germany reported that old people in the region still remember having been told as children to avoid swimming in lakes with greenish water blooms, but that this knowledge has become lost during the past decades, as the public has generally become accustomed to blooms as a meanwhile normal phenomenon. Indeed, intensive bathing activity has been frequent in spite of scums in regions of Germany where the population was uninformed about the potential cyanotoxin hazard and unaffected waterbodies are not available nearby. Summer camps for children and public campsites are not infrequently located at waterbodies with heavy cyanobacterial blooms, particularly in the northern lowland regions of Germany, and intensive water contact occurs over periods of days to weeks.

For assessment of the actual exposure to scums, blooms of *Microcystis* spp. were studied at numerous shoreline sites along the Havel River in Berlin during the summer of 1997 (Fastner et al. 1999). Concentrations of cell-bound microcystins were highly variable (depending on location of scums), and maxima reached more than 10 mg L^{-1} (Fig. 2). Half of these 28 samples showed concentrations of more than 100 $\mu g\ L^{-1}$ cell-bound microcystins, eight of these samples ranged between 100 and 1000 $\mu g\ L^{-1}$, four

Table 1. How much scum could kill a small child? – a theoretical worst-case estimate

Assumptions
• All the microcystins in the scum material are as toxic as microcystin-LR • Acute oral toxicity of microcystin-LR in humans equals that in mice (LD_{50} 5 - 10.9 mg kg^{-1} body weight; Kuiper-Goodman et al. 1999; see also Chapter 5.1) • All the water in scum material has been displaced by the cyanobacterial cells • The toddler exposed weighs 10 kg; thus, 50 mg could be a lethal dose
Data
• Microcystins were found in cells at concentrations up to 3 μg per mm^3 of biovolume, corresponding to 3 mg ml^{-1} of cell material (Chapter 2.1.2).
Consequence
➔17 ml pure cell material may be lethal for a small child.

were between 1000 and 10000 μg L-1 and two samples contained 10500 and 24000 μg L-1. Cell density in these samples was high: measured in terms of chlorophyll-a concentrations, it reached 107 and 156 mg L-1 in the two densest samples. Toxin content of the cells during this bloom in the Havel River was only moderate: the ratio

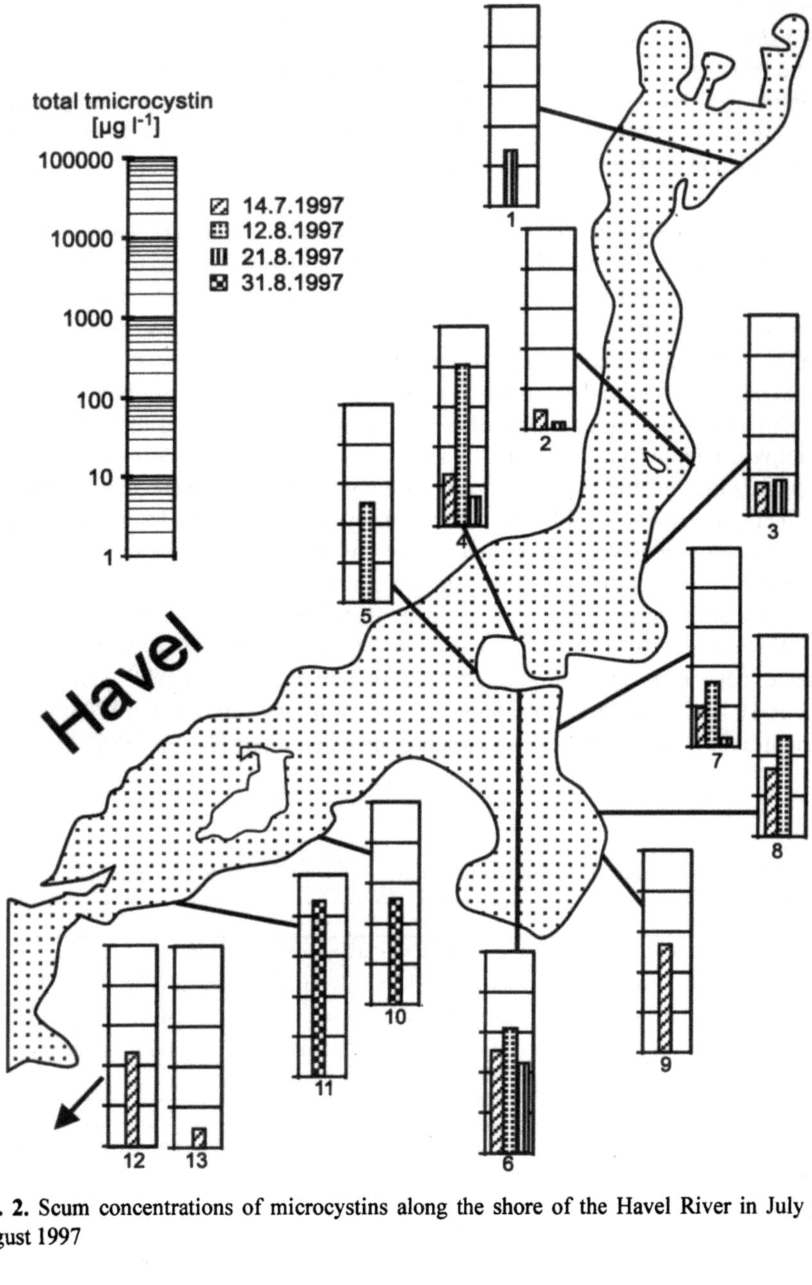

Fig. 2. Scum concentrations of microcystins along the shore of the Havel River in July and August 1997

between microcystins and chlorophyll-a was usually in the range of 0.1 to 0.2, whereas it may reach 1 in *Microcystis* populations with very high microcystin content (see Chapter 2.1.2). The consistency of some of these samples resembled that of pea soup.

At the maximum scum concentration of 24 mg L^{-1} in Fig. 2, a toxic dose for an averagely sensitive 10-kg toddler would require swallowing approximately 2 L of water (Table 2). If a scum is fivefold more toxic (which would be within the range of microcystin/chlorophyll ratios of up to 1 reported by Fastner et al. 1999), 0.4 L could be lethal to a child of 10 kg. If, in addition, cell density of *Microcystis* in the scum is also fourfold higher, 0.1 L could be lethal.

In addition to the possibility of lethal intoxications, there is evidence for health hazards through repeated or chronic exposure to microcystins at sublethal concentrations. Microcystins are currently discussed as tumour promoters (see Chapter 5.1), and cumulative liver damage in consequence of repeated exposure has been demonstrated (Fitzgeorge et al. 1994). Studying effects of recreational exposure to microcystins would require blood samples to investigate serum liver enzyme levels which reflect liver damage, and this has not been undertaken.

For half of the samples taken along the Havel River shoreline during the summer of 1997 (Fig. 2), ingestion of 100 ml by a child of 10 kg would result in a dose of 10-240 µg kg^{-1}. This is in the range of the NOAEL and LOAEL observed in animal experiments (Fawell et al. 1994; Falconer et al. 1994) and leaves no margin for variations in sensitivity between individuals and species after several weeks of exposure. The tolerable daily intake derived from these animal experiments by the World Health Organization was attained already at very low oral uptake amounts – corresponding to the ingestion of 0.2 and 4 ml by a small child for half of the Havel River shoreline samples investigated in 1997. Figure 3 illustrates the fact that this exposure situation is not unique to the Havel River, but rather common at various other sites in Germany.

Table 2. How much scum could cause heath impairment in a small child? – derivations from concentrations measured along the Havel River in July and August 1997

	Water volume to be ingested for corresponding dose					
Criterion	At 100 µg L^{-1} = 0.1 µg ml^{-1}		At 1000 µg L^{-1} = 1 µg ml^{-1}		At 25000 µg L^{-1} = 25 µg ml^{-1}	
	kg^{-1}	For 10-kg child	kg^{-1}	For 10-kg child	kg^{-1}	For 10-kg child
LD_{50}[a]: 5000 µg kg bw^{-1} (Chapter 5.1)	50 L	500 L	5 L	50 L	0.2 L	2 L
TDI[b]: 0.04 µg kg bw^{-1} (WHO 1998)	0.4 ml	4 ml	0.04 ml	0.4 ml	0.0016 ml	0.016 ml

[a] LD_{50} concentration acutely lethal to half of a population of mice in the bioassay.
[b] *TDI* tolerable daily intake, calculated by applying uncertainty factors to the no observed or lowest observed effect levels (NOAEL and LOAEL) in several-week exposure experiments. *bw* body weight

Whereas *Microcystis* spp. cause the highest microcystin concentrations in scums, the widespread filamentous species *Planktothrix agardhii* is notorious for the high filament density it attains in shallow eutrophic water-bodies. These may correspond to several 100 μ L^{-1} chlorophyll-a at transparencies of less than 0.5 m (Secchi disc reading), and microcystin concentrations then may also reach several 100 μ L^{-1}. In contrast to *Microcystis* scums, which are very variable in space and time, high population densities of *Planktothrix agardhii* may be a very persistent phenomenon for weeks on end.

Repeated exposure to scum material may occur during a holiday at one of the many campsites and resorts located along eutrophic waterbodies, or if residents go swimming regularly during the summer. At these concentrations, it could lead to a dose in the range of the dose that caused liver damage in animal studies, with no safety margin, and liver damage in the exposed persons is possible. If humans are more sensitive to microcystins than mice or pigs, if a toddler is more sensitive than the average human being, and if long-term effects (such as tumour promotion) are not sufficiently accounted for in the available studies, liver damage is even more likely.

Cyanobacterial neurotoxins appear to be less relevant in the context of recreational exposure, as neither their toxic mechanisms nor long-term animal experiments indicate them to have chronic effects (Kuiper-Goodman et al. 1999); thus, concentrations well

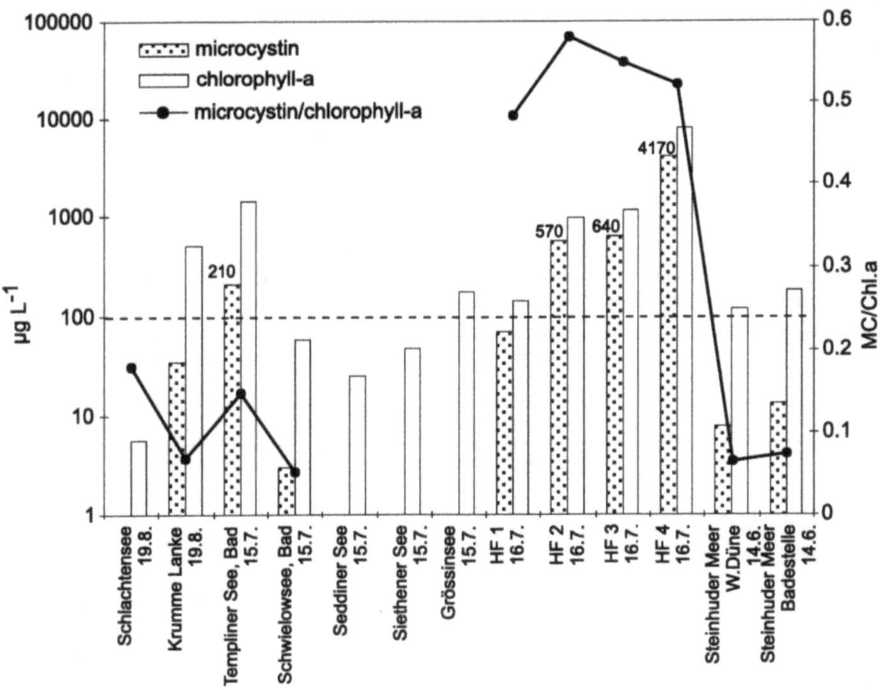

Fig. 3. Scum concentrations of microcystins and chlorophyll a at different sites in Germany in 1997. *Dotted line* German guideline for recreational waters of 100 μ L^{-1} of microcystins

below those causing acute effects may not induce symptoms. The few studies worldwide on their occurrence suggest them to occur less frequently and often only in low concentrations (see Chapter 2.3).

These scenarios lead to two conclusions:

1. Acute intoxications with microcystins with lethal outcome are not impossible at the concentrations encountered in waterbodies used for recreation in spite of cyanobacterial blooms, but they are not very likely.
2. Liver damage through microcystins (with uncertain long-term outcome) is not unlikely if people are exposed to high microcystin concentrations when swimming in scum material, particularly if exposure occurs repeatedly. This is most relevant for children who play extensively in shallow beach areas where the thickest scums tend to accumulate.

Policy for protection of people using recreational areas affected by cyanobacteria should therefore address the risk of liver damage through microcystins because this may be a substantial health hazard. It should further address the general irritative effects of cyanobacteria documented in the study of Pilotto et al. 1997.

Protection Against Recreational Exposure to Cyanotoxins

The Federal Environmental Agency of Germany issued a recommendation to the local authorities responsible for safety and surveillance of bathing sites which addresses both exposure to cyanobacteria in general and specifically exposure to microcystins (Fig. 4). In face of the lack of regulations for recreational waterbody use on the national level, the European Union Directive on the Quality of Bathing Water (EC 1976) was used as point of access. It requires transparency of 2 m (measured with a Secchi disc) as guide value and 1 m as imperative value. Almost half of the sites officially acknowledged as bathing waters in Germany occasionally or regularly do not comply to 1 m Secchi transparency, often because of algal and cyanobacterial proliferation. In many of the shallow recreational waters in Germany, these are not scum-forming cyanobacteria, but rather the homogeneously dispersed *Planktothrix agardhii* populations, which may attain extremely high filament densities, reducing Secchi transparency to 0.3 m or less. In these situations microcystin concentrations may reach levels of 100 $\mu g\ L^{-1}$ or more, even without scum formation (see Chapter 2.1.2).

The cause of cyanobacterial proliferation is eutrophication, in most cases with phosphorus. Thus, for waterbodies with transparencies < 2 m the recommendation for the protection of bathers from toxic cyanobacteria (Bundesgesundheitblatt 1997) suggests assessment of the waterbody's carrying capacity to support cyanobacterial proliferation, thus excluding those with low potential for biogenic turbidity.

The phosphorus threshold for cyanobacterial proliferation depends to some extent upon the size and depth of waterbodies (see Chorus and Mur 1999); in northwestern Europe, concentrations above 20-40 $\mu g\ L^{-1}$ total P are usually required for substantial cyanobacterial blooms. The recommendation therefore suggests that for waterbodies with sufficient carrying capacity for cyanobacterial blooms routine monitoring should be extended to cover cyanobacteria, qualitatively by rapid determination of dominant groups using the microscope, and quantitatively by using the concentration of chlorophyll-a as indicator of phytoplankton biomass. If cyanobacteria are dominant and chlorophyll-a concentrations exceed 40 μg L^{-1}, publishing of warning notices and investigations for remedial measures are suggested (Fig. 3). If chlorophyll-a concentrations exceed 150 μg L^{-1} and cyanobacteria are dominant, temporary closure of the bathing site is recommended. Optionally, microcystin concentrations may be

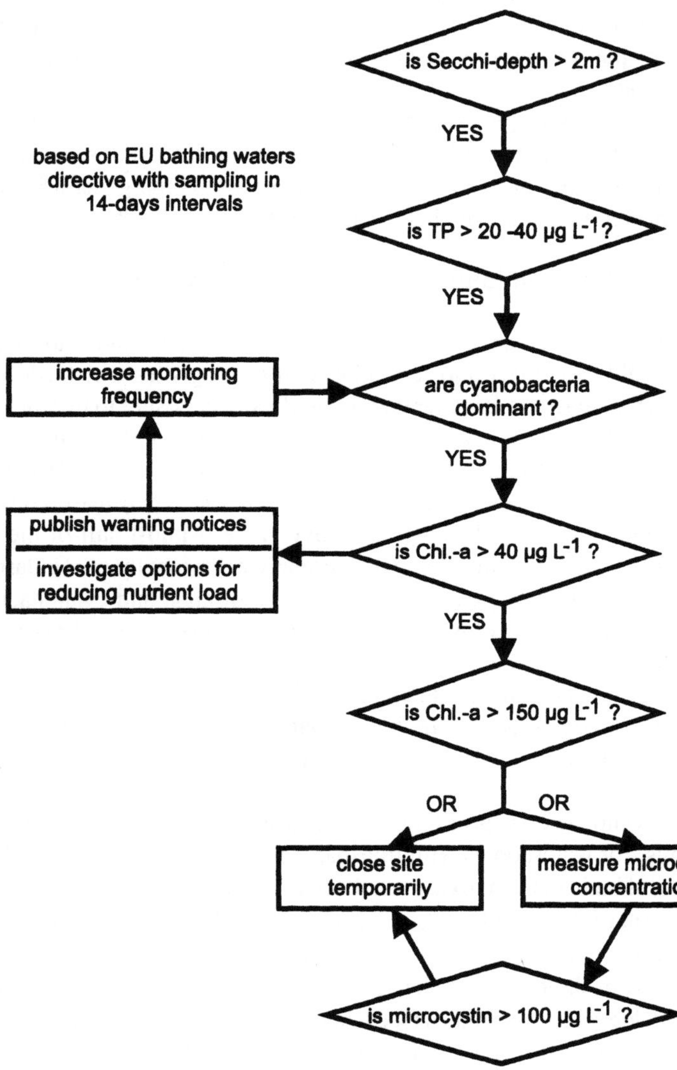

Fig. 4. Structured approach for protection of bathers from toxic cyanobacteria in Germany (Bundesgesundheitsblatt 1997). *TP* total phosphorus

determined. Bathing sites with more than 150 µg L^{-1} chlorophyll-a, but, for example, a bloom of *Aphanizomenon* without microcystins could remain open, but warning notices should be published informing users of the risk of dermal, gastro-intestinal, respiratory, eye or ear symptoms. Sites with microcystin concentrations above 100 µg L^{-1} (sum of all variants) should be closed until the bloom has declined.

The level of 100 µg L^{-1} was considered appropriate in comparison to the WHO drinking-water guideline value of 1 µg L^{-1} because (1) ingestion will be in the range of 100 ml rather than 1-2 L per person and day, and (2) exposure is transient (the cyanobacterial bloom season as well as the bathing season lasts only for weeks to a few months in this part of Europe) rather than life-long. A different toxicological derivation arriving at the same result is given by Fromme et al. (2000).

Current experience with the recommendation is that health authorities responsible for monitoring recreational sites are increasingly aware of cyanobacteria as a concern. Some of the 16 states of Germany have initiated regional programs to assess the situation and to explore options for including cyanobacteria in their routine monitoring. In consequence of press releases on toxin concentrations found at recreational sites, public awareness has also developed in some regions of the country.

Discussions of an expert group for the development of WHO Guidelines followed a slightly different approach, distinguishing between three levels of hazard (see Table 4 in Chapter 5.1): scums (corresponding to > 150 µg L^{-1} Chl.-a or > 100 µg L^{-1} of microcystins in the German recommendation), cell densities in a range in which cell-bound toxins may be a cause for concern (the level of 50 µg L^{-1} Chl.-a is similar to the German level of 40 µg L^{-1} which triggers information and warning), and cell densities in the rather low range that caused increased frequency of symptoms in the epidemiological study by Pilotto et al. (1997) (this level is not addressed by the German recommendation).

On-site information to the users is a key element in both the German and the WHO approach. In areas frequently afflicted by scums or very high cell densitiy (as occurs during dominance of *Planktothrix agardhii*), users of recreational sites are protected best if they learn how to recognize and avoid these situations. A message well received by the public in Germany was: *"if you see scums, or if you carefully wade up to your knees into the water without stirring up sediment, and you cannot see your feet because of a greenish discoloration, don't swim at this locality!"*

5.3 Cyanotoxins in Drinking-Water Supplies

A number of incidents of illness attributed to cyanobacteria in drinking water have been documented worldwide during the past decades (see Chorus et al. 2000 for an overview). Nonetheless, published data on cyanotoxin concentrations in drinking water are scarce. Chapter 5.3 begins with a well-documented incident of human illness due to ingress of river water into the reticulation system, where the causative agent is very likely to have been cyanobacteria (Chapter 5.3.1). The very careful estimation of the microcystin concentration range likely to have prevailed during the incident as well as the symptoms reported strongly point to further bioactive substances in the *Planktothrix agardhii* filaments distributed to the consumers. These results suggest that the current trend of focusing public health attention on microcystins as the most hazardous cyanotoxin may be inappropriate. More likely, as in recreational exposure, short-term and transient symptoms are caused by other components of cyanobacteria, while microcystins are a relevant hazard through long-term exposure.

Results on microcystins found in drinking-water supplies, in most cases in low concentrations, are reported from Canada in Chapter 5.3.2 and from the Czech Republic in Chapter 5.3.3. Chapter 5.3.4 summarizes results from four pilot studies conducted in Germany on the elimination of microcystins in drinking-water treatment and through abstraction procedures.

5.3.1
An Extensive Outbreak of Gastroenteritis Associated with the Toxic Cyanobacterium *Planktothrix agardhii* (Oscillatoriales, Cyanophyceae) in Scania, South Sweden

Helene Annadotter, Gertrud Cronberg, Linda Lawton, Hans-Bertil Hansson, Ulf Göthe and Olav Skulberg

An outbreak of gastroenteritis when 121 persons fell ill occurred in three South Swedish villages in September 1994. The illness coincided with a contamination of the municipal drinking water with untreated water from the River Kävlingeån, the outlet of the eutrophic Lake Vombsjön.

Clinical examination of the patients did not detect any pathogenic bacteria or viruses that could explain the sudden outbreak of gastroenteritis. Therefore, the investigators turned their attention to the presence of the potentially toxic cyanobacterium *Planktothrix agardhii* (Gom.) Anagn. et Kom. in the river. HPLC-analysis of the phytoplankton in Lake Vombsjön 6 days before and 2 weeks after the outbreak showed the presence of microcystins in algal and water samples from the lake. The presence of toxic cyanobacteria in Lake Vombsjön, the absence of secondary cases, the fact that some persons fell ill from having consumed only boiled contaminated water as well as the fact that also cats and dogs fell sick, suggest that the outbreak was caused by cyanobacterial toxins.

Case Description

In September 1994, water from the River Kävlingeån, Scania, southern Sweden, was accidentally mixed with the municipal drinking water. The contamination of the drinking water was due to an erroneous pipe coupling at the sugar refinery in Örtofta, near Lund. At 14:30 h on September 8th, it was noticed that the tap water in the refinery was discoloured. One hour later, it was discovered that a valve, connecting the drinking water with the river water, was open; nobody knew for how long. As soon as the fault had been identified, the valve was immediately closed. Flushing of the drinking water system in the refinery started in the evening and continued throughout the following day.

Those responsible at the sugar refinery assumed that the contamination of the drinking water was confined to the refinery, and that it had not influenced the municipal drinking water system. This was not the case. During the following days, the tap water in the village of Örtofta, as also in the neighbouring villages of Toftaholm and Väggarp, had a musty smell. The time span of intrusion of untreated water from the River Kävlingeån into the municipal drinking water system must have lasted between 1 and 7 h on September 8th.

Between Thursday, September 8th and Tuesday September 13th, a considerable number of people fell ill with gastrointestinal symptoms. The affected persons visited different hospitals and medical care centres. None of the involved physicians suspected a major outbreak. Not until the epidemic was over was the Office of Public Health in Eslöv anonymously contacted by a patient, an employee at the sugar refinery. After communication with some other sick colleagues, he suspected that the incident at the sugar refinery was the cause of their illness. Thursday the 8th was the only day when the tap water at the sugar refinery was contaminated. On Monday 12th, it became apparent that a number of the employees at the sugar refinery were ill with symptoms such as diarrhoea and vomiting. Among the employees who fell ill, the majority had drunk water from the tap on the day when the water was contaminated.

All the employees at the office had drunk coffee or tea, prepared from the same contaminated water. The coffee-drinkers remained healthy while the tea-drinkers fell ill. The coffee was prepared by percolation of hot water through ground coffee in a paper coffee-filter, whereas the tea was made by pouring boiling water onto a tea bag in a cup. All the tea-drinkers at the office lived in villages other than Örtofta, Toftaholm and Väggarp, and thus they consumed the contaminated water only at work as boiled water on September 8th.

The Office of Public Health contacted the sugar refinery, and there received confirmation that the incident with the contaminated water at the refinery was due to the erroneous coupling. The Municipal Services Department in Eslöv was informed, and the water mains for municipal drinking water were flushed.

The information about the outbreak of gastroenteritis first reached the Regional Centre for Public Health at Malmö University Hospital via the local newspaper. A microbiological examination of the water of the River Kävlingeån was initiated while the Office of Public Health in Eslöv conducted an extensive survey of the outbreak by interviewing all those who had drunk the tap water during the days it was contaminated.

Since persons not connected with the sugar refinery got sick, leaflets were distributed by the refinery to the affected households in which information about the accident was given and the households concerned were advised to consume only boiled water until the water was officially declared fit to drink.

Methods of Investigation

A questionnaire was designed in order to map the extent and nature of the incident. The questionnaire was distributed to every household in the villages of Örtofta, Toftaholm and Väggarp.

The River Kävlingeån is the outlet of the eutrophic Lake Vombsjön (12.4 km^2, mean depth 6 m, max. depth 16 m). Lake Vombsjön is a drinking-water reservoir for the towns of Malmö and Lund. In order to detect pathogenic bacteria and viruses, an extensive microbiological analysis of the river water, sampled on September 20th, was carried out. Fortunately, in connection with another investigation, phytoplankton had been sampled in Lake Vombsjön for quantitative analysis of phytoplankton and toxin analysis 6 days before (2 Sept.) and 15 days (23 Sept.) after the incident at the refinery. For quantitative analysis of phytoplankton, samples were taken with a Ruttner sampler and immediately fixed with Lugol solution. Phytoplankton was counted in sedimentation chambers with an inverted microscope according to Cronberg (1982).

For cyanotoxin analysis, phytoplankton material collected from Lake Vombsjön on September 2nd and 23rd 1994 concentrated by a 45-µm plankton net was lyophilized and and extracted in methanol (50 mg DW of cells ml^{-1}) for 1 h. The mixture was then centrifuged and the supernatant analysed for microcystins by HPLC (Lawton et al. 1994). To determine concentrations of microcystin not only in relation to seston dry weight but also to a litre of water, samples were also filtered through GF/C filters and the microcystin content of the filter was examined by HPLC. Peaks were identified as microcystins if their absorption spectra matched those of authenticated microcystins, and the amount of toxin present was calculated by peak area relative to a microcystin-LR standard. The retention time of microcystin-LR was 20.31 min under the HPLC conditions applied.

Further, the phytoplankton from Lake Vombsjön was tested by standard mouse bioassay (Berg et al. 1987).

In order to assess whether ground coffee and coffee filters may absorb cyanobacterial cells, a test was performed with a Lugol-fixed water sample collected on September 2nd from Lake Vombsjön. The lake phytoplankton on that date was dominated by *Planktothrix agardhii*. 5 ml lake water was filtered through normal coffee filters, one filter containing coffee grounds and the other without. The filtrates were collected in plankton chambers and after settling, were inspected with an inverted microscope.

Observations on Illness, Cyanobacteria and Cyanotoxins

The day of the accident (8 Sept. 1994) and several days later, off-flavour (Fig. 1) and discoloration (Fig. 2) of the tap water were experienced in households in the villages of Örtofta, Toftaholm and Väggarp. Of 111 households 22 observed off-flavour and 25 noticed discoloration of the water.

Between Thursday September 8th and Tuesday 13th, 121 of 304 persons who had consumed the tap water fell ill (Fig. 3). Symptoms were abdominal pain, nausea, vomiting, diarrhoea, head-ache, fever and muscular pain (Fig. 4). In Toftaholm, 85% of the persons who drank the tap water fell ill, in Örtofta 68% and in Väggarp 29%.

When the river water was pumped into the refinery, it was not diluted with municipal drinking water. When it was further distributed via the municipal mains, the river water was mixed with municipal drinking water but the proportion of river water to muncipal water differed between the three affected villages. Toftaholm received undiluted river water. Within Örtofta, the people living closest to the refinery received the highest proportion of river water to municipal water. In Väggarp, the proportion of

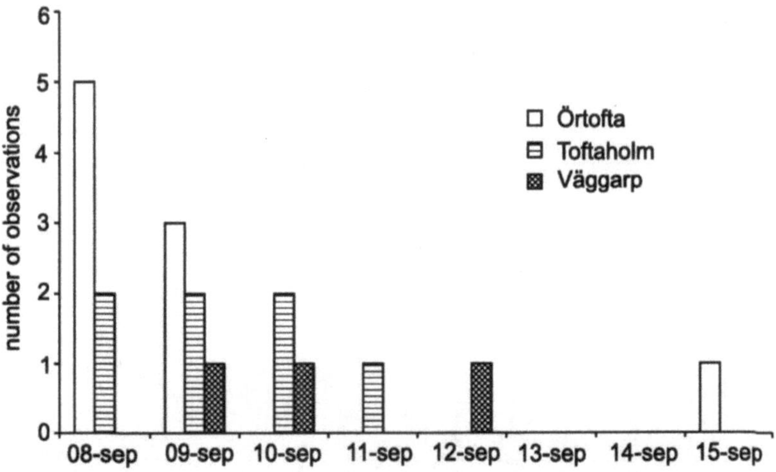

Fig. 1. Number of records of unusual taste and odour in tap water. Örtofta, Toftaholm and Väggarp, September 1994

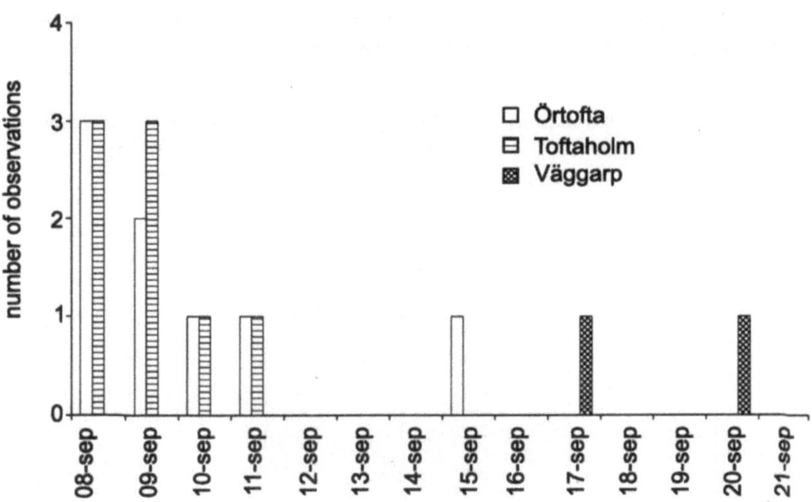

Fig. 2. Number of records of discoloration of the tap water. Örtofta, Toftaholm and Väggarp, September 1994

river water to drinking water was less than in the other villages. This is reflected by the fact that the percentage of persons who fell sick after having consumed the tap water was also highest in Toftaholm (85%), lower in Örtofta (68%) and least (29%) in Väggarp, and the number of complaints of taste and odour or discoloration was substantially lower in Väggarp (Figs. 1-3).

The clinical examination of the patients, carried out by different physicians, detected no pathogenic vectors that could explain the patients' symptoms. The microbiological analysis of the river water revealed no pathogenic microorganisms.

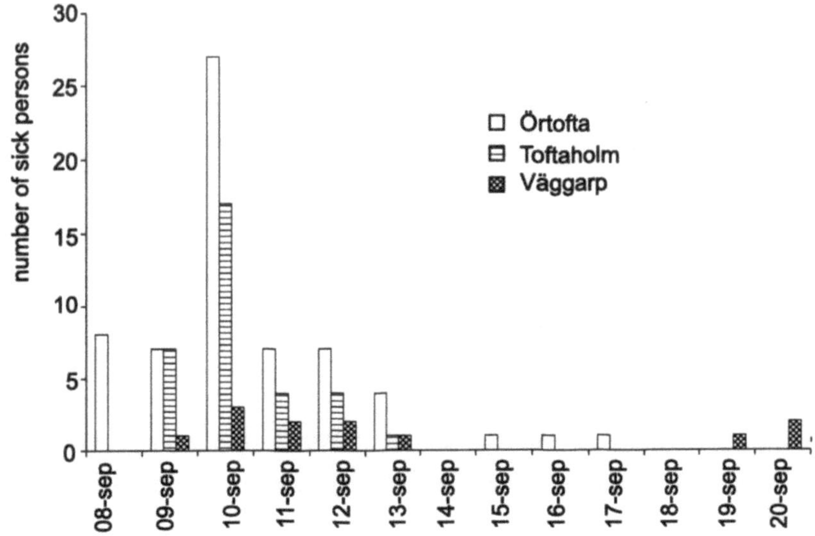

Fig. 3. Development of waterborne illness in Örtofta, Toftaholm and Väggarp, 8-20 September 1994. *Bars* show the number of new cases recorded each day

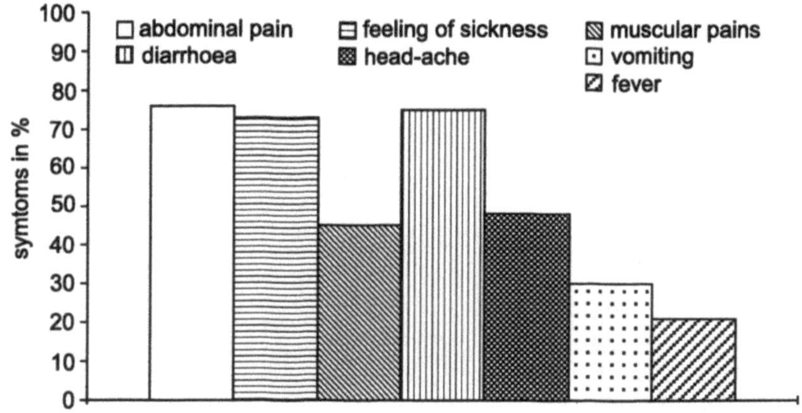

Fig. 4. Symptoms in percent among the 121 persons that fell ill (in total more than 100%, as many patients reported several symptoms)

In Lake Vombsjön, the algal community on all sampling occasions was dominated by *Planktothrix agardhii* (Gom.) Anagn. et Kom. On September 2nd, total phytoplankton biomass was rather high with 22.2 mg L^{-1}, that of *Planktothrix agardhii* amounted to 12.2 mg L^{-1}, and substantial populations of potentially toxic *Microcystis* species occurred (*Microcystis flos-aquae:* 2.3 mg L^{-1}; *M. viridis.* (A. Br.) Lemm.: 2.8 mg L^{-1}). Thus, the potentially toxic species in the sample constituted 78% of the total phytoplankton biomass. On September 23rd, the biomass of *Planktothrix agardhii* was 9 mg L^{-1}.

Microcystin concentrations in the water sample from September 2nd amounted to 0.82 µg L^{-1}. The lyophilized phytoplankton from Lake Vombsjön on September 2nd contained 100 µg microcystins g^{-1} DW and the phytoplankton from September 23rd contained 190 µg microcystins g^{-1} DW.

Acute toxicity tests of lyophilized phytoplankton material from Lake Vombsjön demonstrated the presence of hepatotoxins. The responses included death within 1-4 h after intraperitoneal injection, with massive haemorrhage and necrosis of the liver.

Retention tests for phytoplankton cells on coffee filters with lake water from Vombsjön (sampled on September 2nd) showed that phytoplankton passed through a filter without ground coffee – no retention of the phytoplankton was observed. When ground coffee was added to the filter, all the cells were absorbed to the coffee grounds and no cells went through the filter.

Assessment of Cyanobacterial Toxins as Cause of the Outbreak

No contaminated tap water was sampled before the water mains in the refinery and the municipal mains were flushed. Thus, it was not possible to examine the quality of the drinking water that was suspected to have caused the health problems. The available observations indicate that the accidental ingress of river water into the drinking-water system on September 8th contained cyanotoxins associated with *Planktothrix agardhii.*

The etiology is similar to cases investigated by Dillenberg and Dehnel (1960), who describe the occurrence of toxic water blooms with animal deaths in some lakes in Saskatchewan, Canada, in the summer of 1959. People were warned not to swim in the lakes. However, some persons defied the warnings, and 13 got sick after bathing in the greenish water. They fell ill with symptoms such as headache, vomiting, nausea, stomach pains, muscular pains and diarrhoea. No pathogenic microorganisms were found, whereas cells of the cyanobacteria *Microcystis* sp. and *Anabaena* sp. were detected in the stools of all patients. The amount of lake water swallowed was, however, reported only by one of the patients, a physician. He estimated to have swallowed half a pint (300 ml) after having slipped from the diving board and fallen 5 feet into the water. Three h later, he developed stomach cramps, nausea, and vomited two or three times. Five h after ingesting the lake water, he experienced an attack of painful diarrhoea. The following morning, his temperature was 38.9°C. He further reported a strong headache, weakness, pains in limb muscles and joints, and slimy green stools, similar to the colour of the vomitus. A check for pathogens including viruses was negative, while the microscopic examination revealed high numbers of *Microcystis* spheres and two to three well-preserved curved chains of *Anabaena circinalis* per high-power field. Two days after swallowing the lake water, the patient recovered.

The outbreak around the sugar refinery in Örtofta highlights typical deficits in follow-up investigations due to the common lack of awareness of cyanobacterial health effects in public health authorities. The idea that cyanobacteria could have caused the disease did not emerge until it became clear that the customary microbiological explanation could not be sustained. The possible presence of cyanobacteria was thus not investigated in the stools of the patients.

Further, information on the status of liver enzyme concentrations in the patient's plasma would have been valuable. The microcystin present in Lake Vombsjön 6 days before and 2 weeks after the outbreak is a liver toxin. Toxic injury to the liver, shown as a statistically significant increase in gamma glutamyl transferase (GGT), was observed in an Australian population at the time of a bloom of the cyanobacterium *Microcystis aeruginosa* in the reservoir supplying the persons affected (Falconer et al. 1983). This cyanobacterium is a regular producer of microcystin (see Chapter 2.1.2). However, the physicians whom the 121 patients in Scania consulted were neither aware of any water-borne outbreak, nor did they suspect any hepatotoxic cyanobacterium as the etiological agent. Thus, the liver status of the patients was not checked.

The Scanian outbreak in question was, however, sudden and without secondary cases. This indicates intoxication rather than an infectious disease. This is further strengthened by the fact that all of the seven tea-drinkers in the sugar refinery fell ill. All of them lived in villages other than Örtofta, Toftaholm and Väggarp, but at the refinery, they drank two to three cups of tea, and the tap water consisted of pure river water on September 8th. This implies that the consumed volume of at maximum 1 L boiled river water per person was enough to cause the illness. Microcystins are resistent to heat, whereas boiling water for tea would have inactivated pathogenic microorganisms.

In addition to the 121 persons who fell sick from drinking the tap water, some household pets (dogs and cats) were also affected. Illness of pet animals is rarely reported due to drinking water from lakes and rivers, but reports of human illness as well as animal death from ingesting hepatotoxic cyanobacteria are numerous (Schwimmer and Schwimmer 1964; Lawton and Codd 1991; Kuiper-Goodman et al. 1999).

The role of microcystins in the outbreak requires assessment of the symptoms as well as of the ambient concentrations during the incident. Though the microcystin concentration on the day of the drinking-water contamination was not measured, there is a large body of evidence from ongoing studies at Lake Vombsjön that allow an estimation of the range to be expected: 6 days before the incident it was 0.82 $\mu g\ L^{-1}$. Microcystin analyses of water from Lake Vombsjön 1995-1996 showed similar levels of microcystins. Two years after the incident (in 1995 and 1996) microcystins were detected in 10 out of 32 samples, with a maximum concentration of 1.5 $\mu g\ L^{-1}$ on September 18th, 1996 and an average concentration on the 10 occasions when microcystins were detected of 0.48 μg L^{-1} (Cronberg et al. 1997). Since the River Kävlingeån which contaminated the drinking water is the outflow of Lake Vombsjön, and since the microcystin concentrations in Lake Vombsjön were remarkably similar in 1994, 1995 and 1996, it is likely that the microcystin concentration in the river on September 8th, 1994 were in the same range as in the samples studied from 1994-1996, particularly since both the biomass of potentially toxic cyanobacteria and the concentrations of microcystins in Lake Vombsjön showed normal Lake Vombsjön levels 6 days before the outbreak, and no extreme changes in the weather occurred between September 2nd and September 8th. Thus, the level of microcystins was likely to have been about 1 μg L^{-1} in the river water on the day of the incident. However, higher concentrations cannot be excluded, because microcystin concentration in lake water can vary from one day to another (Kotak et al. 1995), particularly due to the displacement of surface blooms by wind action.

The presence of microcystins in the water that reached the consumers is further supported by mouse bioassays frequently performed: 21 scum samples from Lake Vombsjön sampled from 1986-1992 were examined by mouse bioassay (Annadotter 1993). Hepatotoxic activity was demonstrated in 18 of them, while the rest of them were non-toxic by mouse bioassay. However, though mouse bioassay results reflect the toxicity of particulate material in the samples retained by a plankton net, they do not provide information on the density of these particles in the water and thus on hepatotoxin concentrations. Toxicological information indicates that microcystin concentrations in the range of 1 μg L^{-1} would not cause symptoms (see Chapter 5.1). Further, the symptoms observed during the incident are not typical for microcystin poisoning.

Neurotoxic activity has never been demonstrated in scum samples from Lake Vombsjön. Mouse bioassay of lyophilized phytoplankton material from Lake Vombsjön on September 2nd and 23rd were hepatotoxic without any indications of neurotoxicity. Thus, it is unlikely that the cyanobacteria that were accidentally pumped into the refinery contained cyanobacterial neurotoxins.

Cyanobacterial toxins have been implicated in human illness (i.e., acute non-lethal or chronic toxicity) from municipal water supplies in the USA and Australia (Carmichael 1995; Kuiper-Goodman et al. 1999). In those cases, and in others involving accidental ingestion, the reported symptoms include: abdominal pain, nausea, vomiting, sore throat, dry cough, headache, blistering of the mouth, atypical pneumonia and elevated liver enzymes in the serum. In most of those cases, the cyanobacteria involved and sometimes cyanotoxins were identified. However, in all of these reports it is unclear which substances actually induced the symptoms. The level of a cyanotoxin associated with illness was currently established only for one outbreak in Finland with acute gastric illness in November 1989 in the town of Taalintehtaan (Lepistö et al. 1993). This outbreak was associated with a massive occurrence of microcystin-producing *Planktothrix agardhii* in the raw water supply. Filaments of *Planktothrix agardhii* and 0.1-0.5 μg L^{-1} of microcystins were detected in the drinking water. The water supply was discontinued and the health authoritites banned the use of tap water. Again, the role of microcystins in inducing the symptoms observed is unclear, and the low concentrations detected suggest the presence of other cyanotoxins in the water. This would not be surprising, as evidence for further bioactive substances is increasing for many cyanobacteria, particularly for *Planktothrix agardhii* (see Chapter 7).

In summary, for the Swedish water-borne outbreak reported here, several implications contribute to the assumption that the outbreak was a consequence of intoxication by cyanobacterial toxins and not a bacterial or viral infection: the lack of secondary cases (i.e. the illness was not contagious); the fact that pet animals got sick; that pathogenic bacteria and viruses were found neither in specimens from the patients nor in the river water, and finally, the fact that some people fell sick from having drunk only boiled water, as tea, and no tapwater.

Acknowledgements

The limnological and toxicological investigation was partly supported by Malmöhus läns miljövårdsfond within the project Algal Toxins in Scanian Drinking Water Reservoirs.

5.3.2
Cyanobacterial Toxins in Drinking Water: a Canadian Perspective

Shalini Gupta, Michele Giddings and Marla Sheffer

In Canada, cyanobacterial blooms are common in the Prairie provinces (Alberta, Saskatchewan and Manitoba). Cyanobacterial species known to occur in Canada and most often associated with poisonings are *Microcystis*, *Anabaena* and *Aphanizomenon* (Carmichael 1992). Microcystins are apparently the most commonly encountered cyanobacterial toxin in Canada. The major route of human exposure to cyanobacterial toxins is the consumption of drinking water. Because many rural prairie towns rely on shallow, eutrophic lakes as their drinking water source and a large number of prairie farms depend on dugouts for all or a portion of their domestic water supplies (Kenefick et al. 1992), the possibility of human poisoning from the ingestion of water contaminated with cyanobacterial blooms from lakes and dugouts is currently an issue of concern in Canada.

Occurrence in Canadian Drinking Water Supplies

Few Canadian monitoring data are available on levels of cyanobacterial toxins in raw water supplies. In the summer of 1990, Kenefick et al. (1992) found microcystin-LR to be prevalent in cyanobacterial blooms on three shallow Alberta lakes used as municipal drinking water sources. Microcystin-LR was present in concentrations up to 500 µg g^{-1} algal biomass. More than 70% of over 380 bloom biomass samples from 19 lakes in Alberta between 1990 and 1992 showed detectable levels of toxin (> 1 µg microcystin-LR g^{-1} dry biomass) (Hrudey et al. 1994). Levels of microcystin-LR in Alberta lakes and dugout ponds, measured using high-performance liquid chromatography with ultraviolet detection, ranged from 4 to 605 µg g^{-1} DW of biomass (Kotak et al. 1993) or up to 1500 µg g^{-1} (Hrudey et al. 1994).

The levels of microcystin in the raw intake water for two Alberta drinking water supplies (communities not specified), as measured by the sensitive phosphatase inhibition bioassay, ranged from 0.15 to 4.3 µg L^{-1}, with a large coefficient of variation (59%) for hourly fluctuations over an 11.5-h period; in treated water, levels ranged from 0.09 to 0.64 µg L^{-1}, with a small coefficient of variation (10%). Over a 5-week period, similar coefficients of variation were obtained in the two types of samples (Hrudey et al., 1994).

In the summer of 1993, microcystin-LR was detected (detection limit 0.05 µg L^{-1}) in water samples collected from Shoal Lake, Manitoba, and from within the distribution system following the identification of *Microcystis aeruginosa* blooms in Deacon Reservoir, Winnipeg's main storage facility for water from Shoal Lake (Jones 1996). Water from Shoal Lake is generally considered to be of high quality and requires only disinfection with chlorine prior to distribution and consumption. Maximum microcystin-LR concentrations measured in the raw water of Shoal Lake and in treated tap water were approximately 0.45 and 0.55 µg L^{-1}, respectively.

In 1995, 160 surface water supplies, located mainly in southwestern Manitoba, were chosen for study of cyanobacterial toxins. Microcystin-LR was detected (detection limit 0.1 µg L^{-1}) in 70% of the raw water supplies, 93% of the municipal water supplies, 57% of the dugouts used for domestic and shared domestic plus livestock water consumption, 84% of the dugouts used exclusively for livestock and 44% of the recreational use sites. Treated water samples were analyzed only if the raw water supplies were found to have detectable levels of toxins. Toxin was present in 68% of the treated water samples collected from both the municipal sites and the dugout sites. Toxin concentrations ranged from < 0.1 to 1.0 µg L^{-1} in raw water samples and from < 0.1 to 0.6 µg L^{-1} in treated water samples. Dugouts used extensively for livestock consumption had the highest levels of toxin. Replicate samples taken directly from the algal bloom at two separate livestock sites had toxin levels as high as 1.5-8.0 µg L^{-1}.

Treatment Technology and Management

Good control technology must reflect proper management of the watershed and reservoir to prevent algal and cyanobacterial growth, an appropriate monitoring program and correct treatment technology for both the cyanobacteria and their toxins.

Prevention of Cyanobacterial Growth

Management options are similar to common techniques used to control populations in reservoirs, but with a few exceptions. For example, the use of chemicals that would lead to lysis of cyanobacterial cells must be avoided to prevent the release of their toxins. In the past, chemical treatment with copper sulphate has been the most common technique used to control cyanobacterial blooms in drinking water reservoirs. However,

Kenefick et al. (1993) found that cyanobacterial bloom material from a Canadian prairie lake treated with copper sulphate released most of its toxins during the first 3 days following treatment. Their results indicated that a 99% reduction in microcystin levels would take approximately 3 weeks, suggesting that copper sulphate should not be used to treat potentially toxic cyanobacterial blooms in waters to be consumed by humans within several weeks following treatment.

Monitoring for Cyanobacteria and Their Toxins

An appropriate monitoring program is essential to the overall control of cyanobacteria and their toxins. At present, most municipal water treatment plants in Canada do not regularly monitor their water supplies for cyanobacterial toxins. However, because cyanobacteria have strong smells and tastes and interfere with certain water treatment processes, most municipalities with a history of blooms monitor their surface-water supplies for cyanobacteria.

Monitoring programs must take spatial and temporal variability in microcystin levels into consideration, as there may be large fluctuations in the levels of cyanobacteria and their toxins resulting from the interplay of a variety of physical, chemical and biological factors. Kotak et al. (1995) studied the patterns of occurrence of microcystin-LR (measured as $\mu g\ g^{-1}$ biomass of *Microcystis aeruginosa*) in three hypereutrophic hardwater lakes in central Alberta over three seasons. *Microcystis aeruginosa* was highly variable both temporally and spatially, with differences up to 3 orders of magnitude within each lake over 1 year, between years in an individual lake and between lakes in a year. Seasonal changes in microcystin-LR concentration were positively correlated to the abundance and biomass of the *M. aeruginosa*, total and total dissolved phosphorus concentration, pH and chlorophyll. There was a negative correlation between microcystin-LR concentration and nitrate concentration and no correlation with water temperature. Over a 24-h period, the concentration of microcystin-LR was six times lower at night than during the day.

Treatment Technology

The final step in controlling cyanobacteria and their toxins is the drinking-water treatment process. Although conventional surface-water treatment plants using coagulation, clarification and filtration are effective in removing cyanobacterial cells, the Manitoba study (Jones 1996) illustrated that conventional water treatment methods may be only partially successful in removing the cyanobacterial toxins. This has been confirmed by Lambert et al. (1996), who examined the removal of microcystins from drinking water at two full-scale treatment plants in Alberta that employed coagulation-sedimentation, dual-media filtration and chlorination combined with either granular activated carbon (Ferintosh plant) or powdered activated carbon (Camrose plant) filtration. The Ferintosh treatment plant serves a population of 173 people, while the Camrose plant serves about 15000. The two processes generally removed more than 80% of the microcystin from raw water, particularly when the raw water concentrations were high; however, a residual concentration of 0.05-0.2 µg microcystin-LR equivalents L^{-1} was observed at both treatment facilities (Table 1). The data thus suggest that drinking water consumers could be chronically exposed to low levels of microcystin-LR for the duration of cyanobacterial blooms. Lambert et al. (1996) also found that chlorination was unsuccessful in achieving any reduction in microcystins (Table 2).

Table 1. Removal of microcystin-LR in two full-scale water treatment plants. (From Lambert et al. 1996)

Sampling date	Raw lake water (µg microcystin L^{-1})	Treated drinking water (µg microcystin L^{-1})	Removal from raw water (%)
Ferintosh treatment plant (uses GAC)			
June 9	1.35	0.11	92
June 15	2.28	0.1	96
June 24	0.98	0.11	89
June 30	1.43	0.05	97
July 7	1.31	0.12	91
July 14	0.27	0.11	59
Mean overall removal Median overall removal Range			87% 91% 59-97%
Camrose treatment plant (uses PAC)			
August 26	0.15	0.11	27
August 27	0.87	0.09	90
August 31	0.86	0.13	85
September 2	0.67	0.10	85
September 8	0.62	0.18	71
September 10	0.73	0.17	77
September 14	0.35	0.15	57
September 16	0.29	0.14	52
September 18	0.26	0.14	46
September 22	0.27	0.18	33
September 24	0.15	0.10	33
September 28	0.17	0.14	18
September 30	0.22	0.15	32
October 2	0.15	0.14	7
Mean overall removal Median overall removal Range			48% 49% 7-90%

Table 2. Removal of microcystin-LR at the Ferintosh treatment plant. (From Lambert et al. 1996)

	25 August			1 October		
Unit process	Mean concentration (µg L^{-1})	Cumulative mean removal from raw (%)	Incremental removal by process (%)	Mean concentration (µg L^{-1})	Cumulative mean removal from raw (%)	Incremental removal by process (%)
Raw water	2.9 ± 0.8	-	-	1.2 ± 0.1	-	-
Coagulation-sedimentation	3.1 ± 1.1	0	0	0.7 ± 0.05	39	39
Filtration	1.2 ± 0.2	60	60	0.6 ± 0.07	48	14
GAC	0.5 ± 0.1	83	60	0.3 ± 0.05	71	43
Chlorination	0.5 ± 0.2	82	0	0.3 ± 0.06	71	0

Note: August data based upon 5-8 samples at each treatment step; October data based upon 9-12 samples at each treatment step

Health Canada's Role

Health Canada works with the provinces and territories to establish drinking-water quality guidelines. The finding of microcystin-LR in cyanobacterial blooms in the Deacon Reservoir, Winnipeg's main storage facility for water from Shoal Lake, in the summer of 1993, prompted Manitoba Environment to request that the Environmental Health Directorate of Health Canada initiate development of an Emergency Health Advisory (EHA) guideline for microcystins in drinking water.

The guideline for microcystin-LR is still at the approval stage. The proposed value for this guideline is 0.0015 mg L^{-1}. Once the guideline has been approved, some municipal water treatment plants may be required to monitor for the presence of microcystin-LR in their water supplies, especially if the source is prone to cyanobacterial blooms. Monitoring strategies will vary between provinces.

5.3.3 Dissolved Microcystins in Raw and Treated Drinking Water in the Czech Republic

Ludek Bláha and Blahoslav Maršálek

Freshwater cyanobacteria which occur as the water blooms and surface scum in eutrophic freshwater reservoirs can significantly alter the quality of water (Jacobsen 1994). Besides the effects on oxygen concentrations or trophic state, many cyanobacterial toxins (cyanotoxins) were described (Carmichael 1992). Among them microcystins – cyclic heptapeptides – are the most often studied and reported (Dawson 1998). Awareness of the importance of microcystins for human health increased after a fatal accident in Caruaru, Brazil 1997 where more then 60 people died in the hospital haemodialization centre after microcystin intoxication (Jochimsen et al. 1998; Pouria et al. 1998).

Only some of the processes used in drinking-water treatment plants are able to remove microcystins sufficiently. These are in particular ozonization and/or filtration through granular activated carbon, if properly applied (Lambert et al. 1996; Hrudey et al. 1999). However, both methods are still not often applied in many drinking water treatment plants, and cyanotoxins occurring in the raw water can then be a significant health risk for human populations.

Most of the monitoring reports on environmental cyanotoxin concentrations focused on analyses of microcystins in cyanobacterial biomass (cell-bound microcystins, for review see Sivonen and Jones 1999). On the other hand, current information on environmental concentrations of dissolved (extracellular) microcystins in raw and treated drinking waters is very limited (see also Chapter 2). In spite of the World Health Organization's guideline value for microcystin in drinking water of 1 mmg L^{-1} (WHO 1998), regulatory activities applied concerning toxic cyanobacteria in surface reservoirs of drinking waters are poor or lacking.

In this chapter we tried to answer several questions regarding problems mentioned above: (1) can the extracellular microcystins be detected in raw and treated drinking waters in the Czech Republic?; (2) what are the real environmental concentrations of dissolved microcystins in the waters collected in treatment plants?; (3) can extracellular microcystins in drinking waters represent health risks in the Czech Republic?

Materials and Methods

Samples of raw and treated waters from selected drinking-water treatment plants were collected into the PET bottles and stored frozen at -18°C upon processing. Storage was no longer than 10 days. Concentration of microcystins from the water samples by solid phase extraction (SPE) on C-18 columns was performed according to recommendations of the Environmental Agency, UK (1998) and Moollan et al. (1996): 800 ml of water were filtered through a paper filter, and 8 ml of 10% trifluoroacetic acid (TFA) in methanol and 10 ml of sodium sulphite (10 mM) was added. SPE was performed on Waters Sep-Pac Vac 35cc C-18 (10 g) cartridges, methanolic extracts (30 ml) were dried by vacuum rotary evaporation and dissolved in 200 µl of pure methanol. Extracts were analyzed with HPLC for content of microcystins. Waters 600E HPLC solvent delivery system equipped with Waters 717plus autosampler was used. Analytical column: 150x4.6 mm Supelcosil ABZ+Plus, 5 µm (Supelco). Gradient elution: mobile phases: (A) 0.05% tetrafluoroacetic acid (TFA); (B) acetonitrile + 0.05% TFA. Gradient composition: 0 min 80% A and 20% B, 20 minutes: 48% A and 52% B. Flow rate 1.0 ml min^{-1}, temperature 25°C. Analytically pure microcystin-LR (Calbiochem-Novabiochem Corp., La Jolla, CA) was used as an external analytical standard. A photodiode array detector (Waters PDA 996, wavelength 200-300 nm) was used for scanning UV absorbance and evaluation of spectra. Mean recovery efficacy of the method evaluated with pure microcystin-LR (added into the selected water samples) was 67%. The detection limit of the HPLC method used for analyses of dissolved microcystins may be somewhat variable due to the matrix influence, but for these samples it was in the range of 15-25 ng L^{-1} for microcystin-LR.

The names of drinking-water plants were (according to safety of information law) replaced by numbers. Numbers represent 24 samples from 7 drinking water plants in the Czech republic during the period July-October 1999. Sampling was repeated if the first sample for a given site was positive for microcystins.

Results

In 1999, 24 samples of both raw and treated waters (12 samples each) were collected from 7 selected drinking water plants, whose intake is surface water from freshwater reservoirs where cyanobacterial water blooms developed.

Table 1 summarizes the results of HPLC analyses. The most frequent variant of microcystin – microcystin-LR, the only cyanobacterial toxin for which the WHO drinking water quality safety limit was derived, is depicted. Total microcystin concentrations calculated as microcystin-LR concentration equivalents (Falconer et

Table 1. Concentrations of microcystins in raw and treated drinking waters in the Czech Republic in a 1999 survey

			Raw water			Treated water			Microcystin elimination % of total Microcystins
Sample	Plant	Sampling date	HPLC peaks	MC-LR μg L^{-1}	Total MC	HPLC	MC-LR μg L^{-1}	Total MC	
1	1	22.07.1999	-	-	-	-	-	-	
2	2	11.08.1999	3	-	1.71	-	-	-	100
3	3a	12.08.1999	2	3.04	7.16	2	0.18	0.95	87
4	4a	110.8.1999	2	0.41	1.69	2	-	0.89	47
5	3b	20.09.1999	-	-	-	-	-	-	
6	5a	7.09.1999	2	1.62	2.99	2	0.75	1.03	66
7	4b	12.09.1999	1	0.10	0.10	-	-	-	100
8	5b	24.09.1999	-	-	-	-	-	-	
9	6a	19.08.1999	2	0.30	0.72	-	-	-	100
10	6b	12.09.1999	-	-	-	-	-	-	
11	7a	9.09.1999	3	3.32	8.70	3	1.36	7.79	11
12	7b	12.10.1999	1	0.09	0.09	-	-	-	100

MC-LR microcystin-LR. *Plant numbers* represent particular drinking water plant stations. Repeated sampling on the same plant indicated by letters *a* and *b*

al. 1999), as well as the number of microcystin variants detected in the samples are also reported.

Incidence of dissolved (extracellular) microcystins in raw waters was high: 8 out of 12 samples were positive for microcystins. Concentrations of total microcystin ranged from 0.09 μg L^{-1} up to 8.70 μg L^{-1}. Six samples contained microcystin-LR, for which the WHO guideline value 1 μg L^{-1} was derived. The highest concentration of dissolved microcystin-LR was more than 3 μg L^{-1}. The relative amount of microcystin-LR in the positive samples ranged from 41 up to 100% with a median value of 51%.

The concentrations of total microcystins in positive samples of the treated drinking waters were 953, 893, 1025 and 7785 ng L^{-1} of total microcystins. Although this exceeds the provisional WHO Guideline value (if it is applied to total microcystins) observed concentrations were relatively closely within its range. Overall incidence of dissolved (extracellular) microcystins in both raw and treated waters was 50% (12 out of 24 samples).

Discussion

Our results showed that cyanobacterial hepatotoxins – microcystins – can be found in the Czech reservoirs not only within the cyanobacterial biomass (i.e. as cell-bound microcystin; see also chapter 2.2.3), but also dissolved in both raw and treated drinking waters in the Czech Republic.

Our results correspond well to previously published surveys of dissolved microcystins in freshwater reservoirs worldwide. The concentrations of microcystins, as measured by ELISA, ranged in the waterbodies of China, Germany, Japan and Thailand from 20 ng L^{-1} up to 5600 ng L^{-1} with median values at 200 ng L^{-1} (Ueno et al. 1996). Similarly, Tsuji et al. (1996) reported concentrations of dissolved microcystin from two Japanese lakes ranging from 20 to 3800 ng L^{-1} (based on ELISA). Lahti et al. (1997) reported similar range of microcystin concentrations (60 - 210 ng L^{-1}) from Finland. All the above-mentioned reports were focused on the microcystin concentrations in surface water bodies, thus not exactly matching our samples collected directly in the drinking-water treatment plants. In comparison, the highest concentration in the Czech Republic (8.69 μg L^{-1}) was almost twice as high as the amounts previously reported. Moreover, our concentration was measured directly in the drinking-water treatment plant, which provides a higher degree of relevancy for possible health risks from microcystins in drinking water. The overall incidence of dissolved microcystins in our study was also relatively high (58% positive samples of the raw waters, 50% overall). These indices show that possible risks from microcystins in the drinking waters (hepatotoxicity and liver tumour promotion) should not be underestimated.

Further warning information was also obtained. Total microcystin concentration in the raw water from the treatment plant no. 11 (9/9/99) was eight times higher than the WHO-derived water quality limit (three times higher based on single microcystin-LR). This plant stopped water production and public supply until 12.10. 1999 (sample 12) when no microcystins in treated water were found. Additionally, concentrations in both microcystin-positive samples of treated drinking waters were also relatively high, almost reaching the WHO safety limit.

As known from previous reports, cyanobacterial water blooms, as well as concentrations of cyanotoxins, undergo significant fluctuations during the summer season (Jacobsen 1994; Kotak et al. 1995). Since only limited information on microcystin fluctuations in the reservoirs from which the samples were collected is available, much higher concentrations could be expected during the season. Also the collapse of the bloom at the end of the vegetation season (or herbicide-induced lysis of the cells) can lead to release of extremely high microcystin amounts, which can pose an increased health risk for humans (Lam et al. 1995).

Based on this preliminary survey of dissolved microcystins, several directions are indicated. Future efforts should be focused on detailed relative comparisons of dissolved and cell-bound microcystins in the Czech drinking water reservoirs. Much detailed comparison of the concentrations of dissolved microcystins in the freshwater reservoirs with dominant cyanobacterial blooms, raw and treated drinking water is also urgently required. The efficiency of microcystin-removal processes in the treatment plants should be estimated. Fluctuations of microcystin concentrations during the summer season should also be evaluated.

In conclusion, our study revealed that microcystins can often be detected in both raw and treated drinking water in the Czech Republic, the highest detected concentration (8695 ng L^{-1}) being seven times higher than the WHO-recommended safety limit. Since microcystins were shown to be tumour promoters (Humpage and Falconer 1999) which can have serious human health impacts (Falconer 1998), both research and regulatory activities should be focused on drinking-water reservoirs with occurrence of cyanobacterial water blooms. More importance should be paid to water bloom and microcystin fluctuations over the vegetation season, because extremely high microcystin concentrations can be expected especially at the end of the season when the water blooms collapse.

Acknowledgements

This work was supported by the Project CYANOTOX of EC - DG XII) ENV 4 - CT98 0802.

5.3.4
Microcystin Analysis in Selected Drinking-Water Supplies in Germany

In the context of the German cyanotoxin program, pilot studies were conducted at four very different sites of drinking-water abstraction. The feature they had in common was the occurrence of a high biomass of cyanobacteria containing microcystins in the source water. The aim of these pilot studies was to provide a preliminary assessment of the risk of microcystins in the finished drinking water and to indicate further monitoring needs. Though each study in itself covers only a small number of samples and situations, taken together they support the result of the survey that even when concentrations of cell-bound microcystins are high, the concentrations of dissolved microcystins in the raw water are usually low (see Chapter 2.1.2). Chapter 5.3.4.1 and 5.3.4.2 show successful removal of cell-bound microcystins by drinking-water treatment, and Chapter 5.3.4.3 and 5.3.4.4 show that if intake of cells can be avoided by the choice of offtake depths or by abstraction through bank filtration, muchof the cyanotoxin risk can be avoided.

As all these studies were limited in temporal resolution and extent, further surveillance is needed for generalizations of the risk assessment for these and similar drinking-water supplies, particularly in situations of high cyanobacterial biomass and particularly in order to address the issue of potential microcystin liberation through cell lysis potentially caused by some treatment procedures.

5.3.4.1 Elimination of Microcystins in the Rostock Drinking-Water Treatment Plant

Christine Kruschwitz, Ingrid Chorus, Rita Heinze, Gertrud Schlag and Kerstin Gröbe

The city of Rostock, a port to the Baltic Sea, has relied on the Warnow River as source for its central drinking-water supply since 1867. It currently provides water for 250000 inhabitants. Geo-morphological conditions render the river slow-flowing, influenced by many lakes, and naturally eutrophic. Drinking-water protection zone regulations were implemented in 1969 to protect the catchment and the river, e.g. by prohibition of motor boats in the protection zone and through requirements for specific technology along bridges and roadways to prevent runoff potentially contaminated with fuel from reaching the river. Nonetheless, the Warnow catchment with its area of 3000 km^2 is still rather intensively used for agriculture, and is also a recipient for treated domestic sewage, both of which enhance eutrophication. Annual means of chlorophyll-a concentrations are in the range of 50 µg L^{-1}; maxima reach 150-200 µg L^{-1}. Phytoplankton is dominated by diatoms and green algae, and during the summer months also by cyanobacteria.

Drinking-water treatment was upgraded in 1995 to comprise preozonation with 2.5 mg L^{-1} , flocculation with 8.0 mg L^{-1} polyaluminum chloride and 0.25 mg L^{-1} polyacrylamide, filtration, ozonation with 0.5-1.7 mg L^{-1} ozone and 0.5 mg L^{-1} hydrogen peroxide, filtration through granular activated carbon followed by disinfection with 0.15 mg L^{-1} chlorine-dioxide and 1.0 to 1.5 mg L^{-1} chlorine. With this treatment, it meets the requirements of the German drinking-water ordonance, and consumer acceptance has substantially increased since 1995.

Microcystin concentrations were studied in this system on selected occasions during the summers of 1996 and 1997.

Methods

Microcystin analysis was performed with ELISA (applying the Enviro Gard kit by Coring System, as described in Chapter 8.3). Samples with less than 0.1 µg L^{-1} (detection limit of the ELISA) were concentrated by lyophilization (freeze-drying) in order to detect minor amounts in the range of low concentrations for assessment of treatment performance. Thus, the detection limit could be reduced to 0.004 µg L^{-1}. Losses due to this procedure, assessed at concentrations between 0.1 and 1 µg L^{-1}, amounted to at most 50%. False positives or enhancement of the ELISA response cannot be excluded, so concentrations in finished water were checked with the radioactive protein phosphatase assay as described in Chapter 8.3. For the samples collected in 1996 we are grateful for determinations with the colorimetric protein phosphatase assay by Wayne Carmichael (Wright State University, Dayton, USA).

Results

The situations assessed did not include those with very high cyanobacterial biomass: In the samples investigated in 1996, filamentous cyanobacteria were dominant, but densities were very low. Correspondingly, microcystin levels in the seston samples of the raw water (analyzed by HPLC as described in Chapter 9.1) reached only 2-10 µg g^{-1} DW in May and 200 µg g^{-1} DW in August (total microcystins as sum of all variants). In 1997, *Planktothrix agardhii* formed a low population density of 0.3 to 1.6 mm^3 L^{-1} (biovolume determined microscopically; see Chapter 9.2) at the end of July and in August. Seston microcystin content amounted to 210-250 µg g^{-1} DW on 31 July. *Microcystis* spp. biovolumes were in a similar range with one exception on

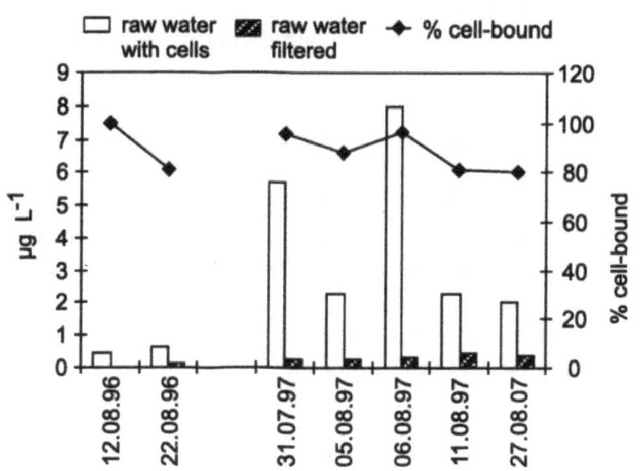

Fig. 1. Microcystin concentrations in raw water before and after filtration of the samples and percent cell-bound microcystin

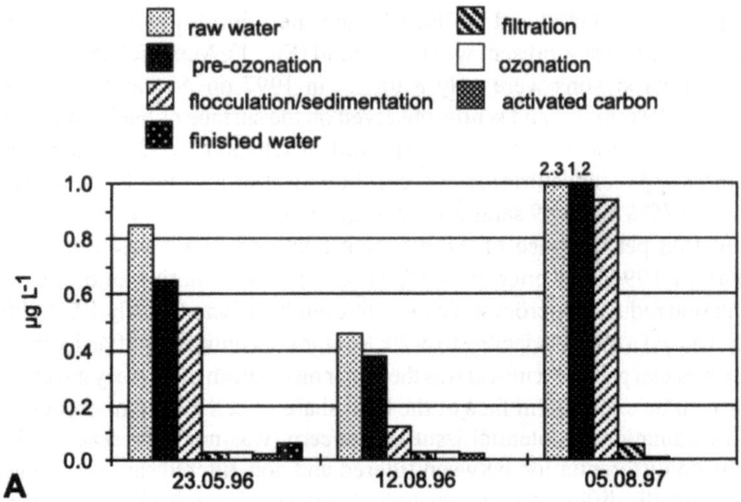

Fig. 2. Microcystin concentration in raw water and after different treatment steps. **A** full range of concentrations. **B** magnification of lower concentration range

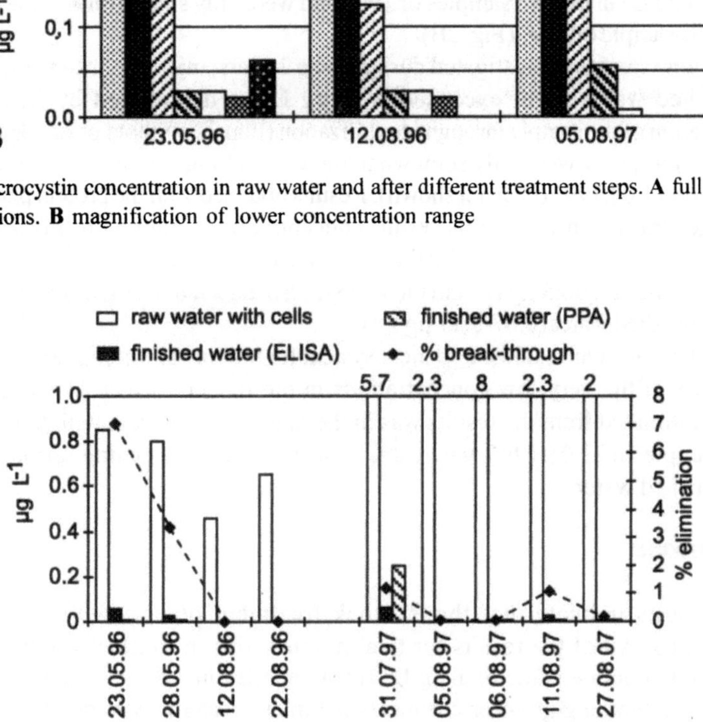

Fig. 3. Microcystin concentrations in raw water and in finished water, determined by ELISA and protein phosphatase assay (*PPA*)

August 6th, when *Microcystis* spp. formed scums on the surface of the intake basins and a maximal biovolume of 19 mm^3 L^{-1} was determined.

Comparison of filtered and unfiltered raw water samples indicated that 80 % or more of the microcystin present was cell-bound (Fig. 1). Maximal concentrations of cell-bound microcystins were only 8 µg L^{-1} in 1997 on August 6th in spite of a biovolume of 19 mm^3 L^{-1} and scums observed on the surface of the intake basin. This agrees well with the observations reported in Chapter 2.1.2 for dominance of *Microcystis* spp.: concentrations of cell-bound microcystins L^{-1} were less than 10 µg L^{-1} for 75% of the 49 samples investigated outside of scums.

Elimination performance of the different treatment steps was assessed on two occasions in 1996 and once in 1997 (Fig. 2). Preozonation and flocculation/ sedimentation reduced microcystin concentrations by a factor of only 1.5-3. Filtration, however, caused a marked decline to 5-20% of the concentrations after sedimentation. This indicates that particle removal was the major mechanism of microcystin elimination, which was to be expected in face of the high share of cell-bound microcystin. Lysis due to preozonation, a potential issue of concern, was not addressed by the study design, as no differentiation between filtered and non-filtered samples occurred for the water from the different treatment steps. However, the high efficacy of filtration in microcystin removal indicates that microcystins reaching the filters were largely cell-bound. Concentrations after ozonation and activated carbon filtration amounted to at most 0.03 µg L^{-1} in the two samples of 1996 and were only slightly above the detection limit in the sample of 1997 (Fig. 2B).

In all nine samples investigated during these 2 years, microcytin concentrations in the finished water never exceeded 0.07 µg L^{-1} as determined by ELISA after concentration of the sample through lyophilization (Fig. 3). Without prior concentration of the samples, they were only somewhat higher, reaching 0.11 µg L^{-1} on 31.07. 1997 and 0.14 on 11.08.1997 (data not shown). Results obtained with the protein phosphatase assays generally confirmed this result: concentrations found with this assay were always lower than 0.01 µg L^{-1} with the exception of the sample on 31.07.1997, where the assay indicated 0.25 µg L^{-1} and the ELISA also detected the highest concentration found in finished water (0.07-0.11 µg L^{-1}).

Overall performance of the treatment system for microcystin elimination is difficult to quantify at the very low concentrations in the finished water. However, break-through estimated from the results was in the range of 1% or less with the exception of one sample on 23.05.1996, where 8% of the raw water concentration was detected in the finished water.

Conclusions

These results indicate that the Rostock treatment plant effectively removes microcystins. All of the results for total microcystins in finished water were well below the Guideline value of 1 µg L^{-1} recommended in 1997 by the World Health Organization Working-group on chemicals in drinking-water for microcystin-LR (WHO 1998). However, the pilot program missed peak concentrations of cyanobacteria and microcystins in the raw water. Further monitoring is necessary to substantiate the results, particularly when scums appear and when non-scum forming cyanobacteria form high population densities in the Warnow River.

Acknowledgements

Special thanks are due to Eurawasser in Rostock for providing the samples, to Katrina Laskus for performing microscopic biovolume determinations and to Jutta Fastner for HPLC analysis of microcystin in seston samples.

5.3.4.2
Elimination of Microcystins at Dörtendorf: Conventional Treatment and a Pilot Experimental Treatment System

Ingrid Chorus, Rita Heinze, Conrad Hübner, Wido Schmidt and Jörg Pietsch

The Dörtendorf drinking-water treatment plant processes 90000 m^3 day^{-1} of raw water from the eutrophic Weida Reservoir, which has substantial phytoplankton and zooplankton populations (see Chapter 2.1.1). Cyanobacteria contribute a high share of phytoplankton biomass, and from autumn to spring, *Planktothrix rubescens* often dominates phytoplankton. Due to its high microcystin content (see Chapter 2.1.2), this taxon is of particular concern in drinking-water sources.

The first drinking-water treatment step is a microsieve which retains on average 98% of the diatoms, 87% of the green algae and 40% of the cyanobacteria in the raw water. This first barrier is followed by conventional treatment of flocculation with aluminum sulphate and rapid sand filtration. The last treatment step is disinfection with chlorine dioxide (dose: 0.15-0.18 mg L^{-1}) and adjustment of pH with calcium carbonate. Concentrations of total microcystins (cell-bound and dissolved; sum of all variants) were assessed on five occasions during February and March 1998 in the raw water, after the microsieve, after flocculation and rapid sand filtration and in the finished water when *Planktothrix rubescens* was abundant in the raw water at biomass densities of 4.1 to 5.2 mm^3 L^{-1}.

Pilot scale experiments for upgrading this plant were conducted in 1998 with two different treatment trains. In one, the effluent of the rapid sand filter was passed through a granular activated carbon filter (nominal diameter of 150 mm with carbon type ROW 0.8 S(supra) produced by Norithã) The other train used both ozone and permanganate alternatively for pre-oxidation followed by a multiple layer filter. This consisted of three layers: 900 mm of Hydro-Anthrazit N (grain diameter 1.4-2.25 mm), 1000 mm of sand (grain diameter 0.71-1.25 mm) and a supporting bottom layer of 100 mm of gravel (grain diameter 2.00-3.15 mm). The filtration rates were 10 m h^{-1}.

Additionally, lab scale batch experiments were performed to assess the efficiency of removal of dissolved microcystins by enhanced treatment trains. For this purpose, the microsieve effluent was spiked with 2 µg L^{-1} of each microcystin-RR, -YR and -LR to assess the removal of these toxins by flocculation and sedimentation as well as by activated carbon. Flocculation tests were carried out with jar test equipment using 6 mg L^{-1} iron (III) chloride-sulphate (Ferrifloc, Kronos Int.) in conjunction with 0.1 mg L^{-1} aid agent Praestol 2530 (Stockhausen). For the activated carbon tests, carbon F 300 (Chemviron Carbon) was added to the test solution at a contact time of 30 min and in two different concentrations.

Analytical methods

Samples of 1 L volume were collected after various treatment steps and stored in a deep freeze (-18 °C) without prior filtration in order to determine total microcystins (dissolved and cell-bound). After thawing and filtration (0.45 µm membrane filters), clean up and enrichment of microcystin concentrations was performed with C18 both for HPLC analysis (see Chapter 9.1) as well as for ELISA (see Chapter 8.3). For raw water samples and some of the microsieve samples, ELISA was also applied to the unconcentrated samples before the C18 enrichment step. For the lab-scale batch experiments, microcystins were determined by HPLC-MSMS (API 2000, Waters, Sciex) as described in Paul (2001).

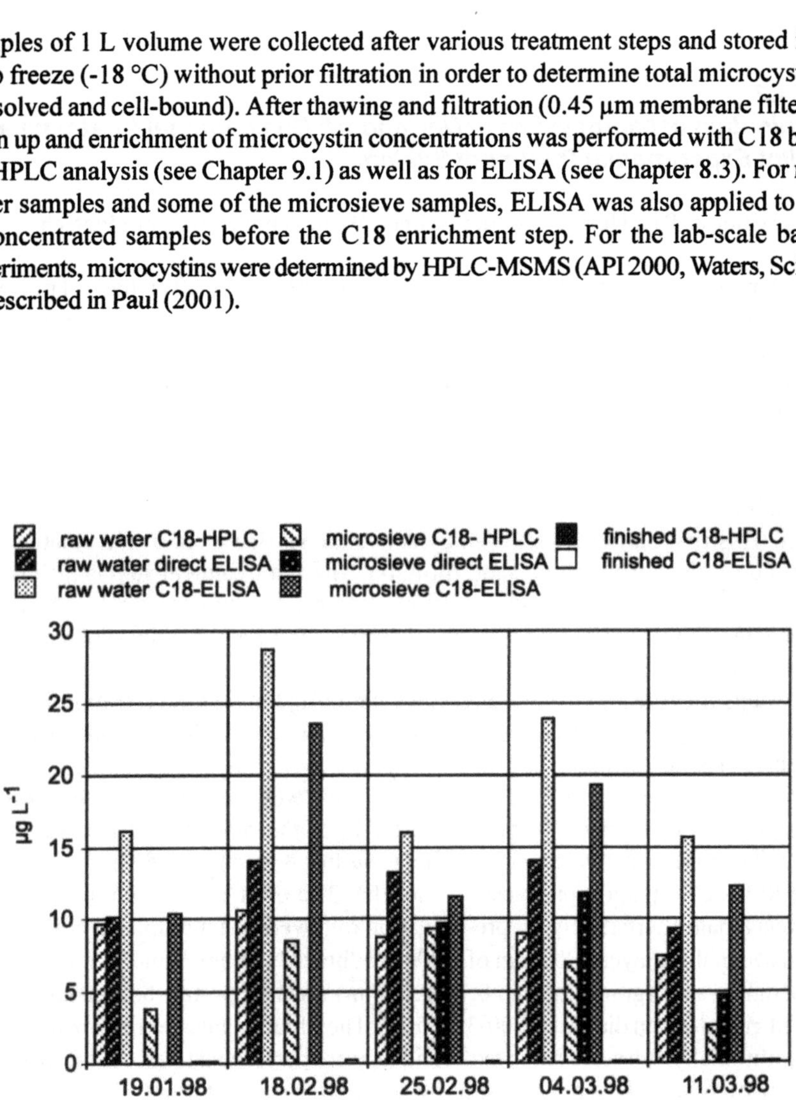

Fig. 1. Microcystin concentrations in raw water, after the microsieve and in finished water, detected with different methods

Results

Concentrations of microcystins (cell-bound and dissolved) in the raw water were in the range of 7.5 to 10 µg L^{-1} determined by HPLC (Fig. 1) on all five occasions investigated in February and March 1998 during the intensive population development of *Planktothrix rubescens*. Results obtained with the ELISA on samples without prior C18 enrichment were in good agreement with those obtained by HPLC, though ELISA results were always somewhat higher – usually only by 20% and by up to 35% on 4th March. Interestingly, results obtained with the ELISA from the samples enriched with C18 always showed the same trends as those obtained by ELISA without prior enrichment or by HPLC, but levels were 1.7 to 2.7-fold higher.

Reductions of microcystin concentrations after microsieve treatment ranged from negligible to at best 65%, thus reflecting the general assessment of performance for cyanobacterial cell removal of 40% (see above). As for raw water samples, concentrations determined by ELISA show the same patterns as those determined by HPLC; values obtained from samples without prior C18 enrichment were slightly higher, and after C18 enrichment they were substantially higher (Fig. 1, columns 4-6).

In spite of these rather high concentrations in the raw water, on all five occasions investigated the concentrations in finished water were below 1 µg L^{-1}, the WHO Guideline value for microcystin-LR in drinking water (Figs. 1 and 2). The filtration step of conventional treatment succeeded in reducing concentrations to less than 0.3 µg L^{-1}, as determined by ELISA after C18 enrichment (Fig. 2). Microcystin concentrations determined by HPLC after the filtration step were below the detection limit of 0.05 µg L^{-1} on 3 occasions and in the range of 0.08-0.1 µg L^{-1} on 2 occasions.

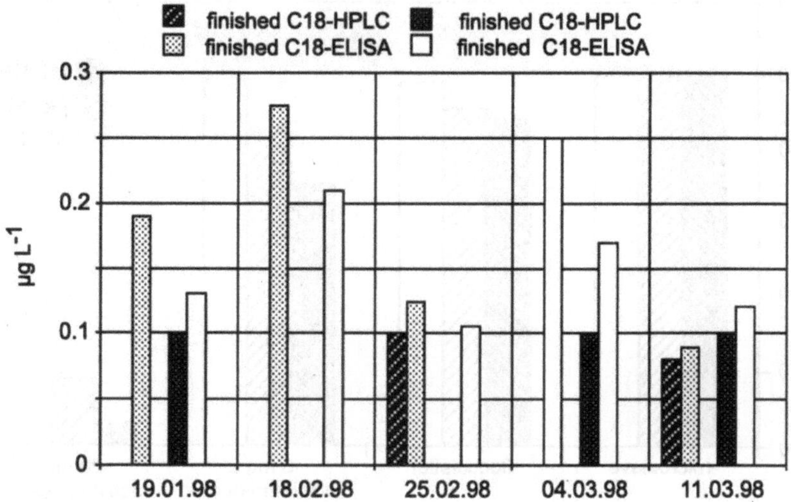

Fig. 2. Microcystin concentrations after rapid sand filtration and in finished water, detected with HPLC and ELISA

The final treatment step of chlorination and pH adjustment had no further effect on microcystin concentrations (Fig. 2).

The experimental pilot plant showed substantial retention of microcystins on the multi-layer filter: microcystins were not detectable in the effluent either with ELISA or with HPLC on two occasions in February, and only 0.1 to 0.3 µg L^{-1} were determined with HPLC or ELISA on the last three sampling occasions (Fig. 3). The following granular activated carbon filter (DN 150) further reduced concentrations to levels in the range of detection limits (Fig. 3).

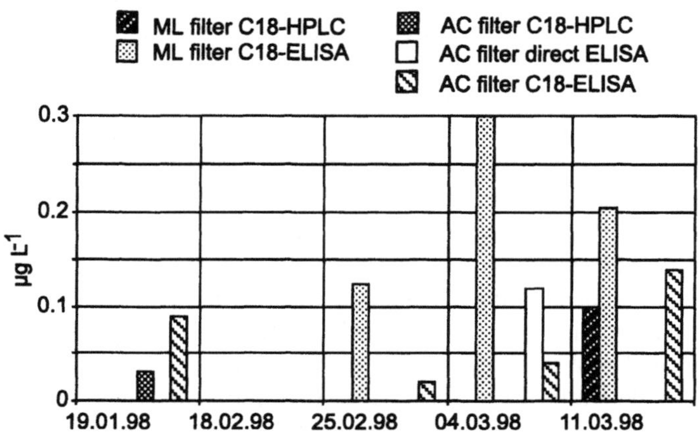

Fig. 3. Microcystin concentrations in the pilot scale experimental treatment system after multilayer filtration (*ML filter*) and after granular activated carbon filtration (*AC filter*), detected with different methods

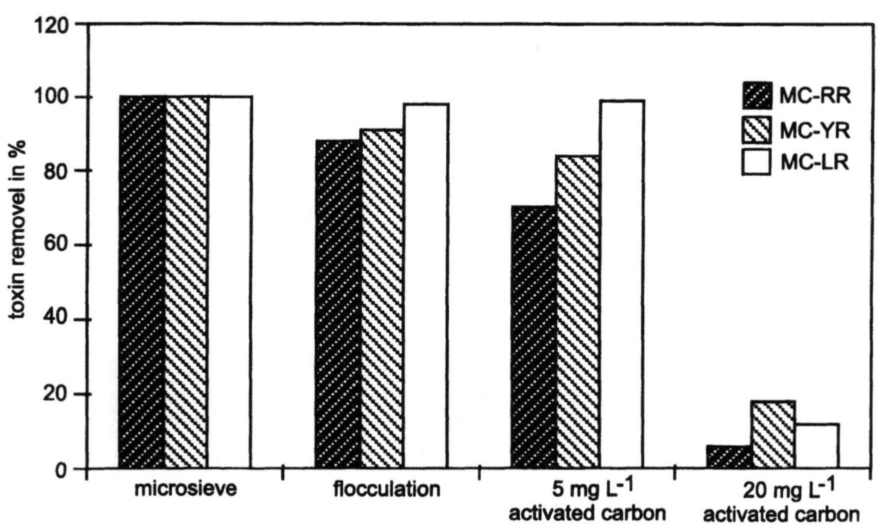

Fig 4. Efficiency of toxin removal by flocculation and activated carbon filtration (lab-scale batch experiments)

Batch experiments for the efficiency of extracellular microcystin removal showed that microcystin removal scarcely occurred in the simulation of conventional treatment through flocculation and filtration (Fig. 4). These results indicate that the use of activated carbon is necessary for sufficient toxin removal. In the batch experiments, more than 90% removal could be achieved, depending on optimization for contact time, the dose of carbon, (Fig. 4) and the choice between powdered activated carbon and carbon filters.

Conclusions

The excellent elimination performance of the conventional rapid sand filter indicates that a high share of total microcystins in the raw water was cell-bound, and that microcystins were presumably retained inside intact cells on the filters. Further investigations over an extended period of time are necessary to assess whether occasions of cell lysis and microcystin break-through may occur on these rapid sand filters. Chlorination at the low concentrations applied had no substantial effect in further reducing microcystin concentrations.

The results gained from the experimental pilot plant demonstrate good performance of the multiple-layer filter and of the granular activated carbon filter in retaining microcystins. Long-term monitoring of performance is necessary for these systems as well, in order to assess saturation and break-through characteristics.

Together, these results show that conventional treatment is an effective barrier against cell-bound toxins. However, the non-calculable risk of cell lysis and increase of dissolved toxins in water highlights the importance of additional treatment steps such as activated carbon filtration.

In comparison to results obtained by HPLC, the ELISA proved reliable, though several-fold higher concentrations determined from C18 enriched samples indicate that interfering substances causing false positive reactions may also be enriched on the C18 material. Nonetheless, this assay proved useful for preliminary assessments and for sensitive detection in the low-concentration range (see also Chapter 8.3).

Acknowledgements

We thank Gertrud Schlag for sample pretreatment, Ingeborg Flieger for HPLC-analysis and Andrea Jäger for performing the ELISA.

5.3.4.3 Elimination of Microcystins Through Bank Filtration at the Radeburg Reservoir

Ingrid Chorus, Gertrud Schlag, Rita Heinze, Klaus Pütz and Ute Kruspe

The Radeburg Reservoir system is situated north of the city of Dresden in the river basin of the Grosse Röder. It consists of the small reservoir Radeburg 1 with a volume of 0.4 mio m^3 and of the larger reservoir Radeburg 2 constructed in shunt with a water volume of 0.6 mio m^3. In the case of flood, Radeburg 1 overflows into Radeburg 2. Flood protection, fishing and augmentation of the low water level were the original functions of this system. Radeburg 2 is moreover part of a national trust property and a valuable bird sanctuary. Since 1986, the reservoir system is used for drinking water abstraction through bank filtration. It is estimated that the drinking-water wells abstract 10% groundwater and 90% bank filtrate of Radeburg 2; thus, bank filtrate has a dominant influence on raw water quality. The mean capacity of the waterworks amounts to 13000 m^3 day^{-1}.

The reservoir Radeburg 2 is shallow and highly eutrophic: total phosphorus concentrations are between 100 and 260 µg L^{-1} and high concentrations of both dissolved phosphorus (usually above 10-50 µg L^{-1}) and nitrogen (usually above 50-100 µg L^{-1}) indicate that phytoplankton in this waterbody is not nutrient-limited. Cyanobacteria and diatoms are usually dominant during the summer months, with *Microcystis* spp. and *Aphanizomenon* spp. as most frequent taxa. The reservoir was regularly monitored for microcystins in the context of the German cyanotoxin survey (see Chapter 2.1). This showed that microcystin content of seston was frequently very high, ranging up to 2.8 mg g^{-1} DW during dominance of *Microcystis aeruginosa*, whose biovolumes reached 15 mm^3 L^{-1} at corresponding chlorophyll-a concentrations of 140 µg L^{-1}. Microcystin concentrations L^{-1} water were often higher than 10 µg L^{-1} during the summer; in some years maxima amounted to 37 and 44 µg L^{-1}. This raised concern about microcystins in the raw water of the treatment plant.

Methods and Study Design

Bank filtration has been proven to retain many contaminants and microorganisms very effectively. To assess whether this applies to microcystins at the Radeburg site, the reservoir and wells collecting bank filtrate were studied from August to December 1997. Samples of the reservoir were taken at weekly intervals and analyzed for both cell-bound and dissolved microcystins. Seepage water collected in a ditch behind a reservoir dam was collected on four occasions to test concentrations after a short distance of soil filtration. Concentrations in selected wells were assessed fortnightly. On each occasion, six wells were selected from a total of ten wells included in the study, depending on their current operation. Cell-bound microcystins were analyzed with HPLC (see Chapter 9.1), and dissolved microcystins were analysed with ELISA as described in Chapter 8.3, but with prior sample enrichment through C18 (see

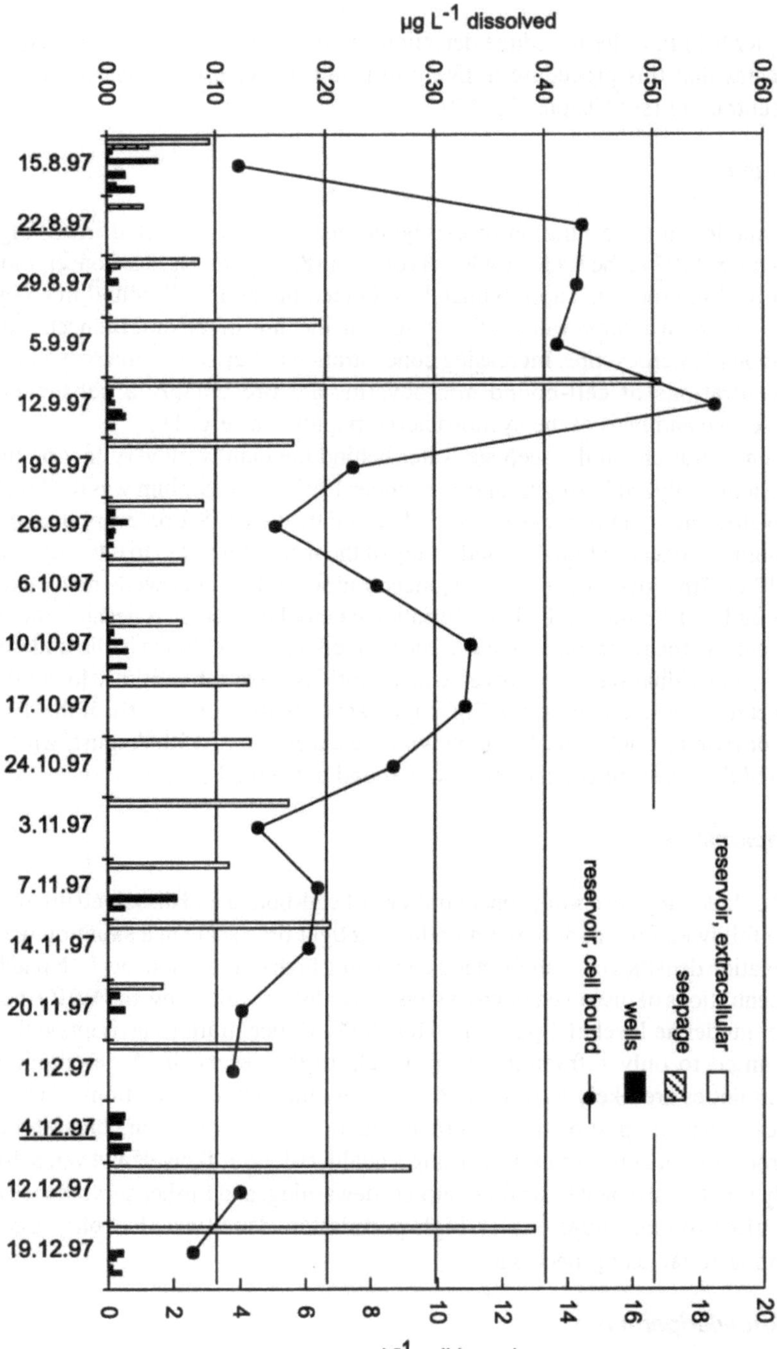

Fig. 1. Microcystin concentrations in the Radeburg Reservoir and its bank filtrate. *Curve* Cell-bound microcystins determined by HPLC in µg L^{-1}; *bars* dissolved microcystins determined with ELISA; *Data underlined* incomplete data

Chapter 9.1) in order to reduce detection limits to 0.01 µg L^{-1}. Current experience indicates that this procedure is likely to result in overestimations of microcystin concentrations (see Chapter 5.3.4.2).

Results

The maximum concentration of cell-bound microcystins detected from August to December 1997 in the Radeburg Reservoir was 18.5 µg L^{-1} (Fig. 1). Concentrations of dissolved microcystins approximately reflected those of cell-bound microcystins, with maximum autumn values of 0.51 µg L^{-1} at the time of maximal concentrations of cell-bound microcystins. Increasing concentrations of dissolved microcystins at low concentrations of cell-bound microcystins are presumably a consequence of senescence and lysis of the cyanobacterial population (Fig. 1).

Concentrations in the seepage water behind the dam were very low on the four occasions analyzed in August and September 1997: the maximum was 0.04 µg L^{-1}. In the wells, traces of microcystins were detected by the ELISA on every occasion, but the highest concentration detected in any of the wells amounted to 0.06 µg L^{-1} in well no. 12 on 7th November (Fig. 1). Concentrations in the other wells never exceeded 0.05 µg L^{-1}. Elimination by bank filtration cannot be calculated because the time of reference in the reservoir is unclear due to uncertain travel times in the underground for the individual wells. However, concentrations were substantially lower than the concentrations in the reservoir. Fig. 1 indicates retention of 75-99% of the dissolved microcystin through bank filtration with the exception of 15th August, when about 40% of the reservoir concentration was found in the wells.

Conclusions

At the Radeburg Reservoir, concentrations of cell-bound and dissolved microcystins were followed for a period of 4 months on a tight time scale in a situation with high population densities of cyanobacteria containing high microcystin levels. Nonetheless, concentrations of dissolved microcystins were always well below the WHO drinking-water guideline level of 1 µg L^{-1} (WHO 1998). Concentrations arriving in the wells amounted to only a fraction of the levels in the reservoir. As methodological uncertainties are likely to result in over- rather than underestimations, these results indicate that in spite of the eutrophic situation in the reservoir, breakthrough of microcystins at concentrations causing a health risk is unlikely. However, a 4-month study is not a sufficient basis for complete dewarning, and further surveillance during and after bloom situations or high population densities of evenly dispersed cyanobacterial taxa is necessary.

Acknowledgements

We wish to thank Ingeborg Flieger for HPLC-analysis and Andrea Jäger for performing the ELISA.

5.3.4.4
Reducing Intake of Microcystins at the Deesbach Reservoir Drinking-Water Abstraction System

Jutta Fastner, Ingrid Chorus, Hartmut Willmitzer and Carol Rabe

The drinking-water treatment plant at Zeigerheim processes 13000 m^3 day^{-1} of raw water from the Deesbach Reservoir. Treatment comprises preozonation; flocculation with aluminum sulfate and filtration at 3.5-5.2 m h^{-1}, pH adjustment and chlorination; optionally, powdered activated carbon may be dosed at 5-50 g m^{-3}).

The Deesbach Reservoir is located in a largely wooded area with only 15% of the catchment being used for agriculture. However, 7% of the catchment area is covered by scattered settlements, and poorly treated domestic sewage led to pronounced eutrophication of the reservoir: total phosphorus concentrations range from 100 to 500 µg L^{-1} and consequently, concentrations of chlorophyll-a are usually above 20 µg L^{-1} and may reach 130 µg L^{-1} (the German working group on drinking-water reservoirs (Arbeitsgemeinschaft Trinkwasser-Talsperren; ATT) recommends no more than 5 µg L^{-1}). At the time of a preliminary investigation in September 1997, a mass development of *Microcystis* spp. was found to contain particularly high concentrations of microcystins of 3.4 mg g^{-1} DW (largely MC-LR and MC-YR). At a microcystin to chlorophyll-a ratio of 0.5 to 1, this would imply 75 to 130 µg L^{-1} of microcystins.

Raw water for drinking-water treatment is abstracted at a depth of 15 or 20 m. It reaches the treatment plant after approximately 2 days and 16.5 km of flow through a system of pipes and tunnels. Effects of this flow on cyanobacterial cells are conceivable; in particular, the potential for lysis during flow in the tunnel was a concern. For long-term alleviation of the situation, sewage treatment measures are being planned. Further, Deesbach Reservoir is intended to serve only as a prereservoir for a large reservoir yet to be constructed, because due to its high hydraulic load per surface area (Table 1), the Deesbach Reservoir has a natural tendency towards eutrophication. However, the current situation required an immediate monitoring program for microcystins during times of cyanobacterial occurrence in the reservoir.

A pilot monitoring program was therefore tested in July and August, 1998, when *Microcystis* spp. occurred in the reservoir.

Table 1. Hydromorphological Characteristics of the Deesbach Reservoir

mean retention time	approximately 0.12 year
mean depth	14.5 m
annual discharge	26.3 mio m^3
volume	3.2 mio m^3
surface area	0.22 km^2
catchment area	49 km^2

Methods

Samples were taken from the surface of the reservoir and from the intake depth at 15 or 20 m, as well as from different sites following the path of water to the treatment plant as well as within the treatment plant. Concentrations of cell-bound and dissolved microcystins were determined by HPLC as described in Chapter 9.1. For dissolved microcystins, sample enrichment and cleanup over C18 were supplemented by a further cleanup step using silica cartridges (Tsuji et al. 1994). This reduced detection limits to 0.01 µg L^{-1}. Samples in the treatment plant were taken from raw water, before and after flocculation, after filtration and from finished water.

Results

Microcystin content of the cyanobacteria was again high in 1998: seston material from the reservoir contained 3 mg g^{-1} DW on 8th July, 1998, chiefly MC-YR and MC-LR. Towards the end of July, seston content declined to 1.8 mg g^{-1} DW. On 5th August, only microcystin-RR was detected, which indicates a shift in species or strain composition within the cyanobacterial population.

Cell-bound microcystin concentrations in surface water reached a peak value of 566 µg L^{-1} on 21st July and then rapidly declined to concentrations below 1 µg L^{-1} (Tab. 2). Dissolved microcystins showed a maximum of 3.6 µg L^{-1} in the week after the maximum of cell-bound microcystins. Surprisingly, in spite of these high levels, concentrations at the offtake depth of 17 m were never higher than 0.3 µg L^{-1}, of both cell-bound and dissolved microcystins. After 16.5 km of flow, concentrations of both cell-bound and dissolved microcystins declined to levels below 0.1 µg L^{-1}, and at the end of the tunnel, only traces of dissolved microcystins and at most 0.09 µg L^{-1} of cell-bound microcystins were detected (Table 2). On all four occasions, no microcystins were detected in the water that reached the drinking-water treatment plant or after any treatment step.

Table 2. Microcystin concentrations in the Deesbach Reservoir 0 and 17 m downstream

Total microcystins (µg L^{-1})		Reservoir surface	Reservoir 17 m offtake	Before tunnel (Hainberg)	After tunnel (Lichtestollen)	Entry to waterworks
14.07.1998	Cell-bound	74.5	0.30	0.04	0.09	n.d.
	Dissolved	0.01	n.d.	n.d.	n.d.	n.d.
21.07.1998	Cell-bound	565.9	0.10	0.05	0.03	n.d.
	Dissolved	0.13	0.02	0.01	0.01	n.d.
28.07.1998	Cell-bound	0.05	0.23	0.16	n.d.	n.d.
	Dissolved	3.63	0.07	0.05	0.02	n.d.
05.08.1998	Cell-bound	0.28	n.d.	0.04	n.d.	n.d.
	Dissolved	0.11	0.02	0.02	0.02	n.d.

Conclusions

The concentrations of cell-bound microcystins were among the highest detected in the entire German survey, with respect to both content in the *Microcystis* cells (analyzed as microcystin per dry weight of seston) and to concentrations L^{-1} of water (see Chapter 2.1.2), yet concentrations of dissolved microcystins in surface water remained low, and at the drinking-water offtake depth, concentrations of both cell-bound and dissolved microcystins were never above the WHO drinking-water guideline level of 1 $\mu g\ L^{-1}$ on the four occasions monitored.

Microcystis spp. form pronounced surface scums but are also capable of vertical migration. Possibly, sampling once per week and only at one time of a day missed higher concentrations in deeper layers as the population migrated. However, if sampling missed substantial concentrations, elevated concentrations further downstream would have been expected, as sampling along four points of flow before the treatment plant and throughout the treatment plant integrates over time. Thus, this preliminary study suggests two conclusions:

1. It is possible to evade high concentrations of cell-bound microcystins by situating offtake depths under the scum horizons.
2. Nonetheless, monitoring of vertical locations of cyanobacterial populations is essential to ensure low concentrations in the raw water.

Ideally, in situations such as the Deesbach Reservoir, continuous monitoring of the cyanobacterial population supported by occasional analysis of cellular toxin content would be desirable. On-line fluorescence detection could be installed for such purposes. If "calibrated" with occasional (e.g. weekly) determination of ratios of fluorescent signals to microcystin content of the cells, elevated fluorescence signals could be employed to indicate elevated microcystin concentrations.

References

Annadotter H (1993) Algtoxiner i dricksvatten. VA-Forsk rapport 1993-03. Svenska vatten- och avloppsverksföreningen, VAV, Stockholm. 20 pp. In Swedish with an English abstract

Baker PD, Humpage AR (1994) Toxicity associated with commonly occurring cyanobacteria in surface waters of the Murray-Darling Basin. Aust J Marine Freshwater Res 45:773–786

Bartram J, Burch M, Falconer IR, Jones G, Kuiper-Goodman T (1999) Situation assessment, planning and management. In: Chorus I, Bartram J (eds) Toxic cyanobacteria in water: a guide to their public health consequences, monitoring and management. Published on the behalf of WHO by E & FN Spon, London, pp. 179–209

Beasley VR, Dahlem AM, Cook WO, Valentine WM, Lovell RA, Hosser SB, Suzuki M, Carmichael WW (1989) Diagnostic and clinically important aspects of cyanobacterial (blue-green algae) toxicosis. J Vet Diagn Invest 1:359–365

Berg K, Carmichael WW, Skulberg OM, Benestad C, Underdal B (1987) Investigation of a toxic water bloom of *Microcystis aeruginosa* (CYANOPHYCEAE) in Lake Akersvatn, Norway. Hydrobiologia 144:97–103

Bourke ATC, Hawes RB, Neilson A, Stallman ND (1983) An outbreak of hepato-enteritis (the Palm Island Mystery Disease) possibly caused by algal intoxication. Toxicon 3(Suppl.):45–48

Bundesgesundheitsblatt (1997) Empfehlung zum Schutz von Badenden vor Cyanobakterien-Toxinen. Bundesgesundheitsblatt 7/97:261

Byth S (1980) Palm Island mystery disease. Med J Aust 2:40–42
Carmichael WW (1992) A status report on planktonic cyanobacteria (blue-green algae) and their toxins. EPA/600/R-92-079, Environmental Monitoring Systems Laboratory, Office of Research and Development, U.S. Environmental Protection Agency, Cincinnati, OH. 141 pp.
Carmichael WW (1992) Cyanobacteria secondary metabolites – the cyanotoxins. J Appl Bacteriol 72:445–459
Carmichael WW (1994) The toxins of cyanobacteria. Sci Am 270:78–86
Carmichael WW (1995) Toxic Microcystis and the environment. In: Watanabe, MF, Harada K, Carmichael WW, Fujiki H (eds) Toxic Microcystis. CRC Press, Inc. ISBN 0–8493–763–9
Carmichael WW, Eschedor JT, Patterson GML, Moore RE (1988) Toxicity and partial structure for a hepatotoxic peptide produced by *Nodularia spumigena* Mertens emend. strain L575 (cyanobacteria) from New Zealand. Appl Environ Microbiol 54(9):2257–2263
Catterall W (1980) Neurotoxins that act on voltage-sensitive sodium channels in excitable membranes. Ann Rev Pharmacol Toxicol 20:15–43
Chiswell R, Smith M, Norris R, Eaglesham G, Shaw G, Seawright A, Moore M (1997) The cyanobacterium, *Cylindrospermopsis raciborskii*, and its related toxin, cylindrospermopsin. Australasian J Ecotoxicol 3:7–23
Chorus I, Falconer I, Bartram J, Salas H (2000): Health risks caused by freshwater cyanobacteria in recreational waters. J Toxicol Environm Health 3:323–347
Chorus I, Mur L (1999) Preventive measures. In: Chorus I, Bartram J (eds) Toxic cyanobacteria in water: a guide to their public health consequences, monitoring and management. Published on the behalf of WHO by E & FN Spon, London, pp. 235–274
Chorus I, Bartram J eds. (1999) Toxic cyanobacteria in water: a guide to their public health consequences, monitoring and management. Published on the behalf of WHO by E & FN Spon, London
Codd GA (1992) Eutrophication, blooms and toxins of cyanobacteria (blue-green algae), and health. In: Keller AZ, Wilson HC (eds.) Proceedings of the 4th Disaster Prevention and Limitation Conference. The changing face of Europe: disasters, pollution and the environment. Vol.4. Aquatic problems, p. 33–62. University of Bradford, UK
Codd GA, Ward CJ, Bell SG (1997) Cyanobacterial toxins: occurrence, modes of action, health effects and exposure routes. Archiv Toxicol 19 (Suppl):399–410
Craig M, Luu HA, Mccready TL, Williams D, Anderson RJ, Holmes CFB (1996) Molecular mechanisms underlying the interaction of motuporin and microcystins with type-1 and type-2A protein phosphatases. Biochem Cell Biol 74:569–578
Cronberg G (1982) Phytoplankton changes in Lake Trummen induced by restoration. Fol Limnol Scand 18:1–119
Cronberg G, Annadotter H, Lindberg M, Lirås V, Lawton L (1997) Undersökning om förekomst av algtoxiner i rå- och renvatten från Vombverket samt påväxtalger i infiltrationsdammarna 1996–1997. Institute of Ecology, Dept. of Limnology Lund University (In Swedish)
Dawson RM (1998) The toxicology of microcystins. Toxicon 36:953–962
De Olivera Araújo F (1995) Effects of cyanobacteria on drinking water and human health: an epidemiological study in Évora, Portugal. In: Reichard EG, Zapponi GA (eds) Assessing and managing health risks from drinking water contamination: approaches and applications. International Association of Hydrological Sciences, Publ. No. 233. IAHS Press, Wallingford, Oxfordshire, pp. 101–110
Dillenberg HO, Dehnel MK (1960) Toxic waterbloom in Saskatchewan, 1959. Canad Med Assoc J 83:1151–1154
Ding WX, Shen HM, Zhu HG, Ong CN (1998) Studies on oxidative damage induced by cyanobacteria extract in primary cultured rat hepatocytes. Environ. Res. 78:12–18

Dube SN, Mazumder PK, Kumar D, Rao PVL, Bhaskar ASB (1996) Cardiorespiratory and neuromuscular effects of freshwater cyanophyte Anabena flos aquae in rats. Def Sci J. 46(3): 135–141

EC Directive 76/160/EWG (1976) Directive on the quality of bathing water, Nr. L 31/1–7

El Saadi O, Cameron AS (1993) Illness associated with exposure to blue-green algae. Med J Aust 185:22

Environmental Agency, UK (1998) The determination of microcystin algal toxins in raw and treated waters by high performance liquid chromatography. In: Westwood D (ed) Methods for the examination of waters and associated materials. Environment Agency, UK, London, pp. 1–12

Falconer IR (1998) Algal toxins and human health. In: Hrubec J (ed.) Quality and treatment of drinking water II. The handbook of environmental chemistry Vol. 5, Part C:53–82. Springer-Verlag, Berlin

Falconer IR (1991) Tumour promotion and liver injury caused by oral consumption of cyanobacteria. Environ. Toxicol. Water Qual. 6(2):177–184

Falconer IR (1993) Algal toxins in seafood and drinking water. Academic Press, London, 224 pp.

Falconer IR, Beresford AM, Runnegar MTC (1983) Evidence of liver damage by toxin from a bloom of the blue-green alga *Microcystis aeruginosa*. Med J Aust 1:511–514

Falconer IR, Buckley TH (1989) Tumour promotion by *Microcystis* sp., a blue-green alga occurring in water supplies. Med J Aust 150:351

Falconer IR, Dornbusch M, Moran G, Yeung SK (1992) Effect of the cyanobacterial (blue-green algal) toxins from *Microcystis* on isolated enterocytes from the chicken small intestine. Toxicon 30:790–793

Falconer IR, Burch MD, Steffensen DA, Choice M, Coverdale OR (1994) Toxicity of the blue-green alga (cyanobacterium) *Microcystis aeruginosa* in drinking water to growing pigs, as an animal model for human injury and risk assessment. J Environ Toxicol Water Qual 9:131–139

Falconer IR, Humpage AR (1996) Tumour promotion by cyanobacterial toxins. Phycologia 35(6 Suppl):74–79

Falconer I, Bartram J, Chorus I, Kuiper-Goodman T, Utkilen H, Codd G (1999a) Safe levels and practices. In: Chorus I, Bartram J (eds) Toxic cyanobacteria in water: a guide to their public health consequences, monitoring and management. Published on the behalf of WHO by E & FN Spon, London, pp. 155–178

Falconer IR, Hardy SJ, Humpage AR, Froscio SM, Tozer GJ, Hawkins PR (1999b) Hepatic and renal toxicity of the blue-green alga (cyanobacterium) Cylindrospermopsis raciborskii in male Swiss Albino mice. Environ. Toxicol. 14:143–150

Fastner J, Neumann U, Wirsing B, Weckesser J, Wiedner C, Nixdorf B, Chorus I (1999) Microcystins (hepatotoxic heptapeptides) in German fresh waters. Env. Toxicol. 14:13–22

Fawell JK, James HA (1994) Toxins from blue-green algae: toxicological assessment of anatoxin-a and a method for its determination in reservoir water. Report No. FR 0434/DoE 3728. Foundation for Water Research, Marlow, Buckinghamshire, 45pp

Fawell J, Mitchell RE, Everett DJ, Hill RE (1999) The toxicity of cyanobacterial toxins in the mouse: I Microcystin-LR. Human and Experim Toxicol 18:162–167

Fawell JK, James CP, James HA (1994) Toxins from blue-green algae: toxicological assessment of microcystin-LR and a method for its determination in water. Report No. FR 0359/2/DoE 3358/2. Foundation for Water Research, Marlow, Buckinghamshire, 46 pp

Fitzgeorge R, Clark S, Keevil C (1994) Routes of intoxication. In: Codd GA, Jefferies TM, Keevil CW, Potter E (eds) Detection methods for cyanobacterial toxins. The Royal Society of Chemistry. ISBN 0–85186–961–0, pp. 69–74

Fitzgerald DJ, Yamasaki H (1990) Tumor promotion: models and assay systems. Teratogen Carcinogen Mutagen 10:89–102

Fitzgerald DJ, Cunliffe DA, Burch MD (1999) Development of health alerts for cyanobacteria and

related toxins in drinking water in South Australia. Environ Toxicol 14:203–209
Fromme H, Köhler A, Krause R, Führling D (2000) Occurrence of cyanobacterial toxins – microcystins and anatoxin-a – in Berlin water bodies with implications to human health and regulations. Environm Toxicol 15:120–130
Fujiki H, Sueoka E, Suganuma M (1996) Carcinogenesis of microcystins. In: Watanabe MF, Harada K-I, Carmichael WW, Fujiki H (eds.) Toxic *Microcystis*. CRC Press, Boca Raton, Florida, p. 203–232
Fujiki H, Suganuma M (1993) Tumor promotion by inhibitors of protein phosphatases 1 and 2A: the okadaic acid class of compounds. Adv Cancer Res 61:143–194
Harada K, Ikuko O, Iwamoto K, Suzuki M, Watanabe M, Watanabe M, Tereo K (1994): Isolation of cylindrospermopsin from a cyanobacterium *Umezakia natans* and its screening method. Toxicon 32:73–84
Hawkins PR, Chandrasena NR, Jones GJ, Humpage AR, Falconer IR (1997) Isolation and toxicity of *Cylindrospermopsis raciborskii* from an ornamental lake. Toxicon 35:341–346
Hawkins PR, Runnegar MTC, Jackson ARB, Falconer IR (1985) Severe hepatotoxicity caused by the tropical cyanobacterium (blue-green alga) *Cylindrospermopsis raciborskii* (Woloszynska) Seenaya and Subba Raju isolated from a domestic water supply reservoir. Appl Environ Microbiol 50(5):1292–1295
Health Canada (1998) Cyanobacterial toxins – microcystins in drinking water. Public Comment Document
Heresztyn T, Nicholson BC (1997) Nodularin concentrations in Lakes Alexandrina and Albert, South Australia, during a bloom of the cyanobacterium (blue-green alga) *Nodularia spumigena* and degradation of the toxin. Environ Toxicol Water Qual 12:273–282
Hermansky SJ, Casey PJ, Stohs SJ (1990a) Cyclosporin A – a chemoprotectant against microcystin-LR toxicity. Toxicol Lett 54:279–285
Hermansky SJ, Wolff SN, Stohs SJ (1990b) Use of rifampin as an effective chemoprotectant and antidote against microcystin-LR toxicity. Pharmacol 41:231–236
Hindman SH, Favero MS, Carson LA, Petersen NJ, Schonberger LB, Solano JT (1975) Pyrogenic reactions during haemodialysis caused by extramural endotoxin. The Lancet ii:732–734
Hitzfeld BC, Höger SJ, Dietrich DR (2000) Cyanobacterial toxins: removal during drinking water treatment, and human risk assessment. Environ Health Perspect 108 (Suppl 1):113–122
Honkanen RE, Zwiller J, Moore RE, Daily SL, Khatra BS, Dukelow M, Boynton AL (1990) Characterization of microcystin-LR, a potent inhibitor of type 1 and type 2A protein phosphatases. J Biol Chem 265:19401–19404
Hrudey SE, Lambert TW, Kenefick SL (1994) Health risk assessment of microcystins in drinking water supplies. In: Toxic cyanobacteria – a global perspective. March 28, 1994. Australian Centre for Water Quality Research, Adelaide, South Australia. pp. 7–12
Hrudey S, Burch M, Drikas M, Gregory R (1999) Remedial measures. In: Chorus I, Bartram J (eds) Toxic cyanobacteria in water: a guide to public their health significance, monitoring and management. E & FN Spon, London, pp. 275–312
Humpage AR, Rositano J, Bretag A, Brown R, Baker P, Nicholson BC, Steffensen DA (1994) Paralytic shellfish poisons from Australian blue-green algal (cyanobacterial) blooms. Aust J Mar Freshwater Res 45:761–771
Humpage AR, Falconer IR (1999) Microcystin-LR and liver tumor promotion: effects on cytokinesis, ploidy, and apoptosis in cultured hepatocytes. Environ Toxicol Water Qual 14:61–75
Humpage AR, Hardy SJ, Moore EJ, Froscio SM, Falconer IR (2000) Microcystins (cyanobacterial toxins) in drinking water enhance the growth of aberrant crypt foci in the mouse colon. J Toxicol Environ Health Part A 61:101–111
Ito E, Kondo F, Harada K-I (1997a) Hepatic necrosis in aged mice by oral administration of microcystin-LR. Toxicon 35:231–239

Ito E, Kondo F, Terado K, Harada K-I (1997b) Neoplastic nodular formation in mouse liver induced by repeated intraperitoneal injections of microcystin-LR. Toxicon 35:1453–1457

Ito E, Kondo F, Harada K-I (2000) First report on the distribution of orally administered microcystin-LR in mouse tissue using an immunostaining method. Toxicon 38:37–48

Jacobsen BA (1994) Bloom formation of *Gloeotrichia echinulata* and *Aphanizomenon flos-aquae* in a shallow, eutrophic, Danish lake. Hydrobiologia 289:193–197

Jochimsen EM, Carmichael WW, An JS, Cardo DM, Cookson ST, Holmes CEM, Antunes MBD, Demelo DA, Lyra TM, Barreto VST, Azevedo S, Jarvis WR (1998) Liver failure and death after exposure to microcystins at a hemodialysis center in Brazil. N Engl J Med 338:873–878

Jones G (1996) Toxic algae study summary. Manitoba Department of Environment, February

Jones GJ, Burch M, Falconer I, Craig K (1993) Cyanobacterial toxicity. Technical Advisory Group Report, Murray-Darling Basin Commission Algal Management Strategy, MDBC, Canberra: 17–32

Kao CY (1993) Paralytic shellfish poisoning. In: Falconer IR (ed) Algal toxins in seafood and drinking water. Academic Press, London, p. 75–86

Katoh F, Fitzgerald DJ, Giroldi L, Fujiki H, Sugimura T, Yamasaki H (1990) Okadaic acid and phorbol esters: comparative effects of these tumor promoters on cell transformation, intercellular communication and differentiation in vitro. Jpn J Cancer Res 81:590–597

Kay D, Fleisher JM, Salmon RL, Jones F, Wyer MD, Godfree AF, Zelenauch-Jacquotte Z, Shore R (1994) Predicting likelihood of gastroenteritis from sea bathing: results from randomised exposure. Lancet 344: 905–909

Keleti G, Sykora JL, Libby EC, Shapiro MA (1979) Composition and biological properties of lipopolysaccharides isolated from *Schizothrix calcicola* (Ag.) Gomont (cyanobacteria). Appl Environ Microbiol 38:471–477

Kenefick SL, Hrudey SE, Peterson HG, Prepas EE (1993) Toxin release from *Microcystis aeruginosa* after chemical treatment. Water Sci Technol 27:433–440

Kenefick SL, Hrudey SE, Prepas EE, Motkosky N, Peterson HG (1992) Odorous substances and cyanobacterial toxins in prairie drinking water sources. Water Sci Technol 25:147–154

Kotak BG, Kenefick SL, Fritz DL, Rousseaux CG, Prepas EE, Hrudey SE (1993) Occurrence and toxicological evaluation of cyanobacterial toxins in Alberta lakes and farm dugouts. Water Res., 27: 495-506

Kotak BG, Lam AKY, Prepas EE, Kenefick SL, Hrudey SE (1995) Variability of the hepatotoxin microcystin-LR in hypereutrophic drinking-water lakes. J. Phycol. 31: 248-263

Kuiper-Goodman T, Falconer I, Fitzgerald J (1999) Human Health Aspects. In: Chorus I, Bartram J (eds) Toxic cyanobacteria in water: a guide to their public health consequences, monitoring and management. Published on the behalf of WHO by E & FN Spon, London, pp. 133–154

Lahti K, Rapala J, Färdig M, Niemelä SI, Sivonen K (1997) Persistence of cyanobacterial hepatotoxin, microcystin-LR in particulate material and dissolved in lake water. Water Research 31: 1005–1012

Lam AKY, Prepas EE, Spink D, Hrudey SE (1995). Chemical control of hepatotoxic phytoplankton blooms – implications for human health. Water Res 29:1845–1854

Lambert TW, Holmes CFB, Hrudey SE (1996) Adsorption of microcystin-LR by activated carbon and removal in full-scale water treatment. Water Res 30:1411–1422

Lawton L, Codd GA (1991) Cyanobacterial (blue-green algal) toxins and their significance in UK and European waters. Water & Environm Man 5:460–465

Lawton LA, Edwards C, Codd GA (1994) Extraction and high-performance liquid chromatographic method for the determination of microcystins in raw and treated waters. Analyst 119:1525–1530

Lepistö L, Lahti K, Rancken H, Salovaara E, Hiisvirta, L (1993) Sinileväongelmat Taalintehtaan vesilaitoksilla 1989 ja 1990. Vesitalous 4:26–30 (In Finnish)

Lippy EC, Erb J (1976) Gastrointestinal illness at Sewickley, PA. J Am Water Works Assoc 68:606–610

Matsunaga S, Moore RE, Niemczura WP, Carmichael WW (1989) Anatoxin-a(s) a potent anticholinesterase from *Anabaena flos-aquae*. J Am Chem Soc 111:8021–8023

McFarren EF, Schafer ML, Campbell JE, Lewis KH, Jensen ET, Schantz EJ (1960) Public health significance of paralytic shellfish poison. Adv Food Res 10:135–179

Moollan RW, Rae B, Verbeek A (1996) Some comments on the determination of microcystin toxins in waters by high-performance liquid-chromatography. Analyst Vol 121: 233–238

Moore MR, Seawright AA, Chiswell RR, Norris RL, Shaw GR (1998) The cyanobacterial toxin, cylindrospermopsin: its chemical properties and toxicology. Human Exp Toxicol 17:503

Nagata S, Soutome H, Tsutsumi T, Hasegawa A, Sekijima M, Sugamata M, Harada K-I, Suganuma M, Ueno Y (1995) Novel monoclonal antibodies against microcystin and their protective activity for hepatotoxicity. Nat Toxins 3:78–86

Negri AP, Jones GJ, Hindmarsh M (1995) Sheep mortality associated with paralytic shellfish poisoning toxins from the cyanobacteria *Anabaena circinalis*. Toxicon 33:1321–1329

NHMRC/Armcanz (1996) Australian drinking water guidelines. National Health & Medical Research Council, and Agriculture & Resource Management Council of Australia and New Zealand, Canberra

Nishiwaki-Matsushima R, Ohta T, Nishiwaki S, Suganuma M, Kohyama K, Ishika T, Carmichael WW, Fujiki H (1992) Liver tumor promotion by the cyanobacterial cyclic peptide toxin microcystin-LR. J Cancer Res Clin Oncol 118:420–424

Ohta T, Sueoka E, Iida N, Komori A, Suganuma M, Nishiwaki R, Tatematsu M, Kim S-J, Carmichael WW, Fujiki H (1994) Nodularin, a potent inhibitor of protein phosphatases 1 and 2A, is a new environmental carcinogen in male F344 rat liver. Cancer Res 54:6402–6406

Ohtani I, Moore RE, Runnegar MTC (1992) Cylindrospermopsin – a potent hepatotoxin from the blue-green alga *Cylindrospermopsis raciborskii*. J Am Chem Soc 114:7941–7942

Paul G (2001) Verhalten von intra- und extrazellularen Algenmetaboliten bei der Trinkwasseraufbereitumg. Diplomarbeit, Technische Universität Dresden

Pietsch J, Schmidt W (1999) Untersuchungen zum Vorkommen und Verhalten ausgewählter Algenmetaboliten bei der Trinkwasseraufbereitung. BMBF-Vorhaben 02WT 9853/0, Zwischenberichte

Pilotto LS, Douglas RM, Burch MD, Cameron S, Beers M, Rouch GR, Robinson P, Kirk M, Cowie CT, Hardiman S, Moore C, Attwell RG (1997) Health effects of exposure to cyanobacteria (blue-green algae) due to recreational water-related activities. Aust N Zealand J Public Health 21:562–566

Pouria S, Deandrade A, Barbosa J, Cavalcanti RL, Barreto VTS, Ward CJ, Preiser W, Poon G K, Neild GH, Codd GA (1998) Fatal microcystin intoxication in haemodialysis unit in Caruaru, Brazil. The Lancet 352:21–26

Rao PV, Bhattacharya R (1996) The cyanobacterial toxin microcystin-LR induced DNA damage in mouse liver in vivo. Toxicol 114:29–36

Ressom R, Soong FS, Fitzgerald J, Turczynowicz L, El Saadi O, Roder D, Maynard T, Falconer I (1994) Health effects of toxic cyanobacteria (blue-green algae). National Health and Medical Research Council, Canberra, 108 pp.

Rodrigue DC, Etzel RA, Hall S, De Porras E, Velasquez OH, Tauxe RV, Kilbourne EM, Blake PA (1990) Lethal paralytic shellfish poisoning in Guatemala. Am J Trop Med Hyg 42:267–271

Runnegar MT, Jackson ARB, Falconer IR (1988) Toxicity of the cyanobacterium *Nodularia spumigena* mertens. Toxicon 26:143–151

Runnegar MTC, Kong S-M, Zhong Y-Z, Ge J, Lu S (1995) Inhibition of reduced glutathione synthesis by cyanobacterial alkaloid cylindrospermopsin in cultured rat hepatocytes. Biochem Pharmacol 49:219–225

Schwimmer M, Schwimmer D (1964) Algae and medicine. In: Jackson DF (ed) Algae and Man, Plenum Publishing Corporation New York, pp. 368–412

Seawright AA, Nolan CC, Shaw GR, Chiswell RK, Norris RL, Moore MR, Smith MJ (1999) The oral toxicity for mice of the tropical cyanobacterium *Cylindrospermopsis raciborskii* (Woloszynska). Environ Toxicol 14:135–142

Sekijima M, Tsutsumi T, Yoshida T, Harada T, Tashiro F, Chen G, Yu S-Z, Ueno Y (1999) Enhancement of glutathione S-transferase placental-form positive liver cell foci development by microcystin-LR in aflatoxin B1-initiated rats. Carcinogen 20:161–165

Sivonen K, Jones J (1999) Cyanobacterial toxins. In: Chorus I, Bartram J (eds) Toxic cyanobacteria in water: a guide to their public health consequences, monitoring and management. Published on the behalf of WHO by E & FN Spon, London, pp. 41–112

Song KY, Lim IK, Park SC, Lee SO, Park HS, Choi YK, Hyun BH (1999) Effect of nodularin on the expression of glutathione S-transferase placental form and proliferating cell nuclear antigen in N-nitrosodiethylamine initiated hepatocarcinogenesis in the male Fischer 344 rat. Carcinogen 20:1541–1548

Soong FS, Maynard E, Kirke K, Luke C (1992) Illness associated with blue-green algae. Med J Aust 156:67

Teixera MGLC, Costa MCN, Carvalho VLP, Pereira MS, Hage E (1993) Bulletin of the Pan American Health Organization 27:244–253

Tereo K, Ohmori S, Igarashi K, Ohtani I, Watanabe M, Harada K-I, Ito E, Watanabe M (1994) Electron microscopic studies on experimental poisoning in mice induced by cylindrospermopsin isolated from blue-green alga *Umezakia natans*. Toxicon 32:833–843

Tsuji K, Naito S, Kondo F, Watanabe MF, Suzuki S, Nakazawa H, Suzuki M, Shimada T, Harada KI (1994) A clean-up method for analysis of trace amounts of microcystins in lake water. Toxicon 32:1251–1259

Tsuji K, Setsuda S, Watanuki T, Kondo F, Nakazawa H, Suzuki M, Harada KI (1996) Microcystin levels during 1992-95 for Lakes Sagami and Tsukui Japan. Nat Toxins 4:189–194

Turner PC, Gammie AJ, Hollinrake K, Codd GA (1990) Pneumonia associated with cyanobacteria. Br Med J 300:1440–1441

Ueno Y, Nagata S, Tsutsumi T, Hasegawa A, Watanabe MF, Park H-D, Chen G-C, Chen G, Yu S-Z (1996) Detection of microcystins, a blue-green algal hepatotoxin, in drinking water sampled in Haimen and Fusui, endemic areas of primary liver cancer in China, by highly sensitive immunoassay. Carcinogen 17:1317–1321

Ueno Y, Makita Y, Nagata S, Tsutsumi T, Yoshida F, Tamura S-I, Sekijima M, Tashiro F, Harada T, Yoshida T (1999) No chronic oral toxicity of a low dose of microcystin-LR, a cyanobacterial hepatotoxin, in female BALB/c mice. Environ Toxicol 14:45–55

Ueno Y, Nagata S, Tsutsumi T, Hasegawa A, Yoshida F, Suttajit M, Mebs D, Puetsch M, Vasconcelos V (1996) Survey of microcystins in environmental water by a highly sensitive immunoassay based on monoclonal antibody. Nat Toxins 4:271–276

Wang H-B, Zhu H-G (1996) Promoting activity of microcystins extracted from waterblooms in SHE cell transformation assay. Biomed Environ Sci 9:46–51

WHO (1998) Guidelines for drinking water quality, Second Edition, Addendum to Volume 2, Health Criteria and other supporting information. World Health Organization, Geneva

Yoshida T, Makita Y, Nagata S, Tsutsumi T, Yoshida F, Sekijima M, Tamura S-I, Ueno Y (1997) Acute oral toxicity of microcystin-LR, a cyanobacterial hepatotoxin, in mice. Nat Toxins 5:91–95

Yoshida T, Makita Y, Tsutsumi T, Nagata S, Tashiro F, Yoshida F, Sekijima M, Tamura S-I, Harada T, Maita K, Ueno Y (1998) Immunohistochemical localization of microcystin-LR in the liver of mice: a study on the pathogenesis of microcystin-LR-induced hepatotoxicity. Toxicol Pathol 26:411–418

Yoshizawa S, Matsushima R, Watanabe MF, Harada K-I, Ichihara A, Carmichael WW, Fujiki H (1990) Inhibition of protein phosphatases by microcystin and nodularin associated with hepatotoxicity. Cancer Res Clin Oncol 116:609–614

Yu S-Z (1989) Drinking water and primary liver cancer. In: Tang Z-Y, Wu M-C, Xia S-S (eds) Primary liver cancer, China Academic Publishers, New York. pp. 30–37

Yu S-Z (1995) Primary prevention of hepatocellular carcinoma. J Gastroenterol Hepatol 10(6): 674–682

Zilberg B (1966) Gastroenteritis in Salisbury European children – a five-year study. Cent Afr J Med 12(9):164–168

6 Effects of *Microcystis* spp. and Selected Cyanotoxins on Freshwater Organisms

Cyanobacterial toxins have adverse effects on mammals, birds and fish and are being increasingly recognized as a potent stress and health hazard factor in the aquatic ecosystem. Microcystins, one main group of the cyanotoxins, are mainly retained within the producer cells during cyanobacterial population growth (see Chapter 2.1.2 and 2.3). However, these toxins are released into the surrounding medium by senescence and lysis of blooms. A wide range of aquatic organisms – phytoplankton, submerged aquatic plants (macrophytes), invertebrates, fish and amphibians – are then directly exposed to dissolved toxin, which may cause diverse effects. Exposure of aquatic organisms may occur both orally by uptake of toxin-containing cells as food, or through the surface tissues of organisms submersed in water containing dissolved toxin.

Previous studies concentrated on the effects of microcystins on fish and daphnia. Data for other aquatic organisms such as amphibians or submerged aquatic plants (macrophytes) as well as for early life stages of fish were lacking.

In most of the published studies on fish, the toxin was chiefly applied by intraperitoneal injection, and in a few cases the organisms (fish or mussels) were fed with cultures of cyanobacteria. Though the results for fish show liver disruption similar to the effects seen in mice, and some evidence associates fish kills with mass developments of toxic cyanobacteria, few studies addressed the effects of environmentally relevant concentrations of cyanotoxins dissolved in the water. This includes the issue of microcystin uptake through membranes without the help of the bile acid carrier, which is also relevant for the exposure of plants to microcystins. Studies on the effect of microcystin on *Daphnia* have suggested these invertebrates to be rather insensitive to dissolved microcystins, but have raised questions regarding the role of microcystins in the selective uptake of cyanobacterial cells as food and the consequences of grazing on microcystin-containing cells.

The following chapter addresses some of these knowledge gaps with results on the impact of dissolved microcystins on the early life stages of fish and amphibians (Chapter 6.1) and on uptake as well as metabolism of microcystin-LR in fish and aquatic plants (Chapter 6.2). Effects of low concentrations of microcystin-LR on the behaviour of fish are investigated in Chapter 6.3, and Chapter 6.4 takes a differentiated look at how ingestion rates, growth and survival of *Daphnia galeata* are affected by microcystin in *Microcystis* cells.

6.1 Effects of Cyanotoxins on Early Life Stages of Fish and Amphibians

Axel Oberemm

The role of cyanobacterial toxins with respect to fish mass mortality following senescence of cyanobacterial blooms is still obscure, since few studies have linked fish mortality and pathology to the presence of dissolved toxins. The study of Rodger et al. (1994), who demonstrated a correlation between a brown trout mass mortality and a hepatotoxic bloom of *Anabaena flos-aquae* gave strong evidence that a combination of mechanical gill injury due to cyanobacterial fibres and presence of dissolved microcystins may cause microcysin-related hepatotoxicity in fish. However, young life stages of fish are known to be generally more sensitive to toxic stress as compared to adults and juveniles, probably due to their thin epithelial layers in combination with a large body surface, a high metabolic rate and the interdependency of early developmental processes easily subject to interference. In addition to these internal factors, the risk of exposure to cyanobacterial compounds is higher than for

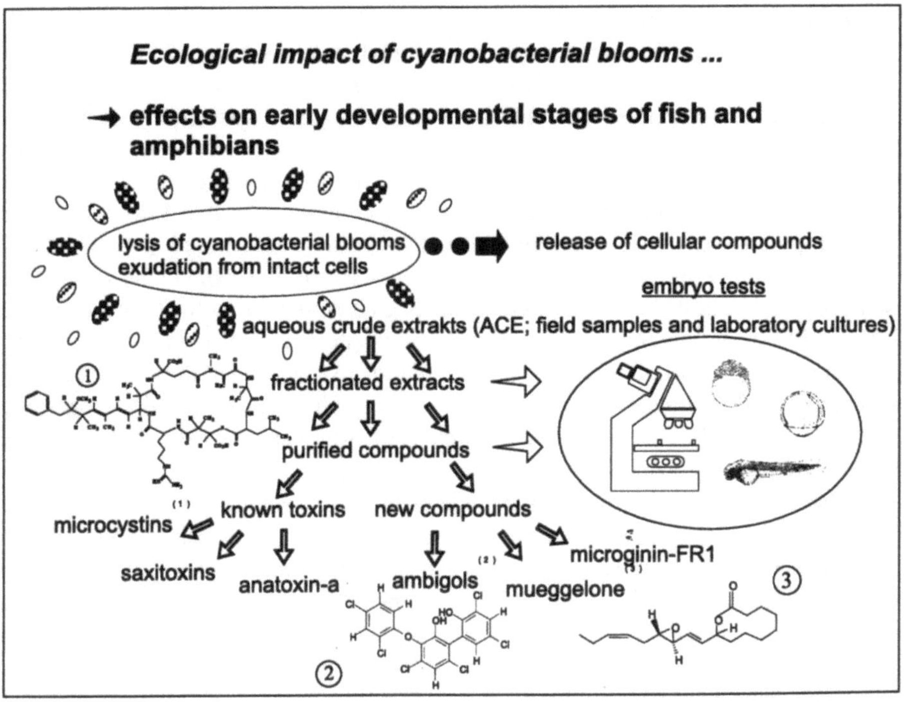

Fig. 1. Release of cellular compounds by cyanobacteria and approaches to evaluate their effects on early life stages of aquatic vertebrates

adults and juveniles, because motility of early developmental stages is limited and thus their ability to avoid contaminations is low. Further, juveniles are often restricted to littoral areas where cyanobacteria frequently form scums or benthic populations.

In order to obtain information on the harmful potential of lysing cyanobacterial mass developments and of compounds released by viable cyanobacterial cells, the effects of common cyanotoxins (selected microcystins, saxitoxin and anatoxin-a), of intact cyanobacterial cells and of aqueous crude extracts from cyanobacteria on different species of fish and amphibians were investigated (Fig. 1). Tests were performed semistatically in petri dishes. Generally, exposure was begun during the end of the cleavage period and maintained up to the end of the embryonic period. (For more information on materials and methods and illustrations of fish development see Chapter 8.4 and Oberemm 2000).

Results for Microcystins (MCs)

The effects of these most common hepatotoxic cyanotoxins were studied comparatively using various species in a range of environmentally relevant concentrations and additionally at high concentrations in zebrafish in order to detect thresholds of acute toxicity (see Table 1). Generally, no acute effects of MC-LR, -RR and -YR at concentrations of 0.5, 5 and 50 µg L^{-1} were observed during embryonic development of the selected species. However, specific effects could be observed.

In rainbow trout, the time course of hatching was altered in dependancy on concentration and structure of MCs (Fig. 2). MC-YR and -RR at all concentrations and MC-LR at 50 µg L^{-1} stimulated hatching, whereas MC-LR at 0.5 and 5 µg L^{-1} led to a minor, but clear delay of hatching causing an increase in synchronicity. This is a quite remarkable biological effect which demonstrates that MCs may interfere with developmental processes or organismic functions. Unfortunately, it is difficult to relate this effect to systemic endpoints, since the regulation of hatching is very complex. However, it may be presumed that the production and release of hatching enzymes was directly affected by MCs. This effect was not observed in any of the other species, which may be due to the slow developmental rate of the trout.

Another effect particularly significant for population dynamics could be demonstrated in some species which were reared in MC-free water after the end of the embryonic period (transition from larval to juvenile phase), after they were exposed during the embryonic development. In zebrafish, MC-LR and -RR led to a comparable decrease of the survival rate and growth (Fig. 3). MC-LR was more effective, since body length was statistically significantly reduced even at a concentration of 0.5 µg L^{-1} MC-LR. Reductions in body length were statistically significant (One-Way ANOVA) in the MC-LR-treated group, but not in the MC-RR-treated fish, which is due to the large variance of control animals in this group. Effects on survival and on growth did not increase with concentrations, except for growth in the MC-RR-treated group (0.5 µg L^{-1} compared to 5 and 50 µg L^{-1}). Experiments with chub (*Leuciscus cephalus*) also indicated effects of MCs on the survival rate during the larval period, since chub embryo survival rates decreased with the increase of concentrations of

Table 1. Test species and cyanobacterial compounds tested: cyanotoxins and ranges of test concentrations and cyanobacterial aqueous crude extracts (ACE)

Species	MC-LR	MC-RR	MC-YR	MC-LF	Saxitoxin	Anatoxin-a	ACE
Fish							
Zebrafish [a] *Danio rerio*	0.5-50 µg L^{-1} 1-10 mg L^{-1}	0.5-50 µg L^{-1} 1-10 mg L^{-1}	0.5-50 µg L^{-1} 1-10 mg L^{-1}	1-10 mg L^{-1}	100- 500 µg L^{-1}	100- 400 µg L^{-1}	*
Rainbow trout *Oncorhynchus mykiss*	0.5-50 µg L^{-1}	0.5-50 µg L^{-1}	0.5-50 µg L^{-1}				
Roach *Rutilus rutilus*	0.5-50 µg L^{-1}						*
Bream *Abramis brama*	0.5-50 µg L^{-1}						
White asp *Leucaspius delineatus*	0.5-50 µg L^{-1}						
Chub *Leuciscus cephalus*	0.5-50 µg L^{-1}	0.5-50 µg L^{-1}	0.5-50 µg L^{-1}				*
Stone loach *Cobitis taenia*	0.5-50 µg L^{-1}						
Amphibians							
Axolotl[a] *Ambystoma mexicanum*	0.5-50 µg L^{-1}	0.5-50 µg L^{-1}	0.5-50 µg L^{-1}				*
Smooth newt *Triturus vulgaris*	0.5-50 µg L^{-1}	0.5-50 µg L^{-1}	0.5-50 µg L^{-1}				*
Marsh frog *Rana ridibunda*	0.5-50 µg L^{-1}						

[a] Not native to Europe

* aqueous crude extract tested at concentrations of 0.01 and 0.1% (dw/vol)

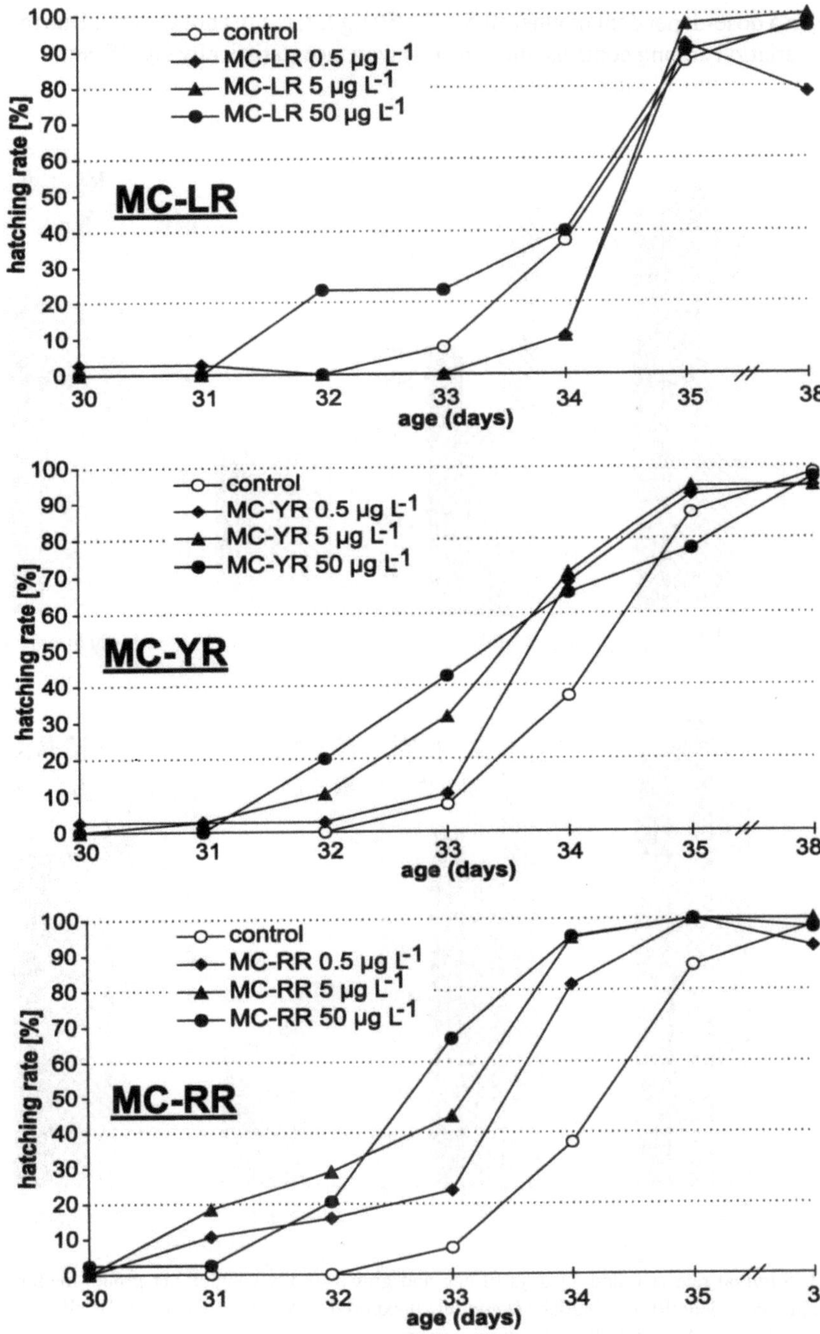

Fig. 2. Hatching course of rainbow trout (*Oncorhynchus mykiss*) exposed to MC-LR, MC-YR and MC-RRs at 0.5, 5 and 50 μg L^{-1}. Time scale is not linear between days 35 and 38. Reprinted from Environ. Toxicol. 14(1),77-88 (1999) with friendly permission from John Wiley & Sons

MCs in a dose-dependent manner (0.5 > 5 > 50 µg L^{-1}; results not shown). Due to the high variation among controls, these results were not statistically significant.

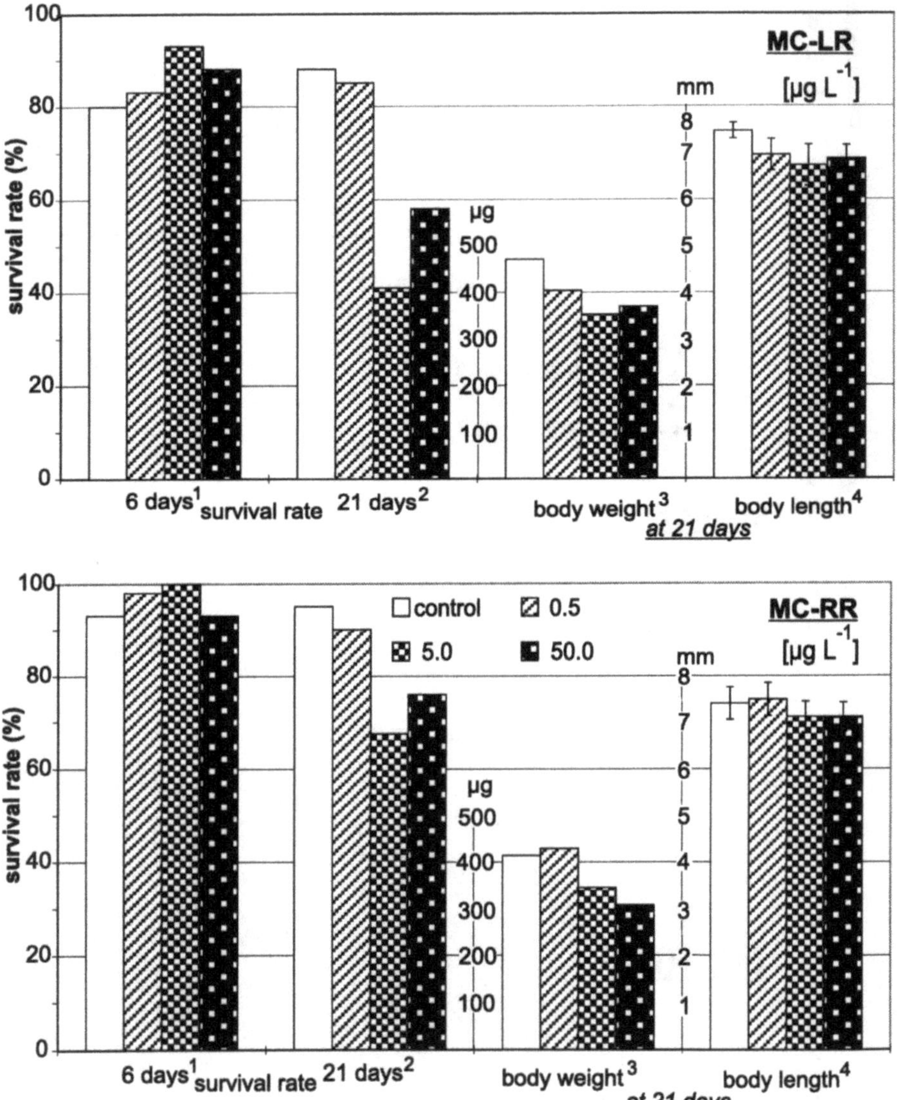

Fig. 3. Survival rate at 6 and 21 days of age and growth (body weights per group, median of individual body length) of zebrafish (*Danio rerio*) exposed to MC-LR (upper panel) and MC-RR (lower panel) at 0.5, 5 and 50 µg L^{-1} during embryonic development. After 6 days of semistatic exposure, larvae were reared in toxin-free water. [1] Related to n at day 1; [2] related to *n* at day 7; [3] median of treatment group (DW); [4] median of individual data. Reprinted from Environ. Toxicol. 14(1),77-88 (1999) with friendly permission from John Wiley & Sons

In contrast to the results with fish, no effects of embryonic exposure to MC-LR, -YR and -RR on the survivalrate of larvae were recorded for axolotls and smooth newt, but other effects were observed in these species: in axolotl, the beginning of food uptake was delayed by about 1 day following an exposure to MC-LR and -YR at 50 µg L^{-1}. After rearing up to an age of 35 days in microcystin-free water, many larvae exposed to MCs during embryogenesis at 50 µg L^{-1} showed less developed forelimbs compared to the control larvae (results not shown). In smooth newt, larvae at an age of 35 days showed a reduction of weight after exposure to MC-YR and -RR at 50 µg L^{-1}.

The influence of humic substances on effects of MC-LR was assessed with zebrafish by addition of Fuhrberg eluate (characterization in Alberts et al. 1992). This humic acid mixture did not increase the effect of MC-LR at a concentration of 50 µg L^{-1} (2 mg L^{-1} DOC). Also, soft water conditions (fivefold dilution of test medium, Chapter 8.3) did not increase the toxicity of MC-LR.

In order to determine acute toxic thresholds of different MCs in fish early life stages, zebrafish embryos were exposed to elevated concentrations of MC-LR, MC-YR, MC-RR and MC-LF (0.1, 1, 5 and 10 mg L^{-1}). MC-LF was included in this series, since it is more lipophilic than MC´s used before (log P_{ow}: MC-LR = 2.2; MC-LF = 3.6) and could therefore accumulate in embryos to a higher degree. No effects on embryonic development were visible under standard exposure conditions (Fig. 4, treatment I), but MC-LR added to eleuthero-embryos at 10 mg L^{-1} led to an

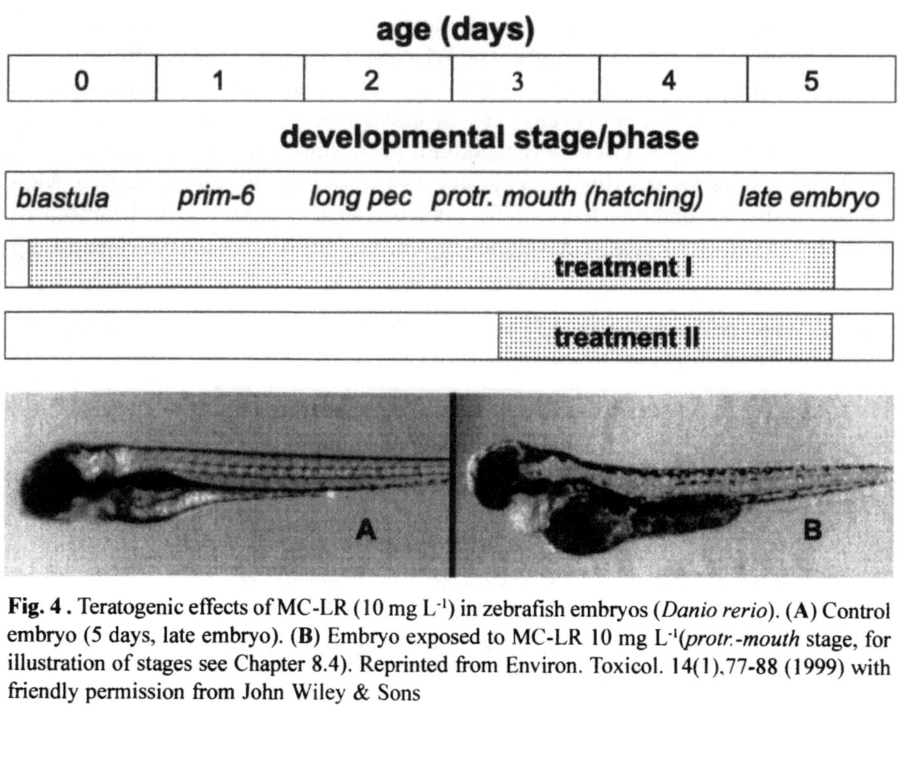

Fig. 4. Teratogenic effects of MC-LR (10 mg L^{-1}) in zebrafish embryos (*Danio rerio*). (**A**) Control embryo (5 days, late embryo). (**B**) Embryo exposed to MC-LR 10 mg L^{-1}(*protr.-mouth* stage, for illustration of stages see Chapter 8.4). Reprinted from Environ. Toxicol. 14(1),77-88 (1999) with friendly permission from John Wiley & Sons

immediate decrease of heart rate. After 24 h, embryos had developed pectoral oedema and showed an enlarged and opaque yolk (Fig. 4B), indicating poor utilization of the yolk at that time. Surprisingly, none of the other MCs applied at high concentrations caused any effect during the embryonic development of zebrafish.

The results demonstrate that no acute toxic effects were apparent in the range of 0.5-50 μg L^{-1} MC-LR, MC-RR and MC-YR. Tests with higher MC concentrations indicated that the acute toxic threshold concentration of MC-LR might be in the same high range in fish embryos as for crustacean zooplankton species (DeMott et al. 1991). However, different chronic effects generated by MCs were detected in concentrations occurring in the field, and these could cause significant disturbances of reproductive success.

The comparative potency of different MCs in inducing a distinct effect in an individual species is presented in Table 2. Remarkably, the succession of potencies related to comparable endpoints (mortality and growth for fish, limb development for amphibians) reversed between fish and amphibians. Thus, it may be concluded that fish and amphibians show different susceptibilities to different MC's although the mechanism(s) of toxic action may be similar. The divergence in susceptibility found at the level of hatching in trout may due to other mechanisms of action. It is not possible to relate detected effects to targets at the organ or molecular level, because molecular markers of potentially affected processes are lacking in the experimental animals. In vivo inhibition of protein phosphatases may contribute to the observed effects. Table 2 shows that the gradations in potencies of different MCs cannot be correlated to molecular features such as hydrophobicity or molecular weight.

Results from related biochemical studies offer some indications to explain the low acute toxicity of MC-LR: Pflugmacher and Wiegand (Chapter 6.3 and 6.4, Pflugmacher et al. 1999, Wiegand et al. 1999) have demonstrated that the activities of detoxication enzymes in plants and zebrafish embryos increased following exposure to MC-LR. In this event, MCs may be effectively metabolized by embryonic tissues and may not reach sensitive target structures at effective concentrations. When exposure of MC-LR began after the embryos hatched, the toxin had an acute effect on the heart rate of the embryos and on successive development, whereas this was not the case if exposure

Table 2. Molecular features of MC-LR, MC-RR and MC-YR and succession of their potencies related to different endpoints (parameters) studied with different test species

Hydrophobicity	*Molecular weight*	
YR > LR > RR	YR > RR > LR	
Mortality: zebrafish	*Growth: zebrafish*	*Body weight: newt*
LR > RR > YR*	LR > RR > YR*	YR > RR > LR*
Limb development: axolotl	*Stimulation of hatching: trout*	*Retardation of hatching: trout*
YR > RR > LR	RR > YR > LR	LR (0.5, 5 μg L^{-1})

Asterisk = NOE (no observed effect)

began at the blastula stage. The increased effect of MC-LR at later begin of exposure could be due to a lack of enzymatic activation. This interpretation is supported by the fact that most of the affected embryos recovered before the end of the exposure period, when detoxication enzymes may have been activated.

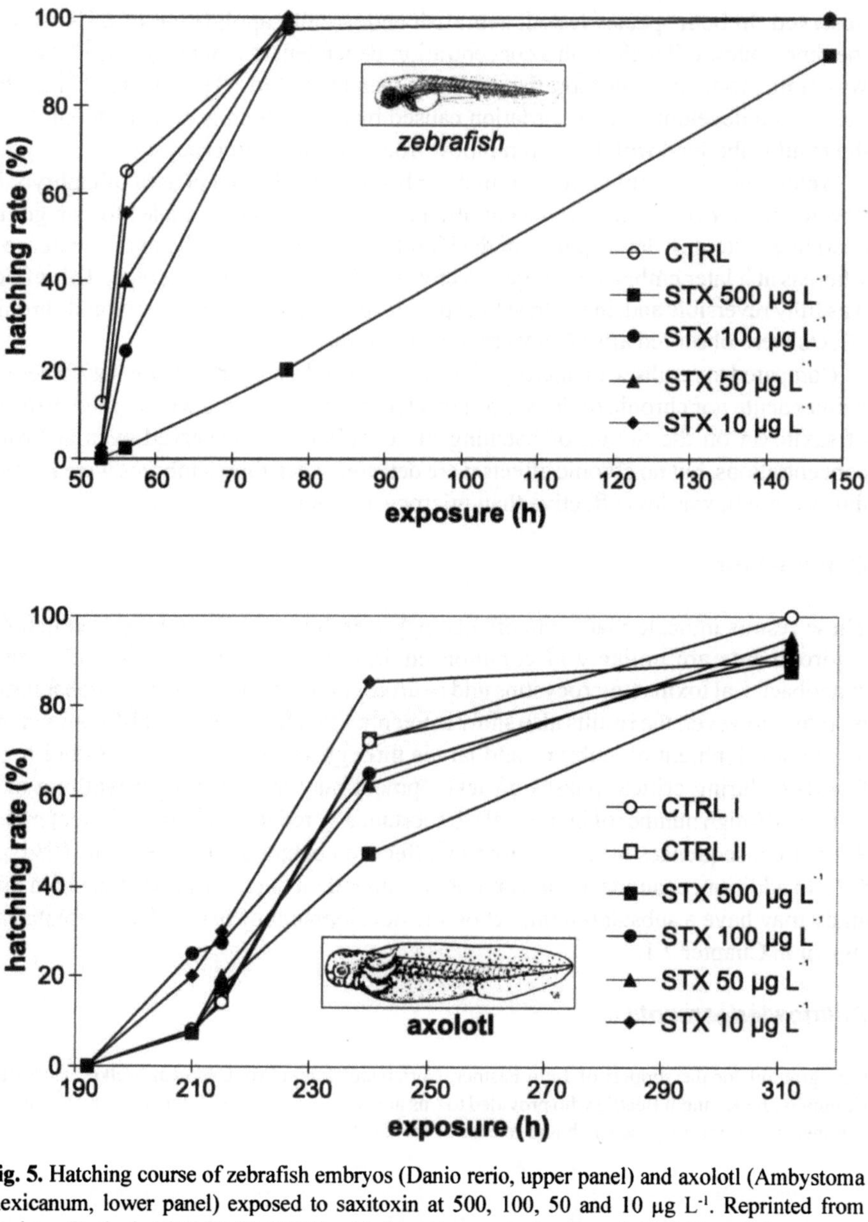

Fig. 5. Hatching course of zebrafish embryos (Danio rerio, upper panel) and axolotl (Ambystoma mexicanum, lower panel) exposed to saxitoxin at 500, 100, 50 and 10 µg L^{-1}. Reprinted from Environ. Toxicol. 14(1),77-88 (1999) with friendly permission from John Wiley & Sons

Results for Neurotoxins (Saxitoxin, STX and Anatoxin-a)

Exposure of zebrafish embryos to STX at a concentration of 500 µg L^{-1} generated malformations in the late phase of embryonic development, chiefly lateral and ventral body curvature and oedema, leading to an increased mortality of about 40% during larval development. At lower concentrations and in axolotl, no malformations were observed. In both species tested, zebrafish and axolotl, application of STX altered the time-course of hatching in a concentration-dependent manner (Fig. 5). This effect was more pronounced in zebrafish embryos than in axolotls. The course of hatching indicates a developmental retardation caused by STX which is quite different from the results obtained with MCs in rainbow trout (cf. Fig. 2, microcystins).

Anatoxin-a was applied to zebrafish embryos only. At two phases of embryonic development incubation in a concentration of 400 µg L^{-1} anatoxin-a led to changes in heart rate. After full development of the blood circulatory system, heart rate increased, whereas at a later embryonic stage a decrease in heart rate was observed. This effect was fully reversible and, in contrast to exposure to STX, no malformations and chronic effects were observed after exposure to anatoxin-a.

Compared to results with microcystins, STX showed a higher acute toxicity, whereas neither acute nor chronic toxicity was detected when anatoxin-a was applied. Effects of saxitoxin on the timing of hatching in zebrafish were observed even at lower concentrations, but no chronic effects were detected when these embryos were reared, thus saxitoxin was less effective than microcystins at the chronic level.

Conclusions

These results indicate that acute effects in fish embryos and amphibians in natural environments are unlikely at common environmental concentrations of known cyanobacterial toxins (microcystins and neurotoxins) associated with cyanobacterial blooms. However, the results also show a potential for chronic effects. Minor impacts on the development of embryos and larvae through short periods of exposure for a few days during critical phases of development may impair the survival rate of a sufficiently high number of individuals to substantially reduce population development. Similar consequences may be caused by effects on behaviour (see Baganz, Chapter 6.2). In addition, cyanobacterial compounds other than the toxins investigated in this study may have a substantial impact on the development of fish and amphibians, as shown in Chapter 7.1.

Acknowledgements

I am grateful for the support of Jutta Fastner, Christiane Bumke-Vogt, Martin Welker, Christian Hummert and Kenneth Beattie who provided toxins and analysis of toxins in batches, and to Angele Schonert for supporting the catch and release of the newts.

6.2 Uptake, Enzyme Effects and Metabolism of MC-LR in Aquatic Organisms

6.2.1 Uptake of Microcystin-LR in Aquatic Organisms

Claudia Wiegand and Stephan Pflugmacher

Most studies investigating effects of microcystins on aquatic organisms have exposed fish by intraperitoneal injection (Råbergh et al. 1991; Carbis et al. 1996; Bury et al. 1997) or orally by feeding the organisms (fish or mussels) with cultures of cyanobacteria (Eriksson et al. 1989; Tencalla et al. 1994; Vasconcelos 1995; Carbis et al. 1997). Only few studies deal with exposure under more appropriate natural conditions as for *Daphnia magna* (Jungmann 1992; Jungmann and Benndorf, 1994).

We studied the uptake of microcystins in aquatic organisms with ^{14}C-labelled microcystin-LR (^{14}C MC-LR), which was purified from *Microcystis* PCC 7820 grown photoautotrophically in batch cultures in the presence of ^{14}C-sodium bicarbonate (Bury et al. 1995).

Methods

For the exposure of aquatic plants, *Ceratophyllum demersum* and *Elodea canadensis* were chosen among the macrophyte species and *Vesicularia dubyana* from the bryophte species. All plants were cultivated in Provasoli medium with supplementary light by daylight lamps (12 µE $m^{-2}s^{-1}$) at a light-dark cycle of 14:10 h and a temperature between 22 and 24°C. Exposure of 5-g whole plants to 2.5 mg L^{-1} ^{14}C-MC-LR was performed in a volume of 250 ml under the same light and temperature conditions. The uptake of ^{14}C-MC-LR in *C. demersum* was investigated in definite exposure times up to 7 days whereas *E. canadensis* and *V. dubayana* were exposed continuously over 7 days.

Exposure of fish was performed using embryos of the zebrafish (*Danio rerio*). For fish maintenance and embryo production procedure, refer to Chapter 8.4 or Oberemm (2000). The adult fish were kept in artificial tank water containing 100 mg L^{-1} sea salt, 200 mg L^{-1} $CaCl_2 \cdot 2 H_2O$ and 103 mg L^{-1} $NaHCO_3$, at 26°C with a light-dark regime of 13:11 h. The fish were fed with a mixture of frozen *Chaoborus* sp., *Daphnia* sp. and *Artemia* ssp. Fish spawned every morning after onset of the light phase. The collected eggs were placed in REKO water (ISO 1984) at 26°C and after a few cleavages, fertilized eggs were distinguishable from non-fertilized and triplicate batches of 50 eggs were selected for exposure in 10 mL REKO water containing 2.5 mg L^{-1} ^{14}C-MC-LR. Uptake during 5 days of development was determined in the following embryonic stages: 32 cell- (1.5 h), 75% epiboly- (6-8 h), prim 6 (approx.

24 h) and long pec stage (approx. 48 h), staging was done according to Kimmel et al. (1995), and up to 3- and 5-day-old eleuthero-embryos.

After termination of exposure, plants and fish eggs were rinsed with MC-LR-free medium to remove non-absorbed ^{14}C-MC-LR. The plants were ground to a fine powder in liquid nitrogen, whereas the fish eggs were homogenized in sodium phosphate buffer 0.1 M using a hand-held glass homogenizer. Extraction was done according to Bligh and Dyer (1959). Radioactivity was determined in the water and organic phases and additionally in the remaining pellets after solubilization in tissue solubilizer NCS-II by liquid scintillation counting (LSC, Wallac 1409) using Optiphase HiSafe 3 cocktail.

Analysis of MC-LR and of *in vitro* enzyme tests were performed by HPLC using a Waters Alliance system 2690 Seperation module with diode array detection according to Plugmacher et al. (1998) with minor modifications, using a Symmetry 5 µ C 18, 3.9 x 150 mm column in an isocratic run with 68 % of solution A (acetonitrile containing 0.1 % (v/v) trifluoroacetic acid) and 32 % of solution B (MilliQ water containing 0.1 % trifluoroacetic acid) over 15 min. Column temperature was maintained at 40 °C and the injection volume was 25 µl.

For characterization of the conjugates MALDI-TOF-MS (matrix-assisted laser desorption/ionization time-of-flight mass spectro-metry) instrument, equipped with a nitrogen laser operating at 337 nm, a two-state ion source operating in the delayed extraction mode (PerSeptive Biosystems, Framingham, MA, USA) was used. All samples were subjected to MALDI-MS using the dried droplet method [14]: 2 µL of a saturated solution of a-cyano-4-hydroxycinnamic acid (Sigma Chemical Co., St. Louis, MO, USA) in 3:2 (v/v) acetonitrile-0.1% trifluoroacetic acid solvent as a matrix, were premixed with 2 µL of the solution of the microcystin-conjugates. From the resulting mixture, 2 µL were applied to the sample plate. Samples were air-dried at ambient temperature (24°C). All measurements were performed at an acceleration voltage of 20 kV (delay 200 ns) using the reflector mode, allowing the determination of monoisotopic mass values. Each spectrum or measurement refers to the sum of 100-200 individual laser shots.

Results

Uptake of ^{14}C-MC-LR in *Ceratophyllum demersum*, after a lag phase of 12 h, increased significantly within the following 6 days (Fig. 1). After 7 days exposure and extraction of the absorbed MC-LR, 0.79 ± 0.21% of the initially applied radioactivity was found in the water phase, 0.14 ± 0.09% in the organic phase and 0.18% in the pellet (Fig. 2). In total, 1.11% ^{14}C-MC-LR was absorbed, which is equivalent to 1.98 µg ^{14}C-MC-LR g^{-1} FW. (Pflugmacher et al. 1999).

Uptake of ^{14}C-MC-LR in different aquatic plants ranged from 1.0 to 3.6 µg g^{-1} FW after 7 days of exposure to 2.5 mg L^{-1} ^{14}C-MC-LR (Fig. 2). The highest uptake was found in the bryophyte *V. dubyana*. The uptake of the toxin by the two macrophyte species *C. demersum* and *E. canadensis* was much lower. The differences in toxin uptake might be due to the different surface area of these plants: *V. dubyana* is a richly brached plant and consequently has the largest ratio of surface area to volume. *C. demersum* with its fine leaves took up twice as much ^{14}C-MC-LR than *E. canadensis*, which has relatively large leaves.

In the zebrafish embryos, MC-LR was detectable in all developmental stages from the 32-cell stage up to 5-day-old larvae (Fig. 3). The concentrations of ^{14}C-MC-LR found before hatching and in the hatched stages were different, but the amounts absorbed per individual (indicated by *) were in the same range. Although the distribution of ^{14}C-MC-LR in embryonic stages before hatching between the embryo itself, the chorion and the perivitelline fluid was not investigated, most of the incorporated ^{14}C-MC-LR may be assumed to be in the embryo, because the chorion pores with a diameter of 0.2 µm are no barrier for the ^{14}C-MC-LR molecules, whose diameter is in the Å range (Rudolph-Böhner et al. 1994; Laale 1977). Exposure up to the 32-cell stage, for about 90 min, led to a ^{14}C-MC-LR concentration of 22.9 ± 4.4%

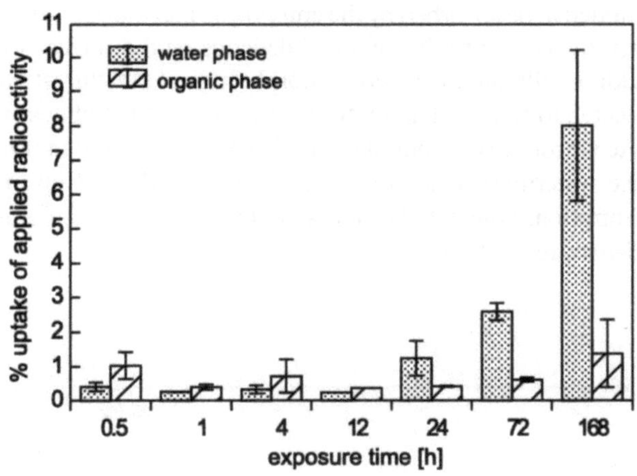

Fig. 1. Time-dependent uptake of ^{14}C-MC-LR by *Ceratophyllum demersum*

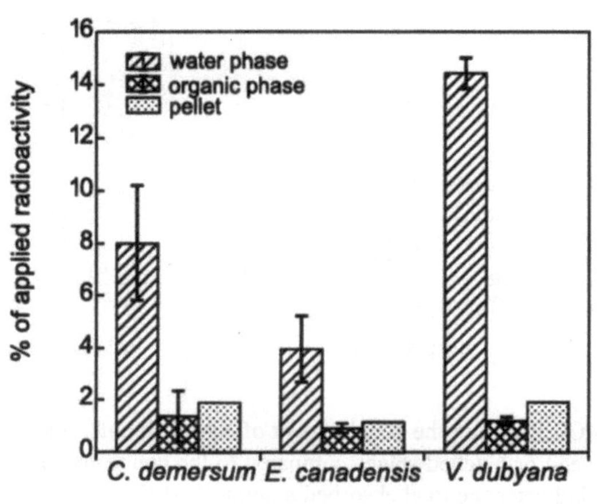

Fig. 2. Uptake of ^{14}C-MC-LR by different aquatic plant species after 7-day exposure to 2.5 mg L^{-1}

of the concentration in the surrounding medium, which is equivalent to 0.44 ± 0.09 ng egg^{-1}, and remained at this level up to the long pec stage. In the 3-day-old, hatched eleuthero-embryos, a ^{14}C-MC-LR concentration of 117.9 ± 59.8% of the medium concentration, corresponding to 0.49 ± 0.29 ng embryo^{-1}, was reached. Extension of the exposure time did not increase toxin accumulation and after 5 days the fish contained 109.7 ± 21.4% of the medium concentration of ^{14}C-MC-LR, i.e. 0.62 ± 0.18 ng fish^{-1}. Radioactivity in the organic and semisolid phases was mostly at background level.

Conclusions

The results show that MC-LR is absorbed by plants as well as by animals. The concentration of the toxin in the hatched-fish stages attained 17% of the concentration of the medium by 17%. Similar microcystin enrichments were found in the egg stages and may be assumed to be absorbed by the embryo, so that the chorion is no protective shield. In plants, the concentration attained depends on the surface to volume ratio and perhaps additionally on the different morphological structures: the macrophyte leaves are protected to some extent by the thin cuticula of the epidermis, in contrast to the bryophyte whose leaves consist of single cell layers without protective cover. Hence, only the watermoss *V. dubayana* concentrates MC-LR up to 144% of the medium concentration, whereas the macrophytes reach 79.2%, *C. demersum* and 40%, *E. canadensis*, respectively.

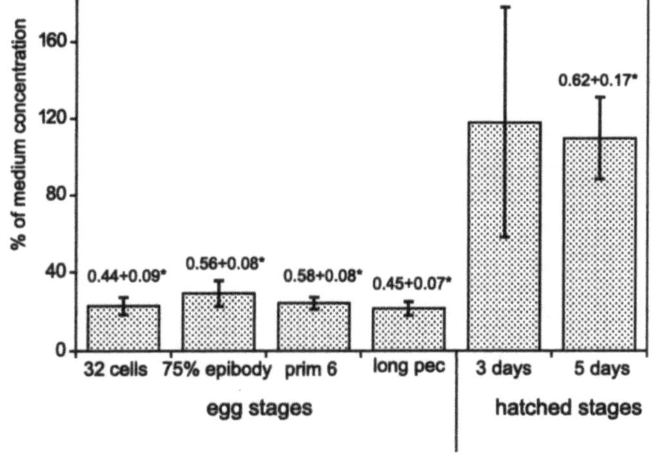

Fig. 3. Uptake of ^{14}C-MC-LR during the development of zebrafish embryos. *Columns* represent the concentration in the egg or hatched stages compared to the concentration in the surrounding medium (2.5 mg L^{-1}); * indicates the total absorbed amount ^{14}C-MC-LR in ng per egg or hatched embryo

6.2.2 Effects of Microcystin-LR on Detoxication Enzymes

Claudia Wiegand and Stephan Pflugmacher

One possibility to examine the impact of microcystins on the physiology of organisms is to measure the activity of detoxication enzymes, because these enzymes are the first to be influenced under xenobiotic or toxic stress (Salaün et al. 1978; Anderson and Gronwald 1991; Pflugmacher and Schröder 1995; Egaas et al. 1993; George 1994). The detoxication of xenobiotics and toxins in most investigated plants and animals proceeds in three phases. Phase I consists of oxidation, reduction and hydrolysis processes mostly performed by cytochrome P-450 monooxygenases, during phase II several conjugation reactions occur by microsomal and soluble glutathione S-transferases, glucosyl- and malonyltransferases, and phase III includes further processing reactions, deposition or translocation, e.g. in the vacuole or cell wall fractions of plant cells or renal excretion in animals (Sandermann 1992; Pflugmacher and Steinberg 1997).

Methods

Effects of microcystin-LR (MC-LR) on the microsomal and soluble gluthatione S-transferase systems (m-sGST, EC 2.5.1.18) were examined in the submerged macrophyte, *Ceratophyllum demersum*, and in embryos of the zebrafish, *Danio rerio*. The plants were precultured for 3 months in Provasoli medium as described in Chapter 6.2.1. Exposure of 5 g FW of the plants to various concentrations of MC-LR was performed in 500 ml volume for 7 days. For the investigations, *D. rerio* embryos were obtained as described in Chapter Chapter 8.4 and Oberemm (2000) and treated as detailed in Chapter 6.2.1. Triplicate batches of 500 eggs were exposed in 150 ml REKO containing 0.5 $\mu g\ L^{-1}$ MC-LR, except for the control, up to the ontogenic stages of 32 cells (aproximately 1.5 h), 75% epiboly (6-8 h), prim 6 (aproximately 24 h), and long pec stage (aproximately 48 h), according to Kimmel et al. (1995). Effects of various MC-LR concentrations were tested using prim 6 stage. After completion of exposure time, the plants and embryos were frozen in liquid nitrogen and enzyme preparation of the plants was performed according to Pflugmacher and Steinberg (1997) and of the embryos with slight modifications according to Wiegand et al. (in press) by homogenization and differential centrifugation to gain the microsomal proteins and ammonium sulphate precipitation to concentrate the soluble proteins. The activity of m- and sGST was measured colorimetrically using 1-chloro-2,4-dinitrobenzene (CDNB), a substrate which is converted by all GST isozymes (Habig et al. 1974). Furthermore, the activity of the glutathione peroxidase (GP-X) was estimated colorimetrically in *D. rerio* embryos according to Drotar et al. (1985). Enzyme activity (katal = conversion rate of 1 M substrate s^{-1}) was calculated in terms of the protein content of the sample (Bradford 1976) and is given in nkat/mg protein.

Results

The activity of the mGST of *C. demersum* towards the substrate CDNB was elevated in *C. demersum* exposed to MC-LR concentrations from 0.12 to 0.50 μg L^{-1} as compared to enzyme extracts from non-exposed plants (Fig. 1A). Higher concentrations of the toxin (> 1.0 μg L^{-1}) caused a reduction in enzyme activity to levels below control plants. In contrast, the activity of sGST increased in a dose-dependent manner in response to MC-LR exposure up to 1.05 μg toxin L^{-1} (Fig. 1B) and no decrease was observed. MC-LR might first come into contact with the mGST of *C. demersum* and an elevation of this enzyme's activity could occur at concentrations lower than those investigated. Breakdown of the enzyme activity at concentrations higher than 1.0 μg L^{-1} indicates that this enzyme or the sorrounding physiological processes are disturbed by the influence of MC-LR on the membranes. In contrast, the sGST responded with higher activities to higher MC-LR concentrations and withstood all the toxin. Hence, at the concentrations tested, the decreasing detoxication capacity of the mGST may be compensated for by increasing sGST activity.

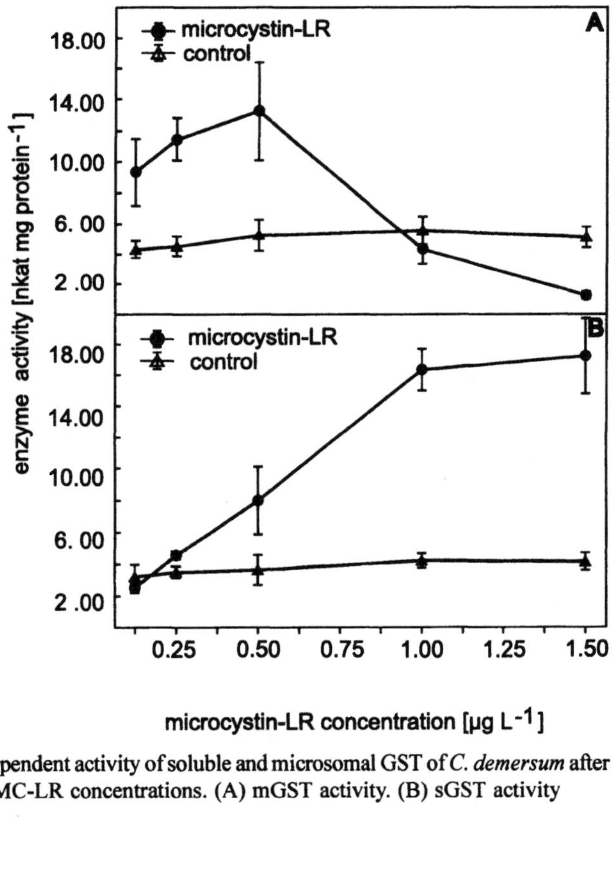

Fig. 1. Dose dependent activity of soluble and microsomal GST of *C. demersum* after 7-day exposure to increasing MC-LR concentrations. (A) mGST activity. (B) sGST activity

The sGST activity, extracted from treated and untreated *D. rerio* embryos decreased with succession of development (from 1.89 ± 0.11 to 0.64 ± 0.02 nkat mg^{-1} protein). Treatment with 0.5 µg L^{-1} MC-LR caused slight, but significant (Students *t*-test; p=0.05) elevation of activities in all developmental stages (from 2.02 ± 0.21 to 1.07 ± 0.33 nkat mg^{-1} protein, Fig. 2A). In the microsomal fraction, the activity was lower at the beginning (0.79 ± 0.20 nkat mg^{-1} protein), slightly decreasing during the first day of ontogenesis (0.44 ± 0.10 and 0.60 ± 0.07 nkat mg^{-1} protein) and increasing to the second day (1.39 ± 0.05 nkat mg^{-1} protein). In contrast to the sGST system and to the mGST system of *C. demersum*, treatment with 0.5 µg L^{-1} MC-LR did not much alter the mGST activity (Fig. 2B).

Exposure of *D. rerio* embryos to increasing MC-LR concentrations from 0.1 to 2.0 µg L^{-1} caused dose-dependent response of sGST activity (Fig. 3). Even a concentration of 0.1 µg L^{-1} MC-LR (which is very low and occurs frequently in natural aquatic environments) caused elevation of sGST activity in the embryos. Higher concentration (5.0 µg L^{-1} MC-LR) supressed sGST activity again to levels slightly beyond control values. The dose-dependent response of sGST activity appears to show that zebrafish embryos are able to withstand MC-LR up to about 2.0 µg L^{-1} but that they cannot compensate for higher concentrations with this enzyme system. These results are consistent with recent developmental studies, where survival and growth were depressed at concentrations of 5.0 µg L^{-1} and higher (Oberemm et al. 1997, 1999).

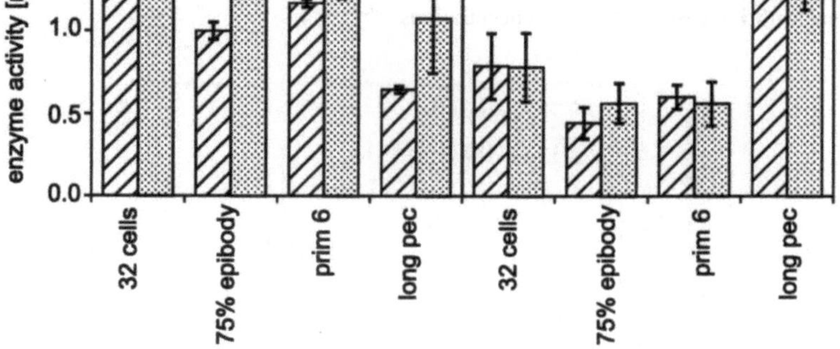

Fig. 2. Activity of sGST (A) and mGST (B) of distinct embryonic stages of *D. rerio* after exposure to 0.5 µg L^{-1} MC-LR

Measurement of another detoxication enzyme, the GP-X, revealed activities more related to the development of the fish embryos (Fig. 4) and also more sensitive to MC-LR. GP-X activity increased with the age of the embryos from 0.43 ± 0.31 nkat mg^{-1} protein at the 32-cell stage to 1.04 ± 0.02 nkat mg^{-1} protein at the long pec stage. Treatment with 0.5 μg L^{-1} MC-LR resulted in slightly elevated activities at all investigated stages, comparable to the elevation of sGST activity.

The observed detoxication reactions indicate that MC-LR might be metabolized in vivo by zebrafish embryos and therefore explain the low acute toxicity of MC-LR. In addition to potential damage of the liver cells, which will be most important for adult fish, these detoxifying enzymatic processes in embryos require additional energy, which is then not available for growth.

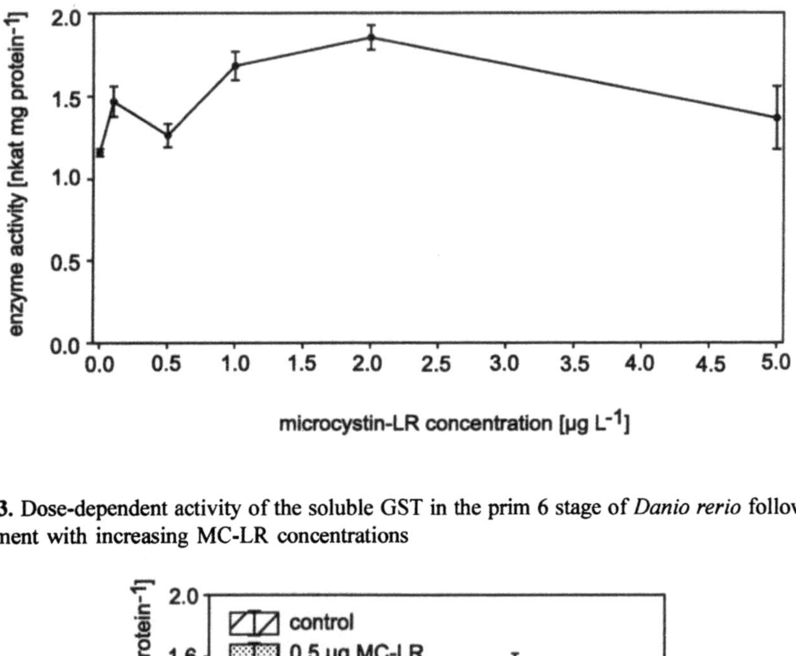

Fig. 3. Dose-dependent activity of the soluble GST in the prim 6 stage of *Danio rerio* following treatment with increasing MC-LR concentrations

Fig. 4. Activity of GP-X of *Danio rerio* embryos during the first 2 days of development after treatment with 0.5 μg L^{-1} MC-LR

Because the detoxication system seems to be comparable in young and adult fish, the energy consumed in these toxin-response reactions may affect all developmental stages of fish perhaps, however, with a stronger influence during early life stages due to their restricted nutritional resources and motility.

The concentrations of MC-LR used in the enzymatic studies are environmentally relevant loads which can be found in many lakes during the summer (Tsuji 1992; Edwards et al. 1994, see also Chapter 2.2, this Volume). In temperate climates, elevated concentrations of microcystins often occur in late summer, when the breeding of fish is almost completed. However, several species mate relatively late in the season (June to August), e.g. stickleback (*Gasterosteus aculeatus*), tench (*Tinca tinca*), bleak (*Alburnus alburnus*) and chub (*Leuciscus cephalus*). Further, populations of toxic cyanobacteria may already occur in spring (Arp 1997), and winter populations of the species *Planktothrix rubescens*, which contains the highest microcystin concentrations of the taxa studied in the survey presented in Chapter 2.2, may reach senescence and lyse in spring. Thus, exposure of fish eggs and embryos to microcystins will not be restricted to the tropics.

Like the fish embryos, aquatic macrophytes are not able to evade contaminations. Hence, the capacity of the detoxication system is surely important for endurance in limnic ecosystems with a meso- to eutrophic nutrient situation and therefore predestined for cyanobacterial growth. Chapter 6.2.3 shows how MC-LR can be metabolized by the GST system in different aquatic organisms.

6.2.3
Metabolism of Microcystin-LR in Aquatic Organisms

Stephan Pflugmacher and Claudia Wiegand

Chapter 6.2.1 and 6.2.2 clearly demonstrate uptake of the hepatotoxin microcytin-LR into fish eggs and different macrophytes. Furthermore, the activation of the glutathione S-transferase system in fish and macrophytes indicates that the detoxication system of the organism reacts under the influence of the toxin; but what happens to the toxin itself?

To clarify this issue, aquatic organisms from different trophic levels were investigated: macrophyte *Ceratophyllum demersum*, two invertebrates *Dreissena polymorpha* and *Daphnia magna* and *Danio rerio* fish eggs and adults.

Investigation of the products of phase II detoxication enzyme system of the soluble glutathione S-transferases (sGST) in extracts of these aquatic organisms revealed metabolites of MC-LR named p1 and p2, which were detected in HPLC in addition to MC-LR (Fig. 1). The assay for the in vitro study of MC-LR conjugation consisted of 100 mM sodium phosphate buffer at pH 6.5, 1 mM GSH in aqueous solution, 3.75 µg ml^{-1} MC-LR and 1.2 to 2.4 mg ml^{-1} protein extract of the investigated species at 28°C. Control reactions were measured with boiled extracts (10 min 100°C) and without protein or GSH to determine the non-enzymatic reaction of GSH with MC-LR. For all tests extractions were performed with Bligh and Dyer solution and the water phase was analyzed for MC-LR and metabolites by HPLC (Pflugmacher et al. 1998b).

These new peaks were more hydrophilic than the parent compound and could be clearly separated from the toxin by HPLC analysis (Fig 1, see Chapter 9.1 for the HPLC method).

The retention times of the new peaks were 4 min (p1) or 5 min (p2) for all investigated organisms, whereas the parent compound MC-LR had a retention time

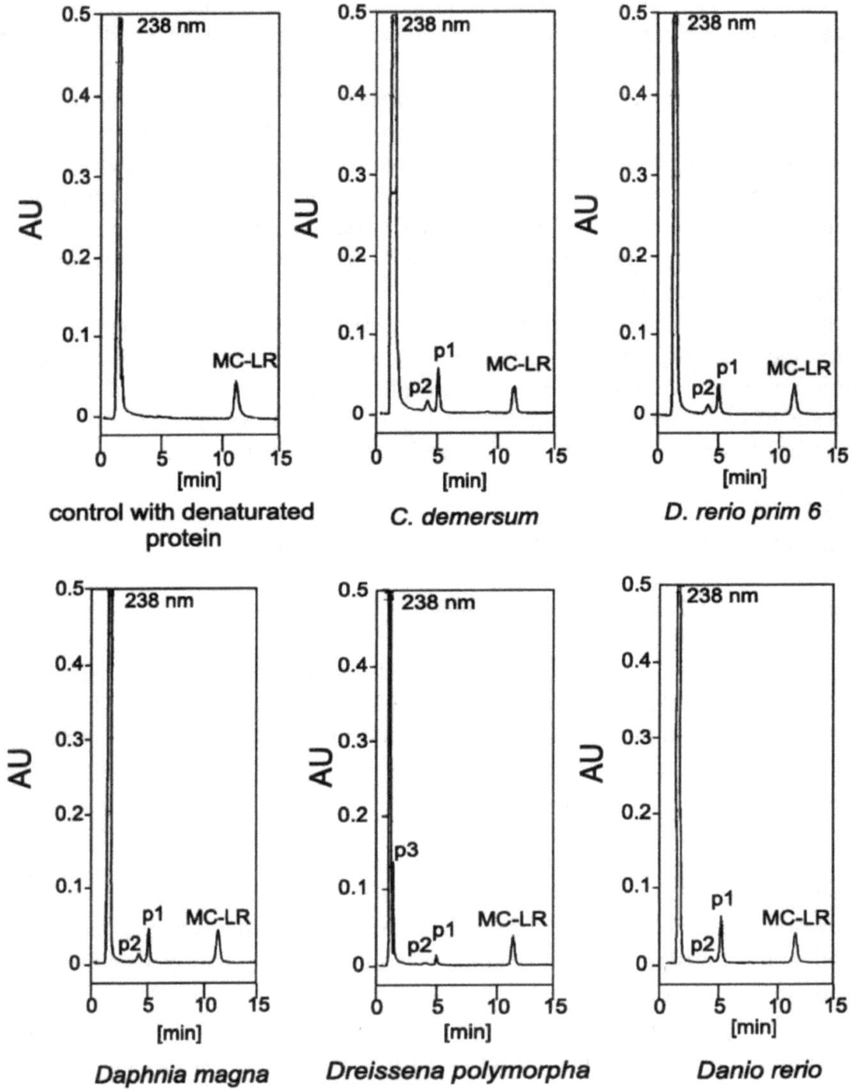

Fig. 1. HPLC chromatograms of *in vitro* enzyme assays for the MC-LR conjugation showing two metabolites, p1 and p2, with retention times of 4 and 5 min. The parent compound, MC-LR, had a retention time of 11 min

of 11 min. In control experiments with boiled extracts no metabolites could be detected, indicating that the conversion of MC-LR was enzyme-dependent; also the chemical formation of the metabolite was slower than conjugation by enzyme catalysis.

Purification of p1, which was the major peak, was carried out by preparative HPLC from the p6 stage of *D. rerio* and from *C. demersum* separately for further characterization. The mass spectrum of the MC-LR peak was monitored at m/z 995, which agreed with the estimated mass for the microcystin. The MALDI-TOF spectrum (provided by E. Krause, FMP) of the metabolite showed a distinct peak at m/z 1302 corresponding to the molecular weight expected for a GSH-conjugate of MC-LR (Fig. 2). All other p1 samples investigated from the other organisms resulted in the same mass.

The UV spectrum showed only minor differences between MC-LR and its GSH conjugate. The absorption maximum of MC-LR was at 238 nm, whereas the absorption maximum of the conjugate occurred at 240 nm; furthermore there was a second maximum at 203 nm which does not occur in the MC-LR spectrum (Pflugmacher et al., 1998).

All data were in correlation with a chemically formed GSH conjugate of MC-LR formed in an aqueous 5% potassium carbonate aqueous solution at room temperature by Kondo et al. (1992) Further studies have shown that this synthetically formed GSH-conjugate of MC-LR can also be detected in extracts of rat and mouse liver after administration of microcystin to the animals in vivo (Kondo et al. 1992,). In conclusion, all the results indicate that the metabolism of microcystin-LR begins with a conjugation reaction to GSH catalyzed by the ubiquitous glutathione S-transferases in plants as well as in animals (Fig. 3). Potential further processing steps could be by transpeptidase and *ciys*-lyase.

Furthermore, this reaction was demonstrated for several aquatic organisms, which might be exposed to microcystins in aquatic ecosystems. Detoxication via GST, which is well known for xenobiotics, may help diverse aquatic organisms to survive under toxic stress caused by toxin release of cyanobacteria.

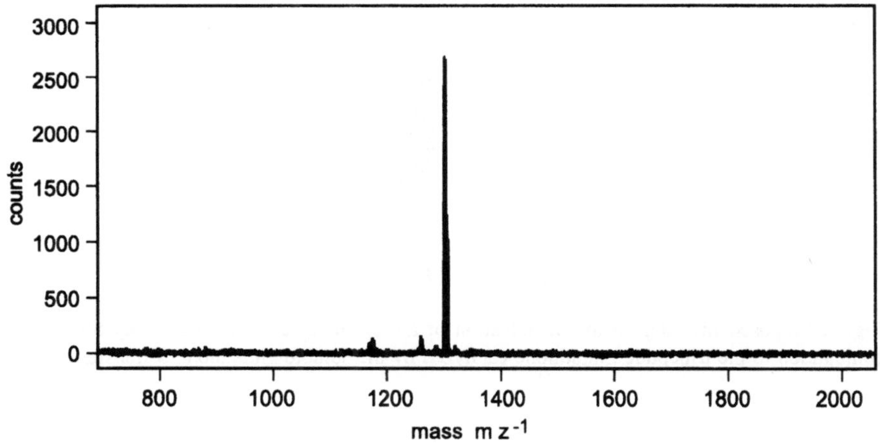

Fig. 2. Mass determination of peak p1 from *Ceratophyllum demersum* in vitro enzyme assay resulting in m/z 1302, which correlates with the mass estimated for a MC-LR conjugate

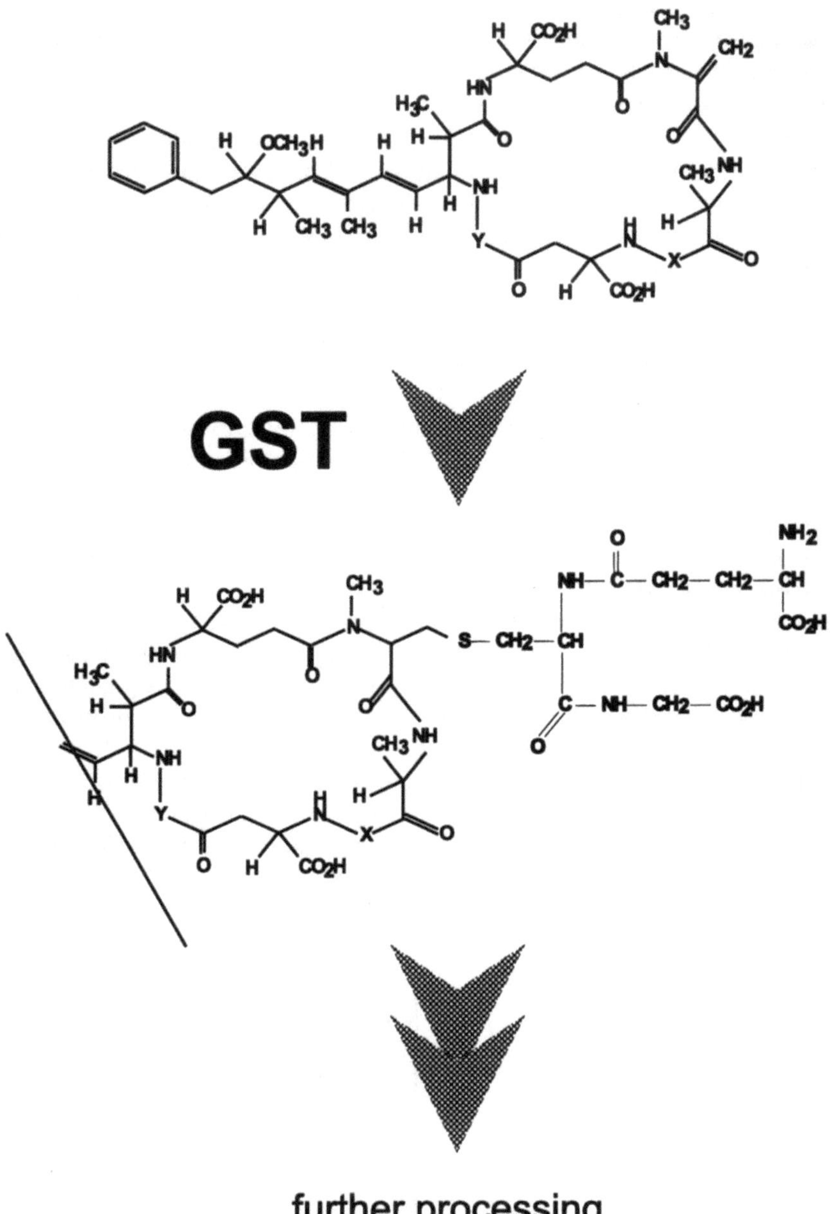

Fig. 3. Proposed first step of the detoxication of the cyanobacterial toxin microcystin-LR: the conjugation to the tripeptide glutathione catalyzed by the soluble glutathione S-transferase

6.3 Changes of Fish Behaviour Affected by Microcystin-LR

Daniela Baganz, Georg Staaks and Christian Steinberg

This study addresses long-term sublethal effects of the cyanobacterial toxin microcystin-LR (MC-LR) on the behaviour of zebrafish, *Danio rerio* (see also Baganz et al. 1998). The question whether MC-LR acts as a stressor for fish was analyzed under laboratory conditions by pattern analysis using an image processing system.

Environmental factors become stressors if they cause harmful effects on organisms. Anthropogenic activity may extend the range of natural stressors, e.g. the eutrophication process may lead to an increased cyanotoxin concentration in water. Biological systems respond to chemical stresses on various levels of aggregation, from molecule to ecosystem.

Behavioral tests addressing whole-organism effects provide primary and very sensitive information about the hazard potential of environmental chemicals for organisms. From the evolutionary point of view, the diversity of mechanisms which control the behaviour of animals is a response to the diversity of ecological problems. There are regulative mechanisms which directly react to changes in the physical or social environment by changing the behavioral disposition of an animal in order to allow adaptability (Alcock 1996).

Many behavioral parameters underlie circadian rhythms. These circadian rhythms are directly connected with the temporal and spatial organization of the organism and with the response to repetitive natural events (Boujard and Leatherland 1992). The behavioral rhythms are based on the temporal coordination of internal processes of organisms with the environment. They are important for the precise temporal fit of the organism into the ecosystem. Time patterns in swimming behaviour show a sensitive response to disturbances. The circadian rhythm is generally highly stable under undisturbed conditions.

The assessment and prediction of long-term sublethal effects of environmental chemicals on the behaviour require quantitative methods with high sensitivity. By using the automated measuring system BehavioQuant (Spieser and Scholz 1992), several test parameters can be recorded and quantified very sensitively: motility (swimming velocity in video-pixels s^{-1}), turns (horizontal changes of swimming direction s^{-1}), swimming height, places of sojourn, and distances between each single fish. In this study, the presented test parameter is the motility which characterizes the swimming activity of fish. To ensure the continuous observation of individuals and to register the movements during periods of particular interest, such as dawn, dusk and night, the system was adapted to cameras which were able to handle normal as well as infrared light.

The scheme of the equipment used for the experiment is shown in Fig. 1. Zebrafish (*Danio rerio*) of 0.25 g mean weight were held in groups of seven individuals (three females, four males) in each of five 15-L glass aquaria. They were exposed to a

12:12 h light/dark rhythm and were fed at 11:00 h and 15:00 h. Tests were carried out in a flowthrough system with a continuous discharge of 10 L aerated tap water day^{-1} from storage tanks. Temperature 26 ± 0.5°C and pH 7.6-7.7 were kept constant.

Four groups of fish were exposed to water containing different concentrations of microcystin-LR. One additional group served as control. The storage tank water was renewed daily and Mycrocystin-LR was added at nominal concentrations of 0.5, 5 and 15 mg L^{-1} for a period of 25 days and 50 mg L^{-1} for a period of 6 days. The continuous water exchange reduced the degradation process described by Cousins et al. (1996).

To carry out our test it was essential to obtain knowledge about the behaviour of fish under undisturbed conditions: daily activity of zebrafish was synchronized with the artificial alternation of light and dark and the feeding times in the 24-h period. Averaged motility curve of all days without MC-LR is shown in Fig. 2. Mating and spawning behaviour occurred every day shortly after the onset of light as described by Westerfield (1989). During this time (around 08:00 h) the maximum daily activity was also registered. Smaller activity peaks were observed at feeding times and during the last hour before dusk. Activity was markedly reduced during the first hour after dark. After this, fairly constant low activity levels were recorded for the rest of the night. The analysis of daily rhythms clearly showed the day-active behaviour of the fish.

The effect of the exposure to MC-LR is illustrated by the daily activity curves, which show dose-effect relationships between behaviour and microcystin concentration (Fig. 3). The reaction to feeding (11:00 h) and to the onset of light

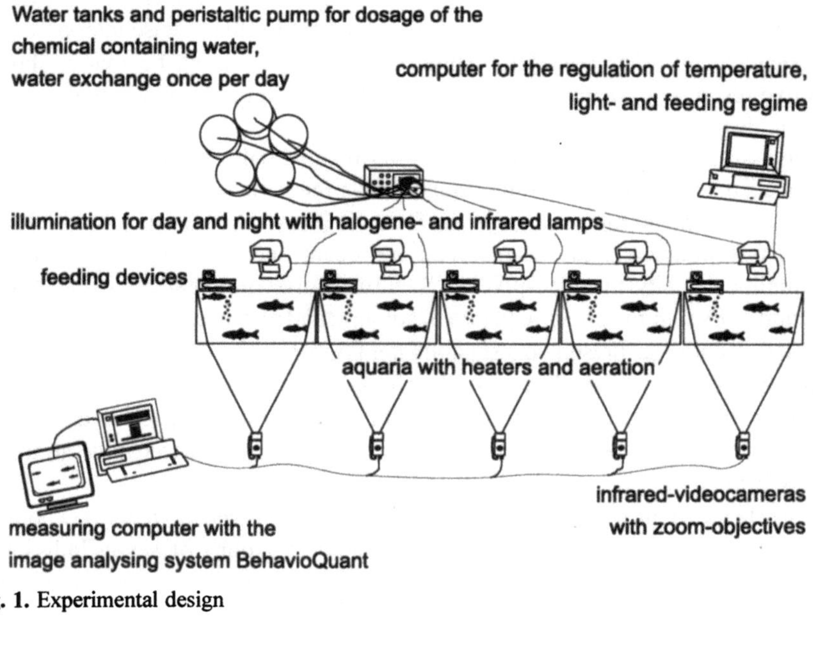

Fig. 1. Experimental design

(08:00 h) was reduced at the higher concentrations of MC-LR (15 and 50 mg L^{-1}) as compared to the control group. At the highest concentration level of 50 mg L^{-1} the activity showed a phase shift with the maximum of activity occurring 11 h later, at 19:00 h rather than at 08:00 h as in the control. As a crucial consequence of the reduced maximum daily activity at the time of mating and spawning behaviour

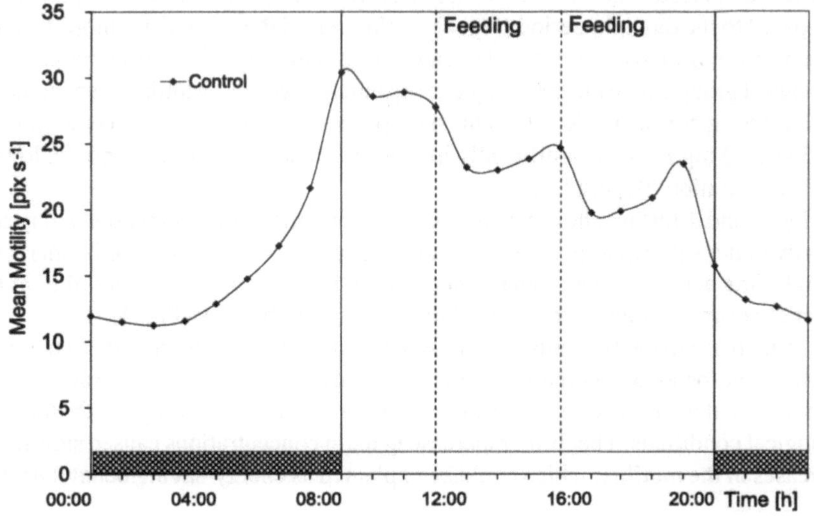

Fig. 2. Smoothed curve of average motility of all days and all groups before exposure to microcystin-LR

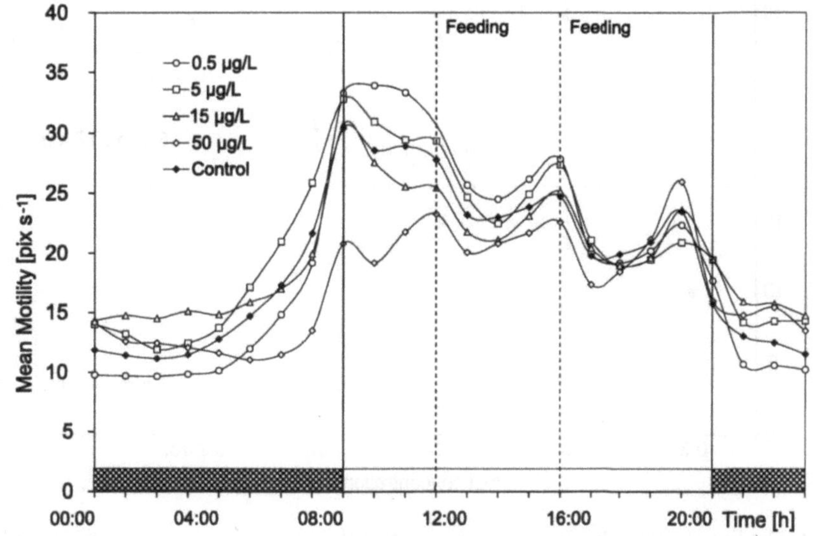

Fig. 3. Smoothed curve of average motility of all days for each test group during exposure to microcystin-LR

(around 08:00 h) the zebrafish stopped spawning and no spawned eggs could be found at this exposure concentration of 50 µg L^{-1}.

The two lower exposure concentrations (0.5 and 5 µg L^{-1}) induced a conspicuous increase in motility between 08:00 h and 12:00 h, whereas elevated exposures led to statistically significantly decreased daytime motilities (Fig. 4). Mean daily motility decreased significantly at the highest test concentration (50 mg L^{-1}).

The dark period from 21:00 h to 04:00 h showed reciprocal behavioral patterns as compared to the daylight period (Fig. 4). In this period the mean daily motility at the lowest concentration (0.5 mg L^{-1}) decreased statistically significantly as compared to the control group and to the three higher concentration groups. Motility then increased with further increase of MC-LR in the dark phase. The motility at concentrations of 5, 15 and 50 mg L^{-1} was significantly higher than that of the control and the lowest test concentration (0.5 mg L^{-1}).

Figs. 3 and 4 further show a clear dose-effect relationship. Decreases in daytime activity under MC-LR exposure are – at least partly – compensated for by increases in nighttime activities. The same applies to the increase in daytime motility at the lower exposures, which corresponded to decreases in the dark (Fig. 4).

The increase in daytime motility at low exposures can be interpreted as an escape reaction and/or as an increased spatial orientation behaviour. Tembrock (1984) described spatial orientation as an active behavioral adaptation to cope with changing ecological conditions. The further increase in toxin concentrations caused significant decreases in the motility, which could be explained as energy-saving behaviour. The

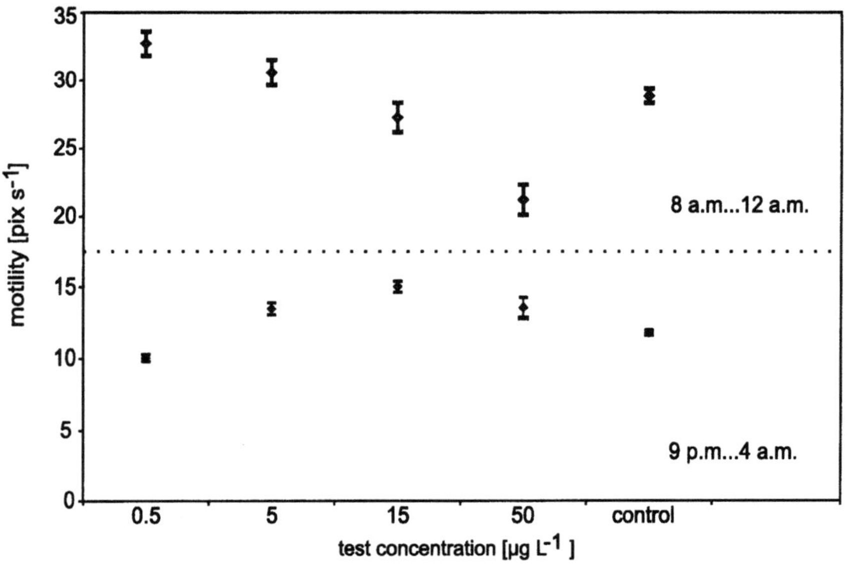

Fig. 4. Motility of zebrafish during the exposure to different concentrations of microcystin-LR; *open squares* indicate the average daily motility between 08.00 h and 12:00 h; *solid squares* indicate the average daily motility between 21:00 h and 04:00 h; *bars* represent the 95% confidence intervals

energy saved by slower swimming would be available as surplus energy at the point of maximum metabolic scope (Staaks 1996).

In addition, the phase shifting of the circadian activity patterns and the elevated activities during nighttimes may be interpreted as sheltering behaviour. This coincides with previous findings of Steinberg et al. (1995) that zebrafish exposed to various concentrations of atrazine showed a pronounced preference for dark habitats. The MC-LR induced phase shift and the reduced reaction on feeding and light-dark changes affected the stability of rhythms and characterized the toxin as a stressor for fish.

Generally, the implications of microcystins for fish are unclear. Whereas acute toxic effects of cyanobacteria toxins on fish were demonstrated unambiguously after intraperitoneal injection and oral application (Phillips et al. 1985; Sugaya et al. 1990; Tencalla et al. 1994; Råbergh et al. 1991; Kent et al. 1994) studies on sublethal and chronic effects are scarce and their results in part contradictory (Råbergh et al. 1991; Oberemm et al. 1997; Pflugmacher et al. 1999; Wiegand et al. 1999; see also Chapters 6.1 and 6.2).

MC-LR diluted in water produced no significant histopathological effects in fish (Phillips et al. 1985; Tencalla et al. 1994). Bury et al. (1995) registered growth inhibition and disturbances of the ion balance in fish exposed to MC-LR diluted in water at concentrations of 41-57 mg L^{-1}. Rodger et al. (1994) found gill and liver damage as well as mortality for brown trout (*Salmo trutta* L.) at concentrations between 16 to 19 mg L^{-1}, but the authors were not able to attribute the effects to microcystin exposure alone. The results of Johnston et al. (1994) showed that the fish respiratory system is unaffected by short-term exposure to microcystin or other biologically active compounds produced by *M. aeruginosa*.

The results of our investigation show effects on the behaviour of fish even with the MC-LR exposure of 0.5 μg L^{-1}. The LOEC (lowest observed effect concentration) appears to be 0.5 mg L^{-1} for behavioral parameters. A LOEC value at the same level of 0.5 mg L^{-1} was also found by Oberemm et al. (1997): larvae of *Danio rerio* reared during the sensitive stages of embryonic development under the influence of MC-LR in the medium showed a decreased survival at 50 mg L^{-1} and a retarded larval growth at concentrations of 0.5; 5 and 50 mg L^{-1} (see also Chapter 7.2).

If further experiments, particularly for native species, confirm these indications of reproductive effects, the ecological consequences would be substantial. Changes in species composition in eutrophicated waters may not only be attributed to losses in spawning sites, but also to direct effects of cyanotoxins on fish reproduction. Further, the findings of reduced motility, increased rates of night activity in relation to day activity, reduced activity during the spawning time period and reduced reaction on feeding and on dusk and dawn are symptoms of the potential danger of MC-LR to fish. For populations competing under natural ecosystems, this reduced overall level of activity will eventually cause disadvantages likely to result in reduced population growth.

6.4 Responses of *Daphnia galeata* fed with *Microcystis* strains With and Without Microcystins

Manfred Henning, Thomas Rohrlack and Johannes-Günter Kohl

Microcystis species is a colonial species, and colonies may attain considerable size. A feature of such colonies is the mucilage in which the cells are embedded. Size, mucilage, and toxin content could individually or jointly affect the ingestion of *Microcystis* as food by zooplankters.

Mass occurrence of *Microcystis* frequently leads to a decrease in large filter-feeding daphnids in lakes (Allan 1977; Gliwicz 1977; Richman and Dodson 1983; Orcutt and Pace 1984). The largest colony size that filter feeders can ingest depends on the animal's body size (Hanazato 1996). If such large colonies go through the carapace gap of daphnids they hinder the filtering process. This leads to an increase in food rejection by the furcal claw (Burns 1968b; Gliwicz 1977; Fulton and Paerl 1987; Gliwicz and Lampert 1990).

Although *Microcystis* in single cells or small colonies can be easily ingested, the utility of these cyanobacteria as food source for *Daphnia* and the role of microcystin content for grazing is under debate: a number of authors (Lampert 1981, 1982; Nizan et al. 1986; Fulton and Paerl 1987; Jungmann et al. 1991; Henning et al. 1991) have shown that *Microcystis*, when growing in single cells, inhibits the ingestion rate of *Daphnia*. Fulton and Paerl (1987) detected a close correlation between toxicity and decrease in the ingestion rate by *Daphnia*. Other authors have found no relationship between cell toxicity (tested on mice) and inhibition of the ingestion rate by *Daphnia* (Nizan et. al. 1986; Jungmann et. al. 1991; Henning et al. 1991).

Regarding the mechanisms of the *Daphnia* food intake, *Microcystis* could possibly influence daphnids at various partial steps of the food intake process. After a *Daphnia* has filtered the particles from the water and collected them in its food groove, the thoracic appendages, maxillules and mandibles bring the particles forward into the oesophagus. Finally, the rhythmic movements of the oesophagus move them into the intestine. When a *Daphnia* has filtered more food than it is able to handle, the furcal claw removes the surplus particles.

McMahon and Rigler (1963) as well as Burns (1968a) studied the food intake process by *D. magna*. This includes the mandibular rate, the swallowing rate and the volume of the swallowed food boluses, which all could affect the ingestion rate.

Apart from all of the above-mentioned influences on the ingestion rate, toxic effects were observed when *Daphnia* was fed with toxic *Microcystis* strains. The toxic properties directly reduced the life-span of daphnids (Porter and Orcutt 1980; Lampert 1981; Reinikainen 1994, Hietala et al 1995). The role of microcystin in intoxication of daphnids is poorly understood. However, DeMott and Dhawale (1995) have shown that microcystin-LR inhibits proteinphosphatases PP1 and PP2A in some *Daphnia* species and in calanoid copepods. In contrast, Matveev et al.(1994) found that toxins of *Microcystis* did not influence *Daphnia carinata*. Jungmann et al. (1991) demonstrated in feeding experiments that the toxic effect of *Microcystis* disappeared when microcystins were removed from cells by extraction.

An independent or combined occurrence of all of the inhibitory effects in *Microcystis-Daphnia* interactions may also result in reduced body growth and egg production by the zooplankter, and thus in reduced population growth (Lampert 1981; Smith and Gilbert 1995). It is difficult to evaluate which of the possible causes are responsible for the inhibition of growth and reproduction in *Daphnia*. The inhibition of food intake as well as sublethal doses of a cyanobacterial toxin will reduce the fitness of daphnids. Further, though easily ingested, *Microcystis* may remain largely unassimilated (Lampert 1981) or be an inadequate food source for *Daphnia* because essential nutrients are absent (Porter and Orcutt 1980).

The following contribution summarizes the results of several laboratory observations concerning the response of *D. galeata* fed with *Microcystis* strains which differ in their morphological and physiological properties.

Material and Methods

Cultures of Microcystis and Daphnia

Altogether five *Microcystis* strains with different morphological and chemical characteristics were used as food for experiments with *Daphnia*. The *Microcystis* strains HUB 5-2-4 and HUB 5-3 were isolated already in 1978 from an intensive water bloom of the Pehlitzsee (Brandenburg, Germany). The water bloom primarily contained *Microcystis aeruginosa* with 95% of total phytoplankton biomass. From several isolations, two unialgal strains were obtained by the washing of single colonies and repeated transfers in culture medium. Both strains have been cultured in the laboratory, but the initial colony characteristics were lost during the culture period and the strains are now mainly growing as single cells. Henning et al. (1991) have reported that, in contrast to the strain HUB 5-3, the strain HUB 5-2-4 was highly toxic in a test with white mice.

All further strains used for feeding experiments descend from isolations gained in 1996 from a *Microcystis* water bloom of the eutrophic Wannsee (Berlin, Germany). The properties of the strains HUB-W75, HUB-W334 and HUB-W368 (shape of colonies, mucilage composition, microcystin content etc.) are described in detail by Rohrlack et al. (Chapter 4.2).

Before being used for the feeding experiments, the *Microcystis* cultures grew in a logarithmic growth phase under a light-dark (12:12 h) regime [see Chapter 3.1 for details on culturing procedures, growth conditions (nutrient limitation, light supply), determination of cell density and analyzes of microcystin content].

In contrast to the strains from Pehlitzsee, the newly isolated cultures from Wannsee grew in colonies of considerable size, of which the larger colonies were not ingestible by *Daphnia galeata*. For these reasons, strains HUB-W75, HUB-W334 and HUB-W368 were reduced to small colonies by pulsed sonification (50 W, 15 times for 0.5 s) and then passed through a 25 μm net. After this procedure the suspension contained only small colonies consisting of a few intact cells surrounded by mucilage. All strains showed a comparable size distribution in the suspension after the size reduction, and no particle greater than 30 µm in diameter was found in the fed cell suspension. Therefore, the suspension tested in the experiments contained only ingestible colonies whose size fell short of the exclusion size for *Daphnia* described by Burns (1968b). Suspensions of single cells without mucilage were obtained from all colony forming strains by washing them with deionized water, following the method of Plude et al. (1991). *Daphnia galeata* was originally obtained from our own laboratory stock culture. *Chlorella* cf. *minutissima* in filtered lake water (membrane filters 0.2 µm) functioned as food for *Daphnia.* Only adult females were used for the experiments. Volume of the daphnids was calculated according to the formulae given by Balushkina and Winberg (1979).

Investigation of Filtering and Feeding Behaviour

A method developed and tested by McMahon and Rigler (1963) was used to determine the filtering behaviour. *Daphnia* were fastened dorsally with their carapace onto the bottom of a flow chamber using grease. The chamber was mounted on a stereo microscope. Filtered lake water (membrane filter 0.2 µm) containing the desired algae at a concentration of 20mm^3 L^{-1} was pumped through the chamber at a rate of 1ml min^{-1} for 1 h. Temperature was kept at 20°C. All reactions of the *D. galeata* were recorded on video tape. Parameters of filtering behaviour were determined after an adaptation time of 50 min using these tapes. In this way the appendage beat rate, the rejection rate, the frequency of the contractions of the maxillules (termed maxillary rate in the following), the frequency of food bolus intake (termed swallowing rate in the following) and the mandibular rate were measured. All of these parameters will be summarized below under the definition food intake behaviour.

To minimise the individual variations and the number of repetitions, several variants of algae (e.g. different strain variants) were tested consecutively on the same *Daphnia.* All data were presented relative to a standard strain without microcystins (HUB 5-3) utilized in all the experiments. This strain showed no inhibitory or toxic effects on *Daphnia* (Rohrlack et al. 1999a) or on white mice (Henning et al. 1991), and hence is suitable as a standard *Microcystis* strain. All experiments were carried out following the same pattern. Daphnids were fed first with standard strain HUB 5-3 for 1 h, then with several variants of the strain to be investigated, and then again with HUB 5-3 to test the reversibility of possible changes in food intake behaviour (see Rohrlack et al 1999a for details).

For comparison of the parameters of food intake behaviour, the *t*-test for paired measurements was applied. The significance of differences between the numerical means of the other values was tested by the Student's *t*-test. The existence of a normal distribution was confirmed by standard skewness and kurtosis. All statistical analyses for these experiments were carried out at the 99% level of significance.

Life Table Experiments

Survival tests were carried out in 100-ml tubes secured at both ends with 200 μm gauze. These tubes were placed into 1.0-L glass bottles with filtered (membrane filter 0.2 μm) lake water containing the desired cyanobacteria at a density of 10 mm^3 L^{-1}. Ten adult *D. galeata* were transferred into the tubes. The suspension was moderately stirred with a magnetic stirrer. Each day, the living animals were counted and transferred into freshly prepared *Microcystis* suspension.

Survival functions were compared with the log-rank test at the 95% level of significance. The time needed to kill 50% of the animals was calculated as median of the survival function estimation (product-limit estimator).

Determination of Body Growth and Birth Rate of D. galeata

For determination of the influence of *Microcystis* on body length, birth rate and reproduction of *D. galeata,* the unicellular strains HUB 5-3 and HUB 5-2-4 were compared with the green alga *Scendesmus* cf. *quadricauda.*

To analyze potential differences in body length and birth rate, 1-day-old *Daphnia* were placed separately into 100-ml vessels filled with a suspension of the chosen food source (HUB 5-3, HUB 5-2-4 or *S.* cf. *quadricauda*) at a temperature of 20°C. The density of the suspensions used was originally adjusted at 10mm^3 algae L^{-1}. A few experiments were carried out with mixed suspensions of these species.

The duration of the experiment amounted at maximum to 13 days, beginning from the first appearance of the neonates. Every 2 to 3 days the animals were transferred into a freshly prepared *Microcystis* suspension. Newborn *Daphnia* were counted and removed from the experiment. After a growth time of 13 days the body length of the *Daphnia* was microscopically measured (from the base of the spine to the end of the head). The average birth rate was calculated for the time interval between the first observation of eggs in the *Daphnia* and the end of the experiment. However, many *Daphnia* fed with *Microcystis* HUB 5-2-4 died during the investigation period. Such events were registered but, due to the mortality of *Daphnia*, the maximal birth rate could not be calculated.

Results

Food Intake Behaviour of D. galeata fed with Different Microcystis Strains

No significant differences were found in appendage beat rate and rejection rate of the *D galeata* when they were fed with different *Microcystis* strains. In contrast, remarkable differences were detected in the mandibular rate, the maxillary rate and the swallowing rate – all of the parameters which determine *Daphnia*'s food intake. In comparison to the standard strain (HUB 5-3), three investigated strains caused a decrease in the maxillary rate and swallowing rate. When *Microcystis* HUB 5-2-4 was fed, both rates were around 30% lower compared to the standard strain. Strains HUB-W75 or

HUB-W334 decreased maxillary rate by only 12%, but swallowing rate in this case was reduced by approximately 30%. All of the observed differences to the standard strain HUB 5-3 were highly significant. Fig. 1 shows that *D. galeata* responds to the different strains in three distinct manners. The strain HUB 5-2-4 causes a decrease in both maxillary- and swallowing rates. Strains HUB-W75 and HUB-W334 caused a smaller reduction of the maxillary rate than of the swallowing rate. HUB-W368 caused only small variances in both maxillary- and swallowing rate

As already described by Rohrlack et al. (1999a) the contractions of the maxillules seem to be mechanically hindered in *D. galeata* when *Microcystis* HUB-W75 or HUB-W334 are fed. The microscopic observation of maxillule-based food-intake behaviour showed that the intensity of *Daphnia*'s maxillule contraction was strongly linked to the amount of cells accumulated in the food groove. The maxillule contracted with normal intensity after food rejection when the food groove was empty. With an increasing amount of accumulated algae, the intensity of maxillules' contractions became weak. Furthermore, the contractions occurred irregularly and developed into an irregular vibration. Sometimes the movement stopped completely until the next rejection occurred.

The mandibular rate tended to decrease when *Daphnia* was fed with strains HUB 5-2-4, HUB-W75 or HUB-W334 (data not shown). Possibly, there is a weak correlation between mandibular rate and maxillary rate; however, the individual variations were pronounced. In consequence, no clear pattern of changes in mandibular rate could be observed. In no case did the mandibular rate fall short of the maxillary rate.

It must be emphasized that all changes in food-intake behaviour occurred immediately when the food was changed from the standard strain HUB 5-3 to HUB 5-2-4, HUB-W75 or HUB-W334. When the strain HUB 5-3 was reintroduced, all observable changes disappeared. Therefore all *Microcystis*-induced changes were completely reversible. In contrast to HUB 5-2-4, HUB-W75 and HUB-W334, the

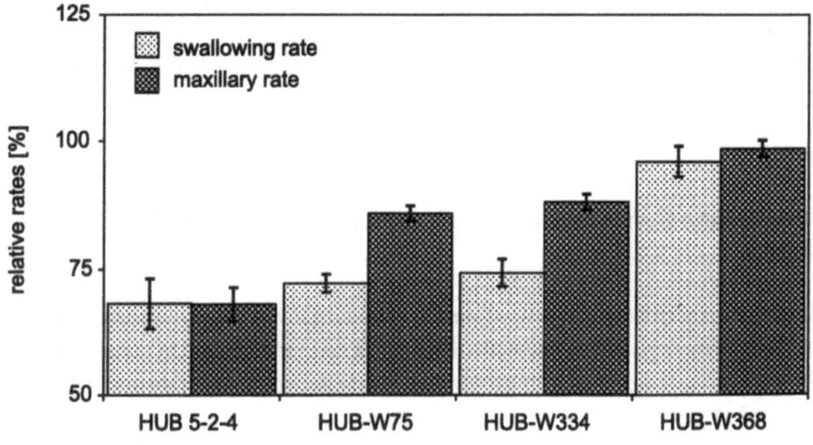

Fig. 1. Relative swallowing and maxillary rates of *D. galeata* when fed with different microcystis strains as % of the respective rates of the strain HUB 5-3

strain HUB-W368 caused no changes in food-intake behaviour compared with the standard strain HUB 5-3.

Daphnia galeata ingests many different *Microcystis* strains at different rates. *Microcystis* HUB 5-3 seems to be a suitable food source for *Daphnia galeata* as indicated by the comparably high ingestion rate of 0.213 $mm^3\ mm^{-3}\ day^{-1}$, and the colony-forming strain HUB-W368 causes no significant reduction in the ingestion rate. *Microcystis* HUB-W75 and HUB-W334 induced significantly lower ingestion rates as compared with the standard strain HUB 5-3. The greatest reduction in the ingestion rate was found when the unicellular strain HUB 5-2-4 was fed.

Nutrient Supply State of Microcystis Strains and the Swallowing Rate of Daphnia

Because variances in nutrient and light during the cultivation of *Microcystis* strains alter the quality of the cells (see Chapter 3.1), a response of *Daphnia* to changes in swallowing rate is possible. In order to test this, *D. galeata* was fed first with the

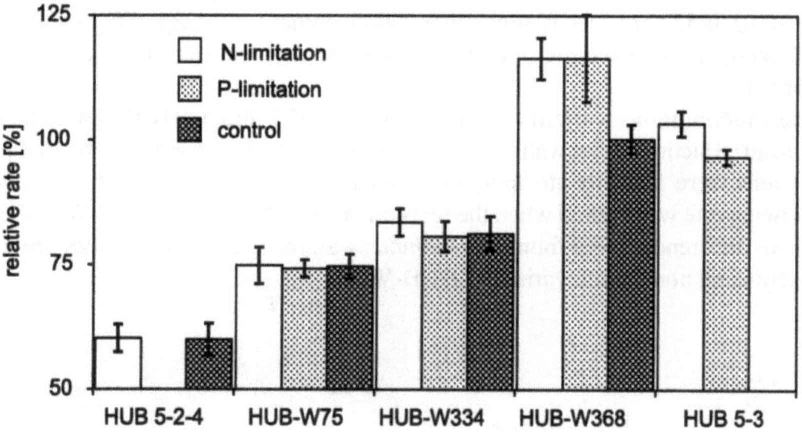

Fig. 2. Relative swallowing rate of *D. galeata* when fed with nutrient-limited grown strains as % of the respective rate of the strain HUB 5-3

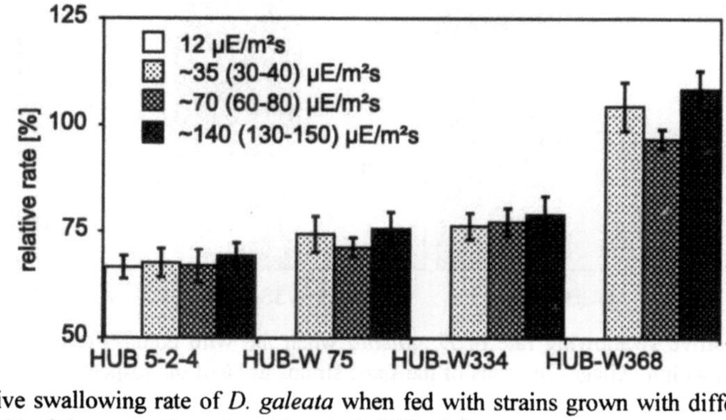

Fig. 3. Relative swallowing rate of *D. galeata* when fed with strains grown with different light intensities as % of the respective rate of the strain HUB 5-3

standard strain HUB 5-3 followed by phosphorus- and nitrogen-limited variants of the strains to be investigated. Additionally, a nutrient-saturated culture was fed to the *Daphnia* as a control variant. The results showed that both P and N limitation in the colony-forming strains HUB-W75 and HUB-W334 caused no further changes in *Daphnia* swallowing rate beyond the reduction induced by nutrient-saturated cultures of these strains as shown in Fig. 1. Only the P-limited culture of unicellular strain HUB 5-2-4 induced an additional decrease in swallowing rate (Fig. 2). For strain HUB-W368, whereas the nutrient-saturated culture caused no changes in *Daphnia* food intake behaviour in comparison to standard strain HUB 5-3 (see above), the swallowing rate tended to increase when P- or N-limited HUB-W368 was offered (Fig. 2). Growth of the strains at varying light intensities did not affect the swallowing rate of *D. galeata* (Fig. 3).

Influence of the Mucilage of Cells on the Swallowing Rate of Daphnia

The possibilty of interference of mucilage with *Daphnia* food intake was tested by feeding first the standard strain HUB 5-3 followed by a colony-forming strain (HUB-W75, HUB-W334 or HUB-W368) either with or without mucilage. Additionally, the swallowing rates of axenic and the non-axenic *Microcystis* HUB-W334 were compared.

The mucilaginous variants of strains HUB-W75 and HUB-W334 caused a significant reduction in the swallowing rate (Fig. 4). This effect disappeared completely when cells were fed after stripping their mucilage envelope, while no changes in swallowing rate were found when the respective variants of strain HUB-W368 were used. No differences were found in *Daphnia* swallowing rate when it was fed with the axenic and non-axenic variant of HUB-W334.

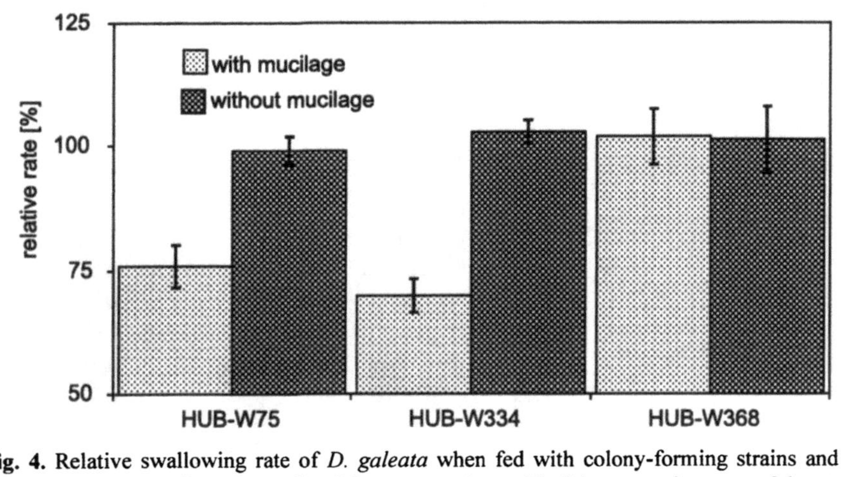

Fig. 4. Relative swallowing rate of *D. galeata* when fed with colony-forming strains and in comparison with mucilage-free cells of the same strains as % of the respective rates of the strain HUB 5-3

Life span of D. galeata fed with Toxic and Non-Toxic Microcystis

All *Microcystis* strains were fed separately in order to test their toxicity on daphnids. The survival curves presented in Fig. 5 show clearly that only HUB-W368, HUB-W334 and HUB 5-2-4 result in a shorter life span compared with starving animals. Differences in life span as compared with starving animals are significant when tested using the Logrank test (a=0.05). Table 1 shows the estimatated 50% percentiles of the life time distribution (length of time for which 50% of the animals survived). The shortest survival time resulted when *D. galeata* was fed with strains HUB-W334 or HUB-W368. *Daphnia* fed with *Microcystis* HUB-W75 resulted in lower mortality rates than those of starving animals, and all *Daphnia* survived no less than 9 days if fed with the strain HUB 5-3 or with *Scenedesmus*. A median survival time of 3.5 days was observed with the strain HUB 5-2-4.

Table 1. Estimated length of time (days) for which 50% of the *Daphnia* survive

No. of experiment	1	2	1	2
Strain	Estimate	Estimate	Standard error	Standard error
HUB-W75	*	7.57	*	0.30
HUB-W368	1.83	2.83	0.26	0.26
HUB-W334	4.33	1.71	0.68	0.22
HUB 5-2-4	3.5	3.5	0.47	0.47
Non-food	9.33	*	0.83	*

* Death rate is lower than 50% within 9 days

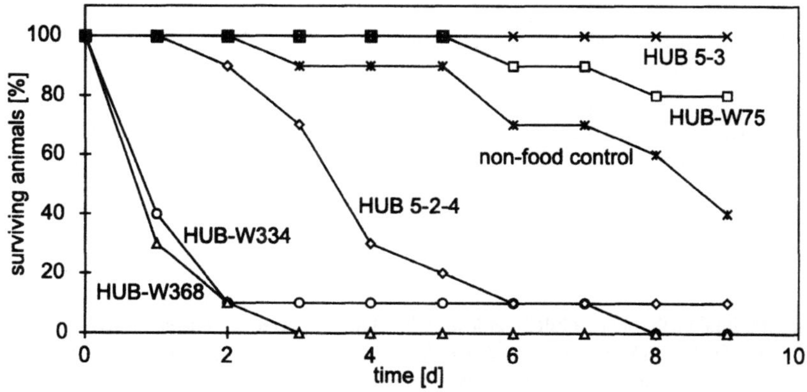

Fig. 5. Survival time of *D. galeata* when fed with different *Microcystis* strains

Growth and Reproduction of D. galeata fed with Microcystis

No significant differences were found in body growth when animals were fed with unicellular strains HUB 5-3 and HUB 5-2-4 or with the green alga *Scenedesmus cf. quadricauda*. The final body length (after 13 days' growth time) ranged between 1.7 and 1.9 mm. Marked differences were found in the maximal birth rate of adult *D. galeata* (Fig. 6). When fed with *Scenedesmus* cf. *quadricauda* or with the *Microcystis* strain HUB 5-3, the birth rate amounted to three to four newborns per day. In contrast, *Microcystis* HUB 5-2-4, as well as a mixture with *Scenedesmus* cf. *quadricauda* (1:1), induced a significant reduction in maximal birth rate by at least 65%.

Discussion

All the experiments showed that some *Microcystis* strains were capable of fundamentally influencing the *Daphnia* food-intake process. Components of the feeding process, appendage beat rate, rate of mandibular movements and rate of postabdominal rejections have previously been described in relation to environmental conditions. Fundamental work on the variations of these rates in relation to the concentration of food has been published by McMahon and Rigler (1963) as well as Burns (1968a). McMahon (1968) noted that the appendage beat rate may also vary with the water quality, and Gliwicz and Sieniawska (1986) assessed this rate as an indicator for cell toxicity.

However, we could not detect that the appendage beat rate varied in *D. galeata* when fed with toxic or non-toxic *Microcystis* strains. Given equal food concentrations, this rate remainded constant over all experiments. Similarly, the mandibular rate and

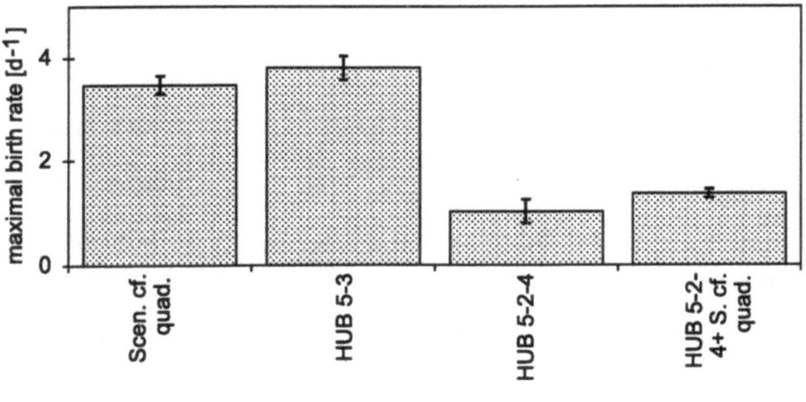

Fig. 6. Maximal birth rate of adult *D. galeata* when fed with *Scenedesmus* and *Microcystis* strains. *Left column* for *Scenedesmus* pure; *middle columns* for non-toxic *Microcystis* strain HUB 5-3 and toxic *Microcystis* strain HUB 5-2-4; *right column* mixed *Microcystis* strain HUB 5-2-4 and *Scenedesmus* in volume relation 1:1

the rejection rate varied only little or not at all. Starkweather (1978) reported that the mandibular rate did not always correlate with the feeding rate. This demonstrates that all of these rates, taken by themselves, do not indicate the toxicity or food quality of a *Microcystis* strain for *D. galeata*.

In contrast, the swallowing rate and, hitherto not maintained, the maxillary rate of *D. galeata*, are influenced by the food quality. The reduction in maxillary rate and in food-transport activity of maxillules seems to be the primary mechanism which inhibits food intake. As a result of the decreased maxillary rate, the swallowing rate and the volume of food bolus decrease. In consequence the ingestion rate of *D. galeata* decreases. A correlation exists between the lower rate of ingestion and the decreased frequency of maxillules movements (Rohrlack et al. 1999a).

Two causes are conceivible for a reduction in the maxillary rate and correspondent food transport activity. The first is a mechanical impediment, the second a sensory perception of an inhibiting factor.

A mechanical failure of the maxillules contractions is likely in colony forming strains HUB-W75 and HUB-W334. The exterior mucilage of cells comes in direct contact with the maxillules. The constitution of the mucilage seems to be important for the action of the maxillules. Hydrolysis of polysaccharides has shown that the strains HUB-W75 and HUB-W334 have high amounts of mannose/xylose (Rohrlack et al. 1999a). These provide many possibilities for cation-based chain-by-chain associations of the envelope polysaccharides, as is confirmed by the relatively high Ca^{2+}-disaggregation endpoint (Angyal 1972; Plude et al. 1991) which may indicate a higher viscosity (Parker et al. 1996). In contrast, the mucilage of the easily ingested strain HUB-W368 is characterized by high amounts of galactose and a very low Ca^{2+}-disaggregation end point (Rohrlack, unpubl. results). This strain does not impede the contractions of the maxillules, and no inhibitory effects could be observed when *Microcystis* HUB-W368 was fed.

Moreover, it could be shown that the inhibition towards food intake disappeared when mucilage of strains HUB-W75 and HUB-W334 were removed. Accordingly, we can conclude for these strains that the factor inducing the hindrance of the maxillules is the mucilage itself, or rather, specific components of the mucilage such as different polysaccharids.

In spite of the absence of mucilage in the unicellular strain HUB 5-2-4, we could observe a similar reduction in maxillules' movement activity. The cells of this strain possibly possess an inhibiting factor that *D. galeata* can perceive. The response of the *Daphnia* to this factor is a decrease in the swallowing rate and a slower ingestion rate (Rohrlack et al. 1999a). The chemical character of this inhibiting factor is not clear, but microcystins may be excluded: strain HUB-W368 did not show this effect in spite of its particularly high microcystin content, and no relationship exists between the ingestion rates of *D. galeata* and microcystin content. This fact was confirmed by alternately feeding *Daphnia* with a microcystin-free mutant and the wild type of the strain PCC-7806. Both variants were ingested at a low rate as compared with the strain HUB 5-3, but no difference occurred between the microcystin-free mutant and the wild type (Rohrlack et al. 1999).

However, microcystins in *Microcystis* do affect *D. galeata*. Survival time of *D. galeata* was substantially impaired when fed with strains that are both readily ingested and contain a high level of microcystins (HUB-368 and HUB-334). Strains that have a lower microcystin content or which were difficult to ingest increased the life span of *D. galeata*, although growth and reproduction experiments with the strain HUB 5-2-4 have shown that *D. galeata* can survive and even reproduce in small numbers in spite of microcystis content in food. However, in the above-mentioned experiment with a microcystin-free mutant, only the wild type reduced the life span of *D. galeata* (Rohrlack et al. 1999).

In summary, these results show that the reactions of *Daphnia* to *Microcystis* with and without microcystin are of a complex nature which cannot be reduced to a single process. Further investigations utilizing a variety of strains with different morphological and physiological characteristics will be necessary in order to understand more fully the *Daphnia-Microcystis* interactions.

References

Alberts JJ, Filip Z, Hertkorn N (1992) Fulvic and humic acids isolated from groundwater: compositional characteristics and cation binding. J Contam Hydrol 11:317–330

Alcock J (1996) Das Verhalten der Tiere aus evolutionsbiologischer Sicht (Behaviour of animals from the evolutionary point of view). Gustav Fischer, Stuttgart Jena New York

Allan JD (1977) An analysis of seasonal dynamics of a mixed population of *Daphnia*, and the associated cladoceran community. Freshwater Biology 7:505–12

Anderson MP, Gronwald JW (1991) Atrazine resistance in a velvetleaf *(Abutilon theophastri)* biotype due to enhanced glutathione S-transferase activity. Plant Physiol 96:104–109

Angyal SJ (1972) Complex formation between sugars and metal ions. In: Doane WM. Carbohydrate chemistry-VI. London: Butterworth, pp. 131–46

Arp W (1997) Blaualgen in Brandenburger Gewässern. In: Chorus (ed) Toxische Cyanobakterien in deutschen Gewässern – Verbreitung, Kontrollfaktoren und ökologische Bedeutung. WaBoLu-Hefte 4/97, pp. 52–57

Baganz D, Staaks G, Steinberg C (1998) Impact of the cyanobacteria toxin, microcystin-LR on behavior of zebrafish, *Danio rerio*. Wat Res 32:948–952

Balushkina EW, Vinberg GG (1979) Dependence between length and body volume from planktonic cladocera. In: Vinberg GG (ed) Laboratory and freewater investigations about the biological fundamentals of the waters productivity (russ.) AN SSSR, Zool Inst Leningrad, pp.58–79

Bligh EG, Dyer WJ (1959) A rapid method of total lipid extraction and purification. Can J Biochem 37:911–917

Bradford M (1976) A rapid and sensitive method for the quantification of microgram quantities of protein utilizing the principle of protein-dye binding. Anal Biochem 72:248–254

Burns CW (1968a) Direct observation of mechanisms regulating feeding behavior of *Daphnia* in lake water. Int Revue ges Hydrobiol 53:83–100

Burns CW (1968b) The relationship between body size of filter-feeding cladocera and the maximum size of particle ingested. Limnol Oceanogr 13:675–8

Bury NR, Eddy FB, Codd GA (1995) The effects of the cyanobacterium *Microcystis aeruginosa*, the cyanobacterial hepatotoxin microcystin-LR, and ammonia on growth rate and ionic regulation of brown trout. J Fish Biol 46:1042–1054

Bury NR, McGeer JC, Eddy FB, Codd GA (1997) Liver damage in brown trout, *Salmo trutta* L. and rainbow trout, *Oncorhynchus mykiss* (Walbaum), following administration of the cyanobacterial heptatoxin microcystin-LR via the dorsal aorta. J Fish Dis 20:209–215

Carbis CR, Mitchell GF, Anderson JW, McGauley I (1996) The effects of microcystins on serum biochemistry of carp, *Cyprinus carpio* L., when the toxins are administered by gavage, immersion and intraperitoneal routes. J Fish Dis 19:151–159

Carbis CR, Rawlin GT, Grant P, Mitchell GF, Anderson JW, McGauley I, (1997) A study of the feral carp, *Cyprinus carpio* L., exposed to *Microcystis aeruginosa* at Lake Mokoan, Australia, and possible implications for fish health. J Fish Dis 20:81–91

Cousins IT, Bealing DJ, James HA, Sutton A (1996) Biodegradation of Microcystin-LR by indigenous mixed bacterial populations. Wat Res 30:481–485

DeMott WR, Zhang QX, Carmichael WW (1991) Effects of toxic cyanobacteria and purified toxins on the survival and feeding of a copepod and three species of *Daphnia*. Limnol Oceanogr 36:1346–1357

DeMott WR, Dhawale S (1995) Inhibition of *in vitro* protein phosphatase activity in three zooplankton species by microcystin-LR, a toxin from cyanobacteria. Archiv Hydrobiol 134:417–24

Drotar A, Phelps P, Fall R (1985) Evidence for glutathione peroxidase activities in cultured plant cells. Plant Science 42:35–40

Edwards C, Lawton LA, Codd GA (1994) Detection of cyanobacterial (blue-green algal) peptide toxins by protein phosphatase inhibition. In: Codd GA, Jefferies TM, Keevil CW, Potter E (eds) Detections Methods for Cyanobacterial ToxinsRoyal Society of Chemistry, Cambridge.

Egaas E, Skaare JU, Sandvik M, Falls JG, Dauterman WC, Collier TK, Netland J (1993) A comparative study of effects of atrazine on xenobiotic metabolising enzymes in fish and insect and of the in vitro phase 2 atrazine metabolism in some fish, insects, mammals and one plant species. Comparative Biochemistry and Physiology C 106:141–149

Eriksson JE, Meriluoto JAO, Lindholm T (1989) Accumulation of a peptide toxin from cyanobacterium *Oscillatoria agardhii* in the freshwater mussel *Anadonta cygnea*. Hydrobiologia 183:211–216

Fulton RSIII, Paerl HW (1987) Toxic and inhibitory effects of the blue-green alga *Microcystis aeruginosa* on herbivorous zooplankton. J Plankt Res 9:837–55

George SG (1994) Enzymology and molecular biology of phase II xenobiotic conjugating enzymes in fish. In: Malins DC, Ostrander GK (eds) Aquatic toxicology. Molecular, biochemical and cellular perspectivesLewis Publishers. Boca Raton, USA, pp. 37–85

Gliwicz ZM (1977) Food size selection and seasonal succession of filter feeding zooplankton in an eutrophic lake. Ekologia Polska 25:179–225

Gliwicz ZM, Sieniawska A (1986) Filtering activity of *Daphnia* in low concentrations of a pesticide. Limnol Oceanogr 31:1132–38

Gliwicz ZM, Lampert W (1990) Food thresholds in *Daphnia* species in the absence and persence of blue-green filaments. Ecology 71:691–702

Habig W, Pabst MJ, Jacoby WB (1974) Glutathione S-transferase: the first step in mercapturic acid formation. J Biol Chem 249:1730–1739

Hanazato T (1996) Toxic cyanobacteria and the zooplankton community. In: Watanabe MF, Harada K, Carmichael WW, Fujiki H (eds.) Toxic *Microcystis*. CRC Press Boca Raton etc. 1996. pp. 79–102

Henning M, Hertel H, Wall H, Kohl JG (1991) Strain-specific influence of *Microcystis aeruginosa* on food ingestion and assimilation of some Cladocerans and Copepods. Int Revue ges Hydrobiol 76:37–45

Hietala J, Reinikainen M, Walls M (1995) Variation in life history responses of *Daphnia* to toxic *Microcystis aeruginosa*. J Plankton Res 17:2307–18

ISO: Internationale Organisation für Standardissierung. Bezugsnummer: 7346/1-1984: Waserbeschaffenheit – Bestimmung der akuten tödlichen Toxizität von Substanzen beim Süßwasserfisch *Brachydanio rerio* Hamilton-Buchanan (Teleostei, Cyprinidae)

Johnston NAL, Campagna VS, Hawkins, PR, Banens RJ (1994) Response of the eastern rainbowfish (*Melanotaenia duboulayi*) to toxic *Microcystis aeruginosa*. Aust J Mar Freshwater Res 45: 917–923

Jungmann D (1992) Toxic compounds isolated from *Microcystis* PCC7806 that are more active against *Daphnia* than two microcystins. Limnol Oceanogr 37:1777–1783

Jungmann D, Henning M, Jüttner F (1991) Are the same compounds in *Microcystis* responsible for toxicity to *Daphnia* and inhibition of its filtering rate? Int Revue ges Hydrobiol 76(1):47–56

Jungmann D, Benndorf J (1994) Toxicity to *Daphnia* of a compound extracted from laboratory and natural *Microcystis* spp., and the role of microcystins. Freshwater Biology 32:13–20

Kent ML, Andersen RJ, Holmes CFB, Mc Cready T, Williams DE (1994) Evidence that microcystin LR is the cause of netpen liver disease of Atlantic salmon (*Salmo salar*). International Symposium on Aquatic Animal Health: Program and abstracts. Univ. of California, School of veterinary medicine, W. 15.4

Kimmel CB, Ballard WW, Kimmel SR, Ullmann B, Schilling TS (1995). Stages of embryonic development of the zebrafish. Dev Dyn 203:253–310

Kondo F, Ikai Y, Oka H, Okumura M, Ishikawa N, Harada KI, Matsuura K, Murata H, Suzuki M (1992) Formation, characterization and toxicity of the glutathione and cysteine conjugates of toxic heptapeptide microcystins. Chem Res Toxicol 5:591–596

Laale HW (1977) The biology and use of zebrafish *B. rerio* in fisheries research. A literature review. J. Fish Biol. 10: 121-173

Lampert W (1981) Inhibitory and toxic effects of blue-green algae on *Daphnia*. Int Revue ges Hydrobiol 66:285–98

Lampert W (1982) Further studies on the inhibitory effect of the toxic blue-green *Microcystis aeruginosa* on the filtering rate of zooplankton. Archiv Hydrobiol 95:207–20

Lampert W (1987) Feeding and nutrition in *Daphnia*. In: Peters RH, De Bernardi R (eds.) Memorie dell'istituto italiano di idrobiologia dott. marco de marchi Vol. 45. *Daphnia*. Verbania Pallanza 1987, pp.143–192

Matveev V, Matveeva L, Jones GJ (1994) Study of the ability of *Daphnia carinata* King to control phytoplankton and resist cyanobacterial toxicity: implications for biomanipulation in Australia. Aust J Mar Freshwater Res 45:889–904

McMahon JW, Rigler FH (1963) Mechanisms regulating the feeding rate of *Daphnia magna* Straus. Canad J Zool 41:321–32

Nizan S, Dimentman C, Shilo M (1986) Acute toxic effects of cyanobacterium *Microcystis aeruginosa* on *Daphnia magna*. Limnol Oceanogr 31:497–502

Oberemm A, Fastner J, Steinberg C (1997) Effects of microcystin-LR and cyanobacterial crude extracts on embryo-larval development of zebrafish. Water Research 31:2918–2921

Oberemm A, Becker J, Codd G, Steinberg C (1999) Effects of cyanobacterial toxins and aqueous crude extracts on the development of fish and amphibians. Environm Toxicol 14:77–88

Oberemm A (2000) The use of a refined zebrafish embryo bioassay for the assessment of aquatic toxicity. Lab Animal 29:32–40.

Orcutt JD Jr, Pace ML (1984) Seasonal dynamics of rotifer and crustacean zooplankton population in a eutrophic monomictic lake with a note on rotifer sampling techniques. Hydrobiologia 119:73–80

Parker DL, Schram BR, Plude JL and Moore RE (1996) Effect of metal cations on the viscosity of a pectin-like capsular polysaccharide from cyanobacterium *Microcystis flos-aquae* C3-40. Appl Environm Microbiol 62:1208–13

Pflugmacher S, Schröder P (1995) Glutathione S-transferase in trees: inducibility by various organic xenobiotics. Z Pflanzenernähr Bodenkunde 158:71–73

Pflugmacher S, Steinberg CEW (1997) Activity of phase I and phase II detoxication enzymes in aquatic macrophytes. J Appl Botany 71:144–146

Pflugmacher S, Wiegand C, Beattie KA, Codd GA, Steinberg CEW (1998a) Uptake of the cyanobacterial hepatotoxin microcystin-LR by aquatic macrophytes. J Appl Botany 72: 228–232

Pflugmacher S, Wiegand C, Oberemm A, Beattie KA, Kruase E, Codd GA, Steinberg CEW (1998b) Identification of an enzymatically formed glutathione conjugate of the cyanobacterial hepatotoxin microcystin-LR: the first step of detoxication. Biochimica et Biophysica Acta 1425:527–533

Plugmacher S, Codd GA, Steinberg CEW (1999) Effects of the cyanobacterial toxin microcystin-LR on detoxication enzymes in aquatic plants. Environm Toxicol 14:111–115

Phillips MJ, Roberts RJ, Stewart JA (1985) The toxicity of the cyanobacterium *Microcystis aeruginosa* to rainbow trout, *Salmo gairdneri* Richardson. J Fish Dis 8:339–344

Plude JL, Parker DL, Schommer OJ, Timmerman RJ, Hagstrom SA, Joers JM, Hnasko R (1991) Chemical characterization of polysaccharide from slime layer of the cyanobacterium *Microcystis flos-aquae* C3-40. Appl Environm Microbiol 57:1696–700

Porter KG, Orcutt JD (1980) Nutritional adequacy, manageability, and toxicity as factors that determine the food qualiry of green and blue-green algae for *Daphnia*. In: Kerfoot WC (ed) Evolution and ecology of zooplankton communities. University Press of New England, Hanover. pp 268–281

Råbergh CMI, Bylund G, Erikson JE (1991) Histopathological effects of microcystin LR, a cyclic peptide toxin from the cyanobacterium (blue-green alga) *Microcystis aeruginosa* on common carp (*Cyprinus carpio* L.). Aquat Toxicol 20:131–146

Reinikainen M (1994) Acute toxicity of the cyanobacterium *Microcystis aeruginosa* 8strain PCC 7820) to *Daphnia pulex* (Cladocera). Algological Studies 75:229–237

Richman S, Dodson SI (1983) The effect of food quality on feeding and respiration by *Daphnia* and *Diaptomus*. Limnol Oceanogr 28:948–56

Rodger HD, Turnbull T, Edwards C, Codd GA (1994) Cyanobacterial (blue-green algal) bloom associated pathology in brown trout, *Salmo trutta* L., in Loch Leven, Scotland. J Fish Disease 17:177–181

Rohrlack Th, Dittmann E, Henning M, Börner Th, Kohl JG (1999) Role of microcystins in poisoning and food ingestion inhibition of *Daphnia galeata* caused by the cyanobacterium *Microcystis aeruginosa*. J Appl Microbiol 65:737–739

Rohrlack Th, Henning M, Kohl JG (1999a) Mechanisms of the inhibitory effect of the cyanobacterium *Microcystis aeruginosa* on *Daphnia galeata's* ingestion rate J Plankton Res 21:1484–1500

Rudolph-Böhner S, Mierke DF, Moroder L (1994) Molecular structure of the cyanobacterial tumor-promoting microcystins. FEBS-Letters 349:319–323

Salaün JP, Benveniste I, Reichart D, Durst F (1978) A microsomal (cytochrome P-450)-linked lauric acid monooxygenase from aged Jerusalem-artichoke tuber tissues. Eur J Biochem 90:155–159

Sandermann H (1992) Plant metabolism of xenobiotics. TIBS 17:82–84

Smith AD, Gilbert JJ (1995) Relative susceptibilities of rotifers and cladocerans to *Microcystis aeruginosa*. Arch Hydrobiol 132:309–336

Spieser OH, Scholz W (1992) Verfahren zur quantitativen Bewegungsanalyse von mehreren Objekten im selben Medium. (A method of quantitative movement analysis of multiple objects in the same medium.) German Patent P 4224750.0

Starkweather PL (1978) Diel variation in feeding behaviour of *Daphnia pulex*. Influences of food densitiy and nutritional history on mandibular activity. Limn Oceanogr 23:307–317

Staaks G (1996) Experimental studies on temperature preference behaviour of juvenile cyprinids. *Limnologica* 26:165–177

Steinberg CEW, Lorenz R, Spieser H (1995) Effects of atrazine on swimming behaviour of zebrafish, *Brachydanio rerio*. Wat Res 29:981–985

Sugaya Y, Yasuno M, Yania T (1990) Effects of toxic *Microcystis viridis* and isolated toxins on goldfish. Jap J Limnol Rikusuiza Tsu 51:149–153

Tembrock G (1984) Verhalten bei Tieren (Behavior of animals). Die neue Brehm-Bücherei, A. Ziemsen-Verlag Lutherstadt Wittenberg, 216 pp.

Tencalla F, Dietrich D, Schlatter C (1994) Toxicity of *Microcystis aeruginosa* peptide toxin to yearling rainbow trout (*Oncorhynchus mykiss*). Aquat Toxicol 30:215–224

Tsuji K, Setsuda S, Watanuki T, Kondo F, Nakazawa H, Suzuki M, Harada K-I (1996) Microcystin levels during 1992–95 for lake Sagami and Tsukui, Japan. Natural Toxins 4:189–194

Vasconcelos VM (1995) Uptake and depuration of the heptapeptide toxin microcystin-LR in *Mytilus galloprovincialis*. Aquatic Toxicol 32:227–237

Westerfield M (1989) The zebrafish book; a guide for the laboratory use of zebrafish (*Brachydanio rerio*). University of Oregon Press, Eugene

Wiegand C, Pflugmacher S, Oberemm A, Meems N, Beattie KA, Steinberg CWE, Codd GA (1999) Uptake and effects of microcystin-LR on detoxication enzymes of early life stages of the zebrafish (*Danio rerio*). Environ Toxicol: 89–95

Wiegand C, Pflugmacher S, Oberemm A, Steinberg CEW (2000) Activity Development of selected detoxication enzymes during the Ontogenesis of the zebrafish (*Danio rerio*). Internat Rev Hydrobiology 85:413–422

7 Toxic Effects and Substances in Cyanobacteria other than Microcystins, Anatoxin-a and Saxitoxins

Cyanobacteria produce a number of biologically active substances. In addition to their well-known alkaloid toxins (see Sivonen and Jones 1999 for an overview), these encompass peptides and depsipeptides. While approximately 60 structural variants of the hepatotoxic microcystins are already known (Sivonen and Jones 1999), the non-hepatotoxic peptides and depsipeptides have been studied to a far lesser extent. These substances have been detected irrespective of whether the object was the study of toxic or of non-toxic strains (Namikoshi and Rinehart 1996). On the other hand, evidence is increasing for biological effects of cyanobacteria which cannot be accounted for by those cyanobacterial metabolites currently understood to encompass "*the* cyanotoxins". This suggests that cyanobacteria contain not only numerous metabolites whose bioactivity is poorly understood, but also further currently unknown toxins. Chapter 7.1 presents a review of peptides and depsipeptides potentially relevant as bioactive substances.

In the context of the project "Toxic Cyanobacteria in German Waterbodies – Occurrence, Causes, Consequences" a number of toxic effects were observed which could not be accounted for by microcystins, anatoxin-a or saxitoxin. Chapter 7.2 summarizes these observations. Further, a number of new cyanobacterial substances with bioactive properties were identified in the context of this project and associated work. These are presented in Chapter 7.3.

7.1 Peptides and Depsipeptides Produced by Cyanobacteria

Heike Bickel, Uwe Neumann and Jürgen Weckesser

Low-molecular peptides with a cyclic structure, with modified amino acids and with D-amino acids are assumed to be formed by non-ribosomal synthesis – as are a multitude of peptide antibiotics (review by Kleinkauf and von Döhren 1990). Owing to their structure, non-ribosomal formation was also assumed for other peptides and depsipeptides (Weckesser et al. 1996), but only for microcystins has this assumption been confirmed by experiments with relevant mutants (Dittmann et al. 1997). In this context it is interesting that ribosomal syntheses have been demonstrated for low-molecular, cyclic peptides such as the tricyclic aborycin from *Streptomyces griseoflavus* (Fréchet et al. 1994) as well as for peptides with D-amino acids such as dermorphin and dermenkephalin from the skin of the frog *Phyllomedusa sauvagii* (Richter et al. 1987; review by Kleinkauf and von Döhren 1996).

The overview given below includes examples of peptides and depsipeptides isolated from cyanobacteria structurally closely related to or identical with ones isolated from marine eukaryotes – poriferans and molluscs. There is growing evidence that the peptides and depsipeptides that have been isolated from eukaryotes originate from grazed, associated or symbiotic cyanobacteria. Some of these metabolites initially isolated from eukaryotes have now also been isolated from cyanobacteria cultures (reviews by Moore 1996; Namikoshi and Rinehart 1996; Weckesser et al. 1996).

Linear and cyclic peptides

The linear pentapeptide microginin inhibits the Zn-metallo-protease angiotensin-converting enzyme (Okino et al. 1993a). Microginin was first isolated from *Microcystis aeruginosa* NIES-100 – a strain which also formed the depsipeptides micropeptins A and B. Neumann et al. (1997) isolated microginin FR1 from a water bloom dominated by *Microcystis* sp.

A description of the cyclic pentapeptide nodularin is given by Namikoshi et al. (1994; see also Sivonen and Jones 1999). Motopurin – a nodularin analogue with a L-Val instead of L-Arg – has been isolated from the marine sponge *Theonella swinhoei* or presumably from its symbiotic cyanobacteria (De Silva et al. 1992).

Lyngbyatoxin A, from *Lyngbia majuscula*, is a modified cyclic dipeptide and a potent tumour promoter. Like microcystin and nodularin, lyngbiatoxin A enhances the phosphorylation of serine and threonine residues of eukaryotic proteins, in that it activates the protein kinase C and thus the phosphorylation of serine and threonine (Cardellina et al. 1979; Fujiki et al. 1984). Microcystin and nodularin, by contrast, inhibit protein phosphatases and thus the dephosphorylation of serine and threonine (MacKintosh et al. 1990; Honkanen et al. 1994).

Westiellamide, a bistratamide-related, cyclic and moderately cytotoxic hexapeptide, has been isolated from the terrestrial *Westiellopsis prolifica* (Prinsep et al. 1992; Patterson et al. 1994).

The cyclic decapeptide calophycin has attracted attention for its broad spectrum of fungicidal activity. It has been isolated from the terrestrial *Calothrix fusca* and contains an unusual (2R,3R,4S)-3-amino-2-hydroxy-4-methylpalmitic acid termed Hamp (Moon et al. 1992). Puwainaphycins from *Anabaena* sp. – likewise cyclic decapeptides – also contain this Hamp unit; some are cardioactive, but they are not fungicidal (Moore et al. 1989; Gregson et al. 1992).

The laxaphycins from the terrestrial *Anabaena laxa* are a family of cyclic undeca- and dodecapeptides (Frankmölle et al. 1991, 1992; Patterson et al. 1994). In terms of structure and biological activity, they resemble the hormothamnins, cyclic undecapeptides from the marine *Hormothamnion enteromorphoides* (Gerwick et al. 1989, 1992). Laxaphycins and hormothamnins exhibit cytotoxic, antimicrobial and fungicidal effects. The cyclic undecapeptides scytonemin A from *Scytonema* sp.

(Helms et al. 1988) and schizotrin A from *Schizotrix* sp. (Pergament and Carmeli 1994) exhibit moderate antimicrobial properties.

The cyclic anabaenopeptin was first isolated from *Anabaena flos-aquae*, and later also from other *Anabaena* sp. strains as well as from *Planktothrix* sp. and *Microcystis aeruginosa* strains (Fujii et al. 1995; Harada et al. 1995; Erhard et al. 1997; Erhard 1999; Murakami et al. 1997a; Shin et al. 1997a). Anabaenopeptins contain a characteristic ureido bond between D-Lys[5] and the varying amino acid in position 6. Another structural peculiarity is the bond between the ω-amide group of the D-Lys[5] and the C-terminal end of the L-Phe[1] (Fig. 1). Anabaenapeptins A and B caused vasodilatations in preparations from the aortas of rats which had been contracted with norepinephrine (Harada et al. 1995). Oscillamide Y, obtained from *Oscillatoria agardhii* and similar to the anabaenopeptins, inhibits chymotrypsin (Sano and Kaya 1995). The ferintoic acids from *Microcystis aeruginosa* (Williams et al. 1996) and the nodulapeptins from *Nodularia spumigena* (Fujii et al. 1997) likewise correspond to the anabaenopeptins in structure. Keramamide A and konbamide from the sponge *Theonella* sp. are also related. Because of this observation, Kobayashi et al. (1991a, b) suspected them to be formed by symbiotic cyanobacteria. Keramamide A and konbamide contain L-Lys in position 5 instead of the D-Lys of the anabaenopeptins (Fig. 1). An overview of the anabaenopeptin derivatives hitherto known is provided in Table 1.

Linear and Cyclic Depsipeptides

Depsipeptides are defined as compounds consisting of α-hydroxy acids and α-amino acids that are interlinked in both ester- and amide-like fashion (Römpp 1995). Strictly speaking, the so-called depsipeptides of the cyanopeptolin type are not depsipeptides, as the ester bond of these depsipeptides is formed between the amino acid's α-carboxy group in position 1 (R1 in Fig. 2) and a β-hydroxy group of threonine.

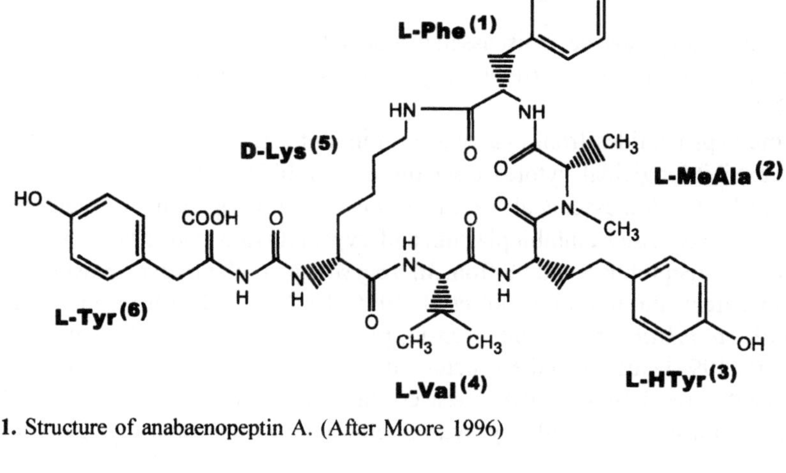

Fig. 1. Structure of anabaenopeptin A. (After Moore 1996)

Table 1. Amino acid variations of anabaenopeptins, oscillamide Y, ferintoic acid, nodulapeptin, keramamide A and konbamide. The abbreviations P1 to P6 stand for positions 1 to 6 of the amino acids

	Reference	P1	P2	P3	P4	P5	P6
Anabaenopeptin A	Harada et al. (1995)	L-Phe	L-MeAla	L-HTyr	L-Val	D-Lys	L-Tyr
Anabaenopeptin B	Harada et al. (1995), Murakami et al. (1997a)	L-Phe	L-MeAla	L-HTyr	L-Val	D-Lys	L-Arg
Anabaenopeptin C	Fujii et al. (1995)	L-Phe	L-MeAla	L-HTyr	L-Val	D-Lys	L-Lys
Anabaenopeptin D	Fujii et al. (1995)	L-Phe	L-MeAla	L-HTyr	L-Val	D-Lys	L-Phe
Anabaenopeptin E	Shin et al. (1997a)	L-Phe	L-MeAla	mHTyr	L-Val	D-Lys	L-Arg
Anabaenopeptin F (Anabaenopeptolin C)	Shin et al. (1997a) Erhard et al. (1997)	L-Phe	L-MeAla	HTyr	L-Ile	D-Lys	L-Arg
Anabaenopeptin G	Erhard (1999)	L-Tyr	MeIle	L-HTyr	L-Ile	D-Lys	L-Arg
Ferintoic acid A	Williams et al. (1996)	L-Phe	MeAla	L-HTyr	L-Val	D-Lys	L-Trp
Ferintoic acid B	Williams et al. (1996)	L-Phe	MeAla	L-HTyr	L-Ile	D-Lys	L-Trp
Oscillamide Y	Sano and Kaya (1995)	L-Phe	MeAla	HTyr	L-Ile	D-Lys	L-Tyr
Nodulapeptin A	Fujii et al. (1997)	Ser(Ac)	MHTyr	HPhe	Met(O_2)	Lys	Ile
Nodulapeptin B	Fujii et al. (1997)	Ser(Ac)	MHTyr	HPhe	Met(O)	Lys	Ile
Keramamide A	Kobayashi et al. (1991a)	L-Phe	MeCht	L-Leu	L-Leu	L-Lys	L-Phe
Konbamide	Kobayashi et al. (1991b)	BhTrp	L-MeLeu	L-Leu	L-Ala	L-Lys	L-Leu

Nonetheless, they have been discussed in association with the group of depsipeptides obtained from cyanobacteria (reviews by Namikoshi and Rinehart 1996; Weckesser et al. 1996).

Some depsipeptides from cyanobacteria inhibit serine proteases or tyrosinases, others exhibit fungicidal, cytotoxic and tumour-inhibiting activities (Namikoshi and Rinehart 1996; Weckesser et al. 1996). For instance, micropeptins A, B and 90 from *Microcystis aeruginosa* inhibit plasmin and trypsin (Okino et al. 1993b; Ishida et al. 1995). Cyanopeptolins S and SS from *Microcystis* sp. inhibit trypsin, plasmin and, to a lesser extent, thrombin (Jakobi et al. 1995, 1996). A90720 from the terrestrial *Microchaete loktakensis* inhibits trypsin, plasmin and thrombin (Lee et al. 1994). Lee et al. (1994) elucidated the structure of a crystallized complex of A90720A and trypsin with a resolution of 1.9 Å. These crystallization results indicate that A90720A interacts with trypsin mainly by hydrophobic interactions such that Leu (R3, cf.

Fig. 2) and MeTyr (R2) occupy the hydrophobic region of the trypsin. A90720A also interacts with trypsin through hydrogen bonds. As the side chain and Arg (R4), the 3-amino-6-hydroxy-2-piperidone unit (Ahp) participates to a significant degree in the formation of the hydrogen bonds and, in addition, prevents the disassociation of A90720A from trypsin.

The following provides an overview of the most important depsipeptides obtained from cyanobacteria:

The linear aeruginosin was first isolated from the toxic strain *Microcystis aeruginosa* NIES-298 (aeruginosin 298-A, Murakami et al. 1994). Aeruginosins from toxic and non-toxic *Microcystis aeruginosa*, *Microcystis viridis* and *Oscillatoria agardhii* inhibit trypsin, thrombin, and (with the exception of aeruginosin 298-A from *Microcystis aeruginosa* NIES-298) plasmin (Murakami et al. 1994, 1995; Shin et al. 1997b; Kodani et al. 1998).

Depsipeptides of the cyanopeptolin type comprise a group of cyclic depsipeptides (reviews by Moore 1996; Namikoshi and Rinehart 1996; Weckesser et al. 1996). Major common features are the ester bond between the hydroxy group of threonine and the amino acid R1 (Val or Ile, see Fig. 2) and a 3-amino-6-hydroxy-2-piperidone unit (Ahp). Dolastatin 13 from the mollusc *Dolabella auricularia* also has this Ahp unit (Pettit et al. 1989). In addition, depsipeptides of the cyanopeptolin type contain an N-methylated amino acid R2, an amino acid R3 (Thr, Leu, Ile, Phe) and a greatly varying amino acid R4 (Fig. 2). The side chain (X) may exhibit the following structural characteristics:

1. The side chain consists of an hexanoyl or octanoyl residue with Asp Glu or Gly, as in the case of e.g. the cyanopeptolins A-D (Martin et al. 1993), the micropeptins A and B from *Microcystis aeruginosa* (Okino et al. 1993b), and the micropeptin 103 from *Microcystis viridis* (Murakami et al. 1997b).
2. The side chain is composed of hydroxyphenyl-lactyl units with one or two amino acids, as in the case of, e.g., microcystilide A (Tsukamoto et al. 1993) and aeruginopeptin from *Microcystis aeruginosa* (Harada et al. 1993).
3. The side chain is a modified glyceric acid with sulphate or methyl groups. This is found, for example, in cyanopeptolins S and SS from water blooms of *Microcystis* sp. (Jakobi et al. 1995, 1996), oscillapeptin from *Oscillatoria agardhii* (Shin et al. 1995), micropeptin 90 (Ishida et al. 1995) and micropeptins 478-A and -B from *Microcystis aeruginosa* (Ishida et al. 1997). The side chain of the component

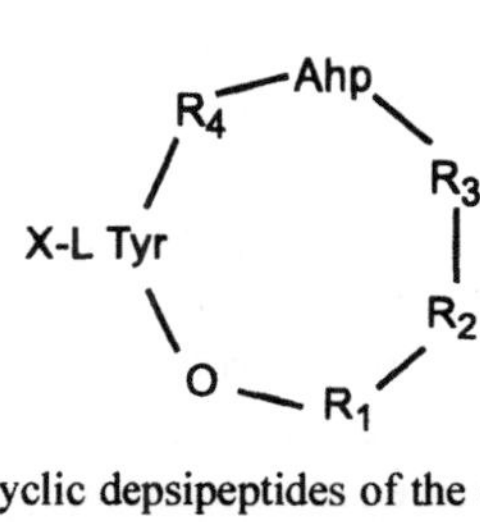

Fig. 2. Common structure of the cyclic depsipeptides of the cyanopeptolin type (Namikoshi and Rinehart 1996). *Ahp* is 3-amino-6-hydroxy-2-piperidone. *X* denotes the side-chain described in the text

A90720A from *Microchaete loktakensis* consists of a Leu-monosulphated glyceric acid (Lee et al. 1994).

4. The side chain is a formyl group with one or two amino acids, as in the case of, e.g., the anabaenopeptilides from *Anabaena* sp. (Fujii et al. 1995).

Other groups of cyclic depsipeptides include:

Microviridins from *Microcystis viridis* and *Microcystis aeruginosa* are tricyclic. Microviridins A and B contain two ester bonds, microviridin C contains only one ester bond. Microviridin A inhibits tyrosinase, and microviridins B and C inhibit trypsin, elastase and chymotrypsin (Ishitsuka et al. 1990; Gerwick et al. 1994; Okino et al. 1995).

The cryptophycins from *Nostoc* sp., which with 25 variants constitute the largest group so far of cyanobacterial depsipeptides, are tumour-selective cytotoxins (Trimurtulu et al. 1994; Golakoti et al. 1995; Moore et al. 1996). Dolastatin 11 from the marine mollusc *Dolabella auricularia* is of a structure similar to the cryptophycins (Pettit et al. 1989, 1993). Cryptophycin-24 is identical with arenastatin from the sponge *Dysidea arenaria* (Kobayashi et al. 1994). Majusculamide C is structurally closely related to cryptophycin, as well.

Majusculamide C has been isolated from *Lyngbya majuscula* (Carter et al. 1984) as well as from the sponge *Ptilocaulis trachys* (or from cyanobacteria accumulated by this sponge) (Williams et al. 1993). This heptacyclodepsipeptide acts as a strong cytotoxin to a broad spectrum of fungal plant pathogens (Moore and Mynderse 1982).

As the last example of a biologically active metabolite from cyanobacteria, mention should be made of hapalosin, which has been isolated from *Hapalosiphon welwitschii* and found to eliminate multidrug resistances in tumour cells (Stratmann et al. 1994).

7.2 Significance of Unidentified Toxic Compounds and Approaches to Their Identification

Axel Oberemm, Rita Heinze, Olaf Papendorf and Jutta Fastner

In the past two decades, many structures and structural variants of cyanobacterial hepato- and neurotoxins have been identified. The most common and diverse group of cyanotoxins are the microcystins, which are well known to destroy the cellular integrity of the liver in mammalians and fish via inhibition of protein phosphatases 1 and 2a (MacKintosh et al. 1990; Runnegar et al. 1995). When this phenomenon was discovered, it was presumed that these toxins are part of an adaptive strategy against cyanobacterial predators and thus may explain the success and dominance of some colony-forming pelagic cyanobacterial families in fresh and brackish waters. However, when the effect of microcystins and microcystin-containing cyanobacteria on the most important group of grazers, the cladocerans, was investigated, most species of cladocerans proved to be rather insensitive to microcystins (DeMott 1991; Jungmann 1992). In contrast, recent results have shown that *Daphnia galeata* is sensitive to microcystins if the toxins are administered orally by feeding intact cells of *Microcystis aeruginosa* (Rohrlack et al. 1999). However, many studies on the effects of microcystins on other aquatic species demonstrated a comparably low toxicity of these potent hepatotoxins if administered dissolved in ambient water or orally to planktonic crustaceans (Anostroca: *Artemia* sp., Feuillade et al. 1996), crayfish (*Pacifastacus leniusculus*, Lirås et al. 1998), insects (Delaney et al. 1995), shellfish (*Mytilus galloprovincialis*, Vasconcelos 1995; *Anodonta grandis simpsoniana*, Prepas et al. 1997), to fish (*Oncorhynchus mykiss*, Johnston et al. 1994; Tencalla et al. 1994; Bury et al. 1995), to fish and amphibian embryos (Oberemm et al. 1999), to frog embryos (Fischer & Dietrich, 2000) or to macrophytes (Casanova et al. 1999). In contrast, studies of the effects of crude extracts of cyanobacteria repeatedly demonstrated toxicity of many cyanobacterial extracts independently of their microcystin content (Jungmann 1992; Fastner et al. 1995; Reinikainen 1995; Feuillade et al. 1996; Oberemm et al. 1997, 1999).

Studies of the effects of cyanobacterial toxins other than microcystins on aquatic organisms are rare. One is the study of Haney et al. (1995), who showed that an extract from *Aphanizomenon flos-aquae,* which is known to produce saxitoxin (STX), inhibited feeding more effectively than pure STX.

Interestingly, the only epidemiological investigation of the impact of cyanobacteria on health of humans exposed during bathing indicated increased frequency of symptoms in relation to cyanobacterial cell density and exposure time, but these effects could not be related to concentrations of microcystins or neurotoxins (Pilotto et al. 1997). Thus, this study also suggests prevalence of further active substances in cyanobacteria.

This chapter presents results of cellular and fish embryo tests which contribute more information to the issue of unidentified toxic factors in cyanobacteria, with respect to both crude extract toxicity and to fractions of cyanobacterial extracts.

Results

Two bioassays were used to detect cyanobacterial toxicity in samples of the screening project: the primary rat hepatocyte (Chapter 8.1) and the zebrafish embryo assay (Chapter 8.4).

The primary rat hepatocyte test was designed and applied in order to detect microcystin-related hepatotoxicity of field samples, and as shown in Chapters 8.1 and 8.2, it does reflect microcystin content of the samples. However, extracts of many samples were much more toxic in the hepatocyte assay than can be accounted for by the concentration of microcystins detected by HPLC. Table 1 compares median microcystin content estimated from the results of rat hepatocyte assays with those obtained from chemical analysis by HPLC. As described in Chapter 8.1, this calculation is based on the assumption that all the microcystins in the sample are as toxic as microcystin-LR, which is an overestimation of toxicity representing a worst-case estimate. Consequently, content of total microcystin calculated from LC_{50} would be expected to be lower than the amounts actually found by HPLC, but the opposite was the case (Table 1).

Table 1. Medians for microcystin content calculated from primary rat hepatocyte LC_{50} compared to microcystin content determined by HPLC

Samples	MC DW^{-1} calculated from rat hepatocyte LC50	determined by HPLC	Factor between result from rat hepatocytes and HPLC
1995 with 100% MeOH; n = 86	830 μg g^{-1} DW	390 μg g^{-1} DW	2.1
1996 with 75 % aqueous MeOH; n = 94	1600 μg g^{-1} DW	1050 μg g^{-1} DW	1.5

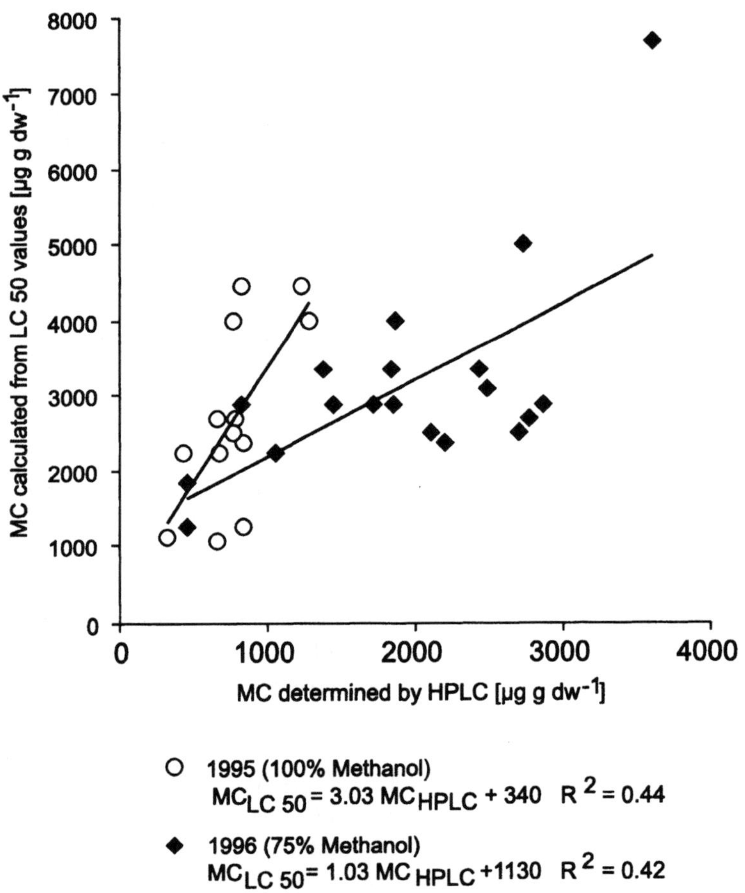

Fig. 1. Microcystin content measured by HPLC (X-axis) and calculated from LD_{50} values of primary hepatocytes (Y-axis) for samples dominated by *Planktothrix agardhii. MC* Total microcystin

Further, the amount of additional toxicity depended on extraction solvents: extraction with pure methanol resulted in toxicity not accounted for by the microcystin content actually found (Table 1). For samples dominated by *Planktothrix agardhii* and extracted with pure methanol, this effect was most pronounced with a slope of the regression indicating that on average, toxicity detected in the primary rat hepatocyte assay was threefold higher than accounted for by the microcystin content (Fig. 1).

Microcystin concentrations estimated from a small number of mouse in vivo assays also indicate toxicity not accounted for by the microcystin content of the samples (see Chapter 8.2 and Hughes et al. 1958, SDF).

For risk assessment, this result has important implications and raises two questions: (1) do these effects also occur in other bioassays, particularly *in vivo* with entire animals, and (2) which substances cause them?

Remarkably, the same crude extracts showed no general cytotoxicity in assays with established cell lines, e.g. CHO-K1 (Chinese hamster ovary) and VERO (African green monkey kidney).

However, fish-embryo tests also demonstrated toxicity not accounted for by known cyanotoxins. Subjected to solutions of pure cyanobacterial toxins, embryos of fish and amphibians showed comparably minor effects (Chapter 6.1), but exposed to suspensions of cyanobacterial matter at concentrations 0.1% DW/vol (approximately 1 mg fresh weight per 100 ml of water, as occurs in bloom situations), they revealed high toxicities. Many field samples dominated by *Aphanizomenon flos-aquae*, as well as cultures of *A. flos-aquae*, *Microcystis* spp. and *P. agardhii*, were highly toxic (Table 2).

Fig. 2 demonstrates the lethal developmental changes and malformations caused by exposure to an 0.1% DW/vol aqueous extract from *Aphanizomenon flos-aquae* (Super-Blue-Green Algae Alpha-Sun, Lake Klamath/ USA). These are typical also for the effects generated by other *Aphanizomenon* samples. In the early phase of development at an age of 8 h, about 75% of the yolk of control embryos is covered by the blastoderm and a streak has appeared as an initial stage of the vertebrate anterior/ posterior axis. In the treatment group exposed to aqueous extract from Lake Klamath *Aphanizomenon*, 30-50% of the embryos showed abnormal epiboly and successive lesions (Fig. 2, individuals A, B). Later, at 26 h, when control embryos have developed a head region, eyes with lenses already visible and a long tail with finfold, surviving embryos of the treatment group showed a considerable, individually differently expressed delay and change in development (e.g. individual C is strongly affected showing a retardation by approximately 12 h and an irregular depression of the tail region, whereas individual D is less affected). All embryos treated with extract from *Aphanizomen flos-aquae*, Super-Blue-Green Algae, died within 28 h of exposure, failing to develop the blood-circulatory system or pigmentation.

Further, an aqueous extract from Super-Blue-Green Algae dragees proved to be toxic to cell cultures, to microcystin-tolerant permanent cell lines as well as to primary hepatocytes (Table 3); this indicates cytotoxicity. As the microcystin content of this sample was low (0.64 $\mu g\ g^{-1}$ DW by ELISA) and all the effects observed are atypical for microcystins, these cyanotoxins can be excluded as their cause.

Table 2. Effects of aqueous extraxts from cyanobacteria on embryonic development of zebrafish. (Legend: see Table 5)

Effect Sample	Exo	EDA	Ret	Mort	Cho	Lor	PYE	Sei	HR	Vit
Mueggelsee sample dominated by *Aphanizomenon flos-aquae*	⊕	⊕			⊕					
Sample from Mueggelsee during lysis of *Aphanizomenon bloom*, (blue discoloration)										
Aphanizomenon flos-aquae lyophilisate from SBGA-Dragees, Lake Klamath	⊕	⊕								
Aphanizomenon flos-aquae SAG 31.87										
Wannsee sample dominated by *Microcystis* sp.			⊕	⊕		⊕	⊕			
Anabaena flos-aquae strain NRC-44-1			⊕	⊕	⊕	⊕	⊕			
Limnothrix redekei strain HUB 010										
Microcystis aeruginosa strain HUB 5-3				⊕		⊕	⊕			
Microcystis aeruginosa strain PCC 7806 (+)				⊕						⊕
Microcystis aeruginosa strain PCC 7806 (-) (0.02 %)				⊕					⊕	
Microcystis aeruginosa strain PCC 7806 (-) (0.002 %)				⊕						⊕
Planktothrix agardhii strain HUB 076 (0.1%)			⊕	⊕	⊕	⊕				
Planktothrix agardhii HUB 076 (0.01%)				∅	⊕					
Planktothrix agardhii strain NIVA Cya 34			⊕	⊕						
Planktothrix rubescens, isolate from Wahnbachtal reservoir				⊕						

Effects of extracts from *Planktothrix* on fish development generally were serious but less pronounced as compared to *Aphanizomenon*, since after a strong retardation of early development, organogenesis was basically performed. However, malformations and subsequent successive mortality occurred later; thus, no viable larvae were present at the end of the embryonic development.

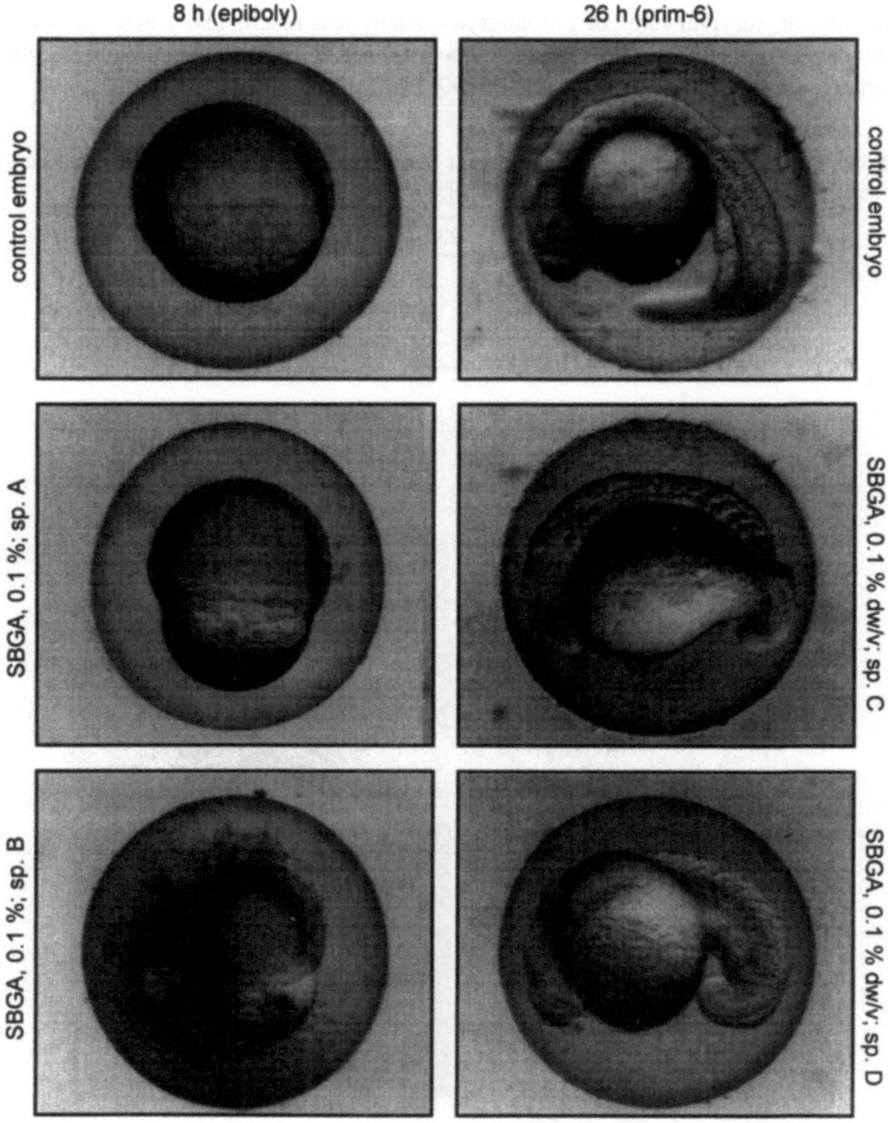

Fig. 2. Inhibition of early development in zebrafish due to exposure to 0.1% DW/vol aqueous crude extract obtained from *Aphanizomenon flos-aquae* (Super-Blue-Green Algae Alpha-Sun). *SBGA* Super-Blue-Green Algae; *sp. A-D* specimen A-D

Another effect was detected especially during exposure to *Planktothrix* extract, which was observed even at a tenfold dilution of the 0.1% DW/vol effect concentration generating malformations. In eggs of both fish and amphibians, ruptures of the egg layers were frequently observed (Fig. 3), which led to premature hatching. This effect is more likely to cause effects also in the field than are malformations, because the effective concentrations are significantly lower and premature hatching may severely reduce the survival rates of the embryos. Analytical results show that the acute developmental effects are not related to the presence of microcystins or other known cyanobacterial toxins (Oberemm et al. 1997, 1999 and unpubl. results).

Table 3. In vitro toxicity of an aqueous extract from *Aphanizomenon flos-aquae* (Super-Blue-Green Algae Alpha-Sun): 20 h LC_{50} (mg DW ml^{-1})

Cell culture	20 h LC_{50} (mg ml^{-1})
rat primary hepatocyte	1.36
CHO-K1 (hamster)	1.36
CaCo2 (human)	1.7

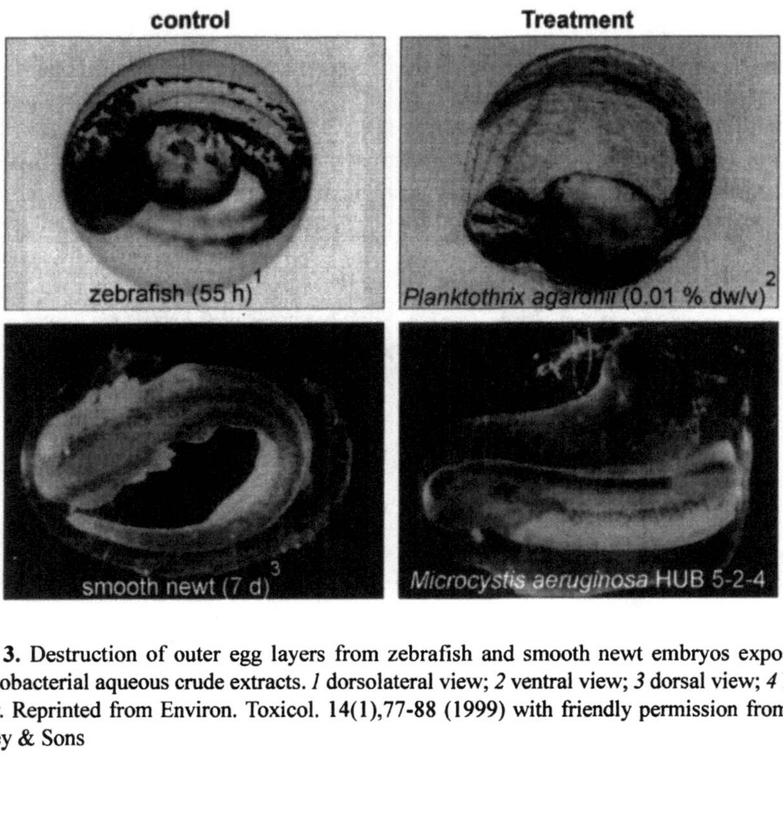

Fig. 3. Destruction of outer egg layers from zebrafish and smooth newt embryos exposed to cyanobacterial aqueous crude extracts. *1* dorsolateral view; *2* ventral view; *3* dorsal view; *4* lateral view. Reprinted from Environ. Toxicol. 14(1),77-88 (1999) with friendly permission from John Wiley & Sons

In order to obtain first insights into the chemical nature of substances that cause toxicity not related to known toxins, cyanobacterial crude extracts were fractionated and the fractions tested. A first approach used methanolic and aqueous extracts from lyophilized *Planktothrix agardhii* HUB 076. Aqueous extracts were subjected to solid-phase extraction using C-18 cartridges, producing an amphiphilic fraction containing microcystins (eluate) and a hydrophilic, polar microcystin-free fraction (permeate). The results of the rat primary hepatocytes assay (Table 4) demonstrate that the aqueous extract is more toxic than the methanolic extract and the microcystin-free fraction of the aqueous extract (permeate) causes the same toxicity as the microcystin-containing fraction (eluate). The higher toxicity of the aqueous extract is consistent with recent findings which demonstrated a higher efficiency of the aqueous

Table 4. Toxicity of crude extracts of lyophilized *Planktothrix agardhii* HUB 076 and its solid-phase C-18 fractions to rat primary hepatocytes, expressed as microcystin-LR concentration (i.e. toxicity equivalents). The values were calculated from calibration tests using pure microcystin-LR

Sample	Calculated microcystin-LR concentration (µg g-1 DW)
MeOH crude extract	322
aqueous crude extract	595
aqueous C18 permeate	217
aqueous C18 eluate	217

Table 5. Developmental effects induced in zebrafish by an aqueous crude extract of lyophilized *Planktothrix agardhii* HUB 076 and its solid-phase C-18 fractions

Sample / Effect	EDA	Ret	Mort	Cho	Lor
Aqueous crude extract (0.1%)	⊕	⊕	⊕	⊕	⊕
Aqueous crude extract (0.01%)	–	∅	–	–	–
Aqueous C-18 permeate (0.01%)	⊕	–	–	⊕	–
Aqueous C-18 eluate (0.017%)	–	–	∅	–	–

Qualitative effect	Abbreviation
Exogastrulation (failure of epiboly)	Exo
Early developmental arrest and mortality	EDA
Retardation in the course of organogenesis	Ret
Mortality in the course of late embryogenesis	Mort
Low viability (decreased survivalrate during larval development)	Vit
Destruction of egg structures and subsequent premature hatching	Cho
Seizures	Sei
Heart rate altered	HR
Malformations: dorsally bent tail (lordosis)	Lor
Pericardial and yolk sac oedema	PYE

⊕ pronounced effect ∅ minor effect

extraction compared to methanolic extraction (Fastner et al. 1998). However, the high toxicity of the C-18 permeate is not consistent with the expectation and must be caused by unknown toxic compounds. However, this result may explain the increased toxicity of cyanobacterial crude extracts to hepatocytes (Fig. 1).

When zebrafish embryos were exposed to these fractions, the microcystin-free permeate produced the strongest effects (Table 5; early developmental arrest), even stronger than the crude extract (cf. effects of 0.01% extracts, Table 5). The delayed toxicity of the C-18 eluate may be attributed to the presence of microcystin-RR (Asp_3), which was detected at 1500-2000 μg g^{-1} DW in the lyophilisate (calculated concentration in crude extract and C-18 eluate approx. 150 μg L^{-1}; for microcystin-related effects refer to Chapter 6.1).

Another approach used a field sample from Lake Mueggelsee strongly dominated by *Aphanizomen flos-aquae* (Berlin, October 1995), which had been shown to produce strong inhibitory effects in early zebrafish development (cf. to Fig. 2). The lyophilisate was eluted successively with methylenchloride (CH_2Cl_2) and methanol. This approach also demonstrated that the most polar fractions have the strongest activity. The highly diluted methanolic fraction 3.2.7.5.2 generated the same pattern of developmental perturbations as the crude extract (Figs. 4, 2). Lack of sufficient amounts of material

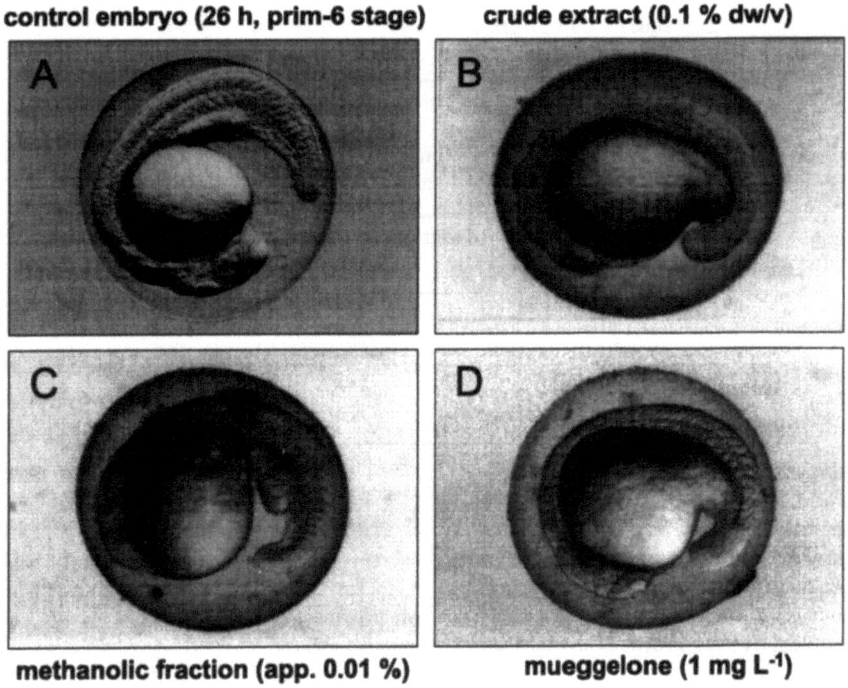

Fig. 4A-D. Inhibition of early development in zebrafish (*Danio rerio*) treated with compounds from *Aphanizomenon flos-aquae*. **A** Control embryo (26 h, prim-6 stage); **B** aqueous crude extract (0.1% DW/vol); **C** methanolic extract 3.2.7.5.2 (0.091 mg L^{-1}); **D** mueggelone 1 mg L^{-1}

from this polar, hydrophilic fraction precluded further analytical processing. This fractionization also produced an active compound which was located in the lipophilic CH_2Cl_2 fraction. However, the determined effect concentration of the isolate mueggelone (Chapter 7.2 and Papendorf et al. 1997) was rather high, so it is unlikely that this lipophilic compound should have contributed to the toxicity of the aqueous crude extract. Fig. 4 demonstrates that inhibition of early development in zebrafish may be caused by different compounds from *Aphanizomen flos-aquae*.

Further work will target step-wise, bioassay-assisted isolation of the substances responsible for these effects and the elucidation of their chemical structures.

7.3 New Cyanobacterial Substances with Bioactive Properties

Ayako Forchert, Uwe Neumann and Olaf Papendorf

Toxic and non-toxic cyanobacterial bloom material from lakes and reservoirs around Berlin, Freiburg and in the state of Thueringen, Germany, was screened for new bioactive secondary metabolites. The bulk of the cyanobacterial material was collected in summer 1994 as a 7-L volume scum of *Microcystis* sp. (120 g lyophilized material) and in late summer 1995 from an algae bloom of *Aphanizomenon flos aquae* (40 g lyophilized material) from the two Berlin lakes Wannsee and Mueggelsee. Further field samples containing almost exclusively *Microcystis* sp. and *Planktothrix rubescens* from Waltershofen / Freiburg i. Br. (750 g lyophilized material) and the Reservoir Weida, Thueringen, were collected in summer 1996 and 1997, respectively.

General Procedure for Isolation of Cyanobacterial Compounds

The isolation of cyanobacterial compounds includes extraction and separation as well as fractionation. Extraction is first carried out for separation into substances of hydrophilic, more hydrophobic and strongly hydrophobic character. Therefore extraction was performed using different solvents to cover a wide variety in polarity of cyanobacterial metabolites. Dichloromethane was used to obtain compounds with strongly hydrophobic properties such as macrolides and lactones, whereas 5% acetic acid was employed for effective extraction of hydrophilic compounds and retained pigment components. Methanol was the most frequently used solvent for extraction of microcystins and related peptide compounds.

Efficient column chromatographic techniques were applied in various forms for the preparation of milligram-amounts of major and minor components needed for both spectroscopic characterization (this Chapter) and biological studies (Chapter 7.2). Chromatographic separations relied on silica-based C-18 reversed phase material which enables separation of many biogenic substances, e.g. amino acids, peptides, polyketides.

The novel Biotage Flash chromatography 75S system (Biotage, Charlottesville, PA, USA), illustrated in Fig. 1, was applied with an ODS packed reversed phase column (9 x 7.5 cm I.D. cartridge) and a stepwise aqueous methanol gradient. The marked advantage of this technique in contrast to normal liquid chromatography with glass columns is its application of approximately 20 psi pressure through cartridges housed in a radial compression module. With a maximum loading of 20 g material this system serves as a very efficient and time-saving cleansing step prior to final purification steps. Edwards et al. (1996) first described the application of the Biotage-Flash system for laboratory-scale purification of microcystins. In our investigation one extract of cyanobacterial material was successfully subjected to a reversed phase Biotage Flash-chromatography system for initial separation.

Reversed-phase high-performance liquid chromatography (RP-HPLC) was used for semi- and preparative purification of biogenic compounds with wide ranges in polarity. In particular, good elution and retention behavior was obtained using water/acetonitrile eluting systems. In addition to reversed phase techniques, vacuum liquid chromatography (VLC) on silica gel and size exclusion chromatography through Sephadex LH20 were employed as an alternative cleansing step. For an efficient separation of pigments, proteins and lipids the latter method with molecular size exclusions of 5000 Da (Amarowicz and Shahidi 1994) was applied.

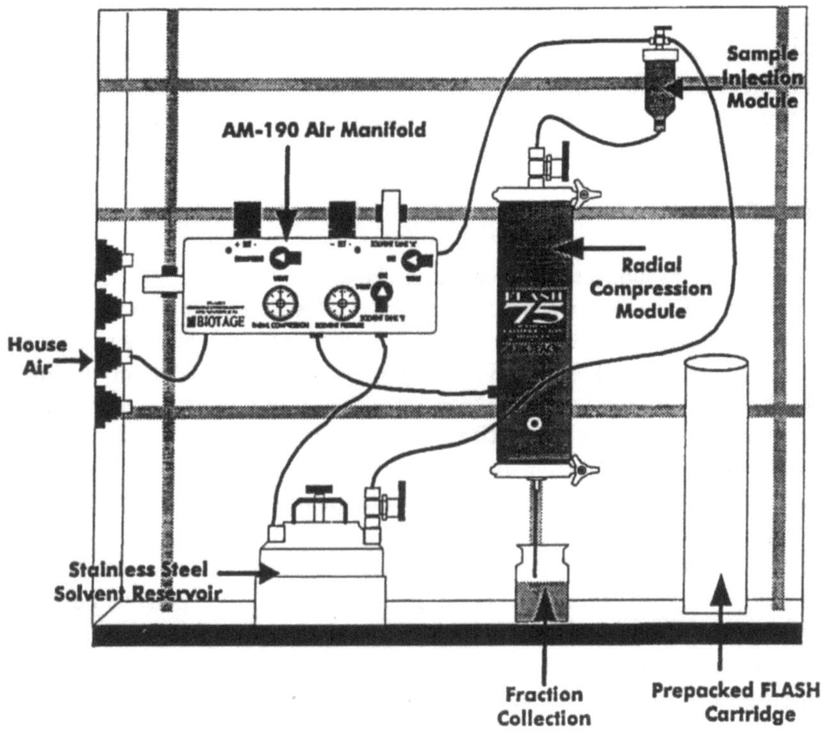

Fig. 1. Scheme of the Biotage RP-flash chromatography system

The detection of various substances during the purification procedure makes use of differences in their physicochemical properties such as retention in chromatographic systems, molecular weight and UV absorption. For example, microcystins exhibit characteristic retention times and UV spectra which are particularly suited for their analytical determination by HPLC. Hence, identification of chromatographed components was achieved by analytical HPLC with high resolution diode array detection. Good retention of microcystins and related compounds was obtained with a water/acetonitrile/0.05% TFA gradient (Lawton et al. 1994). For cleanup steps, analytical HPLC aided in pooling of fractions. In some cases thin-layer chromatography (TLC) on silica plates was employed for rapid and sensitive monitoring of chromatographed compounds. Substances containing chromophores, such as microcystins, were easily detected by characteristic UV spectra.

Furthermore, MALDI-MS was applied for rapid initial molecular mass determination. This very sensitive ionization mass spectrometric method (see also Chapter 8.5) allows assessment of molecular ion peaks of components from intact cyanobacteria, as well as from peaks separated by chromatography.

Spectroscopic Characterization of New Cyanobacterial Compounds

Since the beginning of two-dimensional NMR (2D NMR) 2D measurements has been useful in elucidating chemical structures of cyanobacterial metabolites. A general strategy for identification and assignment of all proton and carbon resonances of moderate complex metabolites (500-2000 Da) used various correlation techniques of 2D NMR experiments (Evans 1995). In the course of our survey of new substances from cyanobacteria, their chemical structures were deduced by 2D H,H and H,C correlated NMR experiments, as well as selected one dimensional NOE difference spectra in combination with mass spectrometric data. All NMR experiments for identification of newly isolated components were measured in deutero-dimethylsulfoxide (DMSO-d_6). The minimum quantity for 2D NMR analyses exceeded 2 mg using a DRX 500 spectrometer (500 MHz) (Bruker, Germany).

The 2D H,H-correlated COSY measurements resulted in cross-peaks of scalar coupling protons within two- or three-bond distances. The positions of the peaks are represented as relative distances to a reference signal (here trimethylsilane) called chemical shifts. The chemical shift of protons depends strongly on their chemical environment revealing information on their characteristic structures. The identification of all detected cross-peaks led to characteristic proton networks used for determination of single structural elements, e.g. amino acid constituents.

The sequence of all constituents was elucidated by 2D NOESY and 1D NOE experiments. The respective cross-peaks and differences are a cause of dipolar-dipolar interactions of neighbouring constituent proton pairs. In addition to NOE resonances, the most reliable heteronuclear H,C HMBC data were successfully used for interresidual correlations of carbonyl carbon atoms to protons of adjacent constituents in order to assess new peptidic structures.

After assigning hydrogen resonances, carbon atoms were detected from the frequently used ^{13}C NMR spectra. As described for proton spectra, the chemical environment of carbon atoms leads to characteristic chemical shifts from which information about carbon functionalities such as carbonyl, aromatic, olefinic or aliphatic carbon atoms can be obtained. Methyl, methylene and methine carbon atoms (multiplicities) were differentiated by ^{13}C DEPT measurements.

In addition to NMR analyses, the molecular formulae of new metabolites were obtained from high resolution FABMS. The high resolution peak matches can be received to an accuracy of ± 5 mmu. Furthermore, structural details could also be determined via FABMS-fragmentation pattern. In some cases all information about sequencing was utilized.

For rapid initial structural classification, MALDI-MS was frequently applied in the post-source decay mode (PSD) for identification of several new analogues which had been isolated in small amounts (> 500 μg). Various chemical structures could be deduced from MALDI-PSD fragmentation patterns which allowed finger-printing identification in comparison with similar patterns of reference material (Erhard et al. 1997).

Finally, the assignment of the absolute configuration of amino acid constituents from hydrolytically degraded peptidic compounds could be determined by GC analyses of the corresponding derivatives using chiral phases of Chirasil-*L*-Val and Lipodex-E. *L*-or *D*-configured amino acids were identified by means of characteristic retention times of reference amino acids.

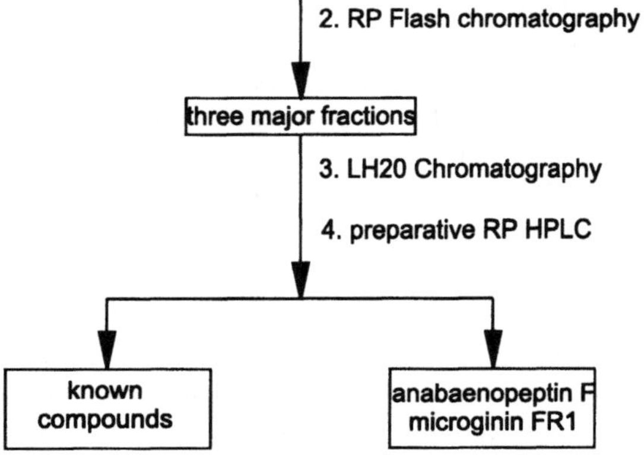

Fig. 2. Isolation procedure applied to algae bloom material consisting almost entirely of *Microcystis aeruginosa*

Isolation Procedures Applied

- For the isolation of the new cyanobacterial compounds microginin FR1 and anabaenopeptin F from *Microcystis* sp. (Wannsee) 60 g of the lyophilized material was extracted twice using 5% acetic acid to obtain more of the hydrophilic compounds. The laboratory-scale precleaning of the crude extract was carried out using a Biotage Flash chromatography system with an ODS reversed stationary phase and a stepwise aqueous gradient from 40 to 80% methanol in 5% increments (500 ml step^{-1}). Up to 80 flash chromatography fractions (100 ml) were collected manually using a flow rate of 50 ml min^{-1} and subsequently screened by TLC and analytical RP-HPLC for the presence of microcystins and related peptidic compounds. Flash chromatography resulted in three major fractions containing hydrophilic, moderately hydrophilic and more lipophilic components. The second purification step permeated the three major fractions, dissolved in methanol, through Sephadex LH20 on a 3 x 35-cm column with methanol and gravity flow. Significant fractions were pooled after monitoring by RP HPLC diode array detection just as employed after flash chromatography. Size exclusion of up to 90% of each fraction was obtained and the three fractions yielded 300-700 mg. The final purification of the three flash and gel chromatographed fractions was achieved by semi- and preparative HPLC using an ODS reversed phase column and various optimized water/acetonitrile gradients. Eluents were monitored at 238 and 214 nm to identify microcystins, as well as other peptidic compounds. The fractionation of repeated HPLC runs led to a number of pure known and unknown components in 0.03-0.01% yield based on the dry weight of cyanobacterial material. In addition to the structurally related compounds already characterized from the substance class of anabaenopeptins (anabaenopeptin a-d, and oscillamide Y: Harada et al. 1995; Sano and Kaya 1995; Shin et al. 1997) and the standard microcystins-RR, -LR and -YR two newly isolated cyanobacterial components were obtained. Hitherto unknown structures of microginin FR1 and anabaenopeptin F were isolated in 4.8 and 2.3 mg, respectively. The purification procedure of cyanobacterial material from Wannsee is summarized in Fig. 2.
- Microginin FR1 was purified from a water bloom of *Microcystis* spp. from Waltershofen / Freiburg i. Br. (Neumann et al. 1997). After methanol extraction a four-step precleaning procedure was applied before fractionation of microginin FR1. The extract was diluted to a final concentration of 10% methanol, applied to ODS silica gel cartridges using 90% methanol containing 0.05% TFA. Collected fractions were then eluted with 50% methanol by LH-20 column chromatography. Using dilutions of aqueous methanol in LH-20 chromatography instead of pure methanol led to a better separation of the medium hydrophobic peptides from strongly hydrophobic pigments. Microginin FR1 containing fractions (HPLC monitoring, Lawton et al. 1994) were applied to ACCELL QMA anion exchange cartridges (Martin et al. 1990). The non-binding fraction was then chromatographed through SA cation exchange cartridges by eluting with 0.25 M ammonium bicarbonate in 30% aqueous methanol. Ion-exchange chromatography has proved to be a powerful method in sample preparation of peptidic substances by separation

due to their charge at a given pH. As final purification step RP-HPLC was repeatedly applied using the same chromatographic conditions as above. The combined fractions (for HPLC profile see Fig. 3) finally yielded pure microginin FR1 in an amount of 1% of cell dry weight.

- Mueggelone was extracted from freeze-dried algae material of *Aphanizomenon flos aquae* (Mueggelsee) using dichloromethane (CH_2Cl_2) to obtain compounds of more hydrophobic properties. The CH_2Cl_2 extract of lyophilized *A. flos-aquae* showed strong inhibitory effects at low concentrations on zebrafish embryonic development (Chapter 7.2), which were not consistent with the microcystin concentration, suggesting that other toxins were present in the sample. Thus, bioassay-assisted secondary metabolite investigation of this extract was undertaken to isolate the principle bioactive compound(s). For this purpose, the crude CH_2Cl_2 extract (2.7 g) was separated by vacuum liquid chromatography over silica gel employing a step gradient from hexane over ethyl acetate to methanol leading to ten fractions of 90 ml each. TLC and ^1H NMR control of all fractions indicated fraction three to be of further interest on the basis of many low field resonances in its NMR spectrum. In addition to the known triterpene lupenyl acetate (Arai et al. 1983), the final HPLC purification of fraction three [Si60, hexane/acetone (96:4) as eluent] gave a hitherto unknown compound, the so-called mueggelone, as clear

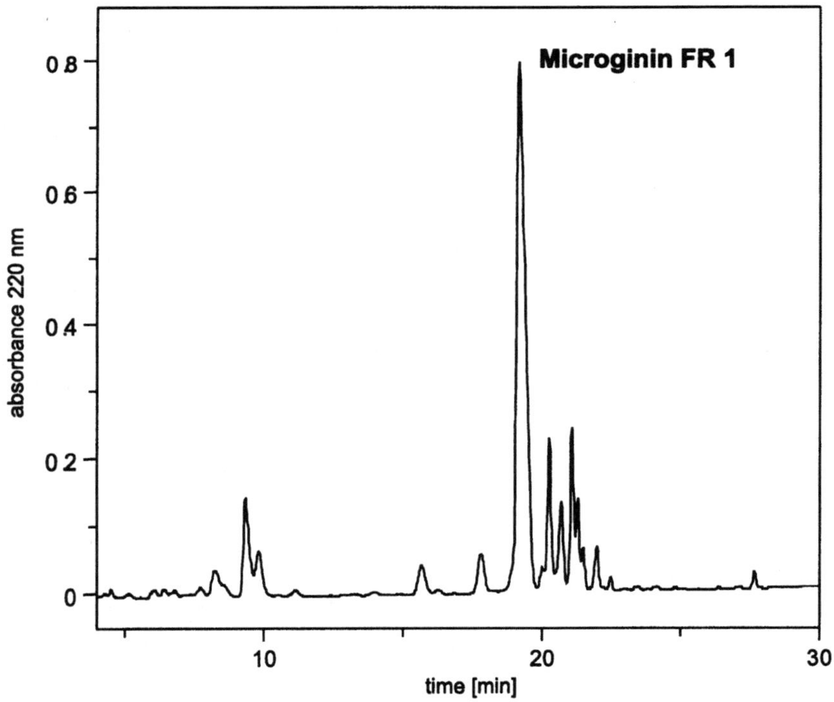

Fig. 3. HPLC elution profile of a microginin FR1 containing extract of a *Microcystis* sp. water bloom

oil in a yield of 0.013% (Papendorf et al. 1997). In 1980, Cardellina and Moore isolated malyngic acid from a sample of *Lyngbya majuscula*. Certainly this compound, if it has the same absolute configuration as mueggelone, could be either its direct precursor or one of its hydrolysis products.

- In the course of the screening program for quantitative analyses of microcystins from a number of eutrophic water bodies (see also Chapter 2.1.2), recovery of microcystins was best using 75% methanol as extractive solvent (Fastner et al. 1998). The same procedure was used to extract microcystin related peptides from algae bloom material consisting mainly of *Planktothrix rubescens* (Reservoir Weida) (Forchert 1999). The extraction of freeze dried algae cells (1.5 g) was thus performed twice with 300 ml 75% methanol for 2 h at ambient temperature. Evaporated supernatants were purified by a two-step reversed-phase HPLC procedure. The first step was performed by a LiChrospher 100, ODS, 5 µm, LiChroCART 250-4 cartridge system using a linear gradient of aqueous acetonitrile containing 0.05% TFA according to Lawton et al. (1994) and a flow rate of 1 ml min^{-1}. The complex HPLC elution profile (Fig. 4) shows signals, including those indicating microcystin. Representative fractions from multiple HPLC runs were combined and screened by

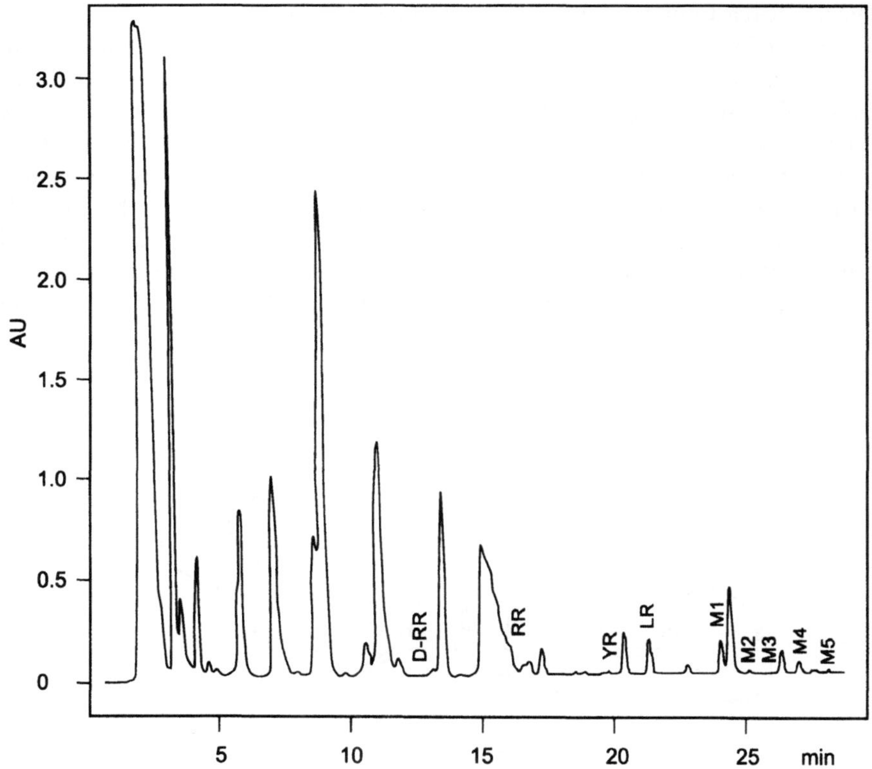

Fig. 4. Analytical HPLC profile of compounds present in a field sample of *Planktothrix rubescens* extracted with 75% methanol. Absorption was measured at 214 nm

MALDI-MS. Mass spectra show hitherto unidentified molecular ion peaks at *m/z* 1130 $[M+Na]^+$, 1148 $[M+Na-H_2O]^+$ and 851 $[M+H]^+$ corresponding to signal at 8.8 min (R_t). This mixture was separated in the course of a second fractionation step by semipreparative RP HPLC using an ODS column and a linear aqueous acetonitrile gradient from 0-45% acetonitrile within 30 min. Combined HPLC runs yielded colourless powders of microcystin-RR (1.2 mg) (Kusumi et al. 1987) and anabaenopeptin F (1.8 mg). The latter was identified as newly discovered cyclic depsipeptide by FAB and MALDI-MS, whose spectrum shows a mass signal at *m/z* 851.

Newly Characterized Compounds

Some newly isolated cyanobacterial metabolites have been successfully characterized by means of extensive NMR analyses, mass spectrometric data and chiral GC analyses. Their structures are members of substance classes of cyclic and linear depsipeptides, as well as lactones.

Mueggelone, a New Ten-Membered Lactone Derivative

The molecular formula of the new lactone mueggelone was found to be $C_{18}H_{28}O_3$, determined by accurate mass measurements. From its ^{13}C NMR data it was evident that mueggelone consists of an ester functionality [74.9 (d), 173.5 (s) ppm], an epoxide group [57.4 (d), 59.9 (d) ppm] and two carbon-carbon double bonds [122.3 (d), 128.6 (d), 132.8 (d), 135.0 (d) ppm]. Based on further NMR resonances, the ester formation was found to be a lactone system. Mass spectrometric analyses which show characteristic signals at *m/z* 275 $[M-OH]^+$, 224 $[M-OH-C_5H_9]^+$ and 194 $[M-OH-C_5H_9-CHO]^+$ support these findings. After all proton resonances were associated with directly bonded carbon atoms via H,C HMQC measurements, H,H COSY cross-peaks indicated a continuous chain of couplings from H_2-6 to H_3-18 and H_2-2 to

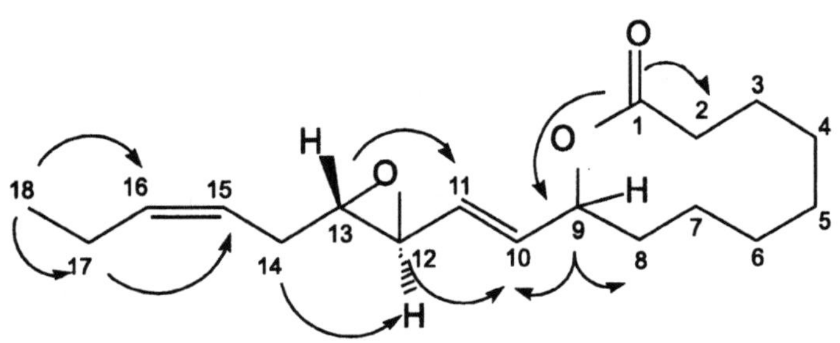

Fig. 5. Chemical structure of the newly characterized lactone derivative mueggelone. The *arrows* indicate diagnostic HMBC correlations

H_2-3. HMBC correlations confirm these molecular fragments of mueggelone. The ring closure of the ten-membered lactone was deduced from HMBC correlations observed between H-9 and C-1, thus completing the basic structure of mueggelone. The double bond $D^{10,11}$ was assigned as E-configured on the basis of its coupling constant $J_{10,11}$ = 15.5 Hz, while $J_{15,16}$ = 8.3 Hz indicates the double bond $D^{15,16}$ to be in *Z*-configuration. Further, NOE differences resulted H-12 and H-13 to be in *trans*-position. The chemical structure of the identified ten-membered lactone mueggelone (Fig. 5), published by Papendorf et al. (1997), consists of *trans* and *cis* double bonds and a C-9 side chain with an epoxide functionality.

Anabaenopeptin F, a Newly Identified Cyclic Depsipeptide

The cyclic depsipeptide anabaenopeptolin F was newly characterized by MALDI-MS fragments (Erhard et al. 1997).

Anabaenopeptins, possess an unusual ureido linkage between two amino acids. The structure of the first anabaenopeptin-type substance was concluded from NMR

Table 1. MALDI-PSD fragments from anabaenopeptin B and F

	Anabaenopeptin B		Anabaenopeptin F
m/z	sequence	m/z	sequence
837	M+H	851	M+H
637	Lys+Phe+Mala+Hty+Val+2H	651	Lys+Phe+Mala+Hty+Ile+2H
---		538	Lys+Phe+Mala+Hty+2H
490	Mala+Hty+Val+Lys+2H	504	Mala+Hty+Ile+Lys+2H
---		476	Mala+Hty+Ile+Lys-CO
460	Val+Lys+Phe+Mala+2H	474	Ile+Lys+Phe+Mala+2H
403	Hty+Val+Lys	417	Hty+Ile+Lys
410	Phe+Mala+Hty+H	410	Phe+Mala+Hty+H
362	Mala+Hty+Val+H	376	Mala+Hty+Ile+H
277	Hty+Val+H	291	Hty+Ile+H
286	Lys+CO+Arg-CN2H2	286	Lys+CO+Arg-CN2H2
263	Mala+Hty+H	263	Mala+Hty+H
233	Mala+Phe+H	233	Mala+Phe+H
231	Mala+Phe-H	231	Mala+Phe-H
201	CO+Arg (side chain)	201	CO+Arg (side chain)
175	Arg+2H	175	Arg+2H
129	Arg-immoniumion	129	Arg-immoniumion
114	Mala+CO+H	114	Mala+CO+H
112	Arg-immoniumion	112	Arg-immoniumion
84	Lys-immoniumion	84	Lys-immoniumion
70	Arg/Lys-related ions	70	Arg/Lys-related ions
58	Mala-immonium ion	58	Mala-immonium ion

data (Harada et al. 1995). In our investigation, Erhard et al. (1997) deduced the chemical structure of a newly isolated anabaenopeptin F from MALDI-PSD fragment analysis in comparison with that similar fragmentation pattern of anabaenopeptin B (for PSD spectra see Figure 6). In addition to the intense molecular ion peak of anabaenopeptin F $[M+H]^+$ *m/z* 851 mass signals at *m/z* 201 $[CO+Arg]^+$ (side chain) and 175 $[Arg+2H]^+$ are subjected to Arg-CO to be attached to the cyclic pentapeptide Lys-Phe-Mala-Hty-Ile which was identified by major mass signal at *m/z* 651 $[Lys+Phe+Mala+Hty+Ile+2H]^+$. Further minor fragment ion peaks at *m/z* 291 $[Hty+Ile+H]^+$, 376 $[Mala+Hty+Ile+H]^+$ and 417 $[Hty+Ile+Lys]^+$ confirmed the structure of anabaenopeptin F bearing Ile instead of Val at position 4 of the corresponding

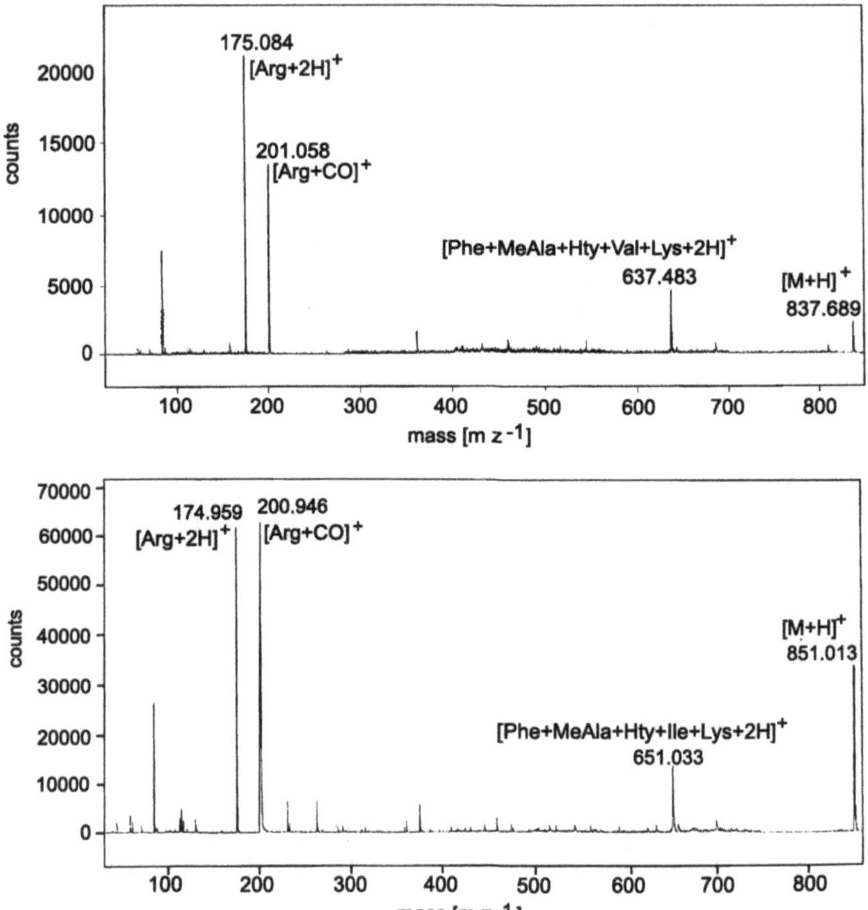

Fig. 6. PSD spectra of anabaenopeptin B and F which are distinguished from each other by fragment ion peaks at *m/z* 277/291 $[Hty+Val/Ile+H]^+$, 362/376 $[Mala+Hty+Val/Ile+H]^+$ and 403/417 $[Hty+Val/Ile+Lys]^+$

structural variant anabaenopeptin B (Figure 7). Interpretation of mass fragments of both anabaenopeptins B and F are listed below (Table 1). Fragment ion peaks at *m/z* 84 [Lys-immonium]$^+$, 263 [Mala+Hty+H]$^+$ and 233 [Mala+Phe+H]$^+$ were detected in both structures which fragments constitute their equal structural units.

Microginin FR1, a Linear Peptide Containing an Unique Structural Constituent

The characterization of the new linear peptide microginin FR1 using ^{13}C DEPT, H,H TOCSY and NOE NMR studies (similar to the aforementioned procedure in determining cyclic depsipeptide planktopeptin) led to the identification of the four common amino acids Ala, Tyr (2x), methylated Leu and one unusual structural element (for H,H TOCSY spectrum see Figure 8) (Neumann et al. 1997, Forchert 1999). Detailed

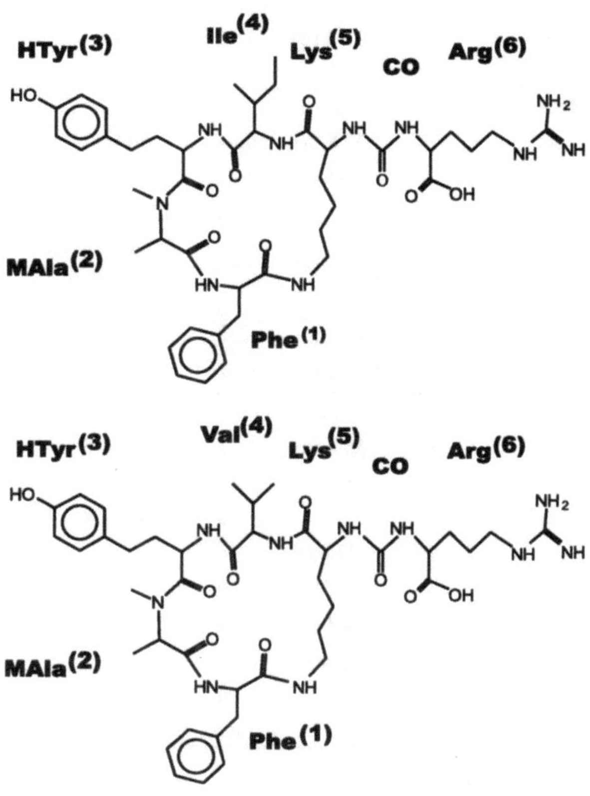

Fig. 7. Chemical structures of variants anabaenopeptin B and F

Table 2. ^{13}C and ^{1}H NMR data of microginin FR1 in DMSO-d_6

Position		13C (mult.)	1H	(mult., J / Hz)
Ahda	1	171.3 (s)		
	2	71.6 (d)	3.94	(br)
	2-OH		6.58	(br)
	3	53.2 (d)	3.02	(m)
	3-NH2		7.80	(br)
	4	31.0 (t)	1.42	(m)
			1.52	(m)
	5	25.2 (t)	1.32	(m)
	6	28.9 (t)	1.23	(m)
	7	28.5 (t)	1.23	(m)
	8	31.2 (t)	1.22	(m)
	9	22.0 (t)	1.23	(m)
	10	13.9 (q)	0.86	(t, 6.6)
Ala	1	172.5 (s)		
	2	44.7 (d)	4.60	(dq, 7.4, 6.9)
	3	17.3 (q)	1.18	(d, 6.9)
	NH		8.10	(d, 7.4)
MeLeu	1	170.1 (s)		
	2	54.8 (d)	4.84	(dd, 6.9, 6.7)
	3	36.6 (t)	1.53	(ddd, 6.7, 6.9, 14)
			1.40	(ddd, 6.7, 6.9, 14)
	4	21.2 (d)	1.28	(ddqq, 6.8, 6.8, 6.0, 6.0)
	5	22.8 (q)	0.75	(d, 6.0)
	5	22.9 (q)	0.85	(d, 6.0)
	N-Me	30.5 (q)	2.56	(s)
Tyr	1	170.9 (s)		
	2	55.8 (d)	4.12	(m)
	3	36.1 (t)	2.80	(m)
			2.96	(dd, 14.3, 4.7)
	4	127.8 (s)		
	5	130.3 (d)	6.90	(d, 8.2)
	6	114.9 (d)	6.63	(d, 8.2)
	7	155.4 (s)		
	OH		8.90	(brs)
	NH		7.69	(d, 7.7)

Table 2. ^{13}C and ^{1}H NMR data of microginin FR1 in DMSO-d_6

Position		13C (mult.)	1H	(mult., J / Hz)
Tyr	1	173.6 (s)		
	2	55.7 (d)	4.35	(m)
	3	33.5 (t)	2.69	(m)
			2.90	(dd, 14.3, 4.7)
	4	128.2 (s)		
	5	129.9 (d)	6.92	(d, 8.2)
	6	114.8 (d)	6.60	(d, 8.2)
	7	155.7 (s)		
	OH		8.55	(brs)
	NH		7.77	(d, 7.7)

s singulett; *d* dublett; *t* triplett; *q* quartett; *m* multiplett; *dd* dublett; *qq* quartett; *br* broad

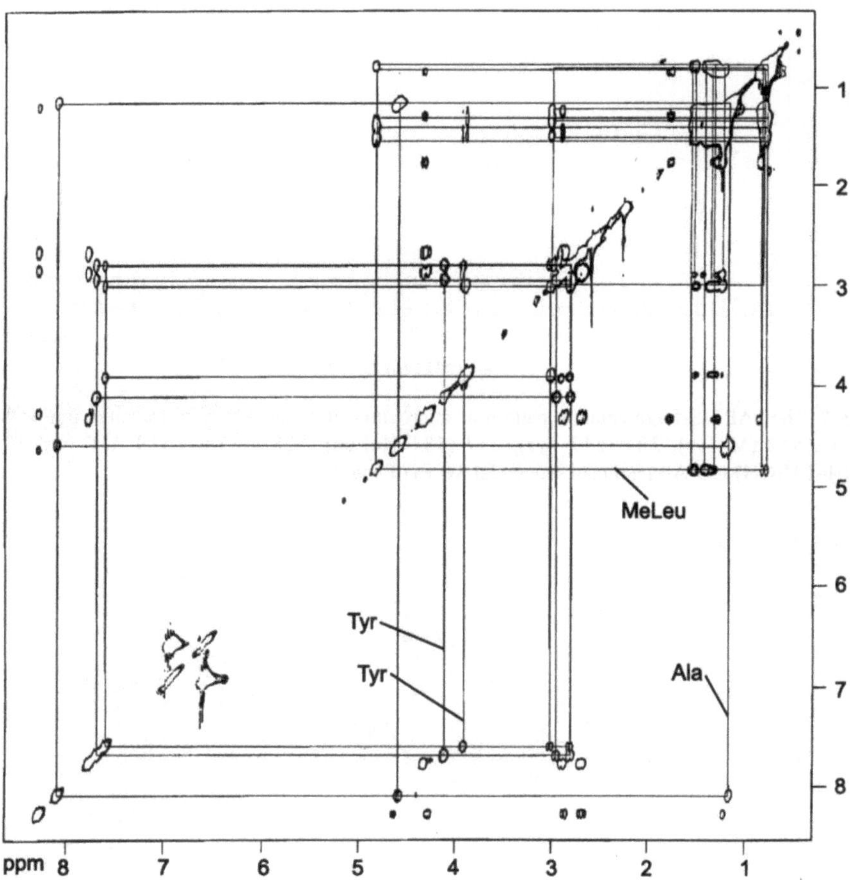

Fig. 8. 2D H,H-TOCSY spectrum of microginin FR1, measured in DMSO-d_6 at 303 K

interpretation of HMQC and ^{13}C NMR spectra indicated the fifth unit as a unique decanoic acid possessing a hydroxy functionality at C-2 [d 3.94 (H-2), 71.6 (C-2)] and an amino group at C-3 [d 3.02 (H-3), 53.2 (C-3)] which was identified as 3-amino-2-hydroxydecanoic acid (Ahda) (for chemical shifts see Table 2).

The amino acid sequence of microginin FR1 was elucidated by HMBC coherences (AhdaCO/AlaNH, AlaCO/MeLeuN-Me, MeLeuCO/TyrNH and TyrCO/TyrNH) and resulted in Ahda-Ala-MeLeu-Tyr-Tyr. Further fragmentation pattern of FABMS (Fig. 9) supported the conclusions from the NMR data, which were consistent with

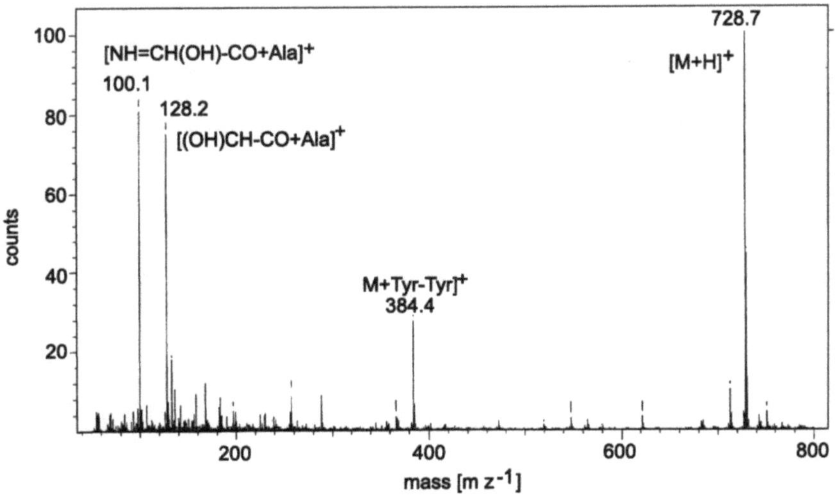

Fig. 9. The FABMS-fragmentation pattern of microginin FR1. The essential fragment ion peaks at *m/z* 547 (M-Tyr), 384 (547-Tyr), 257 (384-MeLeu), 128 (HO-CH-CO-Ala) and 100 (-NH=CH(OH)-CO-Ala) indicate sequential losses of constituents

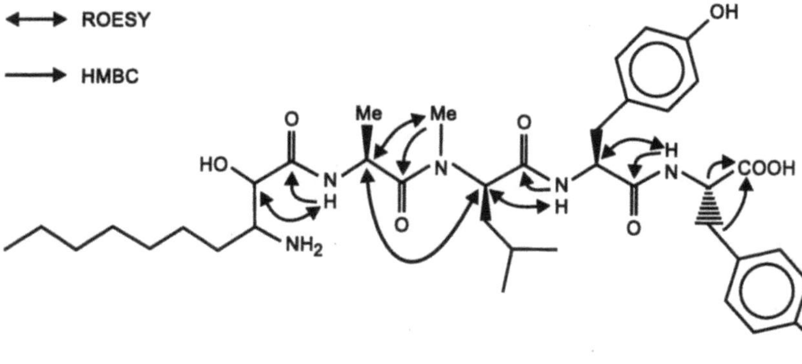

Fig. 10. Chemical structure of microginin FR1. The *arrows* indicate diagnostic HMBC and ROESY correlations

consecutive losses of intense fragment ion peaks at *m/z* 547 (-Tyr), 384 (-Tyr), 257 (-MeLeu), 128 (-HO-CH-CO-Ala) and 100 (-NH=CH(OH)-CO-Ala). The molecular formula $C_{38}H_{57}N_5O_9$ was determined by high resolution FABMS in glycerol [727.4114 M^+; D 4.2 mmu].

The hydrolytic degradation of microginin FR1 (6 M HCl, 115 °C, 20 h) generated the common amino acids Ala, Tyr and MeLeu in a ratio of 1:2:1. The absolute configuration of the amino acids was determined by GC analyses of the N-trifluoroacetyl isopropyl ester derivatives using chiral phases of Chirasil-L-Val and Lipodex-E. Ala and Tyr were shown to be naturally L-configured whereas MeLeu was determined as D-configured amino acid. The stereochemistry of Ahda has not been identified due to the lack of reference material of enantiomeric Ahda. The presented structure of microginin FR1, depicted in Fig. 13, is closely related to the first known Ahda containing linear peptide microginin which possesses angiotensin-converting enzyme inhibitor activity (Okino et al. 1993).

Bioactivities

The newly identified components were tested for their toxic effects on zebrafish embryos (see Chapter 8.4 for methods). One of the tested substances, mueggelone, exhibits inhibitory effects on embryonic development at a concentration of 10 μg ml^{-1} (Papendorf et al. 1997 and Chapter 7.1). Lupenyl acetate was less toxic. It caused pericardial oedema and tail bending after 3 days of treatment at a concentration of 100 μg ml^{-1}. These results suggested that mueggelone and to a lesser extent, lupenyl acetate are likely to be the causative agents of adverse effects on fish development in the CH_2Cl_2 extract (Papendorf et al. 1997).

Underlying enzyme-inhibitor properties of a wide series of known cyanobacterial peptides trypsin and proteinphosphatase-inhibiting activity was measured according to Elfant et al. (1985) and An and Carmichael (1994). Microginin FR1 showed no inhibitory effect on both proteinase trypsin and proteinphosphatase 1A. Microginin FR1, initially solvated in methanol and diluted in reconstituted water and applied to zebrafish during their embryonic development showed no visible effects at concentrations up to 10 mg/L. In addition to the bioassay using zebrafish embryos, microginin FR1 was tested for inhibitory effects on the angiotensin-converting enzyme (ACE) activity using FAPGG as a substrate (Borowitzka et al. 1995). This resulted in a half-maximal inhibition of the ACE by 1.6×10^{-5} M microginin FR1 (Neumann at al. 1997). Current research is directed towards extensive pharmacological research of described compounds.

Conclusions

These results support the assumption that numerous currently unidentified peptides, depsipeptides, lactones and alkaloides occur in cyanobacteria, and that these organisms are both a rich source of potentially useful natural products as well as of yet unknown toxins. Therefore, the isolation and identification of new cyanobacterial metabolites is fundamental to both toxicology and pharmacology. Further extensive efforts are necessary not only for the identification and characterization of hitherto unknown bioactive cyanobacterial compounds, but also for assessment of their occurrence and associated exposure risks with respect to human and environmental health.

Acknowledgements

The authors are grateful to J. Jakupovic for the measurement of a wide range of 2D NMR spectra and for helpful discussions. MALDI-MS and FAB-MS measurements were kindly performed by M. Erhard and G.Höhne. We would like to thank also A. Oberemm for fish-embryo-larval testing and G. Heinrich for chiral analysis.

References

Amarowicz R, Shahidi F (1994) Application of sephadex LH20 chromatography for the separation of cyanogenic glycosides and hydrophilic phenolic compounds of flexsed. J Chromatogr 17: 1291–1299

An JS, Carmichael WW (1994) Use of a colorimetric protein phosphatase inhibition assay and enzyme linked immunosorbent assay for the study of microcystins and nodularins. Toxicon 32:1495–1507

Arai Y, Kusomoto Y, Nagao M, Shiojima K, Ageta H (1983) Yakugaku Zasshi 103:356–359

Borowitzka MA (1995) Microalgae as sources of pharmaceuticals and other biologically active compounds. J Appl Phycol 7:3–15

Bury NR, Eddy FB, Codd GA (1995) The effects of the cyanobacterium *Microcystis aeruginosa*, the cyanobacterial hepatotoxin microcystin-LR, and ammonia on growth rate and ionic regulation of brown trout. J Fish Biol 46:1042–1054

Cardellina J, Moore RE (1980) Tetrahedron 36:993–996

Cardellina JH, Marner FJ, Moore RE (1979) Seaweed dermatitis: structure of lyngbiatoxin A. Science 204:193–195

Carter DC, Moore RE, Mynderse JS, Niemczura WP, Todd JS (1984) Structure of majusculamide C, a cyclic depsipeptide from *Lyngbya majuscula*. J Org Chem 49:236–241

Casanova MT, Burch MD, Brock MA, Bond PM (1999) Does toxic *Microcystis aeruginosa* affect Aquatic plant establishment? Environ Toxicol 14(1):97–110

De Silva ED, Williams DE, Andersen RJ, Klix H, Holmes CFB, Allen TM (1992) Motuporin, a potent phosphatase inhibitor isolated from the Papua New Guinea sponge *Theonella swinhoei* Gray. Tetrahedron Lett 33:1561–1564

Delaney JM, Wilkins RM (1995) Toxicity of microcystin-LR, isolated from *Microcystis aeruginosa*, against various insect species. Toxicon 33:771–778

DeMott WR, Zhang QX, Carmichael WW (1991) Effects of toxic cyanobacteria and purified toxins on the survival and feeding of a copepod and three species of *Daphnia*. Limnol Oceanogr 36:1346–1357

Dittmann E, Neilan BA, Erhard M, von Döhren H, Börner T (1997) Insertional mutagenesis of a peptide synthetase gene that is responsible for hepatotoxin production in the cyanobacterium *Microcystis aeruginosa* PCC 7806. Mol Microbiol 26:779–787

Edwards C, Lawton LA, Coyle SM, Ross P (1996) Laboratory-scale purification of microcystins using flash chromatography and reversed-phase high-performance liquid chromatography. J Chromatogr A 734:163–173

Elfant M, Bryant L, Starcher B (1985) Isolation and characterization of a proteinase inhibitor from marama beans (42184). Proc Soc Exp Biol Med 180:329–333

Erhard M (1999) Matrix-assisted laser desorption/ionisation-time of flight. Massenspektrometrie von Cyanobakterien und Untersuchungen zur Biosynthese und Funktion der Microcystine. Dissertation, Technische Universität Berlin

Erhard M, von Döhren H, Jungblut P (1997) Rapid typing and elucidation of new secondary metabolites of intact cyanobacteria using MALDI-TOF mass spectrometry. Nature Biotechn 15:906–909

Evans JNS (1995) Biomolecular NMR spectroscopy. Oxford University Press, Oxford New York Tokyo

Fastner J, Flieger I, Neumann U (1998) Optimised extraction of microcystins from field samples – a comparison of different solvents and procedures. Wat Res 32:3177–3181

Fastner J, Heinze R, Chorus I (1995) Microcystin content, hepatotoxicity and cytotoxicity of cyanobacteria in some German water bodies. Wat Sci Technol 32:165–170

Feuillade M, Jann-Para G, Feuillade J (1996) Toxic compounds to Artemia from- and isolates of the cyanobacterium *Planktothrix rubescens*. Arch Hydrobiol 138:175–186

Fischer, WJ, Dietrich, DR (2000). Toxicity of the cyanobacterial cyclic heptapeptide toxins microcystin- LR and -RR in early life-stages of the African clawed frog (*Xenopus laevis*). Aquat Toxicol 49:189–198

Forchert A (1999) Isolierung und Charakterisierung neuer Peptidstrukturen aus Cyanobakterien der Gattungen *Microcystis* und *Planktothrix* spp. Dissertation, Technische Universität Berlin

Frankmölle WP, Knübel G, Moore RE, Patterson GML (1992) Antifungal cyclic peptides from the terrestrial blue-green alga *Anabaena laxa*. II. Structures of laxaphycins A, B, D and E. J Antibiot 45:1458–1466

Frankmölle WP, Larsen LK, Caplan FR, Patterson GML, Knübel G, Levine IA, Moore RE (1991) Antifungal cyclic peptides from the terrestrial blue-green alga *Anabaena laxa* I. Isolation and biological properties. J Antibiot 45:1451–1457

Fréchet D, Guitton JD, Herman F, Faucher D, Helynck G, Monegier de Sorbier B, Ridoux JP, James-Surcouf E, Vuilhorgne M (1994) Solution structure of RP 71955, a new 21 amino acid tricyclic peptide active against HIV-1 virus. Biochemistry 33:42–50

Fujii K, Harada KI, Suzuki M, Kondo F, Ikai Y, Oka H, Sivonen K (1995) Novel cyclic peptides together with microcystins produced by toxic cyanobacteria, *Anabaena* sp.. In: 37th Symposium on The Chemistry of Natural Products (Tokushima), Symposium Papers, Tokuschima, pp. 445–450

Fujii K, Sivonen K, Adachi K, Noguchi K, Sano H, Hirayama K, Suzuki M, Harada K (1997) Comparative study of toxic and nontoxic cyanobacterial products – novel peptides from toxic *Nodularia spumigena* AV1. Tetrahedron Lett 38:5525–5528

Fujiki H, Suganuma M, Hakii H, Bartolini G, Moore RE, Takayama S, Sugimura T (1984) A two-stage mouse skin carcinogenesis study of lyngbiatoxin A. J Cancer Res Clin Oncol 108:174–176

Gerwick WH, Jiang ZD, Agarwal SK, Farmer BT (1992) Total structure of hormothamnin A, a toxic cyclic undecapeptide from the tropical marine cyanobacterium *Hormothamnion enteromorphoides*. Tetrahedron 48:2313–2324

Gerwick WH, Mrozek CH, Moghaddam MF, Agarwal SK (1989) Novel cytotoxic peptides from the tropical marine cyanobacterium *Hormothamnion enteromorphoides*. 1. Discovery, isolation and initial chemical and biological characterization of the hormothamnins from wild and cultured material. Experientia 45:115–121

Gerwick WH, Roberts MA, Proteau PJ, Chen JL (1994) Screening cultured marine microalgae for anticancer-type activity. J Appl Phycol 6:143–149

Golakoti T, Ogino J, Hetzel CE, Husebo TL, Jensen CM, Larsen KL, Patterson GML, Moore RE, Mooberry SL, Corbett TH, Valeriote FA (1995) Structure determination, conformational analysis, chemical stability studies, and antitumor evaluation of the cryptophycins. Isolation of 18 new analogs from *Nostoc* sp. strain GSV 224. J Am Chem Soc 117:12030–12049

Gregson JM, Chen JL, Patterson GML, Moore RE (1992) Structures of puwainaphycins A-E. Tetrahedron 48:3727–3734

Haney JF, Sasner JJ, Ikawa M (1995) Effects of products released by *Aphanizomenon flos-aquae* and purified saxitoxin on the movements of *Daphnia carinata* feeding appendages. Limnol Oceanogr 40:263–272

Harada K, Mayumi T, Shimada T, Suzuki M, Kondo F, Watanabe MF (1993) Occurrence of four depsipeptides, aeruginopeptins, together with microcystins from toxic cyanobacteria. Tetrahedron Lett. 34, 6091-6094

Harada KI, Kiyonaga F, Shimada T, Suzuki M, Sano H, Adachi K, Carmichael WW (1995) Two cyclic peptides, anabaenopeptins, a third group of bioactive compounds from the cyanobacterium *Anabaena flos-aquae* NRC 525-17. Tetrahedron Lett 36:1511–1514

Helms GL, Moore RE, Niemczura WP, Patterson GML, Tomer KB, Gross ML (1988) Scytonemin A, a novel calcium antagonist from a blue-green alga. J Org Chem 53:1298–1307

Honkanen RE, Codispoti BA, Tse K, Boynton AL (1994) Characterization of natural toxins with inhibitory activity against serine/threonine protein phosphatases. Toxicon 32:339–350

Hughes EO, Gorham PR, Zahnder A (1958) Toxicity of an unialgal culture of *Microcystis aeruginosa*. Can J Microbiol 4:225–236

Ishida K, Matsuda H, Murakami M, Yamaguchi K (1997) Micropeptins 478-A and -B, plasmin inhibitors from the cyanobacterium *Microcystis aeruginosa*. J Nat Prod 60:184–187

Ishida K, Murakami M, Matsuda H, Yamaguchi K (1995) Micropeptin 90, a plasmin and trypsin inhibitor from the blue-green alga *Microcystis aeruginosa* (NIES-90). Tetrahedron Lett 36:3535–3538

Ishitsuka MO, Kusumi T, Kakisawa H, Kaya K, Watanabe MM (1990) Microviridin: A novel tricyclic depsipeptide from the toxic cyanobacterium *Microcystis viridis*. J Am Chem Soc 112: 8180–8182

Jakobi C, Oberer L, Quiquerez C, König WA, Weckesser J (1995) Cyanopeptolin S, a sulfate-containing depsipeptide from a water bloom of *Microcystis* sp.. FEMS Lett 129:129–134

Jakobi C, Rinehart KL, Neuber R, Mez K, Weckesser J (1996) Cyanopeptolin SS, a disulfated depsipeptide from a water bloom in Leipzig (Germany): structural elucidation and biological activities. Phycologia 35:111–116

Johnston NAL, Campagna VS, Hawkins PR, Banens RJ (1994) Response of the eastern rainbowfish (Melanotaenia duboulayi) to toxic *Microcystis aeruginosa*. Aust J Mar Freshwat Res 45:917–923

Jungmann D (1992) Toxic compounds isolated from Microcystis PCC7806 that are more active against *Daphnia* than two microcystins. Limnol Oceanogr 37:1777–1783

Kleinkauf H, von Döhren H (1996) A nonribosomal system of peptide biosynthesis. Eur J Biochem 236:335–351

Kleinkauf H, von Döhren H. (1990) Nonribosomal biosynthesis of peptide antibiotics. Eur J Biochem 192:1–15

Kobayashi J, Sato M, Ishibashi M, Shigemori H, Nakamura T, Ohizumi Y (1991a) Keramamide A, a novel peptide from the Okinawan marine sponge *Theonella* sp.. J Chem Soc Perkin Trans 1:2609–2611

Kobayashi J, Sato M, Murayama T, Ishibashi M, Walchi MR, Kainai M, Shoji J, Ohizumi Y (1991b) Konbamide, a novel peptide with calmodium antagonistic activity from the Okinawan marine sponge *Theonella* sp.. J Chem Soc Chem Commun 15:1050–1052

Kobayashi M, Kurosu N, Ohyabu N, Wang W, Kitagawa I (1994) The absolute stereostructure of arenastatin A, a potent cytotoxic depsipeptide from the Okinawan marine sponge *Dysidea arenaria*. Chem Pharm Bull 42:2196–2198

Kodani S, Ishida K, Murakami M (1998) Aeruginosin 103-A, a thrombin inhibitor from the cyanobacterium *Microcystis viridis*. J Nat Prod 61:1046–1048

Kusumi T, Ooi T, Watanabe MM, Takahashi H, Kakisawa H (1987) Cyanoviridin-RR, a toxin from the cyanobacterium *Microcystis viridis*. Tetrahedron Lett 28:4695–4698

Lawton LA, Edwards C, Codd GA (1994) Extraction and high-performance liquid chromatographic method for the determination of microcystins in raw and treated waters. Analyst 119:1525–1530

Lee AY, Smitka TA, Bonjouklian R, Clardy J (1994) Atomic structure of the trypsin-A90720A complex: a unified approach to structure and function. Chem Biol 1:113–117

Lirås V, Lindberg M, Nystrom P, Annadotter H, Lawton LA, Graf B (1998) Can ingested cyanobacteria be harmful to the signal crayfish (*Pacifastacus leniusculus*)? Freshwater Biol 39:233–242

MacKintosh C, Beattie KA, Klumpp S, Cohen P, Codd GA (1990) Cyanobacterial microcystin-LR is a potent and specific inhibitor of protein phosphatases 1 and 2a from both mammals and higher plants. FEBS Lett 264:187–192

Martin C, Oberer L, Ino T, König WA, Busch M, Weckesser J (1993) Cyanopeptolins, new depsipeptides from the cyanobacterium *Microcystis* sp. PCC 7806. J Antibiot 46:1550–1556

Martin C, Oberer L, Koenig WA, Busch M, Weckesser J (1990) Rapid purification of peptide toxins microcystin-LR and nodularin. FEMS Microbiol Lett 56:1–5

Moon SS, Chen JL, Moore RE, Patterson GML (1992) Calophycin, a fungicidal cyclic decapeptide from the terrestrial blue-green alga *Calothrix fusca*. J Org Chem 57:1097–1103

Moore RE (1996) Cyclic peptides and depsipeptides from cyanobacteria: a review. J Ind Microbiol 16:134–143

Moore RE, Bornemann V, Niemczura WP, Gregson JM, Chen JL, Norton TR, Patterson GML, Herms GL (1989) Puwainaphycin C, a cardioactive cyclic peptide from the blue-green alga *Anabaena* BQ-16-1. Use of two-dimensional 13C-13C and 13C-15N correlation spectroscopy in sequencing the amino acid units. J Am Chem Soc 111:6128–6132

Moore RE, Corbett TH, Patterson GML, Valeriote FA (1996) The search for new antitumor drugs from blue-green algae. Curr Pharm Design 2:317–330

Moore RE, Mynderse JS (1982) Majusculamide C. US Patent 4342751 Aug

Murakami M, Okita Y, Matsuda H, Okino T, Yamaguchi K (1994) Aeruginosin 298-A, a thrombin and trypsin inhibitor from the blue-green alga *Microcystis aeruginosa* (NIES-298) Tetrahedron Lett 35:3129–3132

Murakami M, Ishida K, Okino T, Okita Y, Matsuda H, Yamaguchi K (1995) Aeruginosins 98-A and -B, trypsin inhibitors from the blue-green alga *Microcystis aeruginosa* (NIES-98). Tetrahedron Lett 36:2785–2788

Murakami M, Shin HJ, Matsuda H, Ishida K, Yamaguchi K (1997a) A cyclic peptide, Anabaenopeptin B, from the cyanobacterium *Oscillatoria agardhii*. Phytochemistry 44:449–452

Murakami M, Kodani S, Ishida K, Matsuda H, Yamaguchi K (1997b) Micropeptin 103, a chymotrypsin inhibitor from the cyanobacterium *Microcystis viridis* (NIES-103). Tetrahedron Lett 38:3035–3038

Namikoshi M, Choi BW, Sakai R, Sun F, Rinehart KL, Carmichael WW, Evans WR, Cruz P, Munro MHG, Blunt JW (1994) New nodularins, a general method for structure assignment. J Org Chem 59:2349–2357

Namikoshi M, Rinehart KL (1996) Bioactive compounds produced by cyanobacteria. J Ind Microbiol 17:373–384

Neumann U, Forchert A, Flury T, Weckesser J (1997) Microginin FR1, a linear peptide from a water bloom of *Microcystis* species. FEMS Microbiol Lett 153:475–478

Oberemm A, Becker J, Codd G, Steinberg C (1999) Effects of cyanobacterial toxins and aqueous crude extracts on the development of fish and amphibians. Environ Toxicol 14:77–88

Oberemm A, Fastner J, Steinberg CEW (1997) Effects of microcystin-LR and cyanobacterial crude extracts on embryo-larval development of zebrafish (*Danio rerio*). Wat Res 31:2918–2921

Okino T, Matsuda H, Murakami M, Yamaguchi K (1993) Microginin, an angiotensin-converting enzyme inhibitor from the blue-green alga *Microcystis aeruginosa*. Tetrahedron Lett 34: 501–504

Okino T, Murakami M, Haraguchi R, Munekata H, Matsuda H, Yamaguchi K (1993b) Micropeptins A and B, plasmin and trypsin inhibitors from the blue-green algae *Microcystis aeruginosa*. Tetrahedron Lett 34:8131–8134

Okino T, Matsuda H, Murakami M, Yamaguchi K (1995) New microviridins, elastase inhibitors from the blue-green alga *Microcystis aeruginosa*. Tetrahedron 51:10679–10686

Papendorf O, Koenig G M, Wright A D, Chorus I, Oberemm A (1997) Mueggelone, a novel inhibitor of fish development from the fresh water cyanobacterium *Aphanizomenon flos aquae*. J Nat Prod 60:1298–1300

Patterson GML, Larsen LK, Moore RE (1994) Bioactive natural products from blue-green algae. J Appl Phycol 6:151–157

Pergament I, Carmeli S (1994) Schizotrin A, a novel antimicrobial cyclic peptide from a cyanobacterium. Tetrahedron Lett 36: 8473–8476

Pettit GR, Kamano Y, Herald CL, Dufresne C, Cerny RL, Herald DL, Schmidt JM, Kizu H (1989) Isolation and structure of the cytostatic depsipeptide dolastatin 13 from the sea hare *Dolabella auricularia*. J Am Chem Soc 111:5015–5017

Pettit GR, Kamano Y, Herald CL, Fujii Y, Kizu H, Boyd MR, Boettner FE, Doubek DL, Schmidt JL, Chapuis JC, Michel C (1993) Isolation of dolastatins 10-15 from the marine mollusc *Dolabella auricularia*. Tetrahedron 49:9151–9170

Pilotto L, Douglas R, Burch M, Cameron S, Beers M, Rouch G, Robinson P, Kirk M, Cowie C, Hardiman S, Moore C, Attewell R (1997) Health effects of exposure to cyanobacteria (blue-green algae) during recreational water-related activities. Austral. New Zealand J Public Health 21:562–566

Prepas EE, Kotak BG, Campbell JC, Evans JC, Hrudey SE, Holmes CFB (1997) Accumulation and elimination of cyanobacterial hepatotoxins by the freshwater clam *Anodonta grandis simpsoniana*. Can J Fish Aquat Sci 54:41–46

Prinsep MR, Moore RE, Levine IA, Patterson GML (1992) Westiellamide, a bistratamide-related cyclic peptide from the blue-green algae *Westiellopsis prolifica*. J Natural Prod 55:140–142

Reinikainen M, Kiviranta J, Ulvi V, Niku Paavola ML (1995) Acute toxic effects of a novel cyanobacterial toxin on the crustaceans *Artemia salina* and *Daphnia pulex*. Arch Hydrobiol 133:61–69

Richter K, Egger R, Kreil G (1987) D-alanine in the frog skin peptide dermorphin is derived from L-alanine in the precursor. Science 238:200–202

Rohrlack T, Dittmann E, Henning M, Börner T, Kohl JG (1999)Role of microcystins in poisoning and food ingestion inhibition of *Daphnia galeata* caused by the cyanobacterium *Microcystis aeruginosa*. Appl Env Microbiol 65:737–739

Römpp Chemielexikon (1995) Vol. 2, 9th edition, p. 891. Edited by J. Falbe, and M. Regitz. Thieme-Verlag, Stuttgart

Runnegar M, Berndt N, Kaplowitz N (1995) Microcystin uptake and inhibition of protein phosphatases: effects of chemoprotectants and self-inhibition in relation to known hepatic transporters. Toxicol Appl Pharmacol 134:264–272

Sano T, Kaya K (1995) Oscillamide Y, a chymotrypsin inhibitor from toxic cyanobacteria. Tetrahedron Lett 36:5933–5936

Shin HJ, Murakami M, Matsuda H, Ishida K, Yamaguchi K (1995) Oscillapeptin, an elastase and chymotrypsin inhibitor from the cyanobacterium *Oscillatoria agardhii* (NIES-204). Tetrahedron Lett 36:5235–5238

Shin HJ, Matsuda H, Murakami M, Yamaguchi K (1997a) Anabaenopeptins E and F, two new cyclic peptides from the cyanobacterium *Oscillatoria agardhii* (NIES-204). J Nat Prod 60: 139–141

Shin HJ, Matsuda H, Murakami M, Yamaguchi K (1997b) Aeruginosins 205A and -B, serine protease inhibitory glycopeptides from the cyanobacterium *Oscillatoria agardhii* (NIES-205). J Org Chem 62:1810–1813

Sivonen K, Jones J (1999) Cyanobacterial toxins. In: Chorus I, Bartram J (eds) Toxic cyanobacteria in water: a guide to their public health consequences, monitoring and management. Published on the behalf of WHO by E & FN Spon, London, pp. 41–111

Stratmann K, Burgoyne DL, Moore RE, Patterson GML, Smith CD (1994) Hapalosin, a cyanobacterial cyclic depsipeptide with multidrug-resistance reversing activity. J Org Chem 59:7219–7226

Tencalla FG, Dietrich DR, Schlatter C (1994) Toxicity of *Microcystis aeruginosa* peptide toxin to yearling rainbow trout (*Oncorhynchus mykiss*). Aquat Toxicol 30:215–224

Trimurtulu G, Ohtani I, Patterson GML, Moore RE, Corbett TH, Valeriote FA, Demchik L (1994) Total structures of cryptophycins, potent antitumor depsipeptides from the blue-green alga *Nostoc* sp. strain GSV 224. J Am Chem Soc 116:4729–4737

Tsukamoto S, Painuly P, Young KA, Yang X, Shimizu Y, Cornell L (1993) Microcystilide A: a novel cell-differentiation-promoting depsipeptide from *Microcystis aeruginosa* NO-15-1840. J Am Chem Soc 115:11046–11047

Vasconcelos VM (1995) Uptake and depuration of the heptapeptide toxin microcystin-LR in *Mytilus galloprovincialis*. Aquat Toxicol 32:227–237

Weckesser J, Martin C, Jakobi C (1996) Cyanopeptolins, depsipeptides from cyanobacteria. System Appl Microbiol 19:133–138

Williams DE, Burgoyne DL, Retting SJ, Andersen RJ, Fahti-Afshar ZR, Allen TM (1993) The isolation of majusculamide C from the sponge *Ptilocaulis trachys* collected in Enewetak and determination of the absolute configuration of the 2-methyl-3-aminopentanoic acid residue. J Nat Prod 56:545–551

Williams DE, Craig M, Holmes CFB, Andersen RJ (1996) Ferintoic acid A and B, new cyclic hexapeptides from freshwater cyanobacterium *Microcystis aeruginosa*. J Nat Prod 59: 570–575

8 Contributions to Toxicity Testing and Toxin Analysis

Toxicity testing is of key importance both for cyanotoxin research as well as for monitoring and can only partially be replaced by chemical analysis of cyanotoxins. Analytical methods were developed to detect selected known substances, whereas toxicological assays aim at determining the effects of all substances, known or unknown, in a sample. Extracts of cyanobacteria may contain a variety of chemical substances, numerous of which may be unknown, and their potential effect can only be detected by toxicological testing complementary to chemical toxin analysis. The mouse bioassay has been the most widely used toxicity test in cyanotoxin research and in the screening programmes published from several countries in the 1980s and early 1990s (Sivonen and Jones 1999). It involves intraperitoneal injection of cyanobacterial material or extracts into mice and determination of the dose that kills 50% of the mice, the classical LD_{50} (Berg and Soli 1985; Falconer 1993). Skulberg et al. (1994) classified cyanobacterial toxins (cyanotoxins) by the response to them as observed in the mouse bioassay as hepatotoxins, neurotoxins and toxins with protracted effects. The time of survival and pathologically observable damage to organs were used as parameters for this classification. For ethical reasons (especially pertaining to LD_{50} tests), because of stringent animal welfare regulations in many countries, and because of time and costs involved, mouse tests are unsuitable for large-scale testing of field samples, and the need to replace the mouse bioassay with alternative methods at the suborganismic level is urgent (another reason being a lack of adequate diagnostic/mechanistic end points). Chapter 8.1 shows that cellular assays are a promising tool for monitoring cyanobacterial toxicity, and Chapter 8.2 provides verification for a cellular microcystin assay by comparing the results to those of the mouse bioassay.

Whereas adequate bioassays for monitoring and research addressing cyanobacterial toxicity require further development, published methods for the chemical detection and quantification of the currently identified cyanotoxins are largely satisfactory. However, research needs are evident in two areas - low levels of detection and identification of microcystin variants as well as rapid identification of substances: analytical methods for low concentrations of dissolved cyanotoxins require further development and verification. The results presented in Chapter 8.3 highlight the applicability of two different enzyme assays and one immunoassay for assessment of microcystin traces in field samples.

The issue of cyanobacterial toxicity further involves the question: toxic to whom? Cyanotoxin research has been triggered by spectacular lethal intoxications of non-aquatic animals and some cases of human illness, and since then has focused on assessing potential human toxicity with mammalian model organisms. The neglect of ecotoxicological approaches discussed in Chapter 6 has in part been due to the very limited experiments with aquatic test animals. Chapter 8.4 shows that the fish-embryo bioassay, a classic widely employed for assessing chemical safety and environmental health (McKim 1977; Birge et al. 1985; van Leeuwen et al. 1990), is invaluable for detecting a variety of toxic effects caused by cyanobacteria.

Monitoring requires rapid identification of microcystin variants as well as of other potentially bioactive peptides. Chapter 8.5 shows how a new spectrometric method – MALDI-TOF – can be employed for the identification of cyanobacterial peptides from small sample volumes without prior sample cleanup.

8.1 Testing Cyanobacterial Toxicity with Primary Rat Hepatocyte and Cell-Line Assays

Rita Heinze, Jutta Fastner, Uwe Neumann and Ingrid Chorus

For the microcystins, a group of hepatotoxic cyanotoxins, supplementing analytical techniques with cellular toxicological assays is especially promising. As shown in Chapter 2, microcystins are the most frequently detected and most widespread cyanotoxins. Their toxicity is characterized by extensive hepatic necrosis and haemorrhages, both in cases of acute animal poisoning and in the mouse bioassay (Sivonen et al. 1992; Falconer 1993). This selective liver damage allows differentiation between the effects of microcystins and those of other toxins by toxicity testing.

Microcystins occur in a multitude of structural variants; Sivonen and Jones (1999) report more than 60 with a wide range of toxicity. However, only four structural variants for use as reference substances in HPLC are available commercially. In consequence, routine procedures of microcystin analysis detect microcystins as such, but do not allow identification of variants other than these four. Toxicological investigations of the effect of the various structural variants are few in number, microcystin-LR (MC-LR) being the most frequently tested variant (Miura et al. 1991; Sahin et al. 1995). Studies of changes in toxicity caused by simultaneous exposition to different microcystin variants are lacking entirely, although cyanobacterial samples frequently contain several structural variants. Toxicity might increase beyond or decrease below purely additive effects. Investigations by Fitzgeorge et al. (1994) showed that prior application of doses of MC-LR below acutely toxic levels enhances the acute toxicity of the neurotoxin anatoxin-a in intranasal application. Therefore, any approach to detect potential harmful effects of cyanobacterial water blooms in lakes and reservoirs cannot rely only on chemical quantification of the single microcystins in the algae or in the water column.

The effects of microcystins are strictly organ- and cell-specific. In line with the selective liver damage found in vivo, in vitro application of these toxins mainly affected primary hepatocytes (Aune and Berg 1986; Aune et al. 1991). The cell specificity is not due to metabolic activation in the liver cells but to a specific pathway of uptake. Microcystins have been shown to be taken up through the multispecific transport system for bile acids which is also an entrance pathway for several xenobiotic substances. The in vitro effects of microcystins on hepatocytes can be inhibited by low concentrations of bile acids or bile acid transport inhibitors (Eriksson et al. 1990; Runnegar et al. 1991; Hermansky et al. 1991; Khan et al. 1995). Only primary hepatocytes and intestinal cells are equipped with this bile acids transport system. Following isolation of hepatocytes for in vitro tests, they dedifferentiate very quickly and also lose this transport system (Petzinger and Frimmer 1988). Experiments of Matsushima et al. (1990) documented that the tissue specificity of microcystins is due to it being difficult for them to penetrate cells without a transport system: after microinjection into primary human fibroblasts morphological changes were induced by dosages which did not induce changes in mouse skin and primary human fibroblasts after incubation.

Intracellularly, microcystins induce an overall increase in phosphorylation of cytosolic and cytoskeletal proteins (Honkanen et al. 1990). Studies by Toivola et al. (1994) and MacKintosh et al. (1995) have demonstrated that microcystins are potent inhibitors of the protein phosphatases 1 und 2A (PP1 und PP2A) by covalently binding to these enzymes. The inhibition of PP1 and PP2A was demonstrated for a variety of animal cells but also for plant cells (MacKintosh et al. 1991).

Primary Rat Hepatocytes as Microcystin Bioassay

In vitro test systems with freshly isolated hepatocytes can be used for the selective investigation of microcystin toxicity of cyanobacterial extracts. Based on this, we have established a toxicity test that uses freshly isolated rat hepatocytes (Heinze 1996). Rat hepatocytes are prepared by a two-step collagenase perfusion technique. After assessment of cell number and viability, cells are seeded into collagenated 96-well plates and cultivated adherently for 2 h. Hepatocytes are exposed to the cyanobacterial extracts (prepared as described in Chapter 9.1 either with 75% aqueous methanol, as recommended by Fastner et al. 1998, or with 100% methanol) for 4 or 20 h. Concentrations of the extracts in the microtitre plates range from 0.003 to 0.83 mg DW ml^{-1} depending on microcystin content and the exposure period. Following exposure, cells are checked for viability using the MTT test (Mosmann 1983). In this test, the soluble yellow tetrazolium salt MTT is converted to insoluble dark-blue formazan crystals by dehydrogenases of the living cells. Formazan (dehydrogenase activity) is measured with a photometer for microtitre plates (ELISA reader, SLT Instr.) The values of the control cells (cells not exposed to cyanobacterial extracts) are set at 100%, the values of the cells incubated with various concentrations of cyanobacterial extracts are calculated and shown as dose-response curves. The LD_{50} values or, more exactly, the LC_{50} values, are determined as the concentration of a cyanobacterial extract at which a 50% reduction in viable cells occurred. LC_{50} values characterize the hepatotoxicity of the cyanobacterial extracts.

Freshly prepared rat hepatocytes respond promptly and sensitively to microcystin exposure. After 1 h exposure to 2.5 μg ml^{-1} (final concentration in the microtiter plate) of the commercially available structural variant microcystin-LR (MC-LR, Calbiochem) viability of the hepatocytes was reduced by 50%. Incubation over a longer period resulted in a lower LC_{50} value, and in exposure for 20 h an MC-RL concentration of 0.05 μg ml^{-1} is sufficient to kill 50% of the hepatocytes (Table 1). In addition to the microcystin concentration, the sensitivity of the test also depends on the exposure time. The longer the exposure period, the lower the microcycstin concentrations shown to induce cytotoxic effects. Limits of determination for MC-LR in seston samples are 240 μg g^{-1} DW after 4 h of exposure and 60 μg g^{-1} DW after 20 h of exposition.

Permanent Cell Cultures as Microcystin Bioassays

Investigations with permanent cell cultures of both animal and human origin, CHO-K1, L-929, VERO, YAC-1, Int 407 and Hep-2, showed no alterations in viability following 20 h exposure to 1 μg ml^{-1} MC-LR. Cytotoxic effects of MC-LR on fibroblasts were described by Wickstrom et al. (1995) for a much higher concentration range of 100 to 200 μM. Investigations by Grabow et al. (1982) and Henning et al. (1992) showed that the effects of cyanobacterial extracts on permanent cell cultures are not attributable to microcystins but to other currently unknown substances. For a

Table 1. LC_{50} values of MC-LR after variable incubation periods with primary rat hepatocytes

Incubation period	LC_{50} values (μg MC-LR ml^{-1})
1 h	2.50
2 h	0.55
4 h	0.20
20 h	0.05

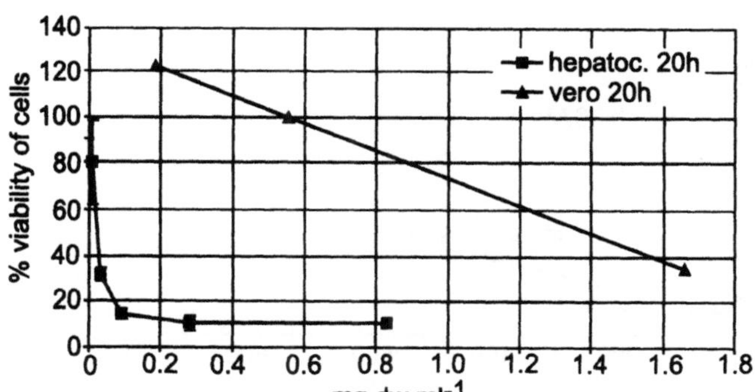

Fig. 1. Hepatotoxicity and cytotoxicity of a methanolic cyanobacterial extract from Lake Langer See (26.06.1996) after 20 h incubation. LC_{50} VERO cells: 1.40 mg DW ml^{-1}; LC_{50} hepatocytes: 0.02 mg DW ml^{-1}

decrease in cell viability to be induced by other cyanobacterial metabolites, higher concentrations and a longer exposure period are necessary (Fig. 1).

Results Obtained in Cyanotoxin-Screening of German Water Bodies

The LC_{50} value of the least toxic samples in the hepatocyte assay was 0.83 mg DW ml^{-1} after 20 h exposure. The most toxic samples investigated killed 50% of the hepatocytes at a concentration of 0.01 mg DW ml^{-1} after 4 h exposure. At exposure times of 4 h, 44% of a total of 430 samples tested were toxic (see Chapter 2.1.2); if exposure time was extended to 20 h, more than half the samples were toxic. The cytotoxicity of the cyanobacterial extracts was tested on CHO-K1, VERO or L 929 cell lines using concentrations twice those used in the hepatocyte assays, and always an exposure period of 20 h. However only a few samples showed cytotoxic effects strong

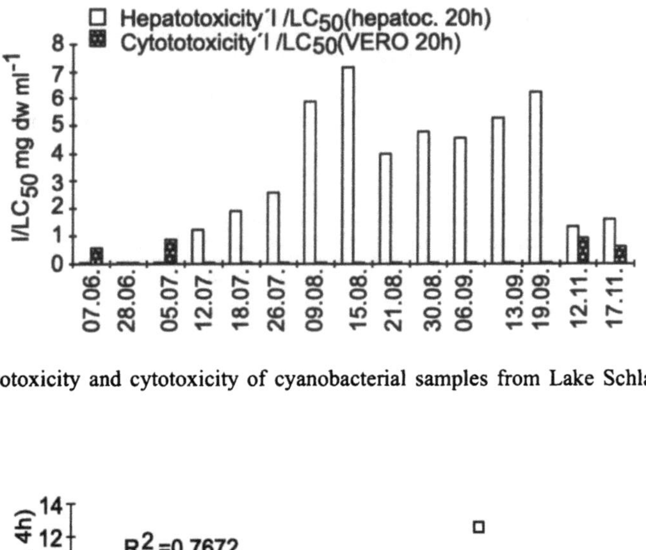

Fig. 2. Hepatotoxicity and cytotoxicity of cyanobacterial samples from Lake Schlachtensee in 1995

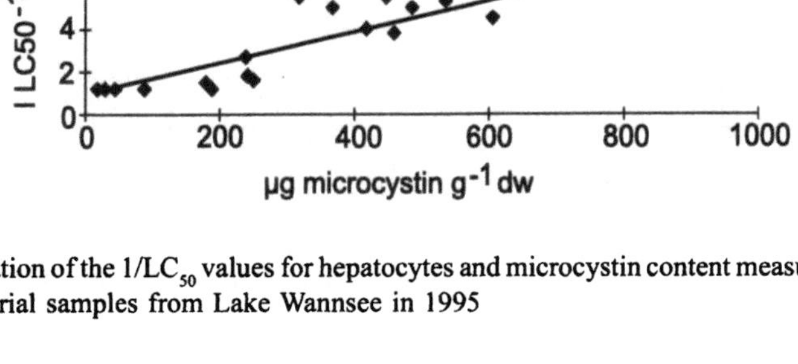

Fig. 3. Correlation of the $1/LC_{50}$ values for hepatocytes and microcystin content measured by HPLC of cyanobacterial samples from Lake Wannsee in 1995

enough to calculate LC_{50} values. There is no correlation between hepatotoxic and cytotoxic effects, as can be seen, for example, for the samples from Lake Schlachtensee (Fig. 2).

However, assessment of cytotoxicity may have been incomplete because the method used to extract the cyanobacteria was optimized for microcystins, and other, unknown cytotoxic substances might have escaped extraction. Also, it is probable that microcystins mask the effect of other cyanotoxins because after their active transport into the cells they may act more quickly and at lower concentrations than cytotoxic substances after passive penetration.

A comparison of the results of the hepatocyte assay with the analytical data from HPLC (see Chapter 2.1.2) generally showed good agreement: of the 255 samples that were not toxic after 4 h of hepatocyte exposure, only 40 contained microcystins at low concentrations; 28 of these contained less than 200 µg g^{-1} DW total microcystins and 12 contained between 200 and 300 µg g^{-1} DW. Further, the analytically determined microcystin concentrations and hepatocyte toxicity, presented as $1/LC_{50}$, correlate well (Fig. 3, see also Chapter 7.2). Thus, hepatocyte toxicity testing generally confirmed the analytical results, i.e. samples with a high measured microcystin content showed high hepatotoxicity and samples found to contain little or no microcystin showed low or no hepatocyte toxicity.

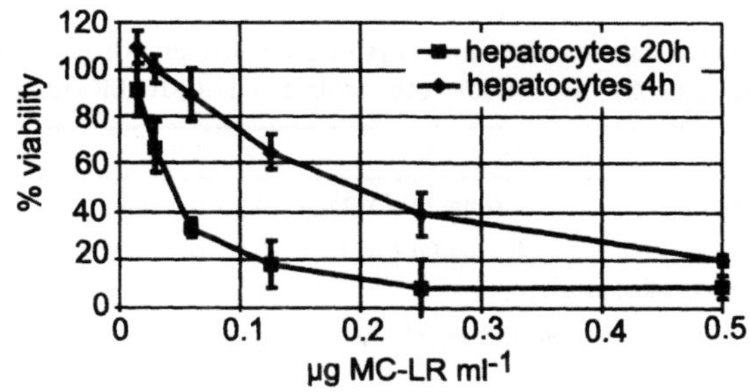

Fig. 4. Toxicity of MC-LR to primary rat hepatocytes after 4 or 20 h incubation. LC_{50} 4 h: 0.20 µg ml^{-1}; LC_{50} 20 h: 0.05 µg ml^{-1}

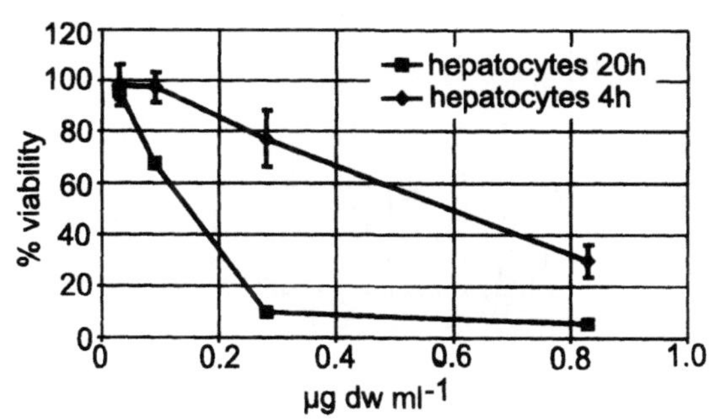

Fig. 5. Hepatotoxicity of a cyanobacterial extract from Lake Oberer See (18.08.1982) to primary rat hepatocytes after 4 or 20 h incubation. LC_{50} 4 h: 0.6 mg DW ml^{-1}; LC_{50} 20 h: 0.15 mg DW ml^{-1}

As discussed above, the comparison of the hepatocyte assay results to the concentrations of microcystins detected with HPLC analysis are of interest for identification of additonal toxic substances, synergistic enhancement of toxicity, or for detection of microcystins with low or no toxicity. This requires expression of toxicity measured in the bioassay and toxin concentration detected analytically in the same units. Therefore, the microcystin content was calculated on the basis of the LC_{50} values of chemically pure MC-LR (Fig. 4), with the assumption that the mg of dry cyanobacterial material that kill 50% of the hepatocytes must contain the mg of microcystins that induce the LC_{50} value of MC-LR. For example, the LC_{50} values for hepatocytes of a cyanobacterial extract from the lake Oberer See (Fig. 5) amount to 0.6 mg DW ml^{-1} after 4 h exposure and to 0.15 mg DW ml^{-1} after 20 h exposure. This is equivalent to the toxicity of 0.2 µg ml^{-1} of MC-LR after 4 h or 0.05 µg ml^{-1} after 20 h exposure (Table 2). For both exposure times this calculates to 333.3 µg MC-LR g^{-1} DW. This calculated microcystin content is a worst-case estimate which equates the toxic effect to only one structural variant of the microcystins – the most toxic one, yet it provides a basis for comparison of the data of analytical investigations and toxicological assays.

For the samples of some water bodies, e.g. those from Tegler See (Fig. 6), the comparison of the analytically measured microcystin concentrations and the

Table 2. Example for the calculation of the microcystin-LR equivalents of a cyanobacterial sample from Lake Oberer See using the LC_{50} values of the sample and of MC-LR (reference substance)

	LC_{50} 4 h	LC_{50} 20 h
Oberer See	0.6 mg DW ml^{-1}	0.15 mg DW ml^{-1}
MC-LR (reference substance)	0.2 µg DW ml^{-1}	0.05 µg DW ml^{-1}
MC-LR equivalents in Oberer See	333 µg g^{-1} DW	333 µg g^{-1} DW

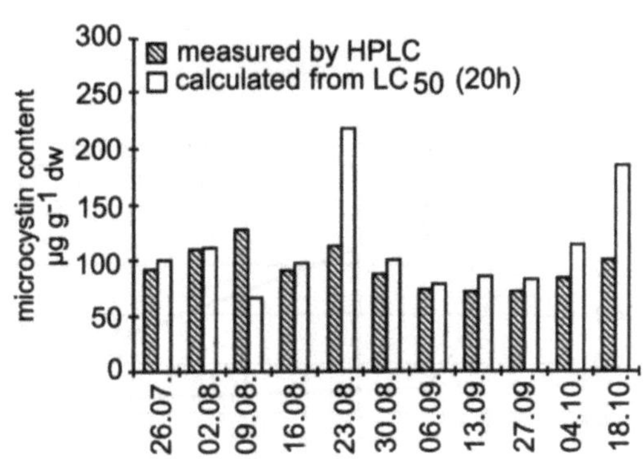

Fig. 6. Microcystin concentration of cyanobacteria from Lake Tegeler See in 1995 analyzed by HPLC and calculated from LC_{50} values of the primary rat hepatocyte assay

microcystin concentrations calculated on the basis of the LC_{50} values revealed good agreement. However, for most samples concentrations derived from the LC_{50} values were slightly higher than those determined analytically by HPLC. As not all of the microcystins in these samples were microcystin-LR and some (such as microcystin-RR) were less toxic, this is surprising – it would be expected that concentrations derived as microcystin-LR equivalents from the LC_{50} values would be somewhat lower than those determined by HPLC. For some water bodies, such as Langer See (Fig. 7), the microcystin concentrations calculated from LC_{50} were even more substantially and regularly higher than those measured by HPLC. Investigations of

Table 3. Microcystin concentrations detected by HPLC and LC_{50} values for rat hepatocytes after 4 h exposure of cyanobacterial samples from Lake Langer See extracted with pure or 75% methanol

Date of sampling	HPLC Extracted by pure methanol µg microcystin g^{-1} DW	HPLC Extracted by 75% methanol µg microcystin g^{-1} DW	LC_{50} values Extracted by pure methanol mg DW ml^{-1}	LC_{50} values Extracted by 75% methanol mg DW ml^{-1}
16.05.1995	517	879	0.1	Missing value
21.06.1995	1067	1557	0.04	0.08
18.07.1995	316	445	0.25	0.35
15.08.1995	555	1274	0.115	0.17
28.08.1995	820	1732	0.045	0.08
13.09.1995	839	1891	0.085	Missing value
10.10.1995	663	2308	0.075	Missing value
23.10.1995	764	2300	0.05	0.07
07.11.1995	1284	2336	0.05	0.04
20.11.1995	853	2284	0.03	0.03
07.12.1995	Missing value	624	0.057	0.14
Mean value	770	1680	0.082	0.120

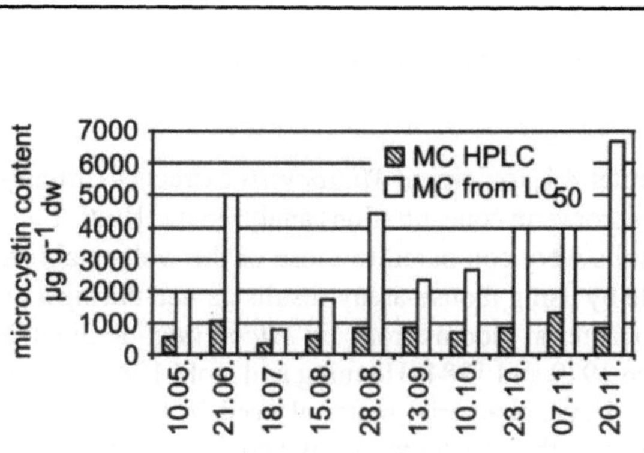

Fig. 7. Microcystin concentration of cyanobacteria from Lake Langer See in 1996 analyzed by HPLC and calculated from LC_{50} values of the primary rat hepatocyte assay

the dominant cyanobacteria in the various water bodies show that this was particularly the case when *Planktothrix agardhii* dominated (see also Chapter 7.2). Samples from both lakes showed no substantial cytotoxicity, deviations in toxicity are caused by unknown hepatotoxic substances or by potentiation of microcystin effects.

Further, the solvent used for extraction proved decisive for the toxicity of the samples. Cyanobacterial samples from Langer See (1995) were extracted with both 75% aqueous methanol and pure methanol. HPLC detected distinctly more microcystins in the samples extracted by 75% methanol (mean value: 1680 µg g^{-1} DW) as compared to those extracted with pure methanol (mean value: 770 µg g^{-1} DW). Nonetheless, most of the samples extracted with 75% methanol were less toxic in the hepatocyte assay than the extracts with pure methanol (Table 3). Correlation between $1/LC_{50}$ and microcystin concentrations determined by HPLC are closer for samples extracted with 75% aqueous methanol (R^2=0.66) than for samples extracted with pure methanol (R^2=0.46). This indicates that extraction with 75% aqueous methanol is optimal for microcystins but other toxic effects are lost. Thus, 75% aqueous methanol is the solvent of choice if bioassays are performed as a measure of microcystin toxicity, whereas if further toxic agents are to be detected other solvents may be adequate, such as pure methanol or water (see Chapter 8.2).

Currently, we have little knowledge of whether the simultaneous effects of various structural variants of microcystins enhance their toxicity or whether the extracts contain other hepatotoxic substances yet unknown. Possibly cyanobacteria, particularly *Planktothrix* spp., contain substances which are not hepatotoxic themselves but enhance the toxicity of the microcystins.

8.2
Can the Primary Rat Hepatocyte Assay Replace the Mouse Assay for Microcystin Testing? Validation with "Historic" Samples from the First Survey on Cyanobacterial Toxicity in Germany

Peter Woitke, Rita Heinze, Marion Dewender, Manfred Henning and Johannes-Günter Kohl

As shown in Chapter 8.1, toxicity of *Microcystis* extracts to primary rat hepatocytes closely reflects microcystin concentrations analyzed by HPLC. For further validation of this assay, results were compared to those of the well-established mouse assay. This was possible by using mouse-assay results gained during a program assessing toxicity of cyanobacterial blooms from six *Microcystis*-dominated lakes in eastern Germany between 1979 and 1984 (Henning and Kohl 1981; Benndorf and Henning 1989). Lyophilized cyanobacterial material used in these mouse assays was still available for testing in the hepatocyte assay and for chemical microcystin analysis, which had not yet been available at the time of this first German screening program on cyanobacterial toxicity.

Cyanobacterial Material and Mouse Assay

The sampling procedure and the mouse bioassay are described in detail in Henning and Kohl (1981). In summary, material of surface blooms of *Microcystis* sp. which were driven to the shore as floating mats was collected using simple scoops. After centrifugation at 2000 g, the cyanobacterial cells were freeze-dried and stored at 20°C. For determination of toxicity, different amounts of the freeze-dried material were suspended in 0.5 ml water and were applied by i.p. injection into male white mice. LD_{50} values were calculated from dose-response curves (Table 1).

HPLC Analysis

Microcystin content of the methanolic extracts was analyzed by a method similar to that of Lawton et al. (1994a; see also Chapter 9 and Woitke et al. 1999) with a binary gradient system on a Merck Hibar LiChrospher RP18 column (4 x 250 mm), protected

Table 1. Mouse bioassay toxicity (aqueous extracts) and microcystin content of water blooms with more than 90% of phytoplankton biovolume being *Microcystis* sp.

Lake	Sampling date	Sample code	LD50 mouse (mg kg-1)	MC-LR (μg g-1 DW)	MC-RR (μg g-1 DW)	MC-YR (μg g-1 DW)	Sum of MC (μg g-1 DW)
Oberer See	23.7.80	OS/1	22.5	1993 ± 184	n.d.	n.d.	1993
Oberer See	5.8.82	OS/2	370	n.d.	44.8 ± 21.3	n.d.	44.8
Oberer See	18.8.82	OS/3	250	90.5 ± 10.8	45.6 ± 3.5	n.d.	136
Talsperre Bautzen	21.9.82	TB	72.5	291.3 ± 41.9	442.9 ± 67.7	n.d.	734.2
Tiefer See	18.8.82	TS	290	17.9 ± 4.9	28.7 ± 8.6	69.5 ± 8.0	116.1
Müggelsee	9.8.82	MS	105	195.5 ± 38.1	152.2 ± 27.4	n.d.	347.7
Radeburg Reservoir	26.7.84	RB/1	Non-toxic		30.4 ± 4.3	n.d.	30.4
Radeburg Reservoir	2.8.84	RB/2	255	89.5 ± 15.5	28.9 ± 6.6	179.7 ± 26.4	298.1
Radeburg Reservoir	9.8.84	RB/3	47.5	213.7 ± 4.8	66.3 ± 10.2	364.0 ± 20.7	644
Radeburg Reservoir	15.8.84	RB/4	47.5	413.2 ± 28.1	152.0 ± 16.6	542.1 ± 27.3	1107.3
Radeburg Reservoir	21.8.84	RB/5	42.5	427.0 ± 42.3	255.3 ± 21.4	578.3 ± 110.1	1260.6
Radeburg Reservoir	29.8.84	RB/6	155	215.4 ± 36.6	178.3 ± 18.1	215.5 ± 34.0	609.2

with a Waters Guard-Pak precolumn (Novapak 18). Results showed a pronounced dominance of the microcystins LR, RR and YR in the more hydrophylic range of the chromatograms.

Most of the extracts contained further substances in the more lipophylic fraction exhibiting the UV spectrum characteristic for microcystins. However, unambiguous identification was not possible. Therefore, microcystins in these samples were determined as the sum of microcystin-LR, -RR and -YR (Table 1).

A comparison of the microcystins present in the samples revealed a different toxin pattern between the investigated lakes, e.g. with MC-YR being found only in the Radeburg Reservoir and in Tiefer See (Table 1). However, across all the lakes toxicity of *Microcystis* is reflected well by the variation of microcystin-content (Table 1) The highest toxin content (only MC-LR) was detected in sample OS/1, the most toxic one with respect to both the mouse bioassay and the hepatocyte test. Changes from late July to late August in the reservoir Speichersystem Radeburg (samples RB/1 - RB/6) also show a congruence between the increase in toxicity (with both the mouse assay and the primary rat hepatocyte assay) and an increase in microcystin content.

Primary Rat Hepatocyte Assay

The assays were performed as described in Chapter 8.1 and by Heinze (1996). Hepatocytes were incubated 20 h with extracts of lyophilized material, and LC_{50} values were calculated from dose-response curves. Lyophilized algal material was extracted

Table 2. Rat hepatocyte bioassay toxicity of methanolic and aqueous extracts of cyanobacterial bloom material. Values are means of three independent determinations

Sample code	LC_{50} rat hepatocytes (methanolic extracts) (mg ml^{-1})	LC_{50} hepatocytes (aqueous extracts) (mg ml^{-1})
OS/1	0.0097 ± 0.004	n.d..
OS/2	0.21 ± 0.011	0.089 ± 0.008
OS/3	0.16 ± 0.035	n.d.
TB	0.053 ± 0.017	0.0315 ± 0.012
TS	0.117 ± 0.025	n.d.
MS	0.26 ± 0.03	0.155 ± 0.02
RB/1	non toxic	0.53 single value
RB/2	0.21 ± 0.015	0.185 ± 0.007
RB/3	0.077 ± 0.015	0.042 ± 0.01
RB/4	0.026 ± 0.009	0.029 ± 0.009
RB/5	0.035 ± 0.021	0.0205 ± 0.0007
RB/6	0.1 ± 0.026	0.019 ± 0.006

For sample code see Table 1.
n.d. not detected

with both methanol and water in order to investigate the influence of the extraction solvent on toxicity (see Woitke et al. 1999).

Most of the methanolic extracts of the freeze-dried material of *Microcystis* blooms sampled in several lakes of the eastern part of Germany were toxic to primary rat hepatocytes. LC_{50} values varied over 1 order of magnitude with sample OS/1 as the most toxic. Only sample RB/1 could be identified as non-toxic with the highest extract concentration of 0.83 mg DW ml^{-1} (Table 2).

The aqueous extracts of these samples showed a grading in their toxicity similar to that found for the methanolic extracts (Table 2); however, for most of the samples the LC_{50} values of the aqueous extracts were slightly lower than those of the methanolic extracts, indicating higher toxicity of aqueous extracts.

Mouse Assay, Primary Rat Hepatocyte Assay and Microcystin Content in Comparison

In a regression analysis of the LD_{50} and LC_{50} values (Fig. 1A and B) both the correlation coefficient and the intercept close to zero indicate the comparability of the mouse test and the rat hepatocyte bioassay, especially in the lower LC_{50} range (i.e. for samples exhibiting a high toxicity). Correlation is better with the hepatocyte values gained from methanolic cextracts than for those from aqueous extracts. A statistical evaluation of the regression function revealed the methanolic and aqueous extracts of the Lake Müggelsee sample (MS) and the aqueous extracts of the sample RB/2 as outliers (a=0.05). Therefore, these samples were rejected in the regression analysis. The reason for the higher toxicity of these samples in the mouse bioassay remains unclear, but possibly other cyanotoxins were present.

Generally, these results point to the rat hepatocyte bioassay as an alternative to the mouse test for assessing the toxicity of cyanobacterial blooms. They confirm the recommendation of Berg and Aune (1987) who suggest the application of freshly prepared rat hepatocytes (with semiquantitative microscopic assessment of cell damage) as a useful model to assess cyanobacterial toxicity.

For comparison of the mouse bioassay and the chromatographically determined microcystin contents in the investigated samples, the inverse LD_{50} values of the mouse test are plotted against the sum of the content of MC-LR, MC-RR and MC-YR (Fig. 1C). The results of both methods show a strong correlation, as was observed by Lawton et al. (1994b). These authors compared the mouse bioassay with several other biological, biochemical and chemical methods for an assessment of cyanotoxicity and found the highest degree of comparability between the mouse bioassay and the HPLC determination of microcystins.

Comparisons of microcystin content determined by HPLC and results of both hepatocyte bioassays (with methanolic as well as with aqueous extracts) show satisfactory correlations (Fig. 1D and E). However, an important indication for an influence of further toxic cyanobacterial metabolites can be derived from the results of the rat hepatocyte bioassay comparing methanolic with aqueous extraction of the algal material. Both extraction methods quantitatively liberated MC-LR, MC-RR, and

MC-YR, because these microcystins could not be detected by HPLC after an additional extraction of the residue with the respective other solvent (data not shown). Therefore, the concentrations of MC-LR, MC-RR and MC-YR in both extracts were comparable. However, a higher toxicity in the rat hepatocyte bioassay was found for the aqueous extracts of almost all investigated samples (see Table 2). This difference in toxicity can only be explained by further substances, more hydrophilic than the microcystins, in the freeze-dried algal material. The poorer correlation suggests variability in this effect.

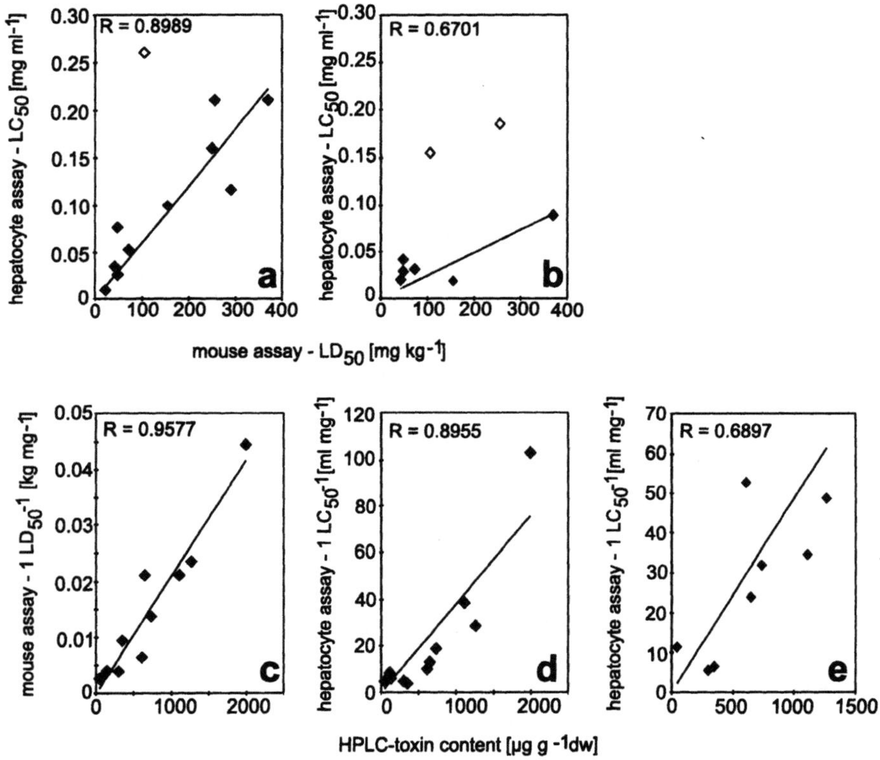

Fig. 1A-E. Comparsion of the LD_{50} values determined using the mouse bioassay with LC_{50} values of the rat hepatocyte bioassay and with microcystin concentrations determined by HPLC (open diamonds: not included in regression).

(**a**) Hepatocyte assay (methanolic extracts) and mouse assay: LC_{50} (rat hepatocytes)=0.000601 * LD_{50} (mouse); r=0.8989 without sample MS (outlier, a=0.05). (**b**) Hepatocyte assay (aqueous extracts) and mouse assay: LC_{50} (rat hepatocytes)=0.000246 * LD_{50} (mouse); r=0.6701 without sample MS and RB/2 (outliers, a=0.05). (**c**) Mouse assay and microcystin concentration: $1/LD_{50}$ (mouse)=2.091 10^{-5} * Toxin content (sum of microcystins MC-LR+MC-RR+MC-YR); r=0.9577. (**d**) Hepatocyte assay (methanolic extracts) and microcystin concentration: $1/LD_{50}$ (rat heparocytes)=0.0380 * Toxin content (sum of microcystins MC-LR+MC-RR+MC-YR); r=0.8955. (**e**) Hepatocyte assay (aqueous extracts) and microcystin concentration: $1/LD_{50}$ (rat heparocytes)=0.0488 * Toxin content (sum of microcystins MC-LR+MC-RR+MC-YR); r=0.6897

In addition to the issue of correlation between bioassay and microcystin content, the question arises to what extent microcystin content is quantitatively reflected by the bioassay, and whether the bioassay reveals further toxicity, as shown for the primary rat hepatocyte test (particularly for samples dominated by *Planktothrix agardhii* or *P. rubescens*; see Chapter 7.1). For this data set, the toxicity of the field samples may be estimated quite precisely from the content of MC-LR, MC-YR and MC-RR, because probably the samples scarcely contained further microcystins, and LD_{50} values from the literature for these as pure substances in the mouse test are available (MC-LR, MC-YR: < 100 µg kg^{-1}, MC-RR: 400-800 µg kg^{-1}, i.p. injection; Harada 1996). Most of the samples revealed a distinctly higher toxicity in the mouse bioassay than was predictable from their microcystin content using an LD_{50} of 100 µg kg^{-1} for the sum of microcystins (Fig. 2A). Results obtained with the primary rat hepatocyte assay using the LC_{50} of 0.05 µg ml^{-1} for MC-LR (see Chapter 7.2 and 8.1) were similar in most cases (Fig. 2B). Only the samples MS and RB/2 showed higher toxicity by the mouse bioassay but not in the hepatocyte assay. The higher toxicity observed in both bioassays may indicate that the peaks in the lipophilic fraction of the chromatograms are indeed microcystins. However, enhanced toxicity of cyanobacterial cells in comparison to their content of microcystins is also conceivable due to synergistic or additive effects of other substances present in the investigated material (see Chapter 7). Therefore, both bioassays indicate that chromatographically determined

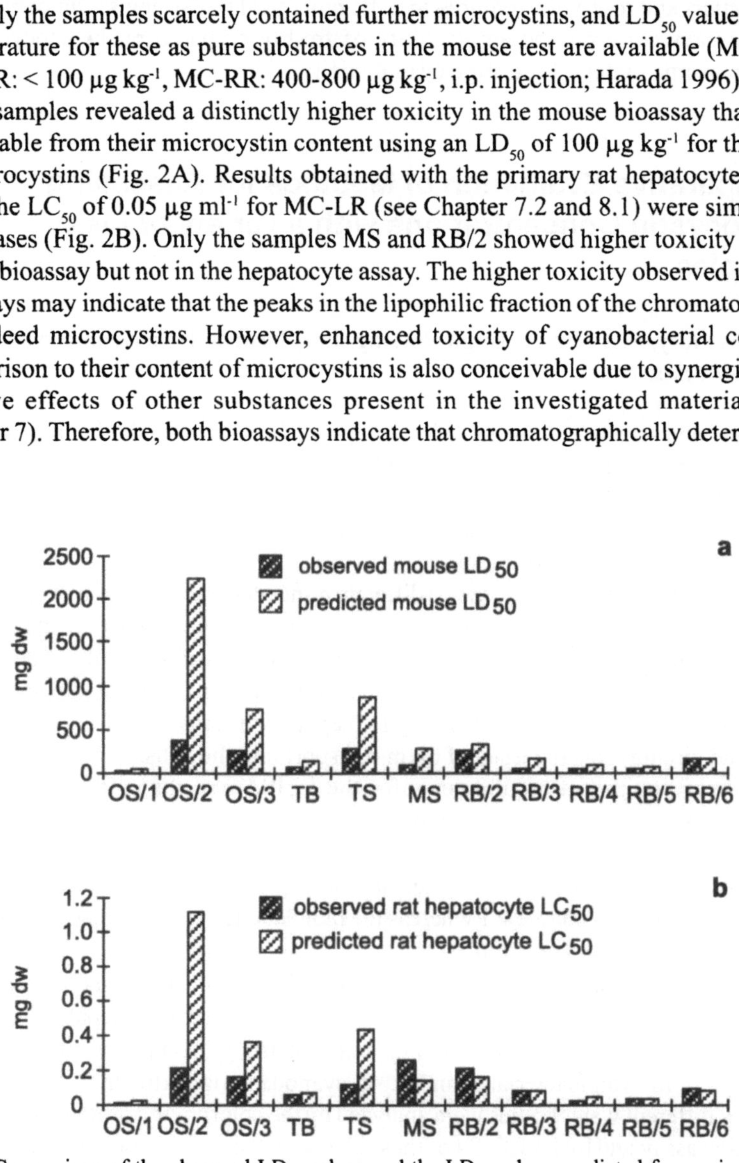

Fig. 2. Comparison of the observed LD_{50} values and the LD_{50} values predicted from microcystin analysis. (**a**) Mouse bioassay (LD_{50}=100 µg kg^{-1} BW). (**b**) Rat hepatocyte assay (LC_{50}=0.05 µg kg^{-1} BW)

toxin concentrations in natural cyanobacterial blooms may underestimate their toxic potential.

In face of the structural variability and different toxicity of microcystins (Harada 1996), the prevailing difficulties in their selective chromatographic determination (Moolan et al. 1996), and the potential for currently unknown cyanotoxins (see Chapter 7), an evaluation of harmful effects of natural cyanobacterial blooms in freshwaters and reservoirs requires specific biological test systems. Our comparison of bioassays confirms the applicability of freshly prepared rat hepatocytes as a suborganismic method for testing cyanotoxicity.

8.3 Comparative Evaluation of Methods for Assessing Microcystin Concentrations with a Variety of Field Samples

Thomas Flury, Rita Heinze, Birgit Wirsing, Jutta Fastner, Uwe Neumann and Jürgen Weckesser

For rapid and sensitive detection of microcystin, the enzyme-linked immunosorbent assay (ELISA, Nagata et al. 1997) and the protein phosphatase inhibition assay (Mackintosh and Mackintosh 1994) are promising. These biological methods are useful for first toxicity screening not only for cyanobacterial samples but also for the extracellular microcystins in water.

These methods of analysis use two different principles: the estimation of microcystin contents by ELISA requires antibodies against microcystin and is based on the structure of the microcystin molecule, while microcystin evaluation with the protein phosphatase inhibition assay is based on toxic effects of microcystins. In contrast to toxicity testing using the mouse assay and the primary rat hepatocyte assay, it acts at the molecular level and circumvents the issue of toxin transport into the cells.

Whereas equipment requirements for the ELISA and the colorimetric protein phosphatase assay are low, the expenses for the sensitive radiolabelled protein phosphatase assay are higher (e.g. for a liquid scintillation counter). A further difference between both protein phosphatase assays is that the colorimetric assay uses only protein phosphatase 1 whereas the radiolabelled assay as used here involves both protein phosphatase 1 and 2A.

Chemical analysis of total microcystins with HPLC (sum of all variants, analyzed as described in Chapter 9.1) and toxicity assays with primary rat hepatocytes served as a background for the evaluation of the colorimetric protein phosphatase assay for microcystin in cyanobacterial samples. 46 cyanobacterial extracts of field samples from six different water bodies were analyzed by the colorimetric protein phosphatase inhibition assay and by HPLC, 34 of these also with the rat hepatocyte assay. Further, for one of the lakes, Langer See, results of the radiolabelled protein phosphatase assay and of ELISA are also available for samples from June to October in 1996.

For the evaluation of methods for detection of extracellular microcystins in very low concentrations, a variety of water samples obtained from the screening program and samples from other sources, particularly from drinking-water supplies, were tested with ELISA and both protein phosphatase assays.

Protein Phosphatase Inhibition Assays

The protein phosphatase inhibition assays use the ability of microcystins to inhibit specifically the catalytic subunits of the serine/threonine protein phosphatases 1 and 2A (Mackintosh et al. 1990; Runnegar et al. 1981; Falconer and Yeung 1992). This inhibitory property leads to cytotoxic effects because protein phosphatases are important regulators of cellular processes.

The Colorimetric Protein Phosphatase Inhibition Assay

A colorimetric protein phosphatase 1 assay which employs p-nitrophenylphosphate as a substrate was reported to be approximately as sensitive as the radiolabelled assay (An and Carmichael 1994), but currently, little is known about its practicality. Recently, a colorimetric protein phosphatase 1 assay was reported to give reliable results in testing cyanobacterial field samples, detection limits being 10-20 µg L^{-1} (Ward et al.1997). Inhibition of protein phosphatase 1 activity by microcystins was measured with the method described by An and Carmichael (1994), taking advantage of the large capacity of microtitre plates. In order to measure different degrees of phosphatase inhibition, dilution steps of the extracts were prepared (10^{-1} to 10^{-5}). In routine assays 10 µl of extract dilution were incubated with 0.05 U protein phosphatase 1 (~0.6 nM, catalytic subunit of protein phosphatase 1, α-isoform, from rabbit muscle, Calbiochem, La Jolla, CA, USA) in 10 µl buffer for 5 min at 30°C. For measuring 100% phosphatase activity 10 µl water or 50% methanol were used instead of extracts. The reaction was initiated by the addition of 180 µl reaction buffer (50 mM Tris-HCl, pH 8.1, 20 mM $MgCl_2$, 0.2 mM $MnCl_2$, 1 mM DTT, 0.5 mg ml^{-1} BSA and 20 mMTris p-nitrophenylphosphate). The assay was run in triplicate for each dilution step of the extract. Microcystin-LR was chosen as a reference for the protein phosphatase 1 inhibitory activity of microcystins, and the microcystin contents of all samples were calculated as µg microcystin-LR equivalents g^{-1} DW. Standard concentrations of microcystin-LR (Calbiochem, La Jolla, CA, USA) were tested in each assay. The microtitre plates were read at 405 nm for 20 min with a Dynatech MR5000UV ELISA reader (Dynatech, Chantilly, Virginia, USA). For the calculation of half-maximal inhibition concentrations (IC_{50}) the GRAFIT software (Erithacus Software Ltd., Staines, UK) was used. IC_{50} values of cyanobacterial samples and microcystin standards were estimated by plotting percent activity of protein phosphatase 1 vs the logarithm of the sample dry weight (DW) contained in different extract dilutions (g ml^{-1}). Since the assay measures the total protein phosphatase 1 inhibition without differentiating between structural variants of microcystin, the concentrations of the toxins were expressed in microcystin-LR equivalents.

The ^{32}P-Radiolabelled Protein Phosphatase Inhibition Assay

For the detection and quantification of microcystin, a protein phosphatase assay using ^{32}P-radiolabelled phosphorylase as substrate proved to be highly sensitive (Lambert et al. 1994). The assay in kit form from Gibco, Life Technologies, Eggenstein, Germany, was employed. This protein phosphatase assay system is designed for the assay of protein phosphatases 1 and 2A. The specific substrate for type 1 and 2A phosphatases, ^{32}P-labelled glycogen phosphorylase-a was prepared by reacting [^{32}P] ATP (NEN, Köln, Germany) with purified phosphorylase-b in the presence of purified phosphorylase kinase. As a source of protein phosphatases 1 and 2A, crude extracts were prepared by homogenizing a rat liver in 50 mM Tris-HCl pH 7.2 containing 0.1 mM EGTA, 0.1% 2-mercaptoethanol and 200 mM mannitol. Phosphatase activity with glycogen phosphorylase-a was determined by the quantification of released [^{32}P] with a liquid scintillation counter (1220 Quantulus Wallac Oy, Turku, Finnland). The inhibition of protein phosphatases was measured after incubation with various concentrations of microcystin-LR (Calbiochem-Novabiochem, Bad Soden, Germany). Microcystin contents were determined by plotting percent inhibition versus the logarithm of microcystin concentration. Then microcystin contents were graphically estimated in the linear range.

ELISA

The 96-well microtitre plate Enviro Gard kit (Coring-System, Gernsheim, Germany) was used as a quantitative test for the detection of microcystins. The kit makes use of polyclonal antibodies which bind either microcystins or a microcystin-enzyme conjugate. Microcystins in a sample compete with the microcystin-enzyme conjugate for a limited number of antibody-binding sites. The colour after enzymatic reaction is inversely proportional to the microcystin concentration. The kit is calibrated with a non-toxic microcystin surrogate at levels equivalent to 0.1, 0.4 and 1.6 $\mu g\ L^{-1}$. Microtitre plates were measured in a microtitre plate reader (SLT,Crailsheim, Germany) at 450 nm. For calculating the results, % B_0 (negative control 100%) were calculated for each set of calibrators or samples and represented graphically against their microcystin concentration at a semilog scale. The microcystin concentration of each sample was read on the graph.

Results for Lyophilized Cell Extracts

Extraction of cells was performed as for HPLC-analysis (see Chapter 9.1). Microcystin contents calculated on the basis of protein phosphatase 1 inhibition in the colorimetric assay showed a close relationship to those determined with HPLC analysis (Fig. 1A), the coefficient R^2 being 0.8782. The slope of 0.88 indicates a slight underestimation of total microcystin content with this assay in relation to the results determined by HPLC.

A similar trend is evident in the results of the rat hepatocyte assay – a measure for microcystin toxicity involving a different level of biological organization. The reciprocals of the LC_{50} values correlate closely to the total microcystin content determined by HPLC analysis (R^2=0.8142 after 4 hs of exposure and 0.8865 after 20 h; Fig. 1B, upper and lower panel).

The results of both assays – the protein phosphatase 1 assay and the rat hepatocyte assay – correlate closely with each other (R^2=0.8108) after 4 h of exposure (Fig. 1C, upper panel), and slightly less closely after 20 h of exposure (R^2=0.7585; Fig. 1C, lower panel), possibly because further effects occur at longer exposure times.

All the cyanobacterial samples containing microcystins as revealed by HPLC analysis also proved to be inhibitory in the colorimetric protein phosphatase assay. Although protein phosphatase 1 inhibitors others than microcystins are well known (Takai and Mieskes 1991; Honkanen et al. 1994) and may occur in cyanobacterial field samples, no false positive results were obtained in the colorimetric protein phosphatase 1 inhibition assays compared to HPLC analysis (above the detection limit of HPLC).

Furthermore, the protein phosphatase 1 inhibition assay may detect quantities below the HPLC detection limit. The microcystin contents of all samples were calculated on the basis of 50% PP-1 inhibition (IC_{50}). For the chosen assay conditions a 50% inhibition of protein phosphatase 1 activity was typically achieved by 0.2 ng microcystin-LR (IC_{50}: 1 ng ml^{-1} reaction volume) resulting in an estimation limit for the 10-µl aliquots of methanolic extracts of 20 ng ml^{-1} (= 20 µg L^{-1}). Compared to HPLC

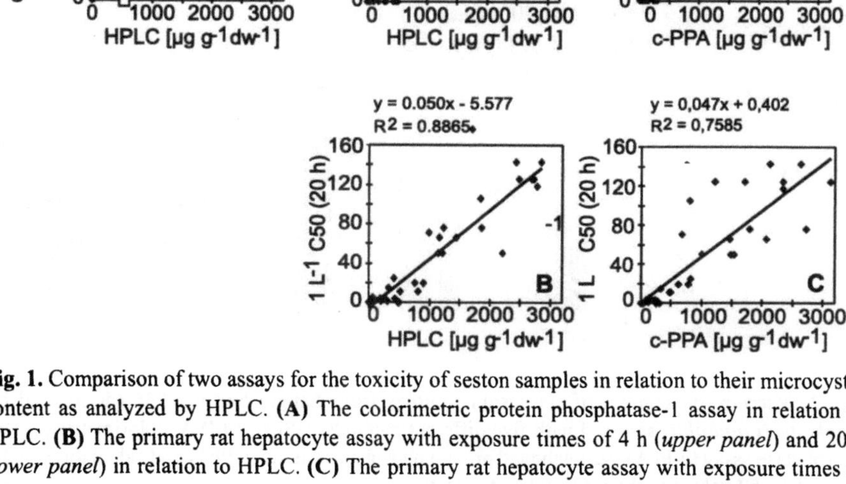

Fig. 1. Comparison of two assays for the toxicity of seston samples in relation to their microcystin content as analyzed by HPLC. **(A)** The colorimetric protein phosphatase-1 assay in relation to HPLC. **(B)** The primary rat hepatocyte assay with exposure times of 4 h (*upper panel*) and 20 h (*lower panel*) in relation to HPLC. **(C)** The primary rat hepatocyte assay with exposure times of 4 h (*upper panel*) and 20 h (*lower panel*) in relation to the colorimetric protein phosphatase1 assay

analysis, the sensitivity of the colorimetric protein phosphatase inhibition assay is about 20-fold higher (injection of 50 µl at a detection limit for field samples of 20 ng per injection results in 400 µg L^{-1} for lysed field sample without concentration steps). Comparative analysis of the six cyanobacterial samples from Langer See during the vegetation period of *Planktothrix agardhii* in 1996 with five different methods, HPLC, ELISA, the colorimetric protein phosphatase assay, the radiolabelled protein phosphatase assay and with the hepatocyte assay, demonstrate the same trend of seasonal development with a maximal microcystin content of seston in early September of several mg g^{-1} DW (Fig. 2). For these samples, correlations are very close between the results of both the ELISA and the rat hepatocyte assay in relation to the colorimetric protein phosphatase 1 assay, and between radiolabelled protein phosphatase assay in relation to HPLC (Fig 3). Particularly in the range of concentrations above 1000 µg g^{-1} DW, comparisons between other methods show more scatter, though coefficients of determination (square of correlation coefficient) are never below 0.6 and all of them are significant at least at the level of 0.03.

Furthermore, the slopes of these regressions indicate enhanced reactions of both the hepatocytes and the ELISA in relation to both protein phosphatase assays and reduced reactions of both protein phosphatase assays in relation to HPLC. However, deviations in both directions were never by more than a factor of 2. Other slopes were close to 1, indicating very similar results of the respective two methods. Thus, these data show that all five methods provided comparable results for estimation of the microcystin content of the cyanobacterial population in Langer See.

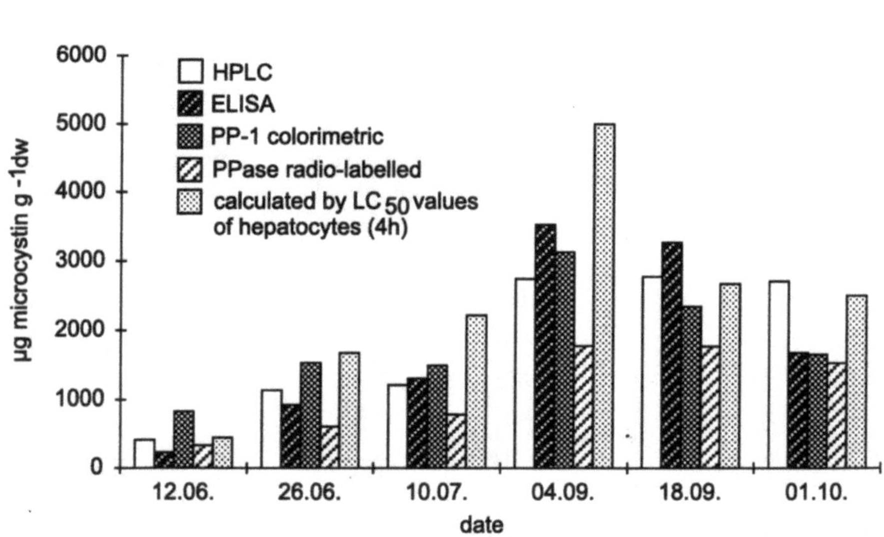

Fig. 2. Microcystin content of six seston samples from Langer See in 1996 (dominated by *Planktothrix agardhii*) assessed with five different methods: HPLC, ELISA, colorimetric protein phosphatase assay (c-PPA), radiolabelled protein phosphatase assay (r-PPA) and calculation on the basis of LC50 values in the rat hepatocyte assay

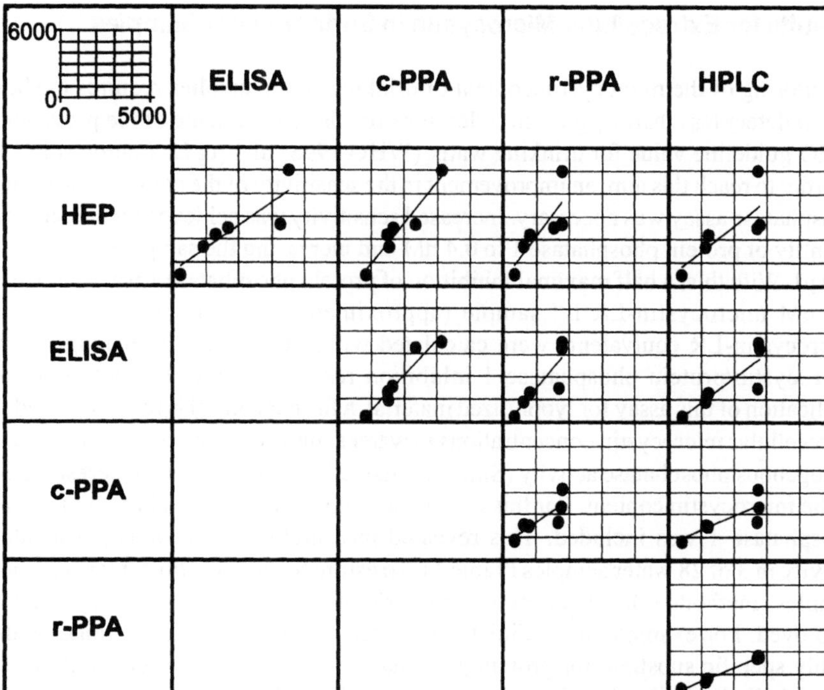

	ELISA	c-PPA	r-PPA	HPLC
HEP	sl: 1.01 ip: 588 R^2: 0.7750 sig: 0.0011	sl: 1.80 ip: 869 R^2: 0.9175 sig: 0.0013	sl: 1.94 ip: 225 R^2: 0.6707 sig: 0.024	sl: 1.14 ip: 335 R : 0.6161 sig: 0.032
ELISA		sl: 1.58 ip: 1094 R^2: 0.9182 sig: 0.0013	sl: 1.76 ip: 242 R^2: 0.7675 sig: 0.011	sl: 1.10 ip: 208 R : 0.7380 sig: 0.0014
c-PPA			sl: 1.00 ip: 675 R^2: 0.6684 sig: 0.023	sl: 0.63 ip: 684 R^2 : 0.6555 sig: 0.025
r-PPA				sl: 0.63 ip: 3 R^2: 0.9903 sig: 0.000018

Fig. 3. Regression analysis for the data shown in Fig. 2. *sl* slope; *ip* x-axis intercept

Results for Extracellular Microcystin in Surface Water Samples

Monitoring of the microcystin content of drinking water supplies requires methods which detect less than 1 µg L^{-1} in order to be reliable in the range of the provisional WHO guideline value for drinking water (WHO 1998) of 1 µg L^{-1} microcystin-LR. In order to reach this aim, an improvement in the sensitivity of the colorimetric protein phosphatase assay was necessary. Increased sensitivity was achieved by reducing the quantity of protein phosphatase 1 to 0.4 nM and increasing the sample volume up to 200 µl. With this, a half maximal inhibition of protein phosphatase 1 was attained by 0.8 nM microcystin-LR L^{-1} sample (approximately equivalent to 0.8 µg L^{-1}). Microcystin-LR equivalents were calculated within the small linear range of the microcystin/protein phosphatase 1 inhibition response curve. In contrast to the application of this assay for lyophilized material, a further control is recommended for extracellular microcystin concentrations in water samples because these may contain endogenous phosphatase activity (Sim and Mudge 1993), causing an underestimation of the microcystin content. For this control, a sample without the addition of protein phosphatase 1 was included. This revealed measurable endogenous phosphatase activity in 3 of 28 water samples (Table 1). To estimate microcystin contents in these samples, methods other than the colorimetric protein phosphatase assay must be employed. For example, the radiolabelled protein phosphatase assay works with a highly specific substrate for protein phosphatases 1 and 2A. The sensitivity of the radiolabelled protein phosphatase assays depends on the dilution of the unspecific

Table 1. Estimation of microcystin contents in water samples with the ELISA and the protein phosphatase inhibition assay (colorimetric PP-1 and radiolabelled PP-1 and 2A)

Waterbody	Date	ELISA (µg L^{-1})	Protein phosphatase assay (PP-1, colorimetric.) (µg L^{-1})	Protein phosphatase assay (PP-1 and 2-A; radio-labelled) (µg L^{-1})
Langer See	10.10.95	1.60	2.4	m.v
	20.11.95	16.00	12.0	10.8
	29.05.96	0.22	<0.3	m.v.
	12.06.96	0.17	<0.3	m.v.
	24.07.96	0.60	m.v.	0.4.
	07.08.96	1.25	m.v.	0.3
	21.08.96	0.83	m.v.	0.3
	18.09.96	0.41	m.v.	<0.3
	01.10.96	0.55	m.v.	0.4
	16.10.96	0.31	m.v.	<0.3
	29.10.96	0.70	m.v.	0.4
	13.11.96	0.55	m.v.	

Table 1. (continued)

Waterbody	Date	ELISA ($\mu g\ L^{-1}$)	Protein phosphatase assay (PP-1, colorimetric.) ($\mu g\ L^{-1}$)	Protein phosphatase assay (PP-1 and 2-A; radio-labelled) ($\mu g\ L^{-1}$)
Zeuthener See	12.07.95	24.00	2.6	15.6
	22.07.96	1.90	2.6	1.6
	22.07.96[+]	7.50	>10.0	3.9
	16.09.96	1.60	2.9	m.v.
	30.09.96	1.50	2.5	m.v.
	06.09.96	16.00	n.d.a	10.2
	04.10.96	>16.00	2.3a	15.2
Weida Reservoir	15.07.96	2.20	<0.3	<1.5
	27.08.96	2.00	<0.3	m.v.
	09.09.96	2.70	<0.3	<1.5
	23.09.96	2.50	<0.3	m.v.
	30.09.96	1.60	<0.3	<1.5
Schmaler Luzin	22.06.95	4.50	n.d.a	<1.5
JFH Steglitz	14.07.97	1.60	4.4	m.v.
Wannsee Strand	14.07.97	2.80	4.4	1.8
Dollgowsee	11.09.97	>1.60	7.0	m.v.
Wolziger See	15.08.95	0.16	<0.3	m.v.
	29.05.96	0.16	<0.3	m.v.
	12.06.96	0.21	<0.3	m.v.
	26.06.96	0.30	0.35	m.v.
	10.07.96	0.18	<0.3	m.v.
	24.07.96	0.34	0.26	m.v.
	07.08.96	0.34	<0.3	m.v.
	21.08.96	0.22	<0.3.	m.v.

[a] endogenous protein phosphatase activity in samples
n.d. not detectable because of further phosphatase activity in the sample
m.v. missing value

source of protein phosphatases 1 and 2a , for example rat liver homogenates. With sufficient dilution, microcystins can be detected in concentrations down to 0.3 μg microcystins L^{-1} water. Using the ELISA, minimal detectable concentrations were 0.1 μg microcystins L^{-1} water (Table 1).

As for the cell material samples, no false positive results were obtained for extracellular microcystin in water samples with either protein phosphatase assay as compared to the ELISA. Table 1). Further, the results for those samples with positive results correlate closely in most cases (Fig. 4), particularly for the radiolabelled protein phosphatase 1 and 2A assay (R^2=0.9946, Fig. 3B). However, the ELISA often indicated higher microcystin concentrations than the radiolabelled protein phosphatase 1 and 2A assay, whereas this effect was usually only slight with the colorimetric protein phosphatase 1 assay. One sample (Zeuthener See, 12. 7. 1995) showed a dramatically higher concentration by ELISA of 24 µg L^{-1} (and was excluded from the regression in Fig. 3A). Whereas the results of the radiolabelled protein phosphatase 1 and-2A assay reflected this rather adequately with 15.6 µg L^{-1}, the colorimetric protein phosphatase assay indicated only 2.6 µg L^{-1}.

Conspicuously, the ELISA did not provide negative results in any of the 28 water samples shown in Table 1, but rather indicated concentrations of approximately 2 µg L^{-1} in all five samples from the Weida Reservoir, even though one or both protein phosphatase assays did not detect microcystin at levels of 0.3-1.5 µg L^{-1}. Here the protein phosphatase assay could help to evaluate potentially false positives of the ELISA. These effects are potentially derived from non-toxic microcystin variants or of a non-specific reaction of the ELISA with structural similar compounds in water.

In conclusion, for the investigation of extracellular microcystins in concentrations lower than 1 µg L^{-1}, the ELISA and both protein phosphatase assays are suitable as

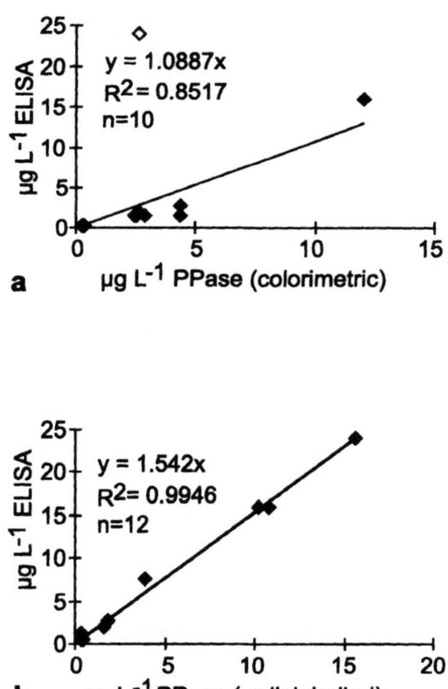

Fig. 4. Comparison of extracellular microcystin content in water samples of a variety of lakes determined with the colorimetric protein phosphatase 1 and the radiolabeled protein phosphatase 1 and 2A assay to results of the ELISA

indicating tests. The ELISA is the most sensitive method and simple in practice, but has a potential for false positive reactions. The colorimetric protein phosphatase assay was improved to be useful for a first screening and rough estimation of microcystin contents in field samples. If the equipment is available, the radiolabelled protein phosphatase assay is sensitive and safe. Both protein phosphatase assays provide preliminary information on the toxicity of the microcystins in comparison to the microcystin content measured by ELISA. Nevertheless, HPLC analysis (confirming test) will remain necessary for confirmation.

8.4 A Fish-Embryo Bioassay for the Assessment of Cyanobacterial Toxicity

Axel Oberemm

Introduction

As mentioned in the preface to this chapter, fish-embryo tests have been widely used for the screening of xenobiotic effects in the aquatic environment. In order to assess hazards of chemicals in aquatic ecosystems, tests with fish as representatives of several trophic levels are generally recommended and accepted (e.g. OECD Guidelines 203, 204, 210, 215). Tests for the assessment of aquatic toxicity and environmental health using early life stages of fish have been applied and refined since decades (McKim 1977; Birge et al. 1985; van Leeuwen et al. 1990). Compared to tests using juvenile and adult fish, embryo assays are more suitable to retrieve information on specific effects, because they offer diverse endpoints to evaluate and are more sensitive against chemical stress (von Westernhagen 1988), which may be due to the large relation between body surface and volume of embryos, their thin epithelia and the vulnerability of developmental processes. Also regarding animal welfare, embryos are preferable to juvenile and adult fish, because chemicals which affect adult fish generally will cause gross effects during early embryonic development, when the central nervous system has not yet developed. In order to obtain eggs in constant quality at every time of the year, the use of tropical fish is required. Zebrafish (*Danio rerio*), originating from northern India, are selected as a suitable species, because this small fish is easy to maintain and breed. Compared to many other species, including most species native to Europe, the experimental handling of zebrafish eggs is much easier, because they are not sticky and translucent, which is essential for the selection and transfer of the eggs into the incubation dishes and for the visibility of the embryonic development. As a member of the family of Cyprinidae, the zebrafish is closely related to the most abundant species of Europe, e.g. roach (*Rutilus rutilus*) and bream (*Abramis brama*), which are the dominating species in lakes and rivers of the lowlands, so that results may have some significance also for fish species native to Europe. Due to the increasing use of zebrafish embryos in molecular and developmental biology in the recent years, much information, e.g. on mechanistic relations of developmental processes is accessible and can contribute to the understanding of toxic mechanisms. Formerly, other aquatic species were used in order to detect cyanobacterial toxicity, e.g. luminescent bacteria (Microtox, Campbell et al. 1994); brine shrimp (*Artemia* spp., Campbell et al. 1994; Reinikainen et al. 1995; Feuillade et al. 1996), *Daphnia* spp. (Jungmann 1992; Reinikainen et al. 1995), *Thamnocephalus platyurus* (Törökné 1998; Thamnotoxkit F) and mosquito larvae (*Aedes aegypti*, Kiviranta and Abdel Hameed 1994). Results concomitantly demonstrated that aquatic species are relatively tolerant to pure microcystins, but are sensitive to extracts from cyanobacteria, more or less independently of the concentration of microcystins (see also Oberemm et al. 1999 and Chapter 6.2 and 7.2).

Materials and Methods

Fish Keeping and Breeding

Since the method described by Creaser (1934), many techniques have been described how to maintain zebrafish and to produce the eggs in order to obtain embryonic stages. A lot of technical and scientific information on zebrafish is accessible from Detrich et al. (1998) and via Internet from ZFIN (Zebrafish Information Network at http://zfish.uoregon.edu/) which is maintained by the Institute of Neuroscience, University of Oregon. For technical information refer to The Zebrafish Book/ University of Oregon, Institute of Neuroscience at http://zfish.uoregon.edu/zf_info/ zfbook/zfbk.html (Z FIN homepage).

Embryo Test

Eggs of zebrafish (*Danio rerio, Brachydanio rerio* Hamilton-Buchanan) are incubated in polystyrene petri dishes (Greiner; Cat. No. 627 102, 35 x 15 mm, 3 ml medium) from blastula stage up to the end of the embryonic period (terminology according to Balon 1975). The incubation is performed semistatically at 26°C, which means that the test medium is changed daily. Reconstituted water according to ISO 7346/1 (1984) is used to keep controls and for diluting the test compounds (11.76 g $CaCl_2$ x $2H_2O$, 4.93 g $MgSO_4$ x 7 H_2O, 2.59 g $NaHCO_3$, 0.23 g KCl, each filled up to 1 L with pure H_2O to make a stock, mix 25 ml of each stock solution and fill up to 1 L pure H_2O). For each test compound a suitable solvent is used when necessary (e.g. MeOH, acetone, DMSO), applied in a safe concentration, which must be demonstrated by using solvent controls. Each dish contains ten eggs and each treatment group plus control consists of four replicates. At distinct phases of development (late epiboly - 8 h, early - 24 h and late - 32 h organogenesis, hatching period - 55 h, posthatch phase, eleuthero-embryo - 75 h and early larva - 144 h (Fig. 1; Table 1; terminology according to Kimmel et al. 1995) suitable endpoints are recorded (developmental speed, malformations,

Table 1. Steps of a zebrafish-embryo test: developmental age, developmental stage and features, and corresponding specific endpoints for evaluation

Age (h)	Normal stage[a]	Description/ features	Specific endpoints
4	Late high blastula	Elevated cap of regular small cells on top of yolk	(Start of exposure)
8	75% epiboly	Late gastrula, epiboly in progression, embryo still spheric, keel-like embryonic shield visible	Progression of growth front, homogeneity of front and tissue
24	Prim 6	Basic vertebrate body organization, beginning of heart beat, melanophores in the eye, spontaneous movements	Heart beat visible
32	Prim 16	Blood circulatory system, melanophores spread in the trunk	Heart rate, malformations of the blood circulatory system, edema
55	Pec fin	Hatching phase, melanophores pronounced, hatching starts	Hatching rate, heart rate, malformations of the blood circulatory system, edema
80	Protruding mouth	Eleuthero-embryo: resting on side, yolk present, xantophores appeared (yellowish gleam)	Heart rate, expression of pigmentation, finfold lesions
144	Early larva	Early larva: swim bladder inflated, swimming, feeding	Swimming performance, feeding success

[a] Kimmel et al. (1995)

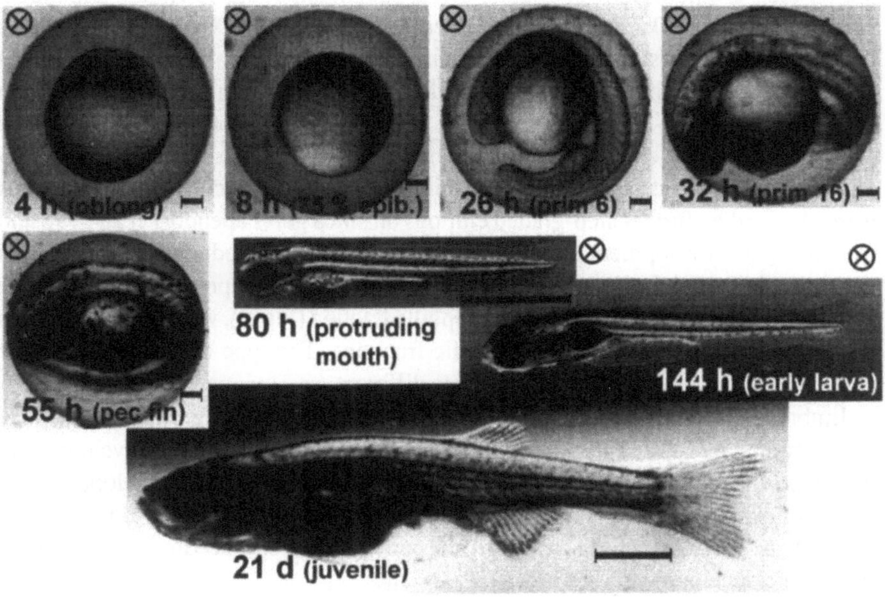

Fig. 1. Steps of a zebrafish-embryo test: developmental age, developmental stage and features, and corresponding specific endpoints for evaluation. Reprinted from Lab Animal 29, p. 35

heart rate, hatching, motility/ swimming ability, feeding, growth, mortality). Exposure is terminated after 6 days, when control embryos have resorbed their yolk, started swimming and are able to feed, which means that the embryonic period is completed. The ability to feed is controlled by adding large *Paramecia* sp. and recorded as a last end point and an essential component of larval viability.

For detection of chronic effects, pretreated larvae are transferred to uncontaminated aquaria to monitor further development and growth. The different treatment groups are kept separate by using meshed chambers. The tank water is circulated and filtered using biofilm coated lava pebbles. Larvae are fed using TetraMin® flakes grounded in liquid nitrogen and newly hatched brine shrimp. At the age of 21 days, when larval development is completed in most specimens (Fig. 1), survival rate, body length and weight are recorded. However, this extension of the short-term test is not applicable for high-throughput screening. In many countries regulatory practices have been defined for the advanced life stages of this fish-embryo test.

Results and Discussion

Results on effects of microcystins and of crude extracts of cyanobacteria on fish and amphibian embryos are presented in Chapter 6.1 and 7.1, respectively. In this chapter, a short example is given how a zebrafish embryo test can be used to detect toxic compounds in fractions of a cyanobacterial crude extract by bioassay-directed fractioning.

Aqueous crude extracts from bloom-forming cyanobacteria have been shown to exhibit adverse effects on aquatic organisms independently of concentrations of known toxins such as microcystins (Chapter 7 and Jungmann 1992; Kiviranta and Abdel Hameed 1994; Reinikainen et al. 1995; Feuillade et al. 1996; Oberemm et al. 1999). In order to obtain first insights into the nature and localization of toxic compounds that must be hypothesized to explain these effects, lyophilized material from *Aphanizomenon flos-aquae* (Kimmel et al. 1995), whose aqueous extracts have been shown to be highly embryotoxic (Oberemm et al. 1997), was eluted successively with methylenchloride and methanol. After this first step of fractioning, subfractions were obtained by further step-graded elutions on silica gel (Papendorf et al. 1998 and Chapter 7.2). Development of zebrafish embryos was inhibited by fractions of both the hydrophilic and the lipophilic solvent. Mueggelone and lupenyl acetate could be identified as active compounds in the lipophilic fraction (Fig. 2), whereas no structures could yet be identified in the hydrophilic fractions. The type of effects caused by mueggelone and lupenyl acetate was quite different.

Embryos exposed to mueggelone at a concentration of 10 $\mu g\ ml^{-1}$ showed 45% mortality after 32 h. The surviving embryos demonstrated considerable developmental retardation, since after 24 and 32 h no blood circulatory system had developed. After 3 days of exposure, the survivors showed edema in the heart region and thrombosis (Fig. 2). At a concentration of 1 $\mu g\ ml^{-1}$, mueggelone had no obvious effects on embryonic development.

In contrast, lupenyl acetate had no effect on early embryonic development. However, after a comparably extended period of 6 days and higher concentration of 100 µg ml^{-1}, embryos had also developed edema in the heart region (Fig. 2) and additionally dorsal tail bending (Lordosis, 30% of specimens). In addition, the yolk resorption was inhibited and the swim bladder was not inflated (Fig. 2). At a concentration of 10 µg ml^{-1}, lupenyl acetate had no visible effects on zebrafish development.

Testing for antimicrobial and antialgal activity did not indicate any effects of the mentioned compounds. This demonstrates the sensitivity of the fish embryo bioassay.

Compared to other common bioassays, fish embryos show much more diverse endpoints for the evaluation of effects and offer more favourable conditions to use them as a diagnostic tool for the assessment of toxicity in aquatic systems. In spite of the potential of fish to have adapted to cyanobacteria, embryos react strongly to their metabolites, and the fish embryo assay is valuable in identifying substances of potential ecotoxicological relevance.

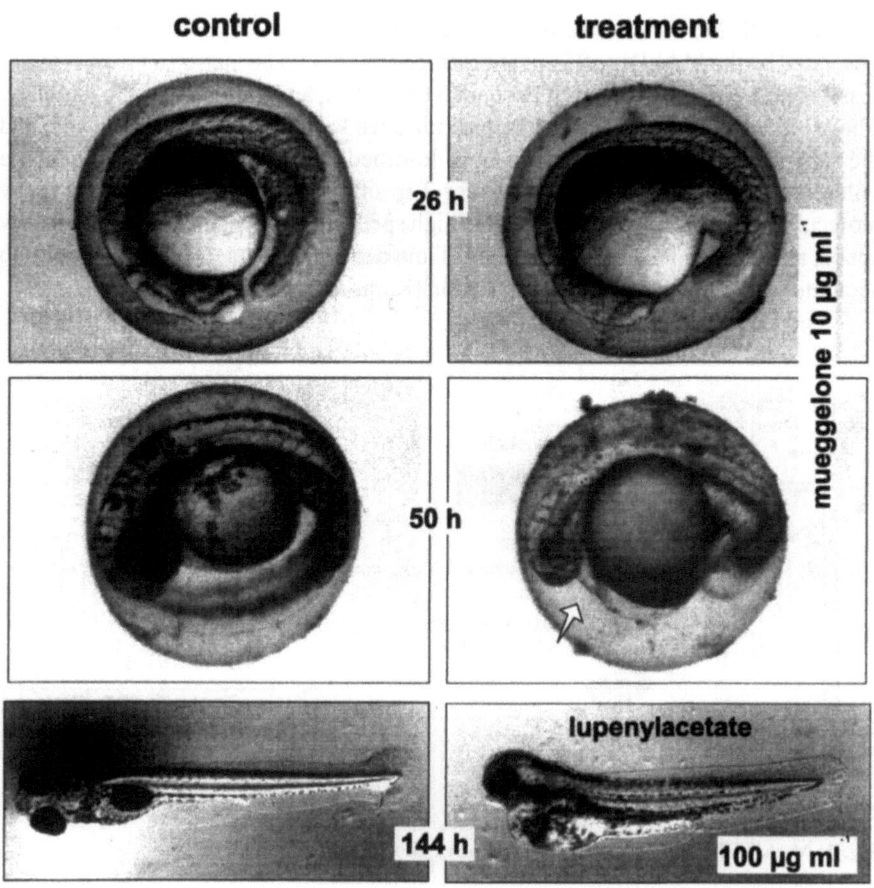

Fig. 2. Developmental disturbances of zebrafish due to exposure to cyanobacterial substances

8.5 Rapid Typing and Structure Determination of Cyanobacterial Peptides Using MALDI-TOF Mass Spectrometry

Marcel Erhard, Hans von Döhren and Peter Jungblut

Matrix assisted laser desorption/ionization – time of flight mass spectrometry (MALDI-TOF-MS) has become an essential tool for the analysis of biomolecules, especially peptides. Accurate molecular weight information can be determined in minutes on extremely small quantities of complete cells, extracts or fractions of further purification steps.

MALDI-TOF-MS Technique

MALDI-TOF-MS makes use of a single principle to determine the mass of ions generated by the MALDI soft ionization process (Fig. 1). The matrix, a low molecular mass organic acid, is mixed with the analyte solution and prepared on the template. The p-electrons absorb the laser photons (nitrogen laser giving a 337 nm output) and the matrix with the analyte molecules is desorbed. The ions are accelerated with a voltage of 20 kV and are detected after 2 m of flight path on the detector. Low molecular weight ions are accelerated to high speed, high molecular mass ions to low speed, and time dependence is detected. Time dependence detection is converted to accurate masses using previous calibriation (Karas and Hillenkamp 1988).

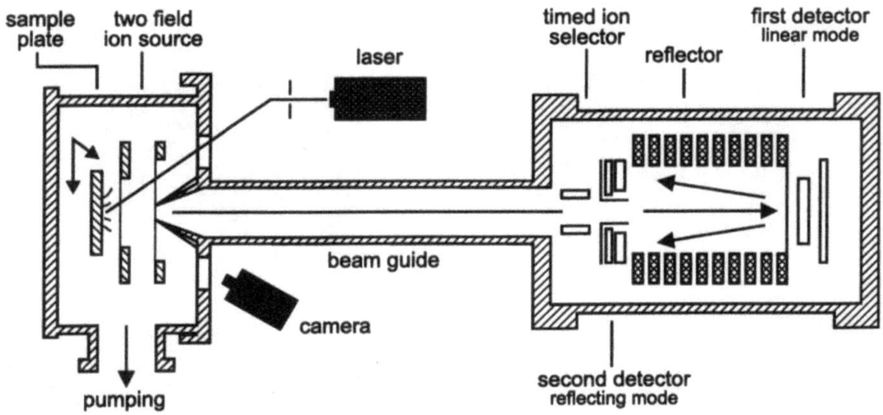

Fig. 1. MALDI-TOF principle (from PerSeptive Biosystems)

Peptide Detection of Intact Cyanobacterial Cells

Cyanobacteria produce many linear and cyclic peptides containing unusual amino acids, fatty acids, polysaccharides, polyketide units, methyl groups, SO_3^- or NO_2^- groups and other derivatives with unusual bonds (e.g. beta- or gamma-amino acid bonds).

Currently more than 60 microcystins (Watanabe et al. 1996), 10 anabaenoeptins (Erhard et al. 1997; Erhard et al. 1999), 10 aeruginosins/spumigins (Fujii et al. 1997), 8 microcolins/mirabimides (Koehn et al. 1992), 10 microginins (Ishida et al. 1997), 30 cyanopeptolins / micropeptins / aeruginopeptins / anabaenopeptilides / nostopeptins (Weckesser et al. 1996) and 8 microviridins (Murakami et al. 1997) are known.

Figs. 2 and 3 show the typical detection of peptides directly from 5 µg of lyophilized cell material or directly from cyanobacterial field samples. All peptides from cyanobacteria were detected as protonated molecular ions $[M+H]^+$ directly from the unfractionated cells.

The spectra illustrate the potential of MALDI-TOF mass spectrometry for the direct detection of known peptides, toxins and additional unknown substances.

Preliminary results indicate that mass developments of cyanobacteria in German water bodies may contain more than 50 peptides. The major compounds can be detected directly from field samples with high cyanobacterial density, particularly from bloom material. For the analysis of compounds in low concentration, purification is necessary. After methanolic extraction and separation by HPLC these compounds can be analyzed by MALDI-TOF mass spectrometry. Because of the high sensitivity of MALDI-TOF, enough material can be collected from a single semipreparative HPLC run to determine the molecular mass of the major peptide peaks. Thus, peptide analysis,

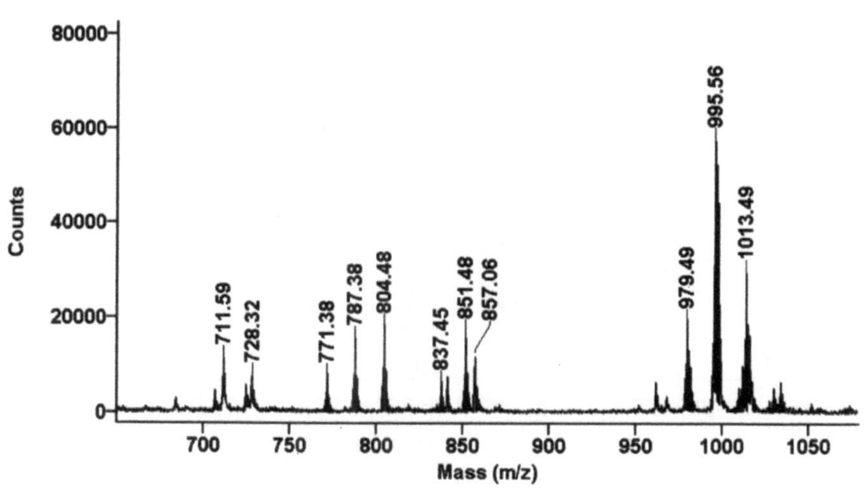

Fig. 2. MALDI-TOF mass spectrum from 5 µg dry water bloom cells from Lake Wannsee 1997, Germany. The water bloom cells produce the microcystin-LR (m/z=995), a new microcystin-LR variant (m/z=1013), a new cyanopeptolin W (m/z=979), anabaenopeptin B (m/z=837) and F (m/z=851) and other unknown compounds (m/z=771, 787, 804), m/z=711 and 728 are fragments

particularly for microcystins, may utilize two different methods: first, purification according physical properties with HPLC and second, analysis of molecular masses by MALDI-TOF. Table 1 shows the analyses of 11 different microcystins with this

Table 1. Analysis of 11 different microcystins by HPLC and MALDI-TOF in a cyanobacterial bloom sample from Wannsee, Germany (4 October 1995), dominated by *Planktothrix agardhii* with some accompanying *Microcystis* spp.

HPLC Retention time (min)	MALDI-TOF $[M+H]^+$	Microcystin
14.3-14.8	1024	$[Asp^{(3)}]$ MC-RR
15.1-15.5	1024	$[Asp^{(3)}, Dhb^{(7)}]$ MC-RR
15.8-16.6	1038	MC-RR
15.8-16.6	1024	$[Dha^{(7)}]$ MC-RR
19.7-20.3	1031	$[Dha^{(7)}]$ MC-YR
19.7-20.3	1045	$[Asp^{(3)}]$ MC-HtyR
20.9-21,3	981	$[Asp^{(3)}]$ MC-LR
21.3-21.8	995	MC-LR
21.3-21.8	967	$[Asp^{(3)}, Dha^{(7)}]$ MC-LR
23.1	1031	$[Asp^{(3)}]$ MC-YR
23.4	1015	$[Asp^{(3)}]$ MC-FR

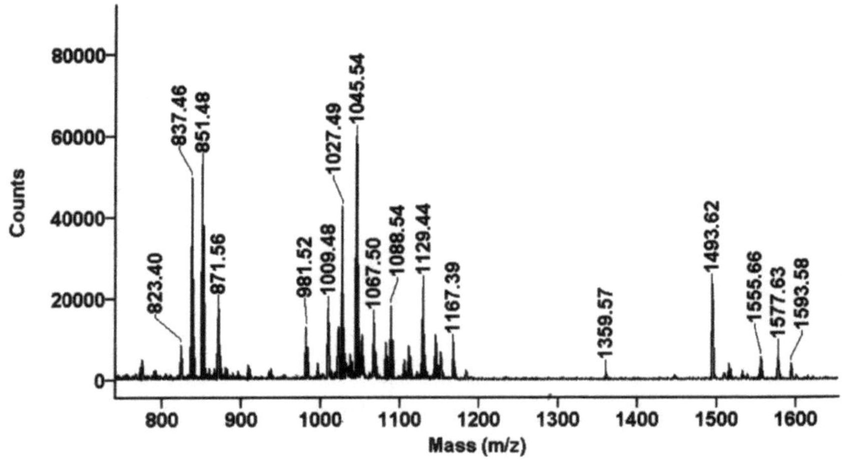

Fig. 3. MALDI-TOF mass spectrum of a *Planktothrix rubescens* srain isolated from Wahnbachtalsperre, Germany. The spectrum shows anabaenopeptin B (m/z=837), anabaenopeptin F (m/z=851), a new anabaenopeptin (m/z=909), chlorophyll-a (m/z=870), $[Asp^{(3)}]$ microcystin-LR (m/z=981), microcystin-LR (m/z=995), $[Asp^{(3)}]$ microcystin-HtyR (m/z=1045) and other unknown signals

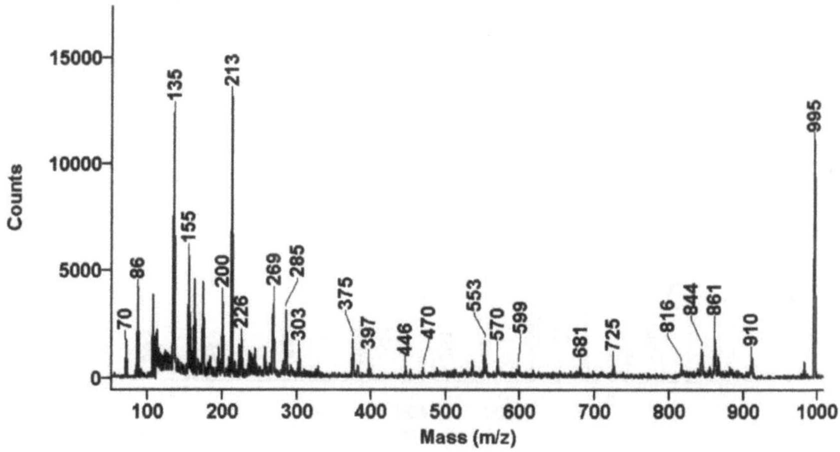

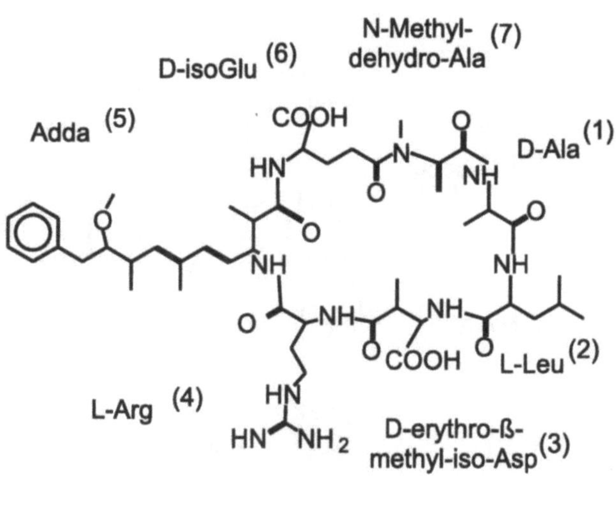

Fig. 4. MALDI-TOF-PSD fragment spectrum from microcystin-LR (m/z=995) directly from 5 µg dry water bloom cells. The typical fragments are: Mdha immonium ion (m/z=56), Arg immonium ion (m/z=70), Leu immonium ion (m/z=86), Adda side chain (m/z=135 and 163), Mdha+Ala+H (m/z=155), Ala+Leu+H (m/z=185), Glu+Mdha+H (m/z=213), Arg+DmiA fragments (m/z=244 and 268), Glu+Mdha+Ala fragments (m/z=329) and Arg+Adda+Glu+Mdha (m/z=680). All fragments show the typical finger print fragments and the complete cyclic structure of the microcystins

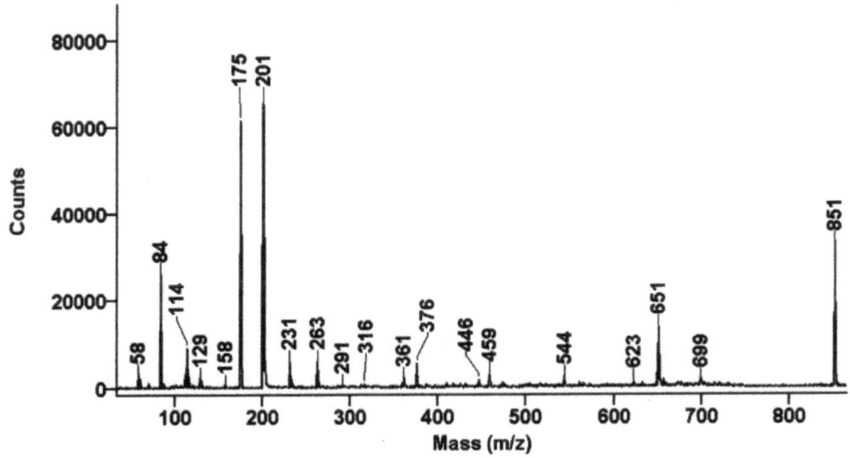

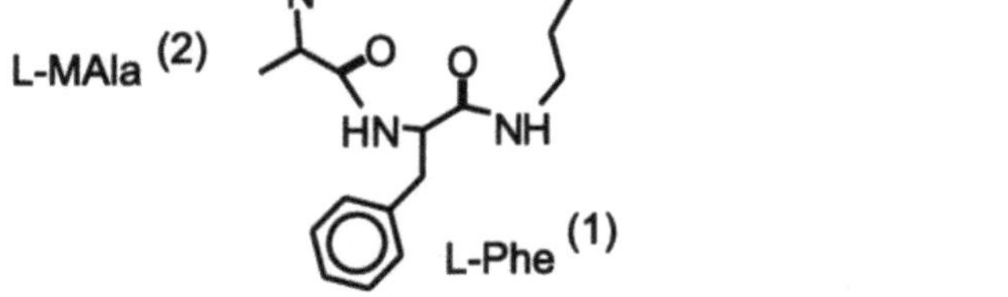

Fig. 5. MALDI-TOF-PSD fragment spectrum from anabaenopeptin F (m/z=851) directly from 5 µg dry cells. The spectrum shows the other typical finger print pattern of the anabaenopeptins with typical fragments: complete ring without side chain (m/z=651), the side chain CO+Arg (m/z=200), Arg from side chain (m/z=175), MAla immonium ion (m/z=58), Arg immonium ion (m/z=70), Lys immonium ion (m/z=84), MAla+Phe fragment (m/z=114 and 231), MAla+HTyr+H (m/z=263), MAla+HTyr+Val+H (m/z=376), Phe+MAla+HTyr+H (m/z=410) and HTyr+Ile+Lys (m/z=417)

procedure. Mircocystins with the same physical properties have different masses (in one HPLC fraction) and microcystins with the same molecular mass may have different retention time (i.e. originate from different HPLC fractions).

Structural Elucidation Using Post-Source-Decay (PSD)

In addition to detection of molecular masses, the PSD mode allows fragmentation of molecules for structural information. For this purpose, the precursor ion of interest is selected by using a timed ion selector which is situated immediately following the source region of the instrument (Fig. 1). With the timed ion selector switched off, the precurser ions are allowed to pass into the mass analyzer. The fragments of precursor ions are analyzed after passing the timed ion selector and the reflector field on the second detector.This selection method is capable of transmitting ions of a 10-Da mass window at a mass of 1000 Da.

This technique was used for the structure determination of cyanobacterial compounds directly from bloom samples or from individual HPLC fractions. Figs. 4 and 5 show two typical MALDI-TOF PSD spectra. Every peptide class (cyclic or linear) produces a typical PSD fragment pattern. The fragments represent amino acids, di-, tri-, tetra-, pentapeptides or other sequences. With the analysis of all the fragments of a molecule, the complete structure can be determined (complete fragment list not shown).

Fig. 4 shows a characteristic microcystin PSD spectrum. Microcystins can be identified by their major fragments m/z = 135, 213 and 375, which provide a typical fingerprint. Fig. 5 shows another typical PSD fragment pattern from the anabaenopeptin class. The major fragments from anabaenopeptins are the side chain (m/z=175 and 200) and the complete ring (m/z=651).

Conclusions

Many secondary metabolites are important because they show toxic effects, or are desired pharmaceutical compounds (e.g. antibiotics). The detection of antibiotics, toxins and related metabolites is generally performed by HPLC methods or bioassays, which, although commercially run in automated procedures, usually require milligram amounts of cell material and time-intensive extraction and purification processes. The procedure shown here employs only microgram quantities of cells, and it enables the rapid detection of metabolites known to reside inside the cell and within the cell membrane. MALDI-TOF mass spectrometry thus has a high potential for the direct identification of known metabolites from whole microorganisms. Simultaneously, it displays unknown metabolites, which can be further characterized in the same experiment by procedures of post-source decay.

References

An JS, Carmichael WW (1994) Use of a colorimetric protein phosphatase inhibition assay linked immunosorbent assay for the study of microcystins and nodularins. Toxicon 32:1495–1507

Aune T, Berg K (1986) Use of freshly prepared rat hepatocytes to study toxicity of blooms of the blue-green algae *Microcystis aeruginosa* and *Oscillatoria agardhii*. J Toxicol Environ Health 19:325–336

Aune T, Yasumoto T, Engeland E (1991) Light and scanning electron microscopic studies on effects of marine algal toxins toward freshly prepared hepatocytes. J Toxicol Environ Health 34:1–9

Balon EK (1975) Terminology of intervals in fish development. J Fish Res Board Can 32:1663–1670

Benndorf J, Henning M (1989) *Daphnia* and toxic blooms of *Microcystis aeruginosa* in Bautzen reservoir (GDR) Int Revue ges Hydrobiol 74:233–248

Berg K, Aune T (1987) Freshly prepared rat hepatocytes used in screening the toxicity of blue-green algal blooms. J Toxicol Environment Health 20:187–197

Birge WJ, Black JA, Westerman AG (1985) Short-term fish and amphibian embryo-larval tests for determining the effects of toxicant stress on early life stages and estimating chronic values for single compounds and complex effluents. Environ Toxicol Chem 4:807–822

Birge WJ, Black JA, Westerman AG (1985) Short-term fish and amphibian embryo-larval tests for determining the effects of toxicant stress on early life stages and estimating chronic values for single compounds and complex effluents. Environ Toxicol Chem 4:807–822

Campbell DL, Lawton LA, Beattie KA, Codd GA (1994) Comparative assessment of the specificity of the brine shrimp and Microtox assays to hepatotoxic (microcystin-LR-containing) cyanobacteria. Environ Toxicol Water Qual 9:71–77

Carbis CR, Waldron DL, Mitchell GF, Anderson JW, McCauley I (1995) Recovery of hepatic function and latent mortalities in sheep exposed to the blue-green alga *Microcystis aeruginosa*. Vet Rec 1:12–15

Carmichael WW, Falconer IR (1993) Diseases related to freshwater blue-green algal toxins, and control measures. In: Falconer IR (ed) Algal toxins in seafood and drinking water. Academic Press London, pp. 186–209

Creaser CW (1934) The technic of handling the zebrafish (*Brachydanio rerio*) for the production of eggs which are favorable for embryological research and are availabe at any specified time throughout the year. Copeia 1934:159–161

Detrich W, Zon, LI, Westerfield M (1998) The zebrafish: Biology. Methods in Cell Biology Series Vol. 59, Academic Press

Erhard M, von Döhren H, Jungblut P (1999) Rapid Identification of the new Anabaenopeptin G from *Planktothrix agardhii* HUB 011 using Matrix-assisted Laser Desorption/ionization Time-of-flight Mass Spectrometry. Rapid Commun Mass Spec 13:337–343

Erhard M, von Döhren H., Jungblut P (1997) Rapid typing and elucidation of new secondary metabolites of intact cyanobacteria using MALDI-TOF mass spectrometry. Nature Biotech 15:906–909

Eriksson JE, Grönberg L, Nygard S, Slotte JP, Meriluoto JAO (1990) Hepatocellular uptake of ^{3}H-dihydromicrocystin-LR, a cyclic peptide toxin. Biochim Biophys Acta 1025:60–66

Falconer IR (1993) Mechanism of toxicity of cyclic peptide toxins from blue-green algae. In: Falconer IR (ed) Algal toxins in seafood and drinking water. Academic Press, London, pp. 177–186

Falconer IR (ed) Algal toxins in seafood and drinking water. Academic Press, London, 224 pp

Falconer IR (1993) Health problems from exposure to cyanobacteria and proposed safety guidelines for drinking and recreational water In: Codd GA, Jefferies TM, Keevil CW, Potter E (eds). Proceeding of the first international symposium on detection methods for cyanobacterial (blue-green algal) toxins, 27-29 September, University of Bath, UK: Detection Methods for

Cyanobacterial Toxins. Special Publication No. 149, The Royal Society of Chemistry, Cambridge (U.K.), pp. 3–10

Falconer IR, Yeung SK (1992) Cytoskeletal changes in hepatocytes induced by Microcystis toxins and their relation to hyperphosphorylation of cell proteins. Chem Biol Interact 81:181–196

Fastner J, Flieger I, Neumann U (1998) Optimised extraction of microcystins from field samples – a comparison of different solvents and procedures. Wat Res 32:3177–3181

Feuillade M, Jann-Para G, Feuillade J (1996) Toxic compounds to *Artemia* from and isolates of the cyanobacterium *Planktothrix rubescens*. Arch Hydrobiol 138:175–186

Fitzgeorge R, Clark S, Keevil C (1994) Routes of intoxication. In: Codd GA, Jefferies TM, Keevil CW, Potter E (eds) Detection methods for cyanobacterial toxins. The Royal Society of Chemistry. ISBN 0-85186-961-0, pp. 69–74

Fujii K, Sivonen K, Adachi K, Noguchi K, Shimizu Y, Sano H, Hirayama K, Suzuki M, Harada K (1997) Comparative study of toxic and non-toxic cyanobacterial products: novel peptides from toxic *Nodularia spumigena* AV 1. Tetrahedron Lett 38:5525–5532

Grabow WOK, Durandt WC, Prozesky OW, Scott WE (1982) *Microcystis aeruginosa* toxin: cell culture toxicity, hemolysis, and mutagenicity assays. Appl Environ Microbiol 43:1425–1433

Harada KI (1996) Chemistry and detection of microcystins. In: Codd GA, Jefferies TM, Keevil CW, Potter E (eds) Detection methods for cyanobacterial toxins. CRC Press, pp. 103–148

Harada KI (1996) Chemistry and detection of microcystins. In: Watanabe MF, Harada K, Carmichael WW, Fujiki H (eds) Toxic *Microcystis*. CRC Press, Boca Raton, pp 103–148

Heinze R (1996) A biotest for hepatotoxins using primary rat hepatocytes. Phycologia 35 Suppl.: 89–93

Henning K, Kremer J, Meyer H (1992) Cytotoxicity of cyanobacterium *Microcystis aeruginosa*. J Vet Med B 39: 307-310

Henning M, Kohl JG (1981) Toxic blue-green algae water blooms found in some lakes in the German Democratic Republic. Int Revue ges Hydrobiol 66:553–561

Hermansky SJ, Stohs SJ, Eldeen ZM, Roche VF (1991) Evaluation of potential chemoprotectants against microcystin-LR hepatotoxicity in mice. J Appl Toxicol 11:65–74

Honkanen RE, Zwiller J, Moore RE, Daily SL, Khatra BS, Dukelow M, Boynton AL (1990) Characterization of microcystin-LR, a potent inhibitor of type 1 and type 2A protein phosphatases. J Biol Chem 265:19401–19404

Honkanen RE, Codispoti BA, Tse K, Boynton AL (1994) Characterization of natural toxins with inhibitory activity against serine/threonine protein phosphatases. Toxicon 32:339–350

Ishida K, Matsuda H, Murakami M, Yamaguchi K (1997) Microginins 299-A nad -B leucine aminopeptidase inhibitors from the cyanobacterium *Microcystis aeruginosa* NIES 299. Tetrahedron 53:10281–10288

Jungmann D. (1992) Toxic compounds isolated from *Microcystis* PCC7806 that are more active against *Daphnia* than two microcystins. Limnol Oceanogr 37:1777–1783

Karas M, Hillenkamp F. (1988) Anal Chem 60:2299–2305

Khan SA, Ghosh S, Wickstrom M, Miller LA, Hess R, Haschek WM, Beasley VR (1995) Comparative Pathology of microcystin-LR in cultured hepatocytes, fibroblasts, and renal epithelial cells. Natural Toxins 3:119–128

Kimmel CB, Ballard WW, Kimmel SR, Ullmann B, Schilling TS (1995) Stages of embryonic development of the zebrafish. Dev.Dyn. 203:253-310Kiviranta J, Abdel Hameed A, Sivonen K, Niemelae SI, Carlberg G (1994) Toxicity of the blue-green alga *Oscillatoria agardhii* to the mosquito *Aedes aegypti* and the shrimp *Artemia salina*. World J Microbiol Biotechnol 10: 517–520

Koehn E, Longley RE, Reed JK (1992) Microcolis A and B, new immunosuppressive peptides from the blue green alga *Lyngbya majuscula*. J Nat Prod 55:613–619

Lambert TW, Boland MP, Holmes CFB, Hrudey SE (1994) Quantitation of the microcystin hepatotoxins in water at environmentally relevant concentrations with the protein phosphatase bioassay. Environ Sci Technol 28:735–755

Lawton LA, Edwards C, Codd GA (1994) Extraction and high-performance liquid chromatographic method for the determination of microcystins in raw and treated waters. Analyst 119: 1525–1530

Mackintosh C, Beattie KA, Klumpp S, Cohen P, Codd GA (1990) Cyanobacterial microcystin-LR is a potent and specific inhibitor of protein phosphatases 1 and 2A from both mammals and higher plants. FEBS Lett 264:187–192

Mackintosh C, Mackintosh RW (1994) The inhibition of protein phosphatases by toxins: implications for health and extremely sensitive and rapid bioassay for toxin detection In: Codd GA, Jefferies TM, Keevil CW, Potter E (eds). Detection methods for cyanobacterial toxins. Special Publication No. 149, The Royal Society of Chemistry, Cambridge (U.K.), pp 90–99

Nagata S, Tsutsumi T, Ueno Y, Watanabe MF (1997) Enzyme immunoassay for direct determination of microcystins in environmental water. J AOAC Internat 80:408–417

MacKintosh C, Coggins J, Cohen (1991) PP Plant protein phosphatases. Biochem J 273:733–738

MacKintosh RW, Dalby KN, Campbell DG, Cohen PTW, Cohen P, MacKintosh C (1995) The cyanobacterial toxin microcystin binds covalently to cysteine-273 on protein phosphatase 1. FEBS Lett 371:236–240

Matsushima R, Yoshizawa S, Watanabe MF, Harada KI, Furusawa M, Carmichael WW, Fujiki H (1990) In vitro and in vivo effects of protein phosphatase inhibitors, microcystins and nodularin, on mouse skin and fibroblasts. Biochem Biophys Res Commun 171:867–874

McKim JM (1977) Evaluation of tests with early life stages of fish for predicting long-term toxicity. J Fish Res Board Can 34:1148–1154

Miura GA, Robinson NA, Lawrence WB, Pace JG (1991) Hepatotoxicity of microcystin-LR in fed and fasted rats. Toxicon 29:337–346

Moolan RW, Rae B, Verbeek A (1996) Some comments on the determination of microcystin toxins in waters by igh-Performance Liquid Chromatography. Analyst 121:233–238

Mosman T (1983) Rapid colorimetric assay for cellular growth and survival: application to proliferation and cytotoxicity assays. J Immunol Methods 65:55–63

Murakami M, Sun Q, Ishida K, Matsuda H, Okino T, Yamaguchi K (1997) Microviridin G and H , two new elastase inhibiting peptides from cyanobacterium Nostoc minutum. Phytochemistry 45:1197-1202

Oberemm A, Fastner J, Steinberg CEW (1997) Effects of microcystin-LR and cyanobacterial crude extracts on embryo-larval development of zebrafish (Danio rerio). Wat Res 31:2918–2921

Oberemm A, Becker J, Codd G, Steinberg C (1999) Effects of cyanobacterial toxins and aqueous crude extracts on the development of fish and amphibians. Environ Toxicol 14:77–88

Papendorf O, König GM, Wright AD, Chorus I, Oberemm A Mueggelone (1998) a novel fish embryo-larval development inhibitor from the fresh water cyanobacterium *Aphanizomenon flos-aquae*. J Nat Prod 60:1298–1300

Reinikainen M, Kiviranta J, Ulvi V, Niku Paavola ML (1995) Acute toxic effects of a novel cyanobacterial toxin on the crustaceans *Artemia salina* and *Daphnia pulex*. Arch Hydrobiol 133:61–69

Rhinehart KL, Namikoshi M, Choi W (1994) Structure and biosynthesis of toxins from blue-green algae (cyanobacteria). J Appl Phycol 6:156–176

Runnegar MTC, Falconer IR, Silver J (1981) Deformation of isolated rat hepatocytes by a peptide hepatotoxin from the blue-green alga *Microcystis aeruginosa*. Naunyn-Schmiedeberg's Arch Pharmacol 317:268–272

Runnegar MTC, Gerdes RG, Falconer IR (1991) The uptake of the cyano-bacterial hepatotoxin microcystin by isolated rat hepatocytes. Toxicon 29:43–51

Sahin A, Tencalla FG, Dietrich DR, Mez K, Naegeli H (1995) Enzymatic analysis of liver samples from rainbow trout for diagnosis of blue-green algae induced toxicosis. Am Vet Res 56: 1110–1115

Sim ATR, Mudge LM (1993) Protein phosphatase activity in cyanobacteria: consequences for microcystin toxicity analysis. Toxicon 31:1179–1186

Sivonen K, Namikoshi M, Evans WR, Carmichael WW, Sun F, Rouhiainen L, Luukkainen R, Rhinehart KL (1992) Isolation and characterization of a variety of microcystins from seven strains of the cyanobacterial genus *Anabaena*. Appl Environ Microbiol 58:2495–2500

Sivonen K, Jones (1999) Cyanobacterial toxins. In: Chorus I, Bartram J (eds) Toxic cyanobacteria in water: a guide to their public health consequences, monitoring and management. Published on the behalf of WHO by E & FN Spon, London, pp. 41–111

Takai A, Mieskes G (1991) Inhibitory effects of okadaic acid on the p-nitrophenyl phosphatase activity of protein phosphatases. Biochem J 275:233–239

Toivola DM, Eriksson JE, Brautigan, DL (1994) Identification of protein phosphatase 2A as the primary target for microcystin-LR in rat liver homogenates. FEBS Lett 344:175–180

Von Westernhagen H (1988) Sublethal effects of pollutants on fish eggs and larvae. In: Hoar WS, Randall DJ (eds) Fish Physiology. Vol. XI: The Physiology of Developing Fish. Part A: Eggs and Larvae, edited by.Academic Press:San Diego, p. 253–346

Ward CJ, Beattie KA, Lee EYC, Codd GA (1997) Colorimetric protein phosphatase inhibition assay of laboratory strains and natural blooms of cyanobacteria: comparisons with high-performance liquid chromatographic analysis for microcystins. FEMS Microbiol Lett 153: 465–473

Watanabe MF, Harada K, Carmichael WW, Fujiki H (eds) (1996) Toxic *Microcystis*, CRC Press, Boca Raton New York, 262 pp

Weckesser J, Martin C, Jakobi C (1996) Cyanopeptolins, Depsipeptides from cyanobacteria. System Appl Microbiol 19:133–138

Wickstrom ML, Khan SA, Haschek WM, Wyman JF, Eriksson JE, Schaeffer DJ, Beasley VR (1995) Alterations in Microtubules, intermediate filaments and microfilaments induced by microcystin-LR in cultured cells. Toxicol Pathol 23:326–337

Woitke P, Heinze R, Dewender M, Henning M, Kohl J-G (1999) Comparison of different methods for an evaluation of the toxicity of natural *Microcystis* blooms. Toxicon, submitted

9 Routine Analytical Methods Applied in the German Cyanotox Project

The following chapter describes the methods used by several partners in the German cyanotoxin survey; reference to these is made in other chapters of the book.

9.1 Microcystin Analysis

Analysis of Cell-Bound Microcystins

Up to 20 mg of lyophilized algal material in a 2-ml Eppendorf tube was extracted three times with either 100% methanol (Lawton et al. 1994) in 1995 or 75% aqueous methanol in 1996/97 (Fastner et al. 1998). The samples were sonicated for 5 min and then shaken for 30 min at each extraction step. After centrifugation the supernatants were pooled and blown to dryness with nitrogen or dried by vacuum centrifugation. Prior to HPLC analysis extracts were resolved in 50% methanol.

The detection of microcystins was performed with a 616 Waters solvent delivery system, 717 WISP autosampler and 991 photo diode array detector (Waters, Eschborn, Germany). Extracts were separated on a LiChrospher 100, ODS, 5 µm, LiChroCART 250-4 cartridge system (Merck, Darmstadt, Germany) using a gradient of aqueous acetonitrile (with 0.05% TFA) according to Lawton et al. (1994) with a flow of 1 ml min^{-1}.

UV spectra were obtained from 200-300 nm, and microcystins identified by their characteristic UV spectrum (UV_{max} 240 nm) (Lawton et al. 1994). Microcystins were quantified at 240 nm using microcystin-LR, -YR and -RR (Calbiochem, Bad Soden, Germany) as external standards. Microcystins other than these were quantified with microcystin-LR. Qualitative standard of microcystin [Dha^7] MC-RR was a gift from Dr. Kaarina Sivonen, University of Helsinki, and qualitative standards of [$D\text{-}Asp^3$] MC-RR, [$D\text{-}Asp^3$, Dhb^7] MC-RR, [$D\text{-}Asp^3$] MC-YR and [$D\text{-}Asp^3$] MC-LR were isolated during this study.

The detection limit for microcystins by HPLC was 1 µg g^{-1} of dry algal material.

For mass spectrometric and amino acid analysis peaks showing the respective microcystin spectra were collected by hand, blown to dryness with nitrogen and stored at -20°C.

During this survey we found that pure methanol does not extract microcystins quantitatively from freeze-dried field samples (Fastner et al. 1998). Therefore in 1996 we used 75 % methanol instead of pure methanol to extract microcystins. Most samples from 1995 were re-extracted with 75% methanol and the results correlated to the results obtained with 100% methanol. Species-specific factors (slope) were found: extraction was enhanced by a factor of 2.1 for samples dominated by *Microcystis* spp., (n=64, r^2=0.91), by a factor of 1.7 for samples dominated by *Planktothrix agardhii* (n=12, r^2=0.8), and 1.2-fold more for samples dominated by *Planktothrix rubescens* (n=12, r^2=0.9). 65 samples from 1995 which could not be reextracted were multiplied with these factors corresponding to the dominant species in the respective samples.

Analysis of Dissolved Microcystins

For the screening program (see Chapter 2.1.2) dissolved microcystins were either analysed from filtered samples directly (i.e. without prior enrichment or clean-up) with ELISA. Alternatively, for specific programs, the commercially available immuno-assay applied was the Enviro Gard kit (Coring-System, Gernsheim, Germany) assessed in Chapter 8.3. Up to 2 L of filtered (0.45-µm membrane filters) water were enriched on preconditioned C18 cartridges (Waters SepPak) and eluted with 90% methanol which was blown to dryness with nitrogen. These eluates were either analyzed with HPLC (detection limit in the range of 1 µg L^{-1}) as described above or with ELISA (detection limit in the range of 0.001 µg L^{-1}), or subjected to further cleanup using silica cartridges (Tsuji et al. 1994; detection limit in the range of 0.01 µg L^{-1}).

Extraction for Toxicity Assays

Aliquotes of the extracts for microcystin-analysis or samples extracted as described above were blown to dryness, resolved in buffer and administered to freshly isolated, primary rat hepatocytes as described in Chapter 8.1 and by Heinze (1999). Results are given in terms of mg DW in 1 ml test solution causing 50% reduction of cell viability within 4 h, the less material needed to induce this effect, the more toxic the sample. Samples requiring more than 0.8 mg DW ml^{-1} for 50% reduction of cell viability were regarded as scarcely toxic.

Structural Elucidation of Microcystins

Amino Acid Analysis

Purified microcystins were hydrolyzed with 6 N HCl in the vapour phase at 150°C for 3 h, derivatized with phenyl isothiocyanate and analyzed by the Waters Pico Tag HPLC system (Thomas and Maug 1984). The derivatives were separated on a C 18 silica gel column (3.9 x 150 mm) using Pico Tag eluents A and B. The eluted compounds were detected by UV absorption at 254 nm.

FABMS

Low-resolution FAB mass spectra were obtained on a ZAB-SE4F mass spectrometer using xenon atoms (8 keV energy) and a matrix of dithiothreitol-dithioerythrol (magic bullet) (Witten et al. 1984). About 5-10 μg of each sample was dissolved in methanol prior to analysis.

MALDI-TOF-MS

Dried HPLC fractions were dissolved in 10 μl water/ethanol/acetonitrile (1:1:1). A mixture of 1 μl sample and 1 μl of matrix (saturated a-cyano-4-hydroxycinnamic acid matrix solubilized in 50% acetonitrile, 0.03% trifluoroacetic acid) was prepared directly on the template. The samples were analyzed using a MALDI-TOF mass spectrometer (Voyager elite, PerSeptive BioSystems, Framingham, MA) with a nitrogen laser giving a 337 nm output. The ions were accelerated with a voltage of 20 kV. Measurements were performed in the delayed extraction mode, allowing the determination of monoisotopic mass values. A low mass gate of 500 improved the measurement by filtering out the most intensive matrix ions. The mass spectrometer was used in the positive-ion detection and reflector mode. Postsource decay (PSD) measurements were performed after peptide mass determination with the same samples on the template. The operating voltages of the reflectron were reduced stepwise to record 12 spectral segments sequentially. See also Chapter 8.5

9.2 Cell Counting, Determination of Biovolume and Chlorophyll-a

Phytoplankton Analysis

Integrated phytoplankton samples were taken from the whole water column or from different depths and preserved with acidic Lugol solution for quantitative, or with formaldehyde solution for qualitative analysis. Species composition was determined with a light microscope (Bourelly 1970; Komárek 1986; Anagnostidis and Komárek 1988; Komárek and Anagnostidis 1989). *Microcystis* and *Anabaena* species were identified to species level, but quantified by genera. Species dominance in the remaining samples was determined semiquantitatively.

Cell numbers of *Microcystis aeruginosa* were determined by counting using an inverted microscope at 400-fold magnification. For cultures (Chapter 3.2), these were differentiated between single cells and cells in division. The method of Utermöhl (1958) was applied with modifications described in Lawton et al. (1999). For *Planktothrix agardii*, the length of filaments within the counting grid was cumulatively assessed rather than counting filaments, as filament length may be extremely variable. Cell and filament diameters were measured at 400-fold magnification concentration.

A cause for variability in cell or trichome counts and biovolumes calculated from them may be uncertainties in trichome counts, e.g. due to irregular sedimentation in the counting chambers: differences between five replicate counts (max/min) were found to be almost 2-fold for *Planktothrix agardhii* and up to 1.4-fold for *Microcysis aeruginosa*.

Chlorophyll-a Concentrations

Both concentrated net samples and unconcentrated water from the same sampling sites were analyzed for chlorophyll-a content according to Deutsche Einheitsverfahren (1985) i.e. ISO (1992).

References

Anagnostidis K, Komárek J (1988) Modern approach to the classification system of cyanophytes: 3 – Oscillatoriales. Arch Hydrobiol Suppl 80:327–472

Bourelly P (1970) Les Algues d'eau doux, Vol. 3. Editions N. Boubee, Paris

Deutsche Einheitsverfahren zur Wasser-, Abwasser- und Schlammuntersuchung (1985) Teil 16, DIN 38412, Verlag Chemie GmbH, Weinheim

Fastner J, Flieger I, Neumann U (1998) Optimised extraction of microcystins from field samples – a comparison of different solvents and procedures. Wat Res 32:3177–3181

ISO 1992 Water Quality Measurement of biochemical parameters. Spectrometric determination of the chlorophyll-a concentrations. ISO 10260, International Organization for Standardization, Geneva

Komárek J (1986) Modern approach to the classification system of cyanophytes: 2 - Chroococcales. Arch Hydrobiol Suppl 73:157-226

Komárek J, Anagnostidis K (1989) Modern approach to the classification system of cyanophytes: 4 – Nostocales. Arch Hydrobiol Suppl 82:247–345

Lawton L, Edwards C, Codd GA (1994) Extraction and high-performance liquid chromatographic method for the determination of microcystins in raw and treated waters. Analyst 119:1525–1530

Lawton L, Marsalek B, Padisák J, Chorus I (1999) In: Chorus I, Bartram J (eds) Toxic cyanobacteria in water: a guide to their public health consequences, monitoring and management. Published on the behalf of WHO by E & FN Spon, London, pp. 348–367

Thomas H, Maugh L (1984) New tool for amino acid analysis. Science 225:42

Tsuji K, Naito S, Kondo F, Watanabe MF, Suzuki S, Nakazawa H, Suzuki M, Shimada T, Harada K-I (1994) A clean-up method for analysis of trace amounts of microcystins in lake water. Toxicon 32:1251–1259

Utermöhl H (1958) Zur Vervollkommnung der quantitativen Phytoplanktonmethodik. Mitt Int Ver Limnol 9:1–38

MIX
Papier aus verantwortungsvollen Quellen
Paper from responsible sources
FSC® C105338

If you have any concerns about our products,
you can contact us on
ProductSafety@springernature.com

In case Publisher is established outside the EU,
the EU authorized representative is:
Springer Nature Customer Service Center GmbH
Europaplatz 3, 69115 Heidelberg, Germany

Printed by Libri Plureos GmbH
in Hamburg, Germany